APPENDIX B: F-TABLE AT 5 PERCENT (UPPER TAIL)

F-TABLE, CRITICAL VALUES, 5 PERCENT IN UPPER TAIL

Degrees of freedom for the numerator along top row

Degrees of freedom for the denominator along side row

	1	2	3	4	5	6	7	8	9	10	12	15	20	24	30	40
1	161	200	216	225	230	234	237	239	241	242	244	246	248	249	250	251
2	18.5	19.0	19.2	19.2	19.3	19.3	19.4	19.4	19.4	19.4	19.4	19.4	19.4	19.5	19.5	19.5
3	10.1	9.55	9.28	9.12	9.01	8.94	8.89	8.85	8.81	8.79	8.74	8.70	8.66	8.64	8.62	8.59
4	7.71	6.94	6.59	6.39	6.26	6.16	6.09	6.04	6.00	5.96	5.91	5.86	5.80	5.77	5.75	5.72
5	6.61	5.79	5.41	5.19	5.05	4.95	4.88	4.82	4.77	4.74	4.68	4.62	4.56	4.53	4.50	4.46
6	5.99	5.14	4.76	4.53	4.39	4.28	4.21	4.15	4.10	4.06	4.00	3.94	3.87	3.84	3.81	3.77
7	5.59	4.74	4.35	4.12	3.97	3.87	3.79	3.73	3.68	3.64	3.57	3.51	3.44	3.41	3.38	3.34
8	5.32	4.46	4.07	3.84	3.69	3.58	3.50	3.44	3.39	3.35	3.28	3.22	3.15	3.12	3.08	3.04
9	5.12	4.26	3.86	3.63	3.48	3.37	3.29	3.23	3.18	3.14	3.07	6.01	2.94	2.90	2.86	2.83
10	4.96	4.10	3.71	3.48	3.33	3.22	3.14	3.07	3.02	2.98	2.91	2.85	2.77	2.74	2.70	2.66
11	4.84	3.98	3.59	3.36	3.20	3.09	3.01	2.95	2.90	2.85	2.79	2.72	2.65	2.61	2.57	2.53
12	4.75	3.89	3.49	3.26	3.11	3.00	2.91	2.85	2.80	2.75	2.69	2.62	2.54	2.51	2.47	2.43
13	4.67	3.81	3.41	3.18	3.03	2.92	2.83	2.77	2.71	2.67	2.60	2.53	2.46	2.42	2.38	2.34
14	4.60	3.74	3.34	3.11	2.96	2.85	2.76	2.70	2.65	2.60	2.53	2.46	2.39	2.35	2.31	2.27
15	4.54	3.68	3.29	3.06	2.90	2.79	2.71	2.64	2.59	2.54	2.48	2.40	2.33	2.29	2.25	2.20
16	4.49	3.63	3.24	3.01	2.85	2.74	2.66	2.59	2.54	2.49	2.42	2.35	2.28	2.24	2.19	2.15
17	4.45	3.59	3.20	2.96	2.81	2.70	2.61	2.55	2.49	2.45	2.38	2.31	2.23	2.19	2.15	2.10
18	4.41	3.55	3.16	2.93	2.77	2.66	2.58	2.51	2.46	2.41	2.34	2.27	2.19	2.15	2.11	2.06
19	4.38	3.52	3.13	2.90	2.74	2.63	2.54	2.48	2.42	2.38	2.31	2.23	2.16	2.11	2.07	2.03
20	4.35	3.49	3.10	2.87	2.71	2.60	2.51	2.45	2.39	2.35	2.28	2.20	2.12	2.08	2.04	1.99
21	4.32	3.47	3.07	2.84	2.68	2.57	2.49	2.42	2.37	2.32	2.25	2.18	2.10	2.05	2.01	1.96
22	4.30	3.44	3.05	2.82	2.66	2.55	2.46	2.40	2.34	2.30	2.23	2.15	2.07	2.03	1.98	1.94
23	4.28	3.42	3.03	2.80	2.64	2.53	2.44	2.37	2.32	2.27	2.20	2.13	2.05	2.01	1.96	1.91
24	4.26	3.40	3.01	2.78	2.62	2.51	2.42	2.36	2.30	2.25	2.18	2.11	2.03	1.98	1.94	1.89
25	4.24	3.39	2.99	2.76	2.60	2.49	2.40	2.34	2.28	2.24	2.16	2.09	2.01	1.96	1.92	1.87
30	4.17	3.32	2.92	2.69	2.53	2.42	2.33	2.27	2.21	2.16	2.09	2.01	1.93	1.89	1.84	1.79
40	4.08	3.23	2.84	2.61	2.45	2.34	2.25	2.18	2.12	2.08	2.00	1.92	1.84	1.79	1.74	1.69
60	4.00	3.15	2.76	2.53	2.37	2.25	2.17	2.10	2.04	1.99	1.92	1.84	1.75	1.70	1.65	1.59
120	3.92	3.07	2.68	2.45	2.29	2.18	2.09	2.02	1.96	1.91	1.83	1.75	1.66	1.61	1.55	1.50
∞	3.84	3.00	2.60	2.37	2.21	2.10	2.01	1.94	1.88	1.83	1.75	1.67	1.57	1.52	1.46	1.39

APPENDIX C: F-TABLE AT 2.5 PERCENT (UPPER TAIL)

F-TABLE, CRITICAL VALUES, 2.5 PERCENT IN UPPER TAILS

Degrees of freedom for the numerator along top row

Degrees of freedom for the denominator along side row

	1	2	3	4	5	6	7	8	9	10	12	15	20	24	30	40
1	648	799	864	900	922	937	948	957	963	969	977	985	993	997	1001	1006
2	38.51	39.00	39.17	39.25	39.30	39.33	39.36	39.37	39.39	39.40	39.41	39.43	39.45	39.46	39.46	39.47
3	17.44	16.04	15.44	15.10	14.88	14.73	14.62	14.54	14.47	14.42	14.34	14.25	14.17	14.12	14.08	14.04
4	12.22	10.65	9.98	9.60	9.36	9.20	9.07	8.98	8.90	8.84	8.75	8.66	8.56	8.51	8.46	8.41
5	10.01	8.43	7.76	7.39	7.15	6.98	6.85	6.76	6.68	6.62	6.52	6.43	6.33	6.28	6.23	6.18
6	8.81	7.26	6.60	6.23	5.99	5.82	5.70	5.60	5.52	5.46	5.37	5.27	5.17	5.12	5.07	5.01
7	8.07	6.54	5.89	5.52	5.29	5.12	4.99	4.90	4.82	4.76	4.67	4.57	4.47	4.41	4.36	4.31
8	7.57	6.06	5.42	5.05	4.82	4.65	4.53	4.43	4.36	4.30	4.20	4.10	4.00	3.95	3.89	3.84
9	7.21	5.71	5.08	4.72	4.48	4.32	4.20	4.10	4.03	3.96	3.87	3.77	3.67	3.61	3.56	3.51
10	6.94	5.46	4.83	4.47	4.24	4.07	3.95	3.85	3.78	3.72	3.62	3.52	3.42	3.37	3.31	3.26
11	6.72	5.26	4.63	4.28	4.04	3.88	3.76	3.66	3.59	3.53	3.43	3.33	3.23	3.17	3.12	3.06
12	6.55	5.10	4.47	4.12	3.89	3.73	3.61	3.51	3.44	3.37	3.28	3.18	3.07	3.02	2.96	2.91
13	6.41	4.97	4.35	4.00	3.77	3.60	3.48	3.39	3.31	3.25	3.15	3.05	2.95	2.89	2.84	2.78
14	6.30	4.86	4.24	3.89	3.66	3.50	3.38	3.29	3.21	3.15	3.05	2.95	2.84	2.79	2.73	2.67
15	6.20	4.77	4.15	3.80	3.58	3.41	3.29	3.20	3.12	3.06	2.96	2.86	2.76	2.70	2.64	2.59
16	6.12	4.69	4.08	3.73	3.50	3.34	3.22	3.12	3.05	2.99	2.89	2.79	2.68	2.63	2.57	2.51
17	6.04	4.62	4.01	3.66	3.44	3.28	3.16	3.06	2.98	2.92	2.82	2.72	2.62	2.56	2.50	2.44
18	5.98	4.56	3.95	3.61	3.38	3.22	3.10	3.01	2.93	2.87	2.77	2.67	2.56	2.50	2.44	2.38
19	5.92	4.51	3.90	3.56	3.33	3.17	3.05	2.96	2.88	2.82	2.72	2.62	2.51	2.45	2.39	2.33
20	5.87	4.46	3.86	3.51	3.29	3.13	3.01	2.91	2.84	2.77	2.68	2.57	2.46	2.41	2.35	2.29
21	5.83	4.42	3.82	3.48	3.25	3.09	2.97	2.87	2.80	2.73	2.64	2.53	2.42	2.37	2.31	2.25
22	5.79	4.38	3.78	3.44	3.22	3.05	2.93	2.84	2.76	2.70	2.60	2.50	2.39	2.33	2.27	2.21
23	5.75	4.35	3.75	3.41	3.18	3.02	2.90	2.81	2.73	2.67	2.57	2.47	2.36	2.30	2.24	2.18
24	5.72	4.32	3.72	3.38	3.15	2.99	2.87	2.78	2.70	2.64	2.54	2.44	2.33	2.27	2.21	2.15
25	5.69	4.29	3.69	3.35	3.13	2.97	2.85	2.75	2.68	2.61	2.51	2.41	2.30	2.24	2.18	2.12
30	5.57	4.18	3.59	3.25	3.03	2.87	2.75	2.65	2.57	2.51	2.41	2.31	2.20	2.14	2.07	2.01
40	5.42	4.05	3.46	3.13	2.90	2.74	2.62	2.53	2.45	2.39	2.29	2.18	2.07	2.01	1.94	1.88
60	5.29	3.93	3.34	3.01	2.79	2.63	2.51	2.41	2.33	2.27	2.17	2.06	1.94	1.88	1.82	1.74
120	5.15	3.80	3.23	2.89	2.67	2.52	2.39	2.30	2.22	2.16	2.05	1.94	1.82	1.76	1.69	1.61
∞	5.02	3.69	3.12	2.79	2.57	2.41	2.29	2.19	2.11	2.05	1.94	1.83	1.71	1.64	1.57	1.48

INDEX

T

U

V

W

Notes

Notes

Notes

Notes

Notes

APPENDIX D: CHI-SQUARED TABLE

Values of χ^2 (Degrees of Freedom, Level of Significance)

Probability in Right Tail

Degrees of Freedom	0.99	0.975	0.95	0.9	0.1	0.05	0.025	0.01	0.005
1	0.000157	0.000982	0.003932	0.0158	2.706	3.841	5.024	6.635	7.879
2	0.020100	0.050636	0.102586	0.2107	4.605	5.991	7.378	9.210	10.597
3	0.1148	0.2158	0.3518	0.5844	6.251	7.815	9.348	11.345	12.838
4	0.297	0.484	0.711	1.064	7.779	9.488	11.143	13.277	14.860
5	0.554	0.831	1.145	1.610	9.236	11.070	12.832	15.086	16.750
6	0.872	1.237	1.635	2.204	10.645	12.592	14.449	16.812	18.548
7	1.239	1.690	2.167	2.833	12.017	14.067	16.013	18.475	20.278
8	1.647	2.180	2.733	3.490	13.362	15.507	17.535	20.090	21.955
9	2.088	2.700	3.325	4.168	14.684	16.919	19.023	21.666	23.589
10	2.558	3.247	3.940	4.865	15.987	18.307	20.483	23.209	25.188
11	3.053	3.816	4.575	5.578	17.275	19.675	21.920	24.725	26.757
12	3.571	4.404	5.226	6.304	18.549	21.026	23.337	26.217	28.300
13	4.107	5.009	5.892	7.041	19.812	22.362	24.736	27.688	29.819
14	4.660	5.629	6.571	7.790	21.064	23.685	26.119	29.141	31.319
15	5.229	6.262	7.261	8.547	22.307	24.996	27.488	30.578	32.801
16	5.812	6.908	7.962	9.312	23.542	26.296	28.845	32.000	34.267
17	6.408	7.564	8.672	10.085	24.769	27.587	30.191	33.409	35.718
18	7.015	8.231	9.390	10.865	25.989	28.869	31.526	34.805	37.156
19	7.633	8.907	10.117	11.651	27.204	30.144	32.852	36.191	38.582
20	8.260	9.591	10.851	12.443	28.412	31.410	34.170	37.566	39.997
21	8.897	10.283	11.591	13.240	29.615	32.671	35.479	38.932	41.401
22	9.542	10.982	12.338	14.041	30.813	33.924	36.781	40.289	42.796
23	10.196	11.689	13.091	14.848	32.007	35.172	38.076	41.638	44.181
24	10.856	12.401	13.848	15.659	33.196	36.415	39.364	42.980	45.558
25	11.524	13.120	14.611	16.473	34.382	37.652	40.646	44.314	46.928
26	12.198	13.844	15.379	17.292	35.563	38.885	41.923	45.642	48.290
27	12.878	14.573	16.151	18.114	36.741	40.113	43.195	46.963	49.645
28	13.565	15.308	16.928	18.939	37.916	41.337	44.461	48.278	50.994
29	14.256	16.047	17.708	19.768	39.087	42.557	45.722	49.588	52.335
30	14.953	16.791	18.493	20.599	40.256	43.773	46.979	50.892	53.672
50	29.707	32.357	34.764	37.689	63.167	67.505	71.420	76.154	79.490
60	37.485	40.482	43.188	46.459	74.397	79.082	83.298	88.379	91.952
80	53.540	57.153	60.391	64.278	96.578	101.879	106.629	112.329	116.321
100	70.065	74.222	77.929	82.358	118.498	124.342	129.561	135.807	140.170

APPENDIX E: CRITICAL VALUES FOR THE DURBIN-WATSON STATISTIC

CRITICAL VALUES FOR THE DURBIN-WATSON STATISTIC (α = 0.05)

	K = 1		K= 2		K = 3		K = 4		K = 5	
n	D_l	D_u	D_l	D_u	D_l	D_u	D_l	D_u	D_l	D_u
15	1.08	1.36	0.95	1.54	0.82	1.75	0.69	1.97	0.56	2.21
16	1.10	1.37	0.98	1.54	0.86	1.73	0.74	1.93	0.62	2.15
17	1.13	1.38	1.02	1.54	0.90	1.71	0.78	1.90	0.67	2.10
18	1.16	1.39	1.05	1.53	0.93	1.69	0.82	1.87	0.71	2.06
19	1.18	1.40	1.08	1.53	0.97	1.68	0.86	1.85	0.75	2.02
20	1.20	1.41	1.10	1.54	1.00	1.68	0.90	1.83	0.79	1.99
21	1.22	1.42	1.13	1.54	1.03	1.67	0.93	1.81	0.83	1.96
22	1.24	1.43	1.15	1.54	1.05	1.66	0.96	1.80	0.86	1.94
23	1.26	1.44	1.17	1.54	1.08	1.66	0.99	1.79	0.90	1.92
24	1.27	1.45	1.19	1.55	1.10	1.66	1.01	1.78	0.93	1.90
25	1.29	1.45	1.21	1.55	1.12	1.66	1.04	1.77	0.95	1.89
26	1.30	1.46	1.22	1.55	1.14	1.65	1.06	1.76	0.98	1.88
27	1.32	1.47	1.24	1.56	1.16	1.65	1.08	1.76	1.01	1.86
28	1.33	1.48	1.26	1.56	1.18	1.65	1.10	1.75	1.03	1.85
29	1.34	1.48	1.27	1.56	1.20	1.65	1.12	1.74	1.05	1.84
30	1.35	1.49	1.28	1.57	1.21	1.65	1.14	1.74	1.07	1.83
31	1.36	1.50	1.30	1.57	1.23	1.65	1.16	1.74	1.09	1.83
32	1.37	1.50	1.31	1.57	1.24	1.65	1.18	1.73	1.11	1.82
33	1.38	1.51	1.32	1.58	1.26	1.65	1.19	1.73	1.13	1.81
34	1.39	1.51	1.33	1.58	1.27	1.65	1.21	1.73	1.15	1.81
35	1.40	1.52	1.34	1.58	1.28	1.65	1.22	1.73	1.16	1.80
36	1.41	1.52	1.35	1.59	1.29	1.65	1.24	1.73	1.18	1.80
37	1.42	1.53	1.36	1.59	1.31	1.66	1.25	1.72	1.19	1.80
38	1.43	1.54	1.37	1.59	1.32	1.66	1.26	1.72	1.21	1.79
39	1.43	1.54	1.38	1.60	1.33	1.66	1.27	1.72	1.22	1.79
40	1.44	1.54	1.39	1.60	1.34	1.66	1.29	1.72	1.23	1.79
45	1.48	1.57	1.43	1.62	1.38	1.67	1.34	1.72	1.29	1.78
50	1.50	1.59	1.46	1.63	1.42	1.67	1.38	1.72	1.34	1.77
55	1.53	1.60	1.49	1.64	1.45	1.68	1.41	1.72	1.38	1.77
60	1.55	1.62	1.51	1.65	1.48	1.69	1.44	1.73	1.41	1.77
65	1.57	1.63	1.54	1.66	1.50	1.70	1.47	1.73	1.44	1.77
70	1.58	1.64	1.55	1.67	1.52	1.70	1.49	1.74	1.46	1.77
75	1.60	1.65	1.57	1.68	1.54	1.71	1.51	1.74	1.49	1.77
80	1.61	1.66	1.59	1.69	1.56	1.72	1.53	1.74	1.51	1.77
85	1.62	1.67	1.60	1.70	1.57	1.72	1.55	1.75	1.52	1.77
90	1.63	1.68	1.61	1.70	1.59	1.73	1.57	1.75	1.54	1.78
95	1.64	1.69	1.62	1.71	1.60	1.73	1.58	1.75	1.56	1.78
100	1.65	1.69	1.63	1.72	1.61	1.74	1.59	1.76	1.57	1.78

HOT ROD volume 2 TECHNICAL LIBRARY

The Best of HOT ROD MAGAZINE

Modifying & Tuning HOLLEY Carburetors

Published by CarTech, Inc.
11481 Kost Dam Road
North Branch, MN 55056 U.S.A.
United States of America
800-551-4754, fax 651-583-2023

ISBN 1-884089-32-1

Book Trade Distribution by
Voyageur Press, Inc.
123 North Second Street
P.O. Box 338,
Stillwater, MN 55082 U.S.A.
612-430-2210, fax 612-430-2211

Distributed in England by
Brooklands Books Ltd
P.O. Box 146, Cobham
Surrey KT11 1LG
England
Tel: 01932 865051, fax 01932 868803

Distributed in Australia by
Brooklands Books Ltd
1/81 Darley St, P.O. Box 199
Mona Vale, NSW 2103
Australia
Tel: 02 9997 8428, fax 029979 5799

Printed in the United States of America

CONTENTS

Jet-Begone

Percy's Holley Metering Blocks Mean No More Jet Changes

By Rob Kinnan

Why is the Holley carburetor the most popular fuel mixer among racers *and* the general public? Ultimate power production is one reason, but it's a lot easier to tune and work on a Holley than a Q-jet or AFB, especially when changing jets. What if we were to tell you that the ritual could be made even easier?

Holley jet changes are fairly easy, but messy. The float bowl must be removed, and that's almost always followed by a torn gasket; inevitably, the fuel spills on a hot engine. Some companies have devised easier ways to get to the jets, from Holley's own quick-change bowl kits to the Carb Shop's SpeedBlocks, but both utilize jets. While jets aren't expensive if you only buy a couple of pairs, purchasing a complete set can be costly. At around $5 a pair, that's $100 for 10 sets of four.

Photography: Rob Kinnan

Percy's High Performance has invented what it considers a better way. The company's Adjust-A-Jet wipes out messy jet changes. The Adjust-A-Jet is an additional metering block sandwiched between the float bowl and the stock metering block, and a tapered needle and seat replaces both jets on either end to meter the fuel. The jets are completely removed from the carburetor. Changes in fuel metering are as easy as loosening a nut and turning the metering screw. Screwing it into the block closes the passage, lessening the amount of fuel to the

This cutaway reveals the Adjust-A-Jet's simplicity. Fuel flows through the passages where the stock jets used to be, then up two channels and over to the metering screw, which threads into the passage shown by the arrow. Metered fuel then flows back down the center passage and into the carburetor body. The plug that seals the horizontal passage is not shown. The metering block has no effect on the emulsion characteristics of the carburetor, and all other tuning procedures remain the same. Percy's claims extensive dyno-testing has shown no perceptible change in the fuel curve over a standard Holley design.

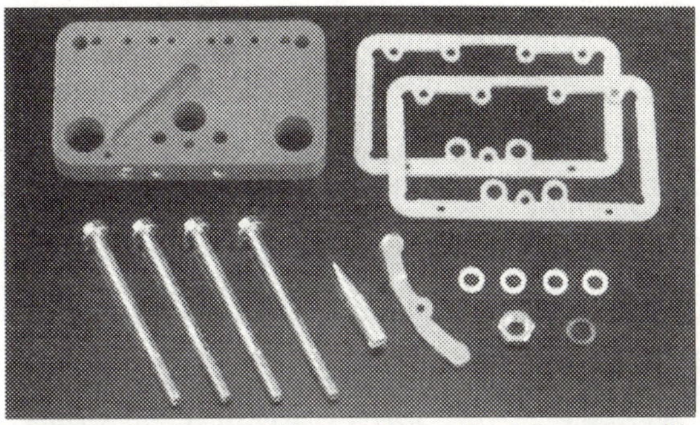

Metering blocks are billet aluminum. The kit comes with block, gaskets, longer bowl screws, and a new accelerator pump arm.

The Percy's block goes between the float bowl and the Holley metering block. The jets are gone, and the passages serve as mere fuel channels; the metering occurs with the tapered screw in the block. The stock metering block needn't be removed.

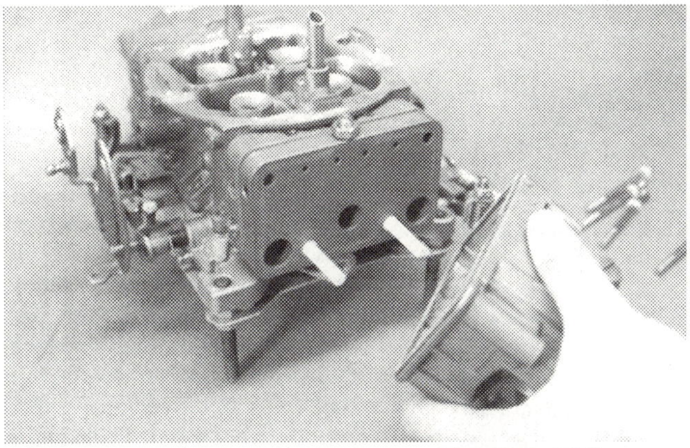

Jet extensions for the secondary bowl ensure adequate fuel delivery under hard acceleration. Make sure that they don't interfere with the float.

The arrow indicates where we had to file the Holley throttle plate to provide clearance.

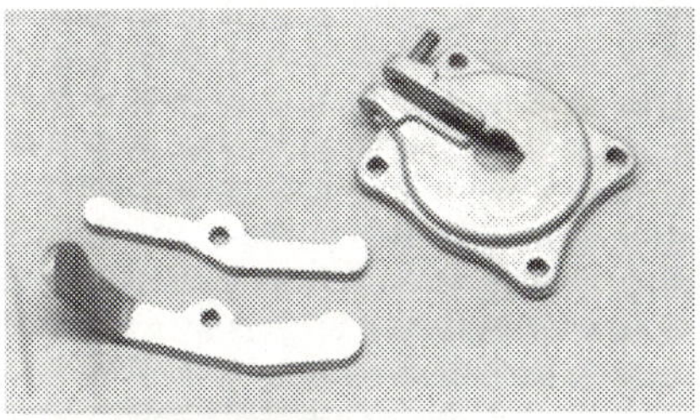

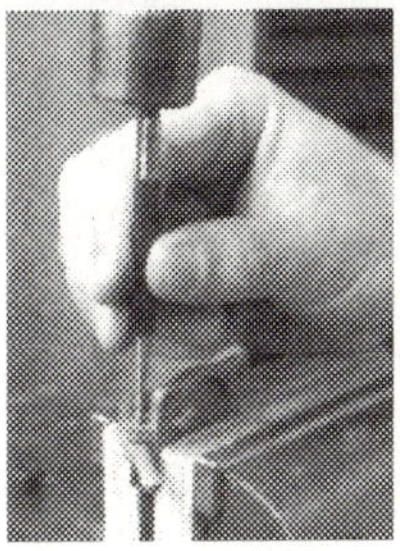

The carb becomes longer with the Percy's kit, so a new accelerator pump arm (bottom) is supplied. Knock out the roll pin with a punch, install the new pump arm, and drive the pin back in.

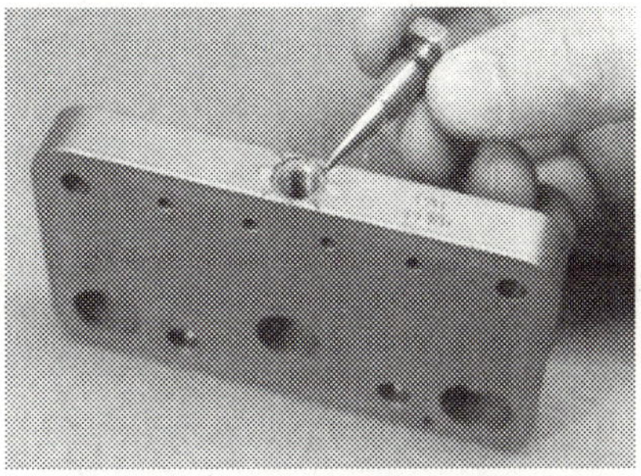

A metering screw controls fuel flow. Turning it in restricts the fuel orifice, thereby decreasing jet size. Turning it out does the opposite. For adjustment from baseline, make sure to note which jets came out of the carb when installing the screw. Gently bottom the screw, then back it out as many turns as the chart indicates to correlate with the jet size.

The air-cleaner mounting flange interfered with the locknut on the metering screw, and creative filing was required for added clearance.

engine, just like installing smaller jets. The blocks have an operating range the equivalent of jets sized from 40 to 100, which covers almost every application known to man.

We installed the kit on a 12-second big-block Camaro in less than two hours (most of which was spent cleaning the carb). Jetting changes literally take only seconds, and we didn't have to deal with torn gaskets or the mess of fuel spilling onto a hot engine. Just turn the screw and go. You can even make changes while the engine is running. Because there is only one metering orifice for each pair of jets, stagger jetting is no longer an option. However, Percy's namesake, Jeff Percival, claims that the metering block eliminates the flow deviance between jets, and therefore compensates for different fuel demands. Stagger jetting, he says, is no longer necessary. One complaint in the past regarding the Adjust-A-Jet was the confusing chart in the instructions that provides a baseline adjustment for the screw. Percy's said by the time you read this, its chart will be user-friendly.

Since the stock metering blocks are retained (unmodified except for removal of the jets), two Percy's blocks add 1 inch to the overall length of the carburetor. A longer fuel-supply line for dual-feed carbs, and possibly a modified air-cleaner base, are required due to the extra length. Otherwise, installation is simple, but there a few places on the carb that need a little more clearance.

The Adjust-A-Jet works on the main metering circuit, and has no effect on the idle circuit. Carbs with fat or lean idle circuits still need attention in those areas, but for quick adjustments or altitude compensation, the Adjust-A-Jet is by far the easiest setup we've ever experienced. Check the captions as we detail installation on an 830-cfm Holley double-pumper. The kits are for Holley 2300-, 3310-, 4150-, 4160-, and 4500-series carburetors, and an alcohol version is also available. They are available from Summit and Jegs, among other places, for $89. **HR**

The kit can be installed on either side of the carb, or both. We put two kits on our double-pumper, one each on the primary and secondary sides.

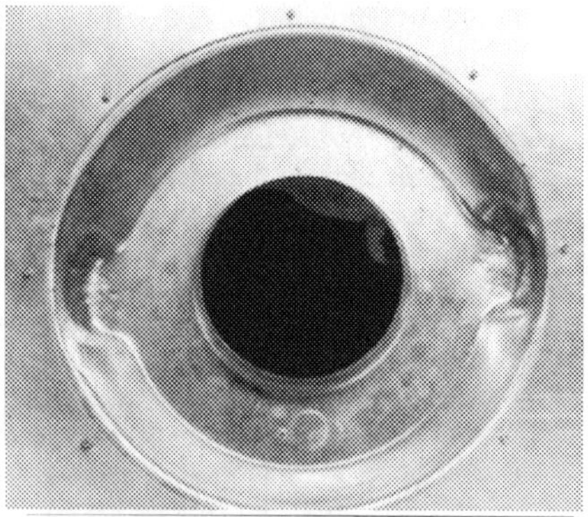

Because carb length grows 1 inch with the Adjust-A-Jet, a dropped air cleaner base may not fit. The base on this Camaro (which was riveted to a fresh-air pan sealed to the hood) needed coercion with a hammer to fit over the carb. A longer fuel line was also required.

Jetting changes are now as easy as loosening a lock nut and turning a screw.

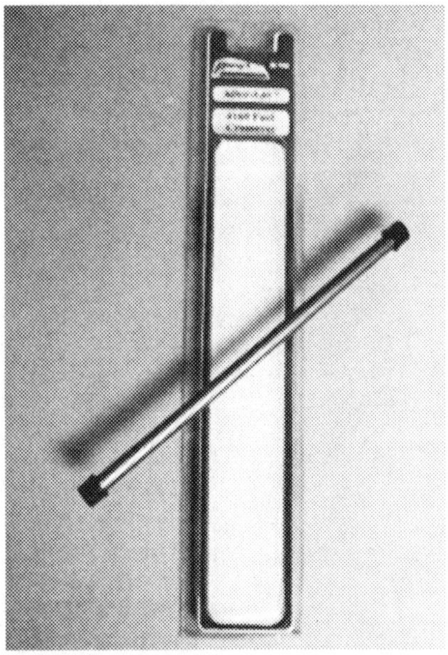

Installing a block on a 4160 single-feed Holley requires a longer Percy's fuel-transfer tube.

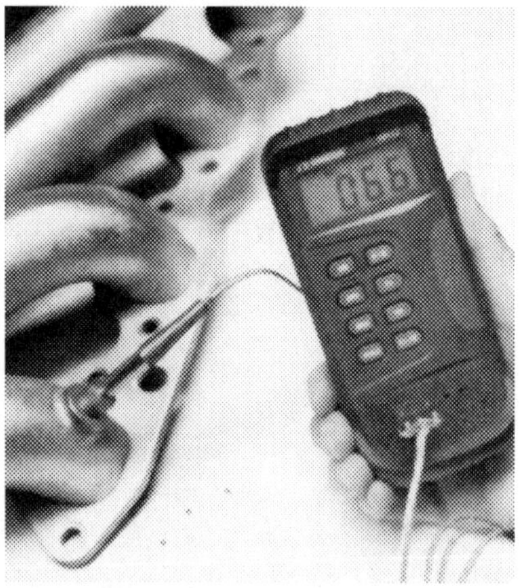

Percy's also provided a digital exhaust-temperature monitor to help with the tuning process. We welded a bung into a header tube to hold a temp probe and read exhaust-gas temperature (EGT). Leaning the carb out raises the EGT, richening cools it. As shown in the photo, the temperature in our shop was 66 degrees Fahrenheit. Percival says that correct jetting will show a temperature of about 1,200 to 1,400 degrees under power on the average street motor.

The accelerator-pump linkage must be readjusted after installation, as per the normal Holley tuning regimen.

SOURCE

Percy's High Performance
Dept. HR12, 5-133 S. Hwy. 5, Camdenton, MO 65020, 573/346-4409

NEW SPEED BLOCKS MAKE JET CHANGES TOO EASY

By Jeff Smith

QUICK-CHANGE JETS

We hot rodders are a lazy lot. We're always looking for ways to make things easier. In the movie *Two Lane Blacktop*, the late Beach Boys drummer Dennis Wilson played a mechanic who spent the entire movie changing jets in the twin Holley carburetors of a big-block–powered '55 Chevy. If he had had a set of Brad Urban's Carburetor Shop Speed Blocks, he might have had more time to go after the girl in the movie.

For the rest of us, Holley carburetor jet changes are a fact of life. You drain the fuel from the float bowl (spilling fuel all over your intake manifold), remove the bowl, change the jets, replace the bowl and then replay the routine for the secondary side. If you're exceptionally gifted, the gaskets tear and you spend half an hour scraping gaskets off the bowl and the metering block. If you haven't been through this drill a few dozen times, don't count yourself as a *true* hot rodder.

Brad Urban has probably changed more carburetor jets than the rest of us combined. After a few thousand repeats of this brain-numbing procedure, he began to think there had to be a better way. Enlisting the help of compatriot Louie Hammel, the two developed the Speed Block, a fully CNC-machined aluminum metering block that duplicates a standard Holley metering block, but with a high-speed twist.

Instead of mounting the jet in the metering block, Hammel created a brass cartridge that screws in from the top of the block. The bottom of the cartridge holds the jet in place inside the metering block in roughly the same location as the original Holley design. The beauty of this is that the cartridge can be removed from the Speed Block without removing the float bowl. This makes for incredibly quick jet changes.

Now you no longer have to drain the

The Speed Block is a direct replacement for the original Holley metering block. With the Speed Block, the jets are easily replaced without removing the float bowl. In fact, the jets can be changed, as shown here, without even shutting down the engine. Note the jet positioned at the bottom of the cartridge.

float bowl to make jet changes, which also means no more spilled fuel over a hot intake manifold. This dramatically reduces the risk of fire.

Taking this one step further, it's possible to change the jets in the carburetor even when the engine is running. Since the idle feed restriction limits the amount of fuel delivered to the idle circuit, there's no change to idle quality if you remove the jet cartridge from the Speed Block while the engine is running. We've tried this on a couple of cars, and it works just as Urban claims. Now that we've played with the Speed Blocks, we're definitely spoiled. We'll certainly never do another dyno test without them.

Over the years, Holley has created several different metering blocks for the 2300 two-bbl, 4150 four-bbl, 4160 four-bbl and Dominator Holley carburetors. Speed Blocks are available for all these applications, including alcohol carbs. You can even use a primary metering Speed Block on the secondary side of any Holley for four-hole idle control with a modification to the throttle plates that Brad Urban's Carburetor Shop can perform for you. The Speed Blocks retail for slightly less than $100 each.

For drag racers, circle burners or street enthusiasts who are looking for ways to make their lives simpler, the new Speed Blocks are the quick-change ticket for any Holley carburetor. **HR**

Fitted with twin Speed Blocks, this 3310 Holley is ready for some ultraquick jet changes. In a quickie test, we were able to change jets in all four corners of this Holley in less than a minute.

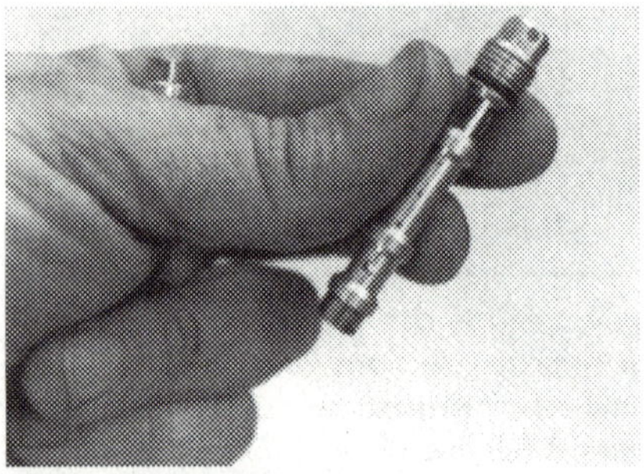

This tight view shows the cartridge removed from the Speed Block. The three holes drilled into the cartridge are the same diameter as the main well-holes in a standard Holley metering block. To make jet changes even quicker, have spare cartridges fitted with the proper jets ready to replace.

One option for the Speed Block is the knurled idle-mixture screw called the Speed Screw. This allows you to adjust the idle mixture without a screwdriver. The tapered portion of the Speed Screw is a duplicate of the stock Holley idle-mixture screw.

This view of the Speed Block (*foreground*) compared to the standard Holley metering block (*background*) illustrates that the stock passages for the idle circuit, the main circuit, the power valve and the air bleeds are all duplicated in the Speed Block. Changing to Speed Blocks is a simple remove-and-replace operation.

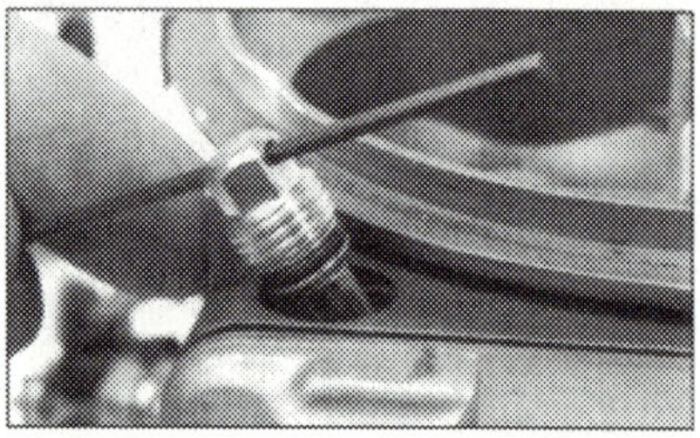

To make pulling the cartridges easier, Urban fitted each cartridge with a small hole that allows the use of an Allen wrench or paper clip to help pull the cartridge from the Speed Block.

SOURCE

Brad Urban's Carburetor Shop
Dept. HR02
8460 Red Oak St.
Rancho Cucamonga, CA 91730
909/947-9722

TRICK NEW HOLLEY BOOSTERS FOR THE STREET

By Jeff Smith

Making power is simple. You combine fuel and air, squeeze it, light it off with a spark and *boom* you have power. Any engine can perform this task rather handily. But hot rodders are always searching for more power. This means tweaking each system within an engine until we make the best power possible.

Brad Urban at the Carburetor Shop is one of those demon tweakers. Lately he's been experimenting with Holley carburetor boosters in the quest for more power and improved driveability. Before we dive into this subject, let's review what a booster is and what it does. In this case, we'll investigate how boosters affect throttle response and power in Holley-carbureted street-driven engines.

Gimme a BOOST!

BOOSTER BASICS

How fuel is introduced into the incoming air and at what ratio the fuel is mixed with air play a big part in power. Each engine requires a particular air-fuel ratio to generate best power. While most contend this ratio is around 12.5:1, that prime air-fuel ratio target changes based not only on the combination of parts used but also varies at different engine speeds.

Boosters are those protrusions found in each carburetor venturi that serve to increase air velocity to create a pressure differential sufficient to generate fuel flow. A venturi is a sleek restriction in the carburetor throat that is designed to increase the velocity of the air passing through the carburetor. When velocity increases, pressure drops. A booster amplifies this reduced pressure, allowing atmospheric pressure in the float bowl to push fuel through the main jet, up through the main well where it is mixed (or emulsified) with air and finally through the booster, where it mixes with the incoming air in the venturi. This amplified pressure drop is commonly called the "signal."

Over the years, Holley has used several different boosters. The two basic designs are the standard, single-outlet booster and the annular-discharge booster. The single-outlet booster introduces fuel into the airstream from one large hole, while the annular-discharge design uses a series of smaller holes to mix fuel with the airstream. The most common single-outlet booster is often referred to as the dogleg booster found in many 4150 Holley carburetors. The annular-discharge concept is generally considered to be more efficient since it conditions fuel into a finer mist.

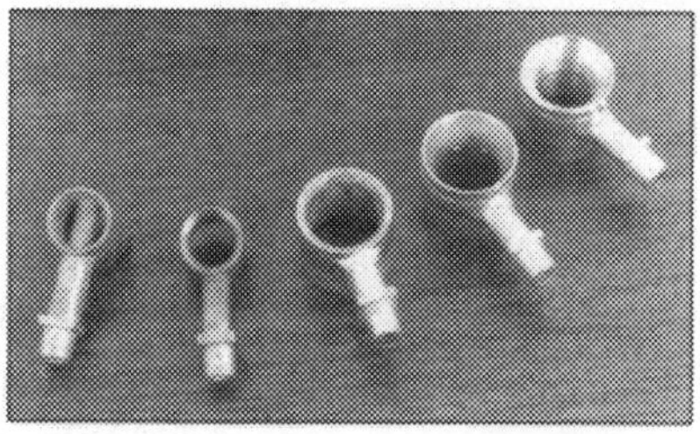

Carburetor boosters can have a dramatic effect on an engine's fuel curve. The Carburetor Shop tested all five of these boosters to evaluate their impact on the fuel curve and engine power. From left to right are the standard straight-leg booster found in most 600-cfm Holleys, the dogleg booster used in the 3310 and many other performance Holleys, the Ford truck annular-discharge booster, the Super Booster 2 (which is a Dominator-style booster with an internal insert) and the original Super Booster annular-discharge with its radiused entry top.

By design, annular-discharge boosters generate a stronger signal to the main jet in comparison with single-outlet boosters. Urban has found that certain annular-discharge boosters can generate a 100 percent increase in signal at the jet over a dogleg booster. This improved signal also flows fuel at a lower rate of airflow. This means that the booster will initiate fuel flow at a lower engine rpm.

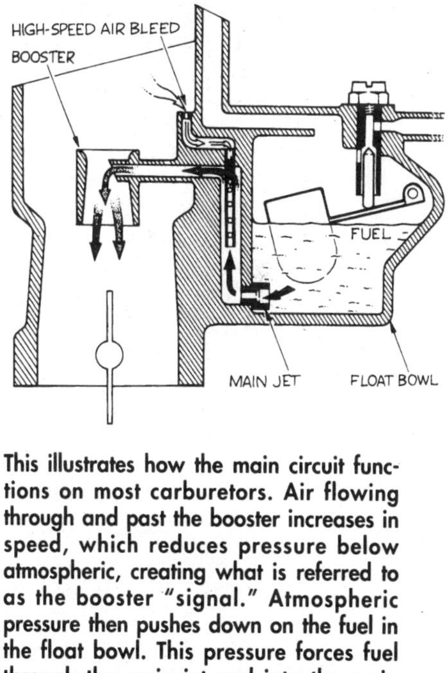

This illustrates how the main circuit functions on most carburetors. Air flowing through and past the booster increases in speed, which reduces pressure below atmospheric, creating what is referred to as the booster "signal." Atmospheric pressure then pushes down on the fuel in the float bowl. This pressure forces fuel through the main jet and into the main well. The main well mixes the fuel with air from the high-speed air bleed before the fuel arrives at the booster.

The theory follows that a booster with a stronger signal will require a smaller jet to flow the same amount of fuel. This should make the annular-discharge booster crisper at part- and full-throttle than a single-outlet booster, while changing the amount of delivered

Annular-discharge boosters introduce fuel into the airstream through eight small holes located around the inside circumference of the booster. Fuel comes down the booster tube, enters an internal channel around the booster, and exits at the holes. Both the Ford truck booster (*left*) and Super Booster 2 (*right*) deliver excellent low- and midrange-rpm fuel delivery at both part- and wide-open throttle. This makes them excellent choices for street-driven carburetors.

The Carburetor Shop tested all of the boosters for signal strength and found that the annular-discharge boosters delivered as much as a 100 percent increase in signal to the main jet. This is why these boosters require a smaller jet to flow the same amount of fuel at the same rpm and load.

DUAL-PLANE/SMALL CAM

The following chart illustrates not only the torque increase of the annular-discharge Super Booster 2 with the small cam and Performer RPM dual-plane intake, but it also reveals the significant change in fuel flow throughout the entire rpm curve. Note that the Super Booster 2 leaned out at the top compared with the dogleg booster. Jetting was established for best overall power. Richer 68 jets were tried with the Super Booster 2, but they hurt power at 3000 (412 lbs-ft) while increasing torque to 379 lbs-ft at 5500.

	DOGLEG BOOSTER 77 jets		SUPER BOOSTER 2 66 jets		DIFFERENCE	
RPM	TORQUE	FUEL (lbs/hr)	TORQUE	FUEL (lbs/hr)	TORQUE	FUEL (lbs/hr)
2500	375	80	396	85	+21	+5
3000	374	78	417	98	+43	+20
3500	433	126	435	121	+2	-5
4000	434	146	439	139	+5	-7
4500	429	169	439	160	+10	-9
5000	415	182	408	175	-7	-7
5500	375	194	374	186	-1	-8
6000	338	203	338	193	0	-10

fuel throughout the engine's rpm range. The amount of fuel delivered at each rpm, if plotted on a graph, is commonly called the fuel curve. One advantage to these more efficient boosters is that you could run a larger carburetor on an engine for best peak power without sacrificing low-speed throttle response.

TEST TIME

Urban decided that theory is great, but he wanted to see what would happen on an engine. First, he commissioned Bob Vrbancic to build a 383 small-block Chevy test engine. Vrbancic assembled a 9:1 TRW short-block along with a set of Dart II iron Sportsman 200cc intake port cylinder heads. The camshaft was a Competition Cams 270 Magnum hydraulic single-pattern cam with 224 degrees of duration at .050-inch tappet lift and .470-inch valve lift. Harland Sharp 1.5:1 roller rockers were used, as was an Edelbrock Performer RPM dual-plane intake. The engine was later fitted with a Victor Jr. intake and longer-duration Comp Cams 292 Magnum hydraulic (244 degrees at .050) to evaluate how the boosters responded to a big-cam/single-plane engine combination. All tests were performed with a 3½-inch Flowmaster muffler-equipped dual-exhaust system.

The carburetor was one of Holley's new Pro-Series 950-cfm double-pumper carburetors. The carburetor is a modified version of the basic 750 double-pumper with an 850-cfm baseplate. Urban also converted the float bowls over to his new needle and seat combination, which increases fuel flow while introducing the fuel below fuel level to reduce aeration. Power valves were used both front and rear.

BOOSTER POWER

Our first major booster discovery was that this engine experienced a sag in the torque curve at 3000 rpm. This is a common occurrence with many street engines, most often in engines with big intake-port heads. This sag was most notable with the factory dogleg boosters.

Comparing the power between the Super Booster 2 annular-discharge booster and the stock dogleg revealed that the Super Booster 2 flowed more fuel at 3000 rpm than the dogleg booster, which pumped an amazing 43 lbs-ft more torque at this rpm. Looking at the fuel flow, the engine wanted more fuel at this rpm, as evidenced by the increase in power. The Super Booster 2 was able to deliver more fuel at this point because of its increased sensitivity. Looking at the power and fuel flow data from the other two annular-discharge boosters revealed that they delivered similar fuel curves, but the Super Booster 2 was the best of the three. We also noticed that the engine went lean at 6000 rpm, which cost a few horsepower. This was due to the large demand placed on the smaller jet at maximum fuel flow. A change to richer jetting improved power at 6000 but hurt the torque at 3000 rpm.

When the engine was changed over to the larger cam and single-plane Victor Jr. intake, we expected to see a similar or even greater difference in power at 3000 rpm, since both the larger-volume plenum and increased overlap camshaft should have contributed to a loss of signal to the carburetor. However, this combination revealed virtually no difference in power between the stock dogleg and the Super Booster 2. Obviously, there is still much to learn about which boosters work the best with certain engine combinations.

POWERFUL POSSIBILITIES

Street engines operate over a very wide rpm range of as much as 4500 rpm or more. This presents an extremely difficult job for the carburetor. Comparing the fuel flow across the entire curve for each of the boosters, the annular-discharge boosters tended to introduce fuel sooner than either the straight-leg or the dogleg boosters. The stock Holley unmodified dogleg booster appeared to be insensitive on the dual-plane manifold below 3500 rpm.

Peak horsepower aside, it seems that street engines fitted with a large cam, single-plane intake and a 750- or larger-cfm Holley could benefit from an annular-discharge booster like the Super Booster 2 or the Ford truck annular booster for better throttle response and driveability.

While our testing was limited only to wide-open throttle, it appears the boosters would compare similarly at part throttle. To test this concept, we bolted this same 950-cfm Holley on a mild 305ci Chevy and drove it outfitted with both the stock dogleg boosters and a set of Super Booster 2s. Amazingly, the 950 Holley performed fairly well with the dogleg boosters, although it was a bit soggy. The Super Booster 2 combination (appropriately leaned out) demonstrated a much crisper throttle response with no wide-open throttle hesitation even when the throttle was hammered at low engine speeds.

There is much more to know about how these boosters perform in the hundreds of different possible street engine combinations. Certainly, these boosters are not intended for every application. The closer the engine is to stock, the less power these boosters may deliver. But if your thumper street engine needs help in the throttle-response department, these boosters could help. Best of all, this is an easy change that the Carburetor Shop can perform on your carburetor for a reasonable price. Urban believes further improvements could follow by matching boosters with modified metering blocks. Either way, these simple items could really give your street stormer a boost. **HR**

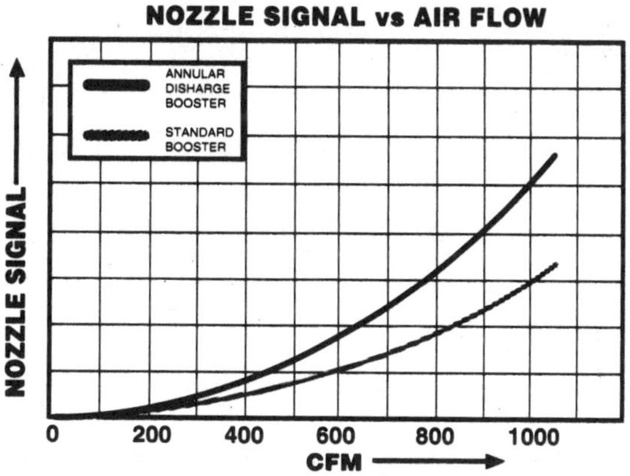

This graph illustrates the annular-discharge booster's signal increase to the jet, and also that it initiates main circuit fuel flow sooner than a standard booster. This is why the Super Booster 2 improved torque at 3500 on Test 1.

MORE POWER WITH LESS FUEL

Booster design will affect how much fuel is used to make power. In this test, Urban placed an undercut on the very bottom of a stock dogleg booster. This is a common trick used by carburetor modifiers. This step shears the fuel as it leaves the booster and improves the fuel distribution to all the cylinders while also reducing the fuel flow slightly. While the power gains are slight, they illustrate how improving the quality of fuel introduced into the engine has a positive effect on power. Note that despite the dramatic 16 lbs-hr reduction of fuel flow at 6000 rpm, the engine still gained 6 lbs-ft of torque (and six horsepower). This test was performed using the tall Victor Jr. intake with the Comp Cams 292 hydraulic Magnum camshaft.

STOCK DOGLEG			MODIFIED DOGLEG		DIFFERENCE	
RPM	TORQUE	FUEL (lbs/hr)	TORQUE	FUEL (lbs/hr)	TORQUE	FUEL (lbs/hr)
2500	366	79	368	75	+2	-4
3000	375	93	374	90	-1	-3
3500	401	118	403	113	+2	-5
4000	406	140	409	135	+3	-5
4500	427	170	427	159	0	-11
5000	431	187	437	186	+6	-1
5500	413	206	415	201	+2	-5
6000	380	223	386	207	+6	-16

All testing was performed at John Baechtel's Westech Performance Group on a SuperFlow 901 dyno. Westech tested five boosters on two different engine combinations that totaled over 100 dyno pulls in three days. Whew!

Bob Vrbancic built this 383 small-block Chevy that served as the testbed. The engine was tested with both a Performer RPM dual-plane and a 270 Comp Cams hydraulic cam, and a Victor Jr. single-plane with a Comp Cams 292 hydraulic cam.

Urban modified a stock dogleg booster (*top*) with an internal step (*bottom*) to help shear the fuel as it leaves the large single fuel outlet hole. This change was worth horsepower even though it *reduced* fuel flow.

Do not try this at home! Urban modified this Holley Pro-Series carburetor to save time during the numerous booster changes by using a set screw to retain each booster. The Carburetor Shop does not recommend you do this on your carburetor because the boosters can easily loosen and fall out. This happened during the testing, but luckily the booster did not find its way into a cylinder. A dropped booster could also jam the throttle open, so you should *never* attempt this.

SOURCES

The Carburetor Shop
Dept. HR05
1457 E. Philadelphia, Unit 24
Ontario, CA 91761
909/947-9722

Westech Performance Group
Dept. HR05
11098 Venture Dr., Ste. C
Mira Loma, CA 91752
909/685-4767

BOWLING FOR HOLLEYS

UPGRADING THE HOLLEY 1850

By David Freiburger

Holley 1850s are inexpensive, factory rebuilds are even cheaper, and swap meet 1850s are nearly free. Their 600-cfm, square-bore, vacuum-secondary design makes them a universal performer for mild small-block applications. But let's face it: those side-hung float bowls are homely, and a secondary metering block isn't stock on the 4160-series 1850. Here we'll show you how to fix these problems to improve the appearance, performance, and tuneability of the venerable 1850. **HR**

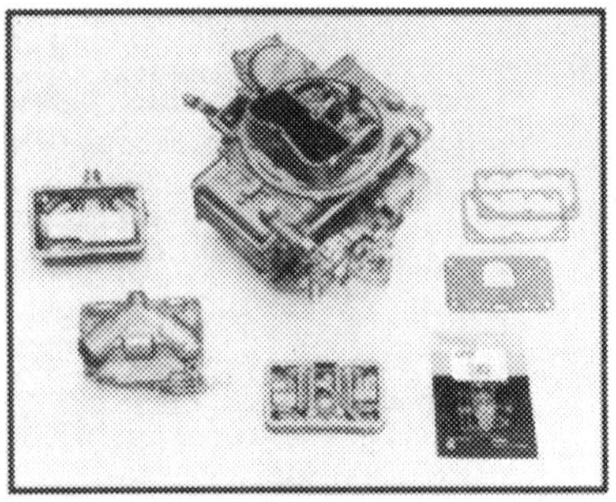

1 Here's what you'll need to install a secondary metering block and center-pivot/dual-feed float bowls on an 1850. The 34-2 bowl kit from Holley includes all the gaskets and screws, or you can use fuel bowls from most vacuum-secondary carbs with center-hung bowls. The metering block can be robbed from a 3310 or a 4776 if your 1850 has list number 1850, 1850-1, or 1850-2, but you need to either use a power valve plug (26-36) or drill out the vacuum passage in the body of the carburetor so a power valve will work. However, the 1850-3 (recognizable by its plastic parts) and the 1850-4 (recognizable by its cast-in bowl-vent tubes) have no provision for a metering plate with a secondary power valve, and a power valve plug won't fit either. These 1989-and-later -3 and -4 carburetors have to use Holley metering block kit 34-6 (with side-hung bowls) or 34-13 (with center-hung bowls). This kit will also work on the earlier carburetors.

2 The secondary metering plate is installed with clutch-head screws, but a ground-down screwdriver will do the job. A single gasket seals both the fuel bowl and the metering plate, so you'll save scraping time if you try to keep the gasket stuck to the body as you remove the fuel bowl. Stock 1850s come with a 134-9 secondary metering plate (stamped "9"), and the equivalent jets are 64s. The 34-13 metering block kit comes with 76 jets, so you'll need to get smaller jets to start with. If you're into stealth tuning a stock-looking 1850, Holley offers secondary metering plates that match jet sizes 53 through 85 and also have different idle-feed restriction sizes.

3 Before assembling the new metering plate and bowl, hold the bowl upside-down and preset the float parallel with the top of the bowl. The bowls in the 34-2 Holley kit come with Duracon (white plastic) floats that can be replaced with the 116-3 Nitrophyl (black plastic) floats. Also remember to put your smaller jets in the metering block. Prevent the gaskets from sticking during your next disassembly by coating them with ChapStick (cherry works best). Locating pins in the metering block will keep you from installing the gasket wrong. Also note that the addition of the metering block requires longer bolts than the ones used with the metering plate *(upper right)*, and that this 1850-4 has no power-valve well *(arrow)*.

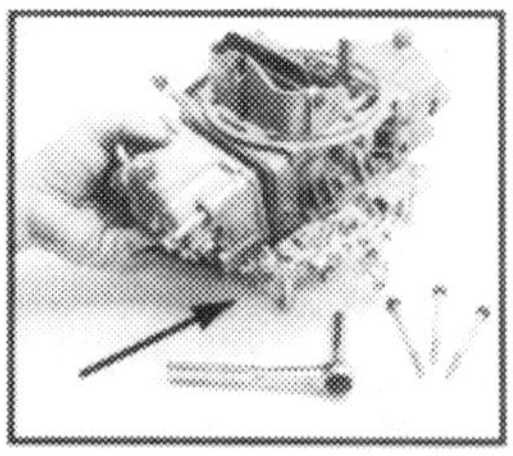

4 Your final step is to remove the front bowl and replace it with the center-hung bowl, again coating the gasket with cherry ChapStick. Holley now uses 5/16-head bolts instead of screws, and don't forget the little bolt gaskets. Also remember to install the accelerator pump lever *(arrow)* as you line up the bowl, and adjust it when you're done. Check out the double-nutted bolts used for the carb stand.

5 The finished carb doesn't look like an 1850, performs better, and is more tuneable. New Holley bowls from the 34-2 kit can be set up for dual-feed from either side, but come with 5/16 inlet fittings. The 26-26 fittings are for 3/8 line, and used bowls can also be upgraded with these larger fittings. You'll find it tough to buy one-piece steel fuel lines for a carburetor with dual metering blocks and 5/16 fittings, so the 3/8 fittings, adjustable fuel lines, or flexible fuel lines may be mandatory for installation. Once you're done, install the carb, double-check the float level, check for leaks, and you're on the road!

SOURCE

Holley Replacement Parts
A Division of Coltec Industries
Dept. HR03
11955 E. Nine Mile Rd.
Warren, MI 48090
313/497-4000

HOW TO TUNE YOUR HOLLEY

By Cole Quinnell

Bolting on high-performance parts is only as helpful in the go-fast game as the extent of your knowledge of how to tune these parts. The carburetor is a prime example of easy bolt-on power because it controls one of the three crucial components necessary to make horsepower. Perhaps the most common and flexible performance carburetor is the Holley four-barrel. This gas-passer has come in hundreds of forms over the years, from O.E.M. status in the '60s to Pro Modifieds today. Best of all, these carbs are easy to work on; follow along and we'll send you in the right direction with screwdriver in hand.

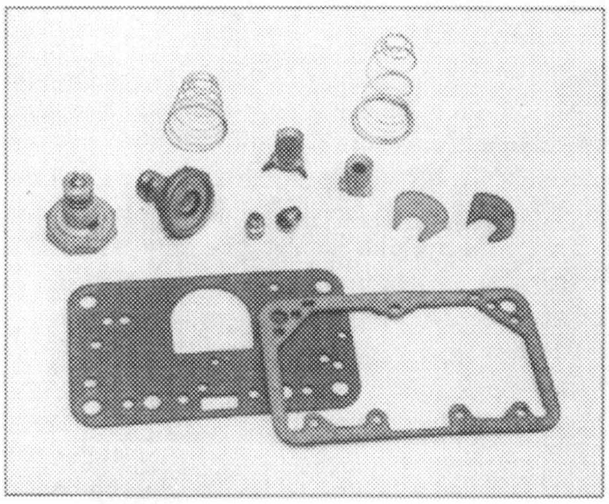

Brad Urban of the Carburetor Shop was kind enough to give us a "begin at the beginning" tuning procedure on common 4150/4160 model high-performance Holley carbs. Tuning means changing components, so some extra parts will be necessary including good, reusable gaskets. Urban recommends using the Holley gaskets that have a brown, paper-bag appearance and a slick, almost waxy finish.

One very common carb problem is vacuum leaks. Often, rubber vacuum plugs can come off or deteriorate due to heat, causing erratic idle. Also, remember that cleanliness is next to godliness on a carburetor. The entire carb should be kept clean, particularly the air bleeds, which are the eight small holes at the top front and rear of the venturis.

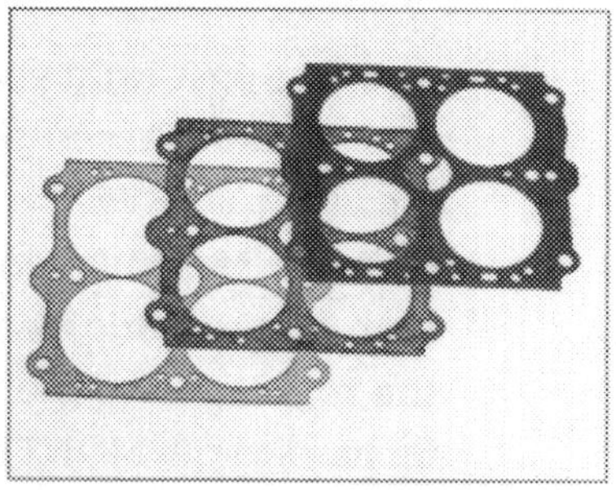

If the base has been removed or replaced, be sure the correct gasket is used when reassembling the carb. There are three different gaskets, all of which will fit, but only one will work correctly. From the left are gaskets for a 800-, 750-, and 600-cfm carburetor. The differences lie in the idle circuit passages, which are the small holes at the top and bottom of the gaskets.

With the engine off, remove the brass sight plug from the bowls. The fuel level should be at the bottom of this sight hole if the float level is set correctly. The Carburetor Shop offers clear sight plugs that don't need to be removed, making this setting much easier, quicker, and cleaner. Be sure to clean up any gas spills immediately, as they pose a fire hazard.

Now the carb is ready to bolt on and run. Set the idle speed so that the engine runs smoothly and allow it to come up to operating temperature. Be sure the car is on level ground so the float level can be set properly. Incorrect settings will cause the carburetor to run slightly rich or lean. Changes in float level are made by loosening the lock screw and adjusting the nut. This moves the seat in relation to the needle to raise or lower the float. Turning the nut clockwise lowers the float level, while counterclockwise raises the float level.

On most 4150 and 4160 Holley carbs, the idle mixture screws are located on the primary metering block. Thread these screws in until the engine stumbles. Then back them out until the highest rpm is reached or the most vacuum is obtained. Most Holleys richen the idle mixture when the screws are turned out, but some emission-application carbs have reverse idle adjustment. Adjust one side, then the other, and then fine tune them both. Most start at 1½ turns from the bottomed position. After the float level and the idle mixture have been set, the idle rpm can be readjusted.

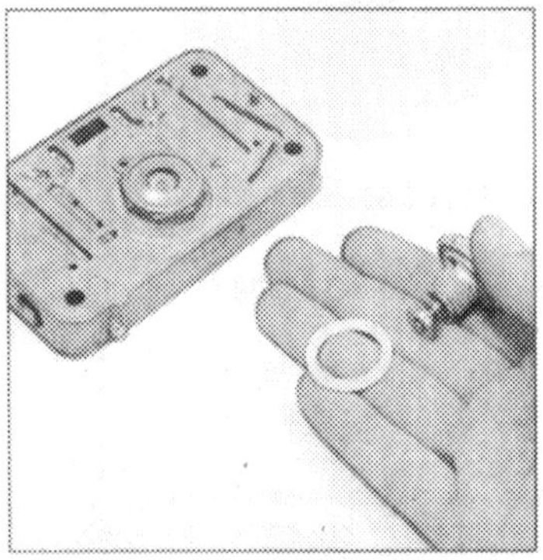

The power valve provides more fuel during high-load/high-speed operation, opening at a predetermined vacuum point that is stamped on the valve. A power valve marked 65 opens at 6.5 inches of mercury. New Holley power valves have the number spread out on the face, but read the same as the old units. Power valve tuning considerations dictate that the power valve only open at wide open throttle (WOT) and not at idle. First, use a vacuum gauge to check manifold vacuum at idle and WOT. The key is to ensure the idle vacuum does not drop below the power valve's opening point. For example, erratic idle and an over-rich condition is caused by an 8.0-inch idle vacuum with a power valve rated at 10.5 inches of manifold vacuum.

Once all the other metering systems have been dialed in, attention can be turned to the secondaries. On vacuum-secondary models, a spring within a vacuum diaphragm controls the opening rate of the secondaries. Holley offers a quick-change kit (part No. 20-59) which makes secondary springs quick and easy to change. Holley also offers a spring kit (part No. 20-03) containing six different secondary springs. During transition into the secondaries, a bog or stumble indicates that the secondaries are opening too soon; slow acceleration could indicate the secondaries are opening too late. When the secondaries are opening at the correct time, the only noticeable difference will be an increase in power and a change in the exhaust tone. A lighter spring will open the secondaries sooner, while a heavier spring will open them later.

During part-throttle operation, the main jets handle the fuel-metering chores. An HC/CO meter is extremely helpful when tuning part-throttle metering, but performance experimentation can be just as helpful, as is checking spark plugs. Dark carbon deposits on the porcelain of a spark plug indicate a rich condition, while a completely white or damaged spark plug can mean a lean air/fuel ratio. A correct air/fuel ratio will leave a light brown ring on a plug's porcelain. Jet changes may be necessary to compensate for different engine modifications, altitude, temperatures, or other conditions.

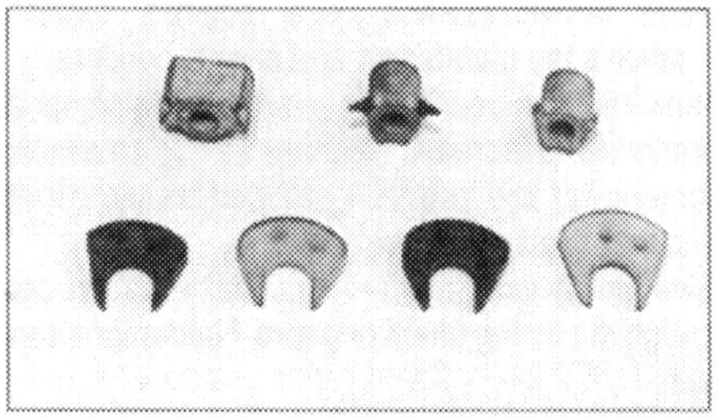

The accelerator pump circuit is controlled by a combination of the plastic pump cam located on the primary throttle shaft and the accelerator pump "shooter" located between the venturis. Hesitation on acceleration can be caused by too much or, more often, too little fuel introduced by the accelerator pump circuit. Holley carb books will reference the different pump cams while numbers stamped on the pump squirters determine their size. Larger squirters deliver more fuel in a shorter period of time. Holley offers a kit (part No. 20-12) that contains an assortment of cams. Finally, there are different styles of squirters; the one in the middle has discharge nozzles to better direct the flow.

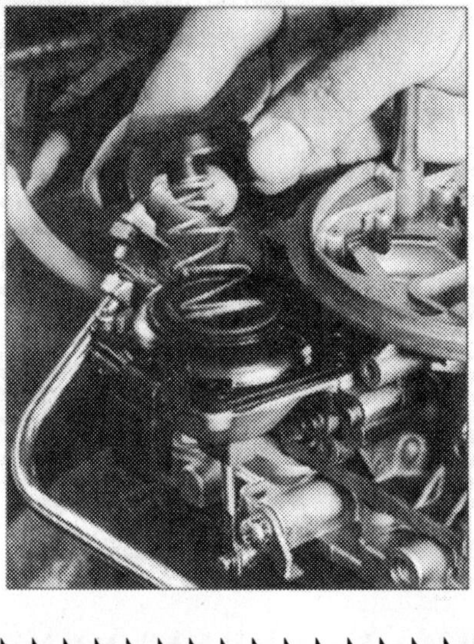

SOURCES

The Carburetor Shop
Dept. HR11
1457 Philadelphia, Unit 24
Ontario, CA 91761
714/947-7744

Holley Replacement Parts
Dept. HR11
11955 E. Nine Mile Rd.
Warren, MI 48090
313/497-4000

HARD-CORE HOLLEY

BUILDING THE 3310 HOLLEY FOR MAXIMUM STREET PERFORMANCE

When factory musclecars stalked the streets in the halcyon days of the '60s, Holley's 3310 4-barrel carburetor fueled the best of them. Big-block Corvettes, Z/28 Camaros, and all the heavy hitters were Holley equipped. Rated at 780 cfm, the 3310 was a terrific carburetor for these applications. With vacuum secondaries and more-than-adequate flow capacity, it delivered good power and excellent driveability on engines that often ran on the ragged edge of reality.

Through the years, the 3310 remained a stable hallmark of the high-performance industry. In both street and racing applications, it was distinguished time and again, but in the '90s, the 3310 has grown long in the tooth. As the world of electronic fuel injection overtakes us, carburetors still remain the primary fueling systems for street and racing applications. The demands and expectations placed on carburetors have increased dramatically, however, and carburetor experts have taken on the challenge of making the 3310 meet modern street and racing requirements.

At The Carburetor Shop in Ontario, California, Brad Urban is building his own "high-spec" 3310 Holleys, with a bag full of tricks intended to make the 3310 a full-service carburetor for the most demanding jobs. Brad performs a host of modifications to the 3310, most of which are unavailable even in the current Holley Premium series.

Brad cites five reasons why his carburetors are a performance asset: 1) 4-corner idling circuits; 2) flow-balanced venturis; 3) high-flow boosters; 4) trimmed throttle shafts; and 5) redyeing for original color, with a process The Carburetor Shop calls ColoRestore. On closer investigation, you'll find a lot of intelligent thought in his other modifications. At this writing, modified carburetors right out of the box will run about $550, while a good, original core can be fully modified for under $200, as long as it doesn't need any major new parts.

With a new carburetor or a thoroughly cleaned original model, Brad begins by removing the boosters from the main body. Then he mills off the air horn, removes the casting flash in the venturis, and selectively hones each venturi until all four bores flow the same on the flow bench. Both ends of the main body are milled in parallel, and the bottom is surfaced. If the carb is a new or late-model version, it comes with restrictive boosters identified by a bar across the throat of the booster. These boosters

By John Baechtel

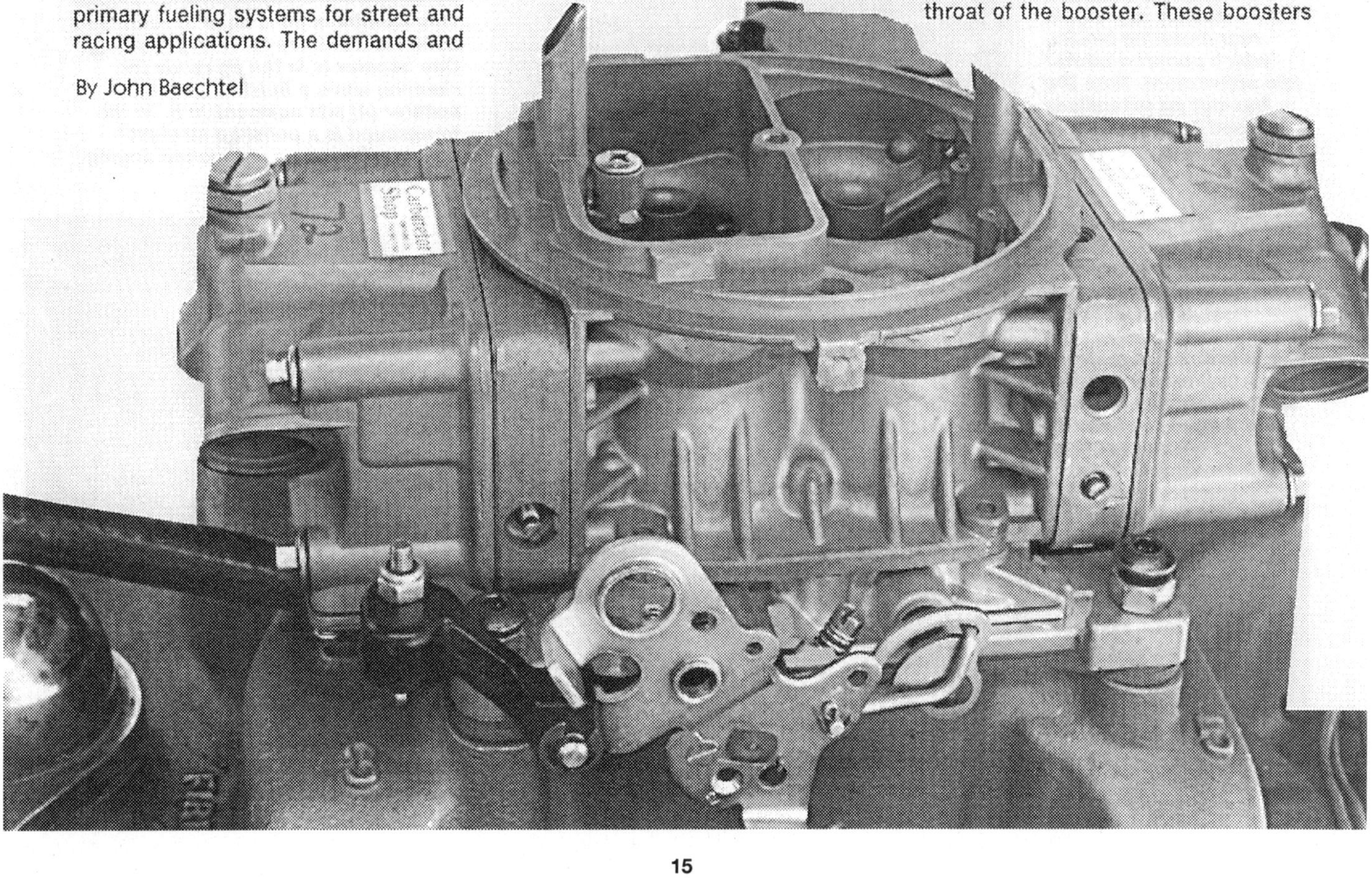

have a straight attachment shaft and a .125-inch feed hole. Brad prefers earlier dogleg boosters with .160-inch feed holes, and he built a special fixture to hold these dogleg boosters while they are modified. The casting flash is removed from the outside of the booster, and the inner venturi is enlarged with a reamer. The entire booster is polished and radiused to prevent as little restriction as possible. Once a set of boosters has been prepped, Brad flows them in the main body and rehones them until they all flow the same. These booster modifications generally yield an additional 30 cfm of airflow.

Next, Brad begins the conversion to 4-corner idling circuits. He prefers to use a new primary metering block with a power valve on the secondary side of the carburetor. By using the exact same primary metering block on the secondary side Brad keeps the carburetor balanced. To accommodate the power valve, the main body is placed in a fixture and a quarter-inch-deep opening is bored in the required location with an end mill.

The baseplate is also modified with a pair of new idle-air circuit holes drilled in another special fixture. Then the front and rear idle circuits are separated for increased control by blocking the idle-air

Brad's initial step is to knock out the boosters and flow each venturi separately. The venturis are then individually honed until they all flow the same.

Modified boosters are installed and the venturis are reflowed to check for consistency. Any variation in flow caused by a booster is corrected by honing the individual booster until all venturis and boosters are balanced. When the boosters are flowed and the metering plate is installed, the pressure drop across the entire system is measured. By installing jets, Brad can even flow-match a set of jets so the whole system is balanced.

To accept a standard primary metering block and power valve, a quarter-inch-deep hole is bored in the secondary side of the main body.

This aluminum jig is used to ream and hone the boosters for flow balancing. One booster is in the jig ready for reaming while a finished dogleg booster (A) sits adjacent to it. In the foreground is a polished straight booster (B) and an untouched dogleg booster (C).

New 3310s come with a nonadjustable rear metering plate, as shown on the left. Brad's modified carburetors are fitted with fully adjustable rear metering blocks, which permit 4-corner idle adjustment. Note the hex-nut jet extensions used to keep the jets covered with fuel during acceleration.

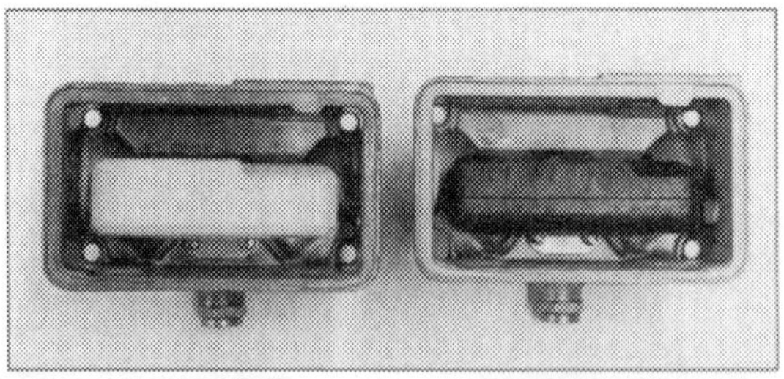

New 3310 carbs come with a white plastic float (left). Brad's carburetors use a modified Nitrophyl float with notches to clear the rear jet extensions. He matches the front (brass) and rear float weights by adding a small amount of shot to the rear float.

Brad plugs the secondary feed hole (see arrow), causing the secondary diaphragm to quicken secondary action. Because the primary feed hole is too small, Brad installs a modified brass tube cut at a 45-degree angle (inset A) in the primary bore to pick up a better air-speed signal. The soft lead in the channel is pushed past the primary hole so that it only blocks the secondary hole (inset B). Now, secondary action is controlled by primary-bore air

The throttle plate fits into one of Brad's special jigs to drill the secondary idle-air passage at the proper angle.

Brad uses clear epoxy to secure the new, modified venturi boosters after installing them with a special booster installation tool that locks the boosters solidly in place. This is a critical step in preventing internal vacuum leaks and the possibility of hanging the throttle open if a booster were to fall out.

Brad installs a small plug in the end of the throttle shaft bore to prevent leakage and to keep dirt out of the throttle shafts.

Clear Holley sight plugs are used to make the float level observable. The secondary diaphragm housing is outfitted with Holley's removable cap to permit quick and easy spring changes.

While this practice freaks out Holley reps, Brad likes to remove the secondary-throttle stop screw and reinstall it from the top of the throttle plate. This allows you to adjust it without taking the carburetor off the engine.

The Carburetor Shop's ColoRestore process restores the original Holley colors on all the carburetors on which it is applied.

CARBURETOR SHOP 3310 HOLLEY
PRIMARY MODIFICATIONS

✔	4–CORNER IDLE CIRCUITS
✔	FLOW BALANCED VENTURIS
✔	HIGH-FLOW BOOSTERS
✔	THINNED THROTTLE SHAFTS
✔	COLORESTORE FOR ORIGINAL COLOR

SOURCE

The Carburetor Shop
Dept. HR
1457 Philadelphia Ave.
Unit 24
Ontario, CA 91761
(714) 947-7744

channel in the middle on each side with lead shot. Additional machining on the main body is required, but Brad doesn't recommend doing it yourself because you can cause vacuum leaks. The throttle shafts are fitted with four No. 172 throttle blades and then milled with the screws in them to thin their profiles at wide-open throttle. This is good for an extra 25 cfm. The end of the throttle shaft is shortened 3/8 inch to allow room for a small plug in the end of the shaft bore. This prevents leaks and keeps dirt out of the throttle shaft bore.

The primary float bowl is fitted with standard Holley brass floats like those used on earlier carburetors. Late-model carbs come with white plastic floats that don't provide balanced float action because they are 4.5 grams lighter. Since the rear floats have to be notched to clear the rear jet extensions, Brad uses a Nitrophyl float with small bits of shot epoxied into it to match the weight of the brass float in the front bowl.

Once the machining operations are completed, Brad uses ColoRestore on the carburetor. His unique process recreates the original Holley color found on early carburetors. The carburetor is assembled with all-nylon gaskets because Brad doesn't want you tearing any paper gaskets if you work on it. A smaller .125-inch squirter and an accelerator pump cam with a larger ramp are included for tuning purposes. A .128-inch squirter is installed during assembly.

With Brad's modifications, these Holleys are capable of delivering greater performance than ever before. We swapped Brad's carburetor for a previously optimized 3310 on our 383-cubic-inch small-block Chevelle and picked up a tenth right out of the box on a 90° day. Optimizing the jetting and changing the squirter gave us another .05 second and the car ran its quickest pass ever at 12.60, and the smoothness and driveability gained from the 4-corner idle circuit was dramatic. On a 70° day there is probably another tenth still lurking in there.

Brad says the flow capacity can be as much as 870 cfm, but he stresses that the balanced flow capacity and critical control of 4-corner idle circuits are more important. Each carburetor is run on a test engine to set float levels and to check for leaks prior to shipping. These are ready-to-run-right-out-of-the-box, hard-core Holleys for hard-core racers and street machiners! HR

POWDER TO THE PEOPLE

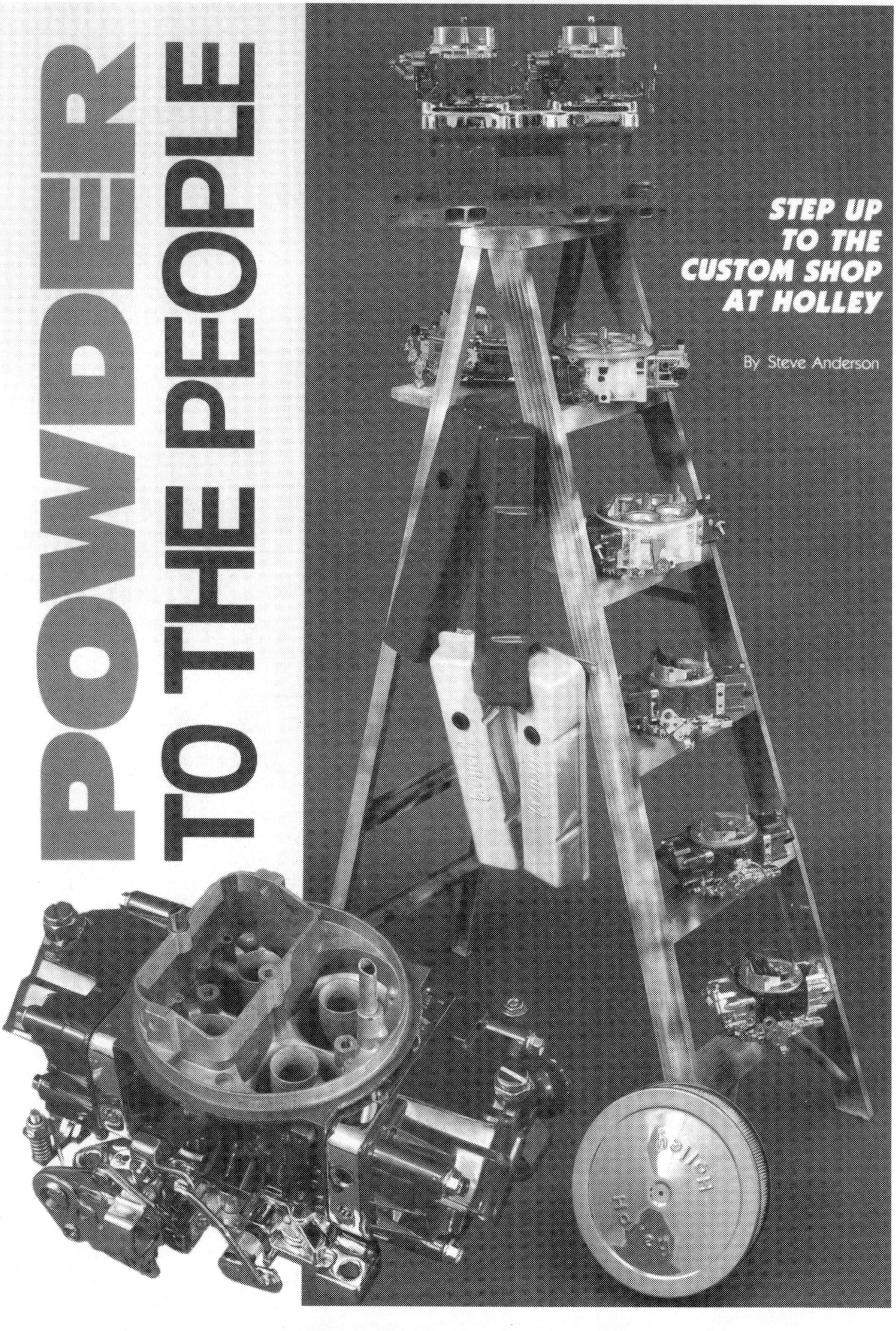

STEP UP TO THE CUSTOM SHOP AT HOLLEY

By Steve Anderson

There's no doubt about it: Today's hot rods are looking better than ever! Detail is the key to everything, including the paint, the interior, and especially in the engine compartment.

What better time for Holley to introduce its neatly detailed line of carburetors and related components—products aligned with the specific needs of today's enthusiasts? Virtually any combination can be created utilizing chrome, black chrome, or 24-karat gold plating with any of 11 different shades of powder coating, including three translucent candy-apple colors. Whatever the application, there's a coating, or combinations of coatings, that is well suited to it, and without the problems once associated with other plating processes. Not only are these various coatings great looking, but they also are resistant to scratches and chips, high temperatures, and harmful elements such as solvents, brake fluid, gasoline, and battery acid. In addition, a variety of steel and aluminum valve covers, air cleaners, intake manifolds, and electric fuel pumps is available in a number of matching hues.

Whatever the application, Holley's Custom Shop can cover it with a minimum of hassle. Simply call or write to order your carburetor of choice and to determine exactly how it will be detailed. If you have already purchased a new Holley, it too can benefit from the same procedures. In addition, plated screws, bolts, and nuts are available to finish off the update. Unfortunately, the plating and coating processes cannot be applied to used carburetors or related components.

For those interested in replicating an O.E.M. look, Holley's Custom Shop also offers its original zinc dichromating service. Restoration buffs will now be able to send their larger carburetor components (float bowls, metering blocks, main body, choke, and secondary diaphragms) to be cleaned and refinished in their original, gold-tone finish.

For those less inclined to rework their carburetors, Holley's Custom Shop also offers a complete rebuilding service. The procedure consists of a visual inspection for worn components, which are replaced, and stripped threads, which are retapped. All components are then cleaned both chemically and with a vibratory process. From there the various castings are redichromated; steel components are cadmium chromated or plated to resist corrosion. All gaskets are replaced along with the choke and secondary diaphragms, the accelerator pump assembly, the needle and seat assembly, and the power valves. Once the carburetor has been assembled, the thermostat, choke, and overall operation are verified during flow-bench testing.

With such a wide variety of platings and coatings offered on virtually all of Holley's carburetors, hot rodders are sure to find the combination that's right for them. The highest levels of quality are maintained throughout the entire process, including the bench testing of every Holley carburetor before it's shipped to the customer.

Another specialty of the Custom Shop is its first-rate line of blueprinted carburetors, from the 2-barrel 500-cfm 4412 series to the awesome 1050-cfm Dominator 4-barrels. Each has been carefully matched to the requirements of NASCAR or NHRA (depending on the application), with several sizes for each. In addition to fine tuning performed on the boosters, idle feed, transfer slots, air bleed, and jets, each carburetor is maximized within various class regulations. Like all carburetors reworked by Holley, each is flow-bench tested to ensure ideal operation, performance, and reliability.

As we said, detailing has become one of the most—if not *the* most—important considerations for today's hot rodder. How an engine looks is equally as important as how it runs. And, in giving attention to both areas, Holley's Custom Shop has made it possible to add greater detail in areas once considered purely functional. So before you bolt down your new Holley, consider a variation on an age-old theme. Then you too can give your Holley the powder!

To order, write to Holley Replacement Parts, Custom Shop, Dept. HR, 601 Space Park North, S. Cartwright St., Goodlettsville, TN 37072, or call (615) 859-3124, 7:30 a.m. to 3:30 p.m., CST. HR

In addition to its wide range of coated carburetors, Holley's Custom Shop also offers a complete line of intake manifolds, valve covers, air cleaners, and electric fuel pumps in several matching coatings. Whether alone or in combination with other visual enhancements, the results are eye-popping.

MIX MASTERS

DECK YOUR ENGINE WITH NEW CARBS FROM HOLLEY

By Marlan Davis

New Holley high-tech polished aluminum carbs combine good looks with ease of manufacturing and maintenance. Selected models will incorporate special mods (such as anti-slosh "J"-shaped bowl vents, right) that make them fully qualified for Marine use.

The original Holley Model 4150 carburetors were introduced over 30 years ago. While the now-classic design has withstood the test of time and outlasted most other competitive 4-barrels, inevitably the 30-year learning curve and modern technological advances have shown there's a need for design refinements. In that vein, Holley Replacement Parts Division is introducing all-new, high-tech universal carburetors that incorporate significant improvements over previous Holley 4-barrels, while still retaining all the ease of tuneability and parts interchangeability for which traditional Holleys are famed.

Why a new carburetor model at this time? There are several reasons: With manufacturing and material costs continually rising, Holley needs an easier-to-manufacture, less-labor-intensive design with fewer separate parts and castings. Traditional Holleys look somewhat dated in the modern 1980s billet, polished, "high-tech" engine compartment. Existing Holley carburetors are not recommended for Marine applications since their basic design doesn't conform with current Coast Guard safety guidelines. Finally, Holley engineers thought that (good as it is compared to the competition) they could design a carb that's even easier to tune and service.

The new carburetors are intended to complement, not replace, the existing Holley carburetor line: Both standard/Holley flange (basic Model 4010) and spread-bore/Q-jet flange (basic Model 4011) designs will be offered, with a choice of vacuum or mechanically actuated secondaries (for specific part numbers, see chart). Someday there may be a 2-barrel version.

The Model 4010/4011's basic design is a radical departure from previous Holley practice. Up to now, Holleys have basically disassembled "horizontally." In other words, you'd loosen the float bowl screws, then disassemble the fuel bowls and metering blocks or plates from the main body, which in turn was unscrewed from the base plate. This system used lots of gaskets, increasing the potential for leaks. The float bowl and metering block gaskets also had to "fight" gravity, which was constantly trying to "push" gasoline in a downward direction. A jet change required draining the float bowl first. All these separate castings and gaskets, the fasteners to retain them, and the people to assemble them, are quite costly in today's world. By contrast, the new Holleys have but two main components: the main body and the airhorn. The main body encompasses the throttle body, main body and fuel bowls in one complete, integral unit, while the airhorn casting acts as a "cover." The result: Only one major gasket (between the airhorn and the main body) is now required, and because it installs flat, high up on the carb's body, it works with gravity instead of fighting it. The two major castings simplify manufacturing and assembly. And the float bowls no longer need to be drained to remove jets—a definite advantage in the "between rounds thrash."

The 30cc accelerator pump housing (left) and diaphragm is the same as used on "traditional" Holley 4-barrels. So is the power valve, but it's relocated within a separate housing on the carb bottom.

The jets themselves are completely interchangeable with existing Holley

The standard "Holley" flange carb (basic Model 4010, left) will be offered in 600- and 750-cfm ratings; the spread-bore (basic Model 4011, right) comes in a choice of 650 or 800 cfm. Holley plans to offer both vacuum secondary and double-pumper versions. Vacuum ports are provided for timed distributor spark (a), timed EGR (b), manifold vacuum (c), and PCV (d).

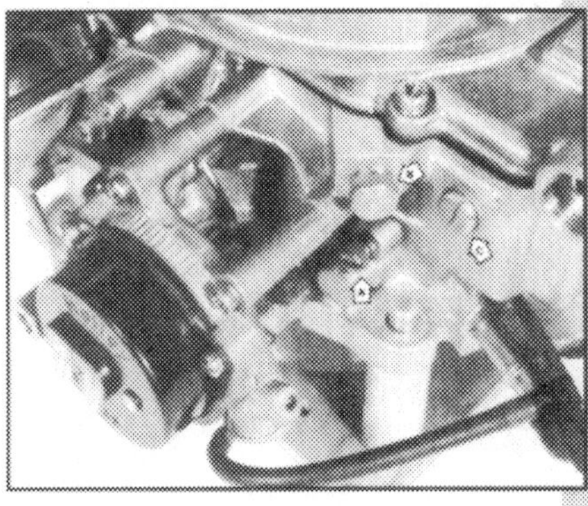

Except for List Nos. 0-84021 and -84022, all the new carbs have manual chokes. Electric choke conversion kits are available separately. Note location of idle mixture adjustment screws (one side shown, a). Marine carbs have provisions for venting mechanical fuel pumps directly to venturi (fitting, b), thereby ensuring against accumulation of harmful vapors in the typical restricted, Marine enclosed engine compartment. As with existing Holleys, a float sight-level gauge is provided on the side of the bowls (c).

Accelerator pump cams and linkage are interchangeable with traditional Holley carbs. The new plastic arm shown here will eventually be incorporated on all Model 4150/4160/4165/4175 Holley carbs, except for the musclecar replacement series. It has been criticized by some hard-core professional racing carb builders, who often bend the old metal pump arms to optimize accelerator pump tip-in. (Holley says they will continue to offer metal "arm kits" separately for those who demand them.)

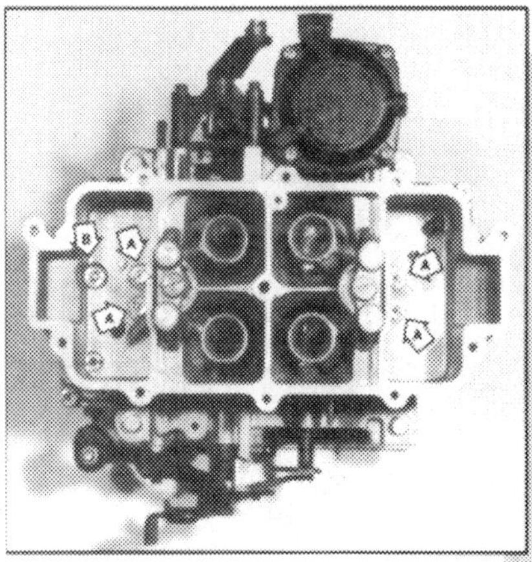

The new carbs have only two major castings—a main body incorporating the throttle body, main body, and fuel bowls in one integral unit is sealed by the airhorn "cover." Integral float bowls have the same fuel capacity as traditional center-hung bowls. Jets (arrows a) are easily accessed by removing cover; it's no longer necessary to drain float bowl. Power valve (b) screws in from bottom (see photo, page 80). Production versions will have power valves on both primary and secondary sides to enhance tuning flexibility. Vacuum secondary diaphragm and linkage is the same as on earlier Holley carbs. The plastic cover is found on most new Holleys.

HOLLEY MODEL 4010/4011 AVAILABILITY

MODEL 4010/HOLLEY FLANGE	
LIST NO.	**DESCRIPTION**
0-84010	600 cfm vacuum secondary
0-84011	750 cfm vacuum secondary
0-84012	600 cfm mechanical secondary
0-84013	750 cfm mechanical secondary
0-84018	750 cfm vacuum secondary, Marine
0-84020	600 cfm vacuum secondary, Ford kickdown lever
MODEL 4011/SPREAD-BORE FLANGE	
LIST NO.	**DESCRIPTION**
0-84014	650 cfm vacuum secondary
0-84015	800 cfm vacuum secondary
0-84016	650 cfm mechanical secondary
0-84017	800 cfm mechanical secondary
0-84019	800 cfm vacuum secondary, Volvo Marine
0-84021	650 cfm vacuum secondary, electric choke
0-84022	800 cfm vacuum secondary, electric choke, Marine

4150/4160/4165/4175 carburetors. So, too, are the power valves; accelerator pump actuating arm, linkage, cam, and diaphragm; needle and seat; floats; secondary vacuum diaphragm and diaphragm springs; and chokes. That means tuning parts are readily obtainable. One "tuning" part that's unique is the accelerator pump shooter(s)—there isn't any. Traditional Holleys have separate bolt-in shooters of various orifice sizes and design configurations. They install above the booster venturis, which in turn require special tools for removal and installation. On the long oval tracks or in severe off-road applications, the boosters sometimes vibrate loose. In the new 4010/4011, the shooters' function is assumed by restrictors pressed into a bolt-in cluster that combines the "shooters," booster venturis, main well, and idle tubes in a single integral unit—one each for the primary and secondary side. The unique clusters are easily removable, allowing the custom carb tuner easy access for polishing the bores, and/or substitution of custom venturis of varying designs. All the new carbs will come with factory-installed annular discharge booster venturis; similar to the ones used in the latest Holley racing carbs, they bring the main system in quicker, while offering better fuel atomization.

With their anti-leak design, the new carbs are much better suited to the Marine environment than earlier Holleys. However, Holley also plans to produce dedicated Model 4010 and 4011 carbs with additional special components designed to fully conform to all Coast Guard requirements: "J"-style fuel bowl vent tubes will be incorporated to resist fuel slosh. Special throttle shafts with grooves to direct fuel back into the bore eliminate yet another path of potential fuel seepage. Finally, provisions are included to allow venting mechanical fuel pumps directly to the venturis.

While traditional Holleys are built from zinc alloys, the new Model 4010/4011 4-barrels will be manufactured from aluminum, with major components put through a vibratory milling (tumbling) process for a polished appearance. The bright polished aluminum is set off by

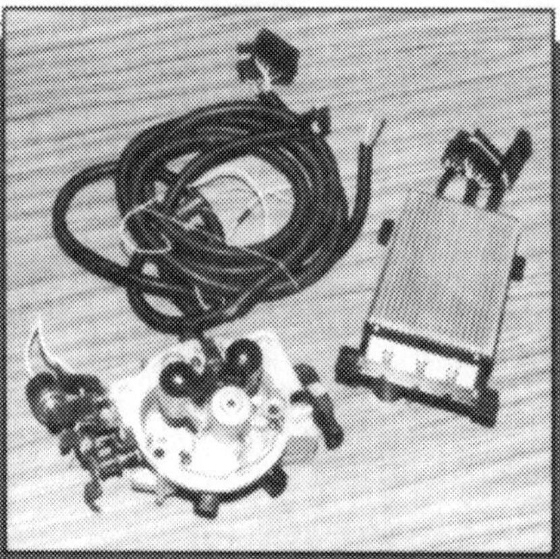

HARD BODIES

While most hot rodders are still into carburetors, Detroit has switched over almost entirely to fuel-injection systems, which usually offer superior driveability over the typical stock carburetion systems they've superseded. Unfortunately, the new FI systems don't readily lend themselves to the usual performance modifications, and are also expensive to repair. Holley plans to solve these problems with their stand-alone throttle body injection (TBI) system, under development for possible introduction by fall 1988. Components of the new system include the throttle body, throttle position sensor (TPS), inline electric fuel pump, computer, engine temperature sensor, intake manifold adapters, and complete wiring harness.

The new throttle body features dual 2-inch o.d. bores; without the typical restrictions imposed by a conventional carburetor's venturis, the units flow 650 cfm at 1½-inch vacuum drop (the same as an equivalent Holley 4-barrel). All system components are manufactured by Holley, down to and including the injector nozzles. With currently available injector nozzle sizes, one throttle body is capable of supplying the fuel/air needs of a 300-hp engine. Holley plans to offer a dual throttle body system capable of feeding up to a 600-hp engine. These could be used to make a tunnel ram much more streetable than is usually the case with conventional carburetors. The throttle body is fed fuel under 12 to 15 psi by the supplied inline electric pump. A return line to the fuel tank is required.

The new stand-alone computer is fully adjustable—merely "dial-in" the desired fuel curve characteristics via its external adjustment knobs. The computer determines fuel delivery based on input from the TPS and tachometer. "Choke mode" is supplied by the engine temperature sensor. A prefabricated wiring harness with moisture- and fluid-resistant "Weather-Pak" terminals connects the various sensors with the computer.

Holley reports that, installed on a '79 350-powered Z/28 *(photo)*, the throttle body has produced up to ½-second quicker quarter-mile e.t.'s when compared to a 0-1850 600-cfm 4-barrel (which itself is quicker than the stock Q-jet). I drove the test car and found low-end response definitely superior, with no loss of power on top, either. In addition to the stand-alone unit, Holley plans to offer 650-cfm TBI units to replace the smaller stock GM units used on late-model pickup trucks; they'll bolt right on and hook up to the stock electronics. A smaller unit for S10s (that's still larger than the GM stocker) is also under development. And further down the road there are rumors of a Holley port fuel-injection setup.

blackened screws, nuts and hardware, all resulting in a custom "high-tech" appearance. Individual enthusiasts who desire a unique stylistic approach can opt to have the aluminum components locally anodized in any one of a virtual rainbow of hues offered by most such companies (color-coordinated engine compartments, anyone?).

All this sounds as if Holley has indeed succeeded in meeting their design goals without sacrificing quality. Are there any drawbacks to the new design? Maybe, depending on your attitude toward "plastic." Many parts on these new prototype carbs—including the floats, secondary vacuum diaphragm cover, accelerator pump arm, and parts of the choke linkage—are made from various plastic compounds. In addition to their use on the new Model 4010/4011, these parts will eventually be installed on most traditional Holley carbs, too. While highly controversial among some enthusiasts, Holley assures us that all components have been thoroughly tested. For example, the new one-piece plastic float is said to be resistant to all fuels and solvents; it eliminates the porosity problems encountered with nitrophyl floats or the occasional leakage through the brass float design's brazed seam. Nevertheless, the plastic choke throttle blade shaft conceivably could sustain damage induced by an engine backfire and resultant flash fire. (Holley is aware of this, and the latest word is that choke blade shafts will continue to be made from metal.)

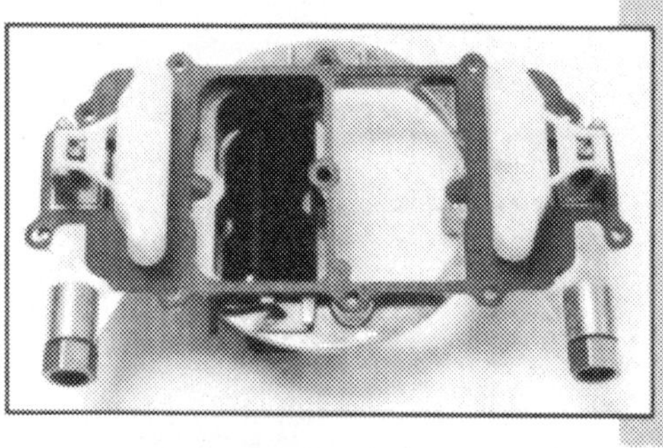

The airhorn cover contains the float, needle and seat, choke blade and shaft, and dual-feed fuel inlets. Externally adjustable needle-and-seat assemblies and center-hung plastic floats interchange with parts used on existing Holleys. (Traditional Holleys are switching to the new plastic floats.) The 5-inch-diameter airhorn mounting flange accepts standard air cleaner assemblies.

Annular discharge booster venturis are incorporated as part of bolt-in (and, therefore, easily removable) "cluster casting," which also houses accelerator pump discharge ports (arrows), *main well, and idle tubes. Holley eventually plans to offer a "restrictor kit" consisting of discharge port restrictors with different orifice sizes, but in the meantime you can increase pump discharge size by drilling the orifices oversize.*

Minor quibbles aside, on the whole it appears that Holley has another winner. Holley plans to price the new units somewhere in between the no-frills, low-cost universal models and the more expensive competition units. The Model 4010 should be available this month, with the Model 4011 to follow in July. While primarily intended for the discriminating street enthusiast, street rodder, or hot boat owner, the new carb's leakproof design and ease of disassembly could find it a home in selected competition applications as well. Then Holley will really be able to take a "bough," as enthusiasts deck their engines with their latest fuel metering product! **HR**

ADJUST-A-CARB

HOLLEY BORROWS FROM THE COMPUTERIZED CAR TO BUILD THE ENGINE-TUNER'S DREAM CARBURETOR

By Pat Ganahl

Why do you suppose carburetors are so hard to adjust? Even a simple jet change requires major disassembly. It's my hunch they're made this way just so the average idiot can't screw them up. But knowledgeable performance enthusiasts seeking the perfect engine tune on any given day—especially race day—are resigned to the ordeal of draining float bowls, tearing gaskets, mending leaks, and all the rest that goes with typical jet changing.

Ever dreamed about a carburetor . . . a high-performance carb, no less . . . that you could tune by simply twisting a knob on the dashboard? It's no dream. Holley has had one (actually dozens) for a few years now. We have no idea why it's been such a well-kept secret.

They're called Mile-Dial and Quarter-Mile-Dial carburetors, and they are a very logical outgrowth of our new computer car generation. As we explained in our treatise on new-car computer systems ("Computer Class," HRM February '85), the typical computer-controlled "feedback" carburetor uses a solenoid-actuated enrichment valve to add extra fuel to the standard circuits in the carburetor. The solenoid lifts a plunger, or needle, off its seat 10 times every second. The seat is located in the carburetor float bowl, and it has a hole in it just like a main jet. When the needle lifts, fuel flows from the bowl, through the seat, and into a passage connecting to the

What's so unusual about this Holley 4-barrel? Other than the odd-shaped front float bowl, the two wires coming out of it are the only clue. They lead to a small control box on the dash that allows you to tune the carb 10 jet sizes at the twist of a knob!

Typical Quarter-Mile-Dial kit contains primary and secondary float bowls with electronic solenoids, metering blocks, the signal generator control box, and hook-up wire. Some applications, and all Mile-Dials, fit primary side only. Complete carbs are also available.

ADJUSTO

main jet wells. In a computerized car, the ECM monitors exhaust gas oxygen content (among other things), and sends a signal to the carb solenoid 10 times a second, telling it how long to keep the needle on the seat during that tenth of a second. The "on-seat time" of the needle is called the duty cycle of the solenoid, which determines how much fuel enrichment the solenoid will add to the mechanical carb circuits, thereby either richening or leaning the fuel mixture. If the needle stays on the seat 100 percent of the time (100-percent duty cycle), it would be in its leanest adjustment, since no extra fuel would be allowed to pass.

The computer constantly adjusts the fuel mixture in new cars to keep emissions to a minimum. But a perfect stoichiometric air-fuel ratio (14.7:1) isn't necessarily best for either performance or economy driving. But a manually—or electronically—adjustable carburetor is. Since Holley already had computer-adjustable carburetors (commonly called "feedback" carbs)—that is, they had float bowls with electronically controlled enrichment solenoids—they figured, "Why not hook one of these up to a control knob, instead of a computer?" That's just what they did.

The control knob operates a small integrated-circuit signal generator contained in a black box about the size of a cigarette lighter. Depending where you set the knob, it sends a signal to the carburetor solenoid 10 times a second, telling it how long to keep the needle on the seat per cycle. There are 10 settings (clicks) on the knob, each click being equivalent to a mixture change of approximately one main jet size (number). The signal generator mounts anywhere within easy driver reach, hooks up with four wires (12-V in, ground, and two leads to the carb), and lets you tune your carburetor up to five jet sizes richer and five leaner, as you drive, just by turning a knob.

Holley's Mile-Dial carburetor is designed primarily to optimize fuel economy and allow in-car adjustment for altitude driving and other variable conditions. The Quarter-Mile-Dial is a performance carb. Both can be ordered as complete units, or as an add-on kit to fit most common Holley 2 and 4-barrels. If you have a Holley on your car, chances are you can convert it to dashboard adjustability without even removing the carb from the manifold.

Besides having a slightly smaller orifice in the solenoid seat, Mile-Dials channel enrichment fuel to the power valve circuit. Thus fuel enrichment is limited to the capacity of the power valve channel restriction orifices, and is superfluous once the power valve opens (at low vacuum; i.e., wide-open throttle).

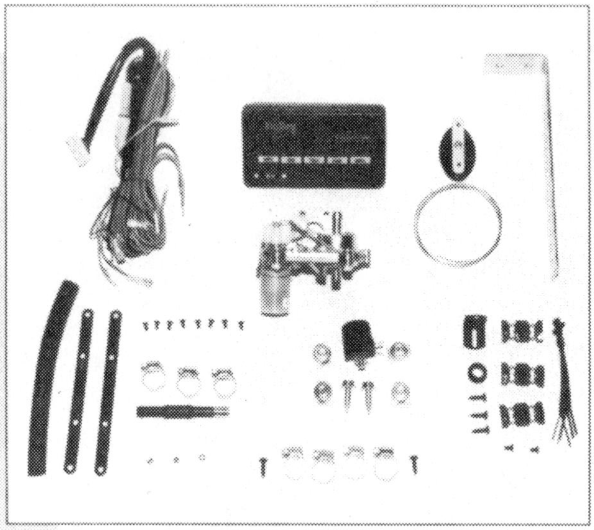

If you install Holley's slick trip computer along with a Mile-Dial carb, you can read mpg fuel consumption instantaneously, and adjust the carb for best mileage.

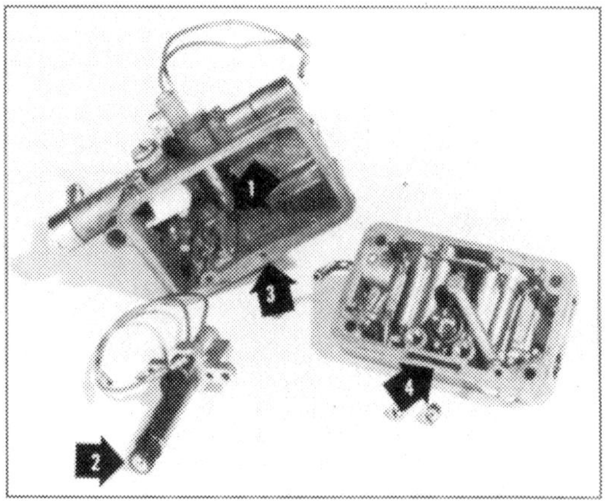

In the electronic carb the fuel enrichment solenoid (shown separately, foreground) ***takes fuel from bowl through slot*** (1)***, meters it through seat orifice*** (2) ***when needle lifts, sends it through passage*** (3) ***to slot in metering block*** (4) ***which sends it to main wells or power valve, depending on carb type.***

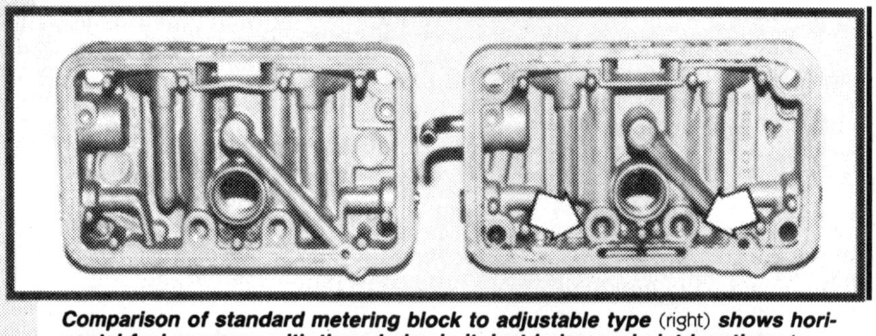

Comparison of standard metering block to adjustable type (right) ***shows horizontal fuel passage, with three holes in it, just below main jet locations*** (arrows)***. In Quarter-Mile-Dials, middle hole is plugged, routing extra fuel from solenoid through end holes, directly into main wells. On Mile-Dials, end holes aren't drilled, and fuel is sent through middle hole to power valve above it.***

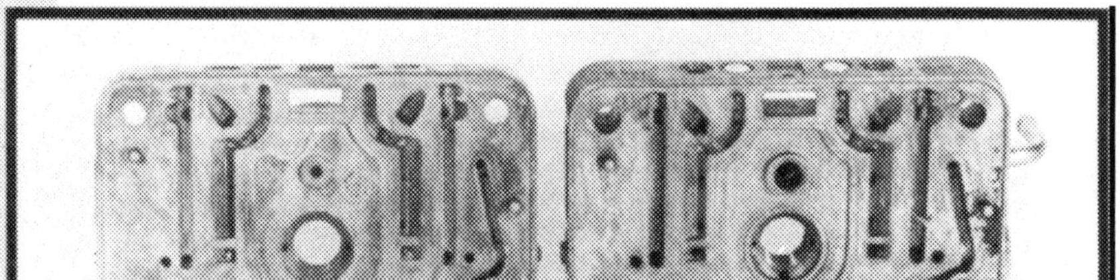

Back side of Mile-Dial metering block (right) ***shows extra hole*** (arrow) ***in bottom of power valve well. Enrichment fuel enters here and is fed to main wells through the two normal power valve orifices. Thus maximum enrichment is limited to power valve restriction hole capacity, and is "over-ridden" when power valve opens.***

ADJUSTO

That is, the Mile-Dial's adjustability primarily affects part-throttle operation, such as highway cruising. The Quarter-Mile-Dial, however, channels enrichment fuel directly to the main jet wells, so it is effective throughout the driving range above idle. Several models of the Quarter-Mile-Dial have enrichment solenoids on both the primary and secondary sides, for full top-end adjustability. To set primary and secondary jets separately, you could use two control boxes.

For a real-world test of Holley's adjusto-carb, we mounted a 600-cfm model 4160 (PN 1850) Mile-Dial carb on a 351-C 2-barrel Ford engine, fitted with an Edelbrock Performer intake manifold and a mild Crane hydraulic cam, installed in a '56 F-100 pickup. The only installation glitch was that the bottom of the solenoid housing in the front float bowl hit the raised lettering on the manifold (grinding a letter down fixed it); plus finding an aesthetically pleasing and accessible location for the stick-on control box was difficult (it was relegated to the glovebox, on a fabricated bracket).

Once the unit was intstalled, it was apparent that this carburetor was too rich, overall, to allow any adjustability. This is an important point. Since the adjustable carb adds extra fuel to the main jet circuits, it stands to reason that it must use smaller main jets than a comparable non-adjustable model. For instance, a standard No. 1850 carb comes with No. 66 main jets. The Quarter-Mile-Dial version comes with No. 64's (whether you get a complete carb or a replacement kit, the main jets in the metering block will be two to three sizes smaller than a non-adjustable version of the same carb). For this particular application, however, we had to change the primary jets to No. 62's to get the carb jetted properly at the 50-percent setting on the knob. This accomplished, we could really feel the effect of just two or three clicks to either the "lean" or "rich" sides of the dial.

For around-town driving, we found that leaving the knob in the middle position was most practical. To find the best setting, you could lean it down to where the engine pings or stumbles on acceleration, and then richen it one or two stops. Then, if you want to maximize fuel economy, you can crank it down a few notches leaner when you're cruising on the freeway. Most helpful, we found, was being able to richen the mixture to reduce pinging when pulling a load up a grade, even on regular gas. And, of course, being able to tune the carb from the dashboard whenever you decide to go racing (especially at a high-altitude track, like ours) is a welcome relief. Why didn't they do this sooner? HR

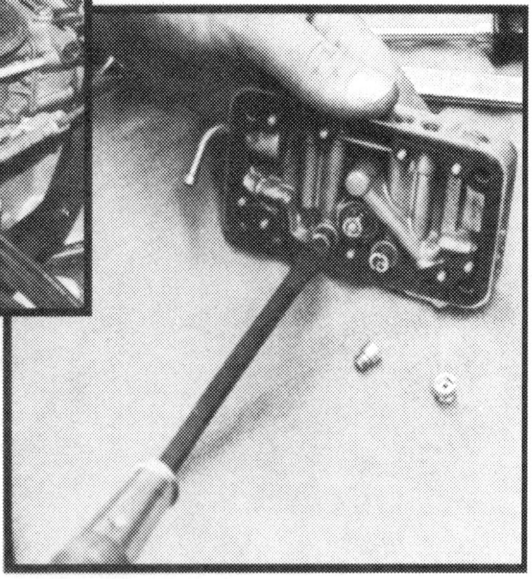

To get the electronic carb to work properly, you must jet it (the old way) for correct air-fuel mixture under normal driving conditions with the knob set in the middle position. This is important. Our test application required primary main jets two sizes smaller.

But once you've got the carb dialed-in for your specific combination, you can do the rest of your dialing, for most driving and racing conditions, with a dash-mounted knob.

POSITIVE FEEDBACK

Holley's electronically adjustable carburetors are stand-alone systems. You can't interface one (as they say) with a computer-controlled vehicle. Holley does not make a computer-compatible adjustable carb. But they do make a computer-controlled "feedback" 4-barrel that will bolt on (and plug in) in place of the feedback Quadrajet on your late-model Camaro, Firebird, Monte Carlo, or other Chevy V8. Listed as part No. 80073, it is a 650-cfm Model 4175 spreadbore, having small primaries and large secondaries like the Q-jet.

Why swap? Holley claims a slight performance improvement. Some people just like Holleys better. And the Holley will allow you to make jet changes (affecting wide-open throttle operation) on race day. However, the real bottom line is that the Holley costs a couple hundred bucks less than a replacement feedback Q-jet, should the original carb go kaput. Now if they could just attach a tuning knob to it. . .

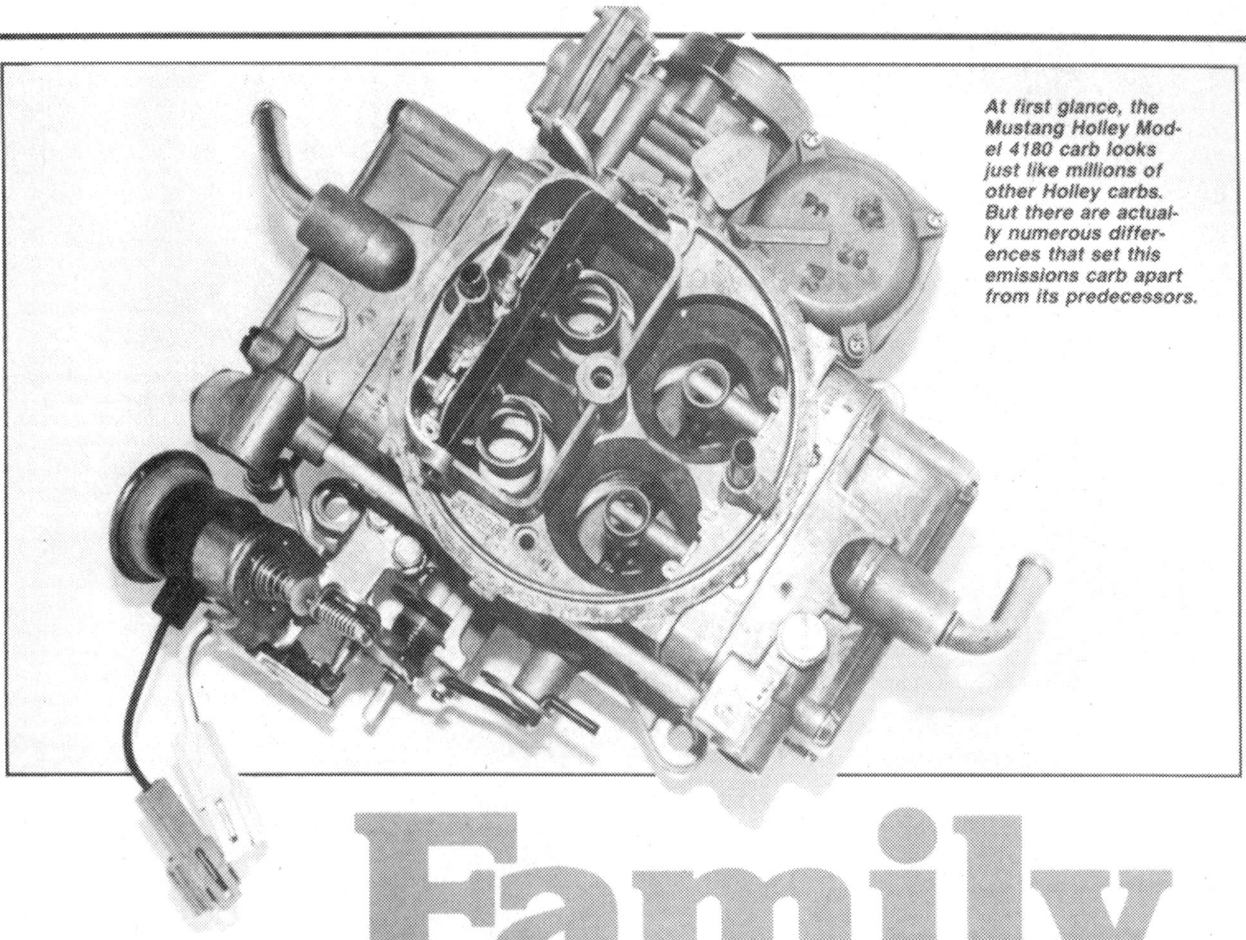

At first glance, the Mustang Holley Model 4180 carb looks just like millions of other Holley carbs. But there are actually numerous differences that set this emissions carb apart from its predecessors.

Family TIES

INVESTIGATING THE LATE MUSTANG/CAPRI HOLLEY 4-BARREL CARB

By Marlan Davis

At first glance, the Mustang Holley Model 4180 carb looks just like millions of other Holley carbs. But there are actually numerous differences that set this emissions carb apart from its predecessors.

Family TIES

The most visual external difference is the relocated idle mixture screws. Instead of the previous primary metering block location (A, arrow), they're now found in the main body, hidden behind "tamper-proof" block-off plugs (B, arrows). Corresponding internal fuel delivery passage redesign precludes use of easier-to-modify old-style metering blocks—could be a hassle with rad cams in all-out race applications.

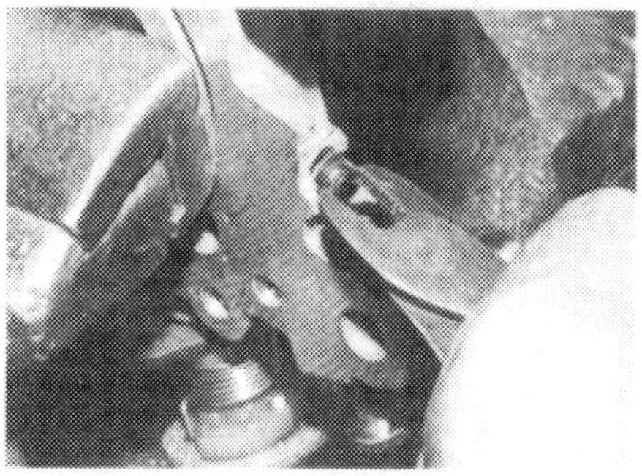

Hidden screws are accessed by sawing or chiseling away casting surrounding the plugs, then pulling plugs with pliers. Recessed screws then adjust conventionally (turn out to richen) with Allen wrench.

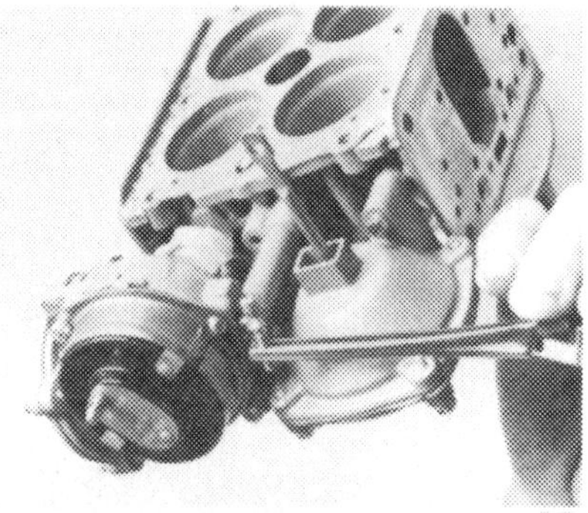

Factory-pre-set choke also has anti-tamper provisions, with choke cover retained by breakaway-head screws. To remove the special screws, carefully file heads to permit gripping them with conventional pliers; replace with standard screws.

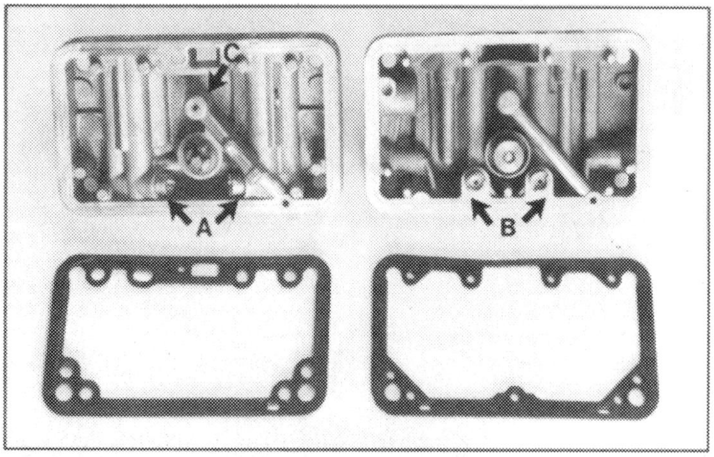

As delivered, 4180 metering block (left) comes with close-limit main jets (A). Compared to conventional jets (B), close-limit design has less percentage flow variation between jets of like number size, yet are completely interchangeable with standard design. Note how 4180's jets are installed at an angle—making some internal race passage mods harder. Block dowels are different sizes, too. Accelerator pump bleed hole (C) reduces hot-soak and percolation problems. If running big cam and/or 50cc accelerator pump, this hole should be plugged to avoid lean condition—but hot start characteristics will suffer. The 4180 requires unique float bowl-to-block gasket, part No. 108-56 (lower left).

PROOF TEST

Naturally, whenever major engine mods are performed, the carburetor usually requires recalibration. We wondered how good the stock Mustang carburetor calibration was on the stock, unmodified vehicle. Would a production carb benefit from some Fuel Curve Engineering fine-tuning? To find out, we borrowed an '84½ Mustang from Fred Colley, owner of Colley Ford (655 S. Grand Ave., Glendora, CA 91740, 818/963-4131), the largest Ford Motorsports parts dealer west of the Mississippi. The rare 5-speed 302-H.O. GT-350 convertible had only 2000 miles on the clock, having been in storage virtually from Day One.

The car was driven to 2500-foot high Los Angeles County Raceway, and baselined in stock form with all emissions control equipment hooked up. A best (corrected to sea level) e.t. of 15.75 seconds @ 87.19 mph was recorded. Bob Szabo of Fuel Curve Engineering then installed his reworked duplicate of the stock carburetor, List 50265 (Ford B5ZE-9510-GA). Various jet combinations, power valves, and secondary springs were tried, with the optimum combination (on that particular day) listed in the chart below, along with the factory's baseline calibration. The relatively light Mustang proved insensitive to secondary spring changes, until the lightest (white) or heaviest (black) springs were tried. The secondary diaphragm check ball was removed for all modified carb tests. Also keep in mind that at an actual sea level dragstrip, slightly richer jetting would be needed.

Well, how much did we gain? Slightly over two-tenths and 1 mph, actually: 15.515 e.t. @ 88.70 mph. The factory calibration is pretty good, but you can always gain more power from a little fine-tuning. Still, Ford did their homework—*with* a Holley, and *without* a computer.

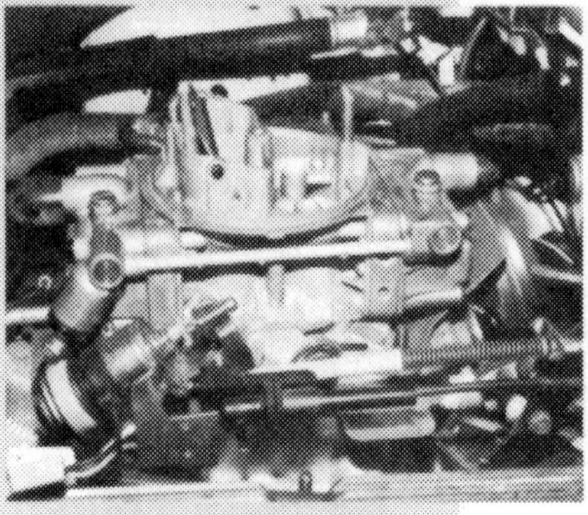

Secondaries were converted to Model 4150-style metering block with replaceable main jets. Side pivot bowl was retained to clear factory heat isolator plate and EGR valve. Center hung bowls aid cornering, will fit if EGR is deleted.

The modified carb would have made a greater difference if we could have gotten the overpowered, tire-limited Mustang to hook up. As it was, the car had to be feather-footed off the line.

TEST CARBURETOR CALIBRATION

Holley Model 4180/List 50265

	Primary Jet	Secondary metering plate or jet	Primary power valve opens	Secondary diaphragm spring
Stock	622	(Plate) 34R-10465-B	11"/8" hg	Purple (med. light)
Best Modified	622	(Jet) 68	6.5" hg	Long yellow (light)*

Note: Modified carb also included manually adjusted idle mixture, reworked accelerator pump leakage adjusted for best tip-in performance.

* Check ball removed.

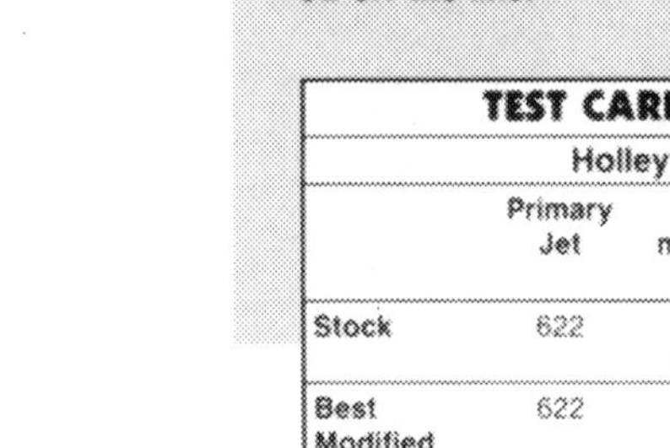

The 4180 comes with two-stage power valve (left)***. High-capacity "window-style" single-stage valves are recommended for performance use. Be sure to use plain round power valve gasket 8R-1597 with window valves having internal shoulder.***

(continued)

Special thanks to: Colley Ford/Glendora, California; Fuel Curve Engineering/Palm Desert, California; Holley Replacement Parts Div./Warren, Michigan.

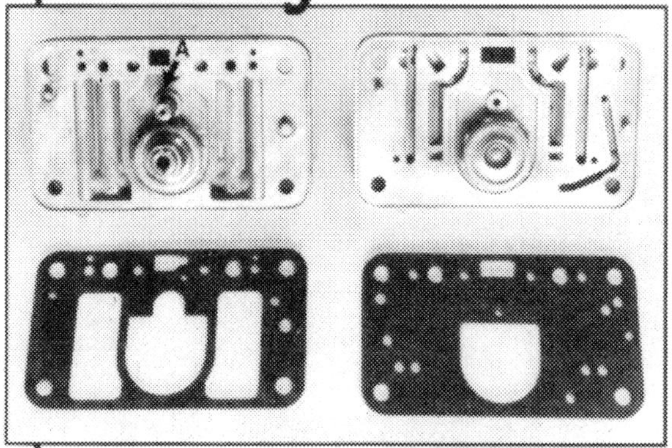

Main body side of primary metering block graphically illustrates internal design changes. Model 4180 (left) *uses separate accelerator pump transfer tube with O-ring* (A). *Most non-emissions Holleys don't have a separate tube, which was designed to correct previous internal leakage problems. Non-transfer-tube metering blocks tended to heat-warp under long-term usage, causing loss of gasket tension and consequent leakage. Use gasket 108-55* (lower left). *Fix cures the problem with no performance drawbacks.*

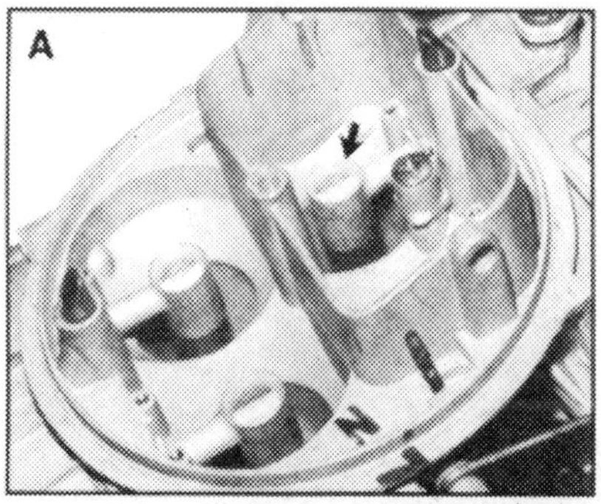

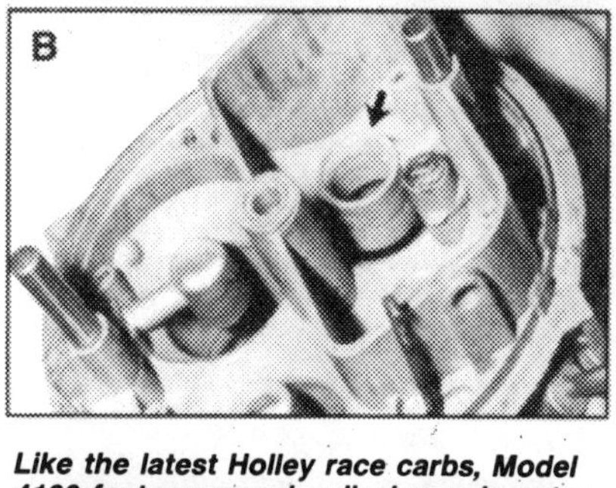

Like the latest Holley race carbs, Model 4180 features annular discharge booster venturis (B, compared to conventional design, A). *Annular discharge boosters improve overall fuel metering efficiency and broaden low-end response because of a stronger signal felt at the booster throughout the operating range, as well as improved fuel atomization.*

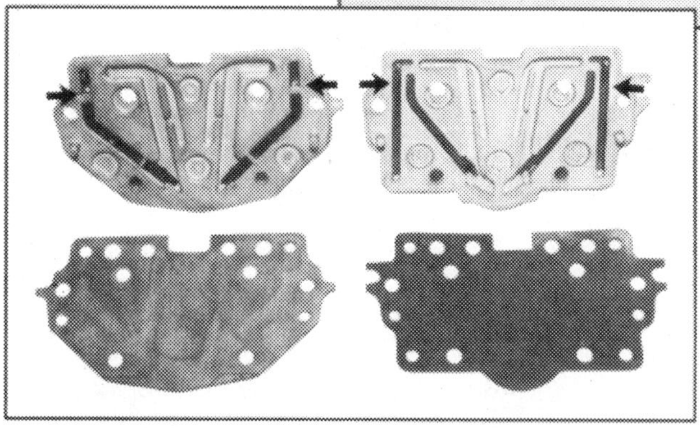

Like the earlier Model 4160's, 4180 carbs use a metering plate in the secondary side in lieu of main jets. To change secondary calibration requires either different metering plates or conversion to Model 4150 secondary metering block with replaceable jets. Model 4180 secondary plate (left) *doesn't interchange directly with most 4160 plates* (right)*—the only exceptions are metering plates used on Holley Model 2300 2-barrel Chrysler 3x2 Six-Pak end carbs, and certain Model 4160 Chrysler OEM 4-barrel installations. All other 4160 plates can be used on the 4180 only if their fuel delivery passages are blocked off to match 4180 plate block-off locations as shown by arrows. Metering plate gasket must match metering plate design—use 108-13 for 4180 plates; 108-27 for 4160 plates.*

Reason for different metering plate design: 4180 (left) *transfers fuel directly through main body, compared to 4160* (right)*, which has transfer passage as part of metering plate. The 4180 design prevents internal leaks.*

Metering plate hassles can be avoided by changing secondary metering to 4150 configuration. Kit 34-6 includes metering block, gaskets, screws, and all other necessary parts. Main jets must be purchased separately. On Mustang, use No. 70 secondary jets as sea-level baseline

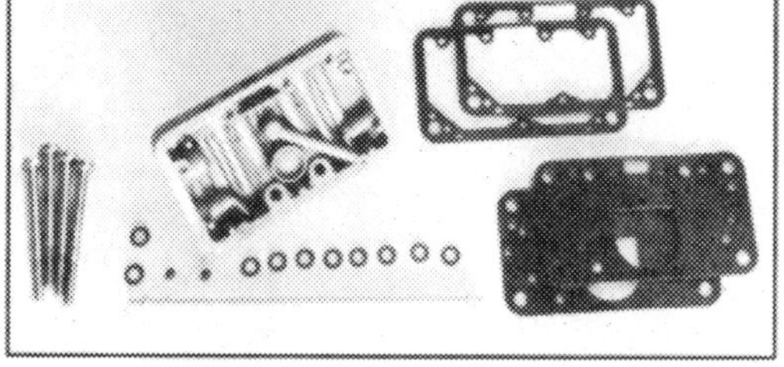

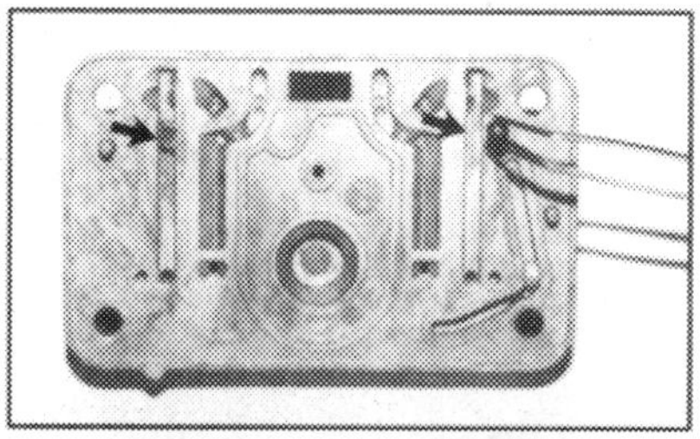

Because the 4180 main body meters secondary fuel internally when converting rear to 4150 style, use the two lead balls contained in Kit 34-6 to block off new metering block's fuel passages, as shown.

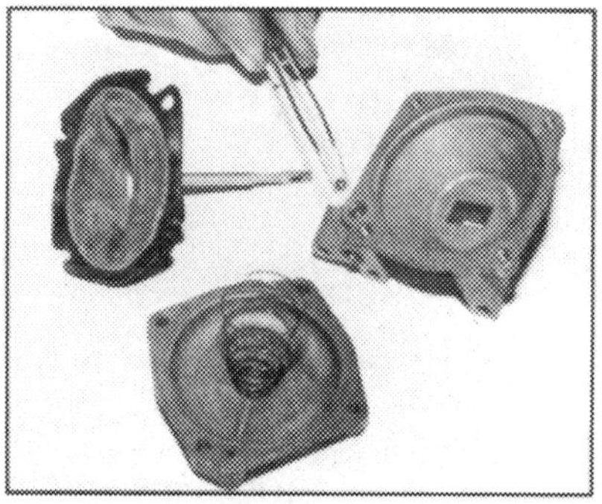

The 4180 carbs use vacuum secondaries, whose rate of opening is varied by changing to different tension springs, as contained in Kit 20-13. The lighter the spring tension, the faster the secondaries will open. Use the lightest spring that does not cause the carb to bog. Removing check ball also quickens opening rate. On light cars like the Mustang, Fuel Curve recommends first removing this ball before changing springs.

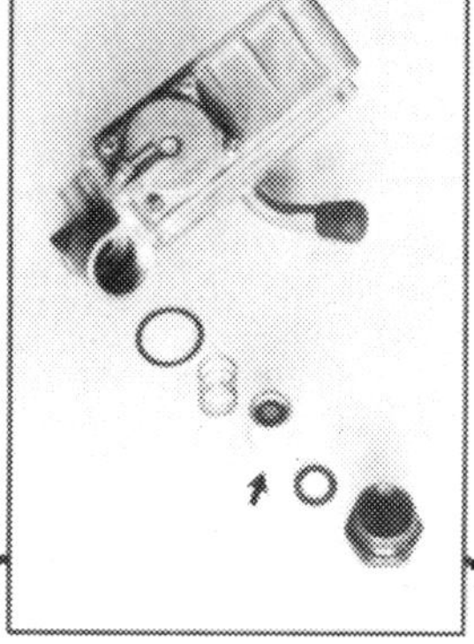

Replace restrictive, clog-prone bronze in-bowl fuel filter (arrow) *with inline fuel filters for high-performance work.* **HR**

WEBERIZE YOUR HOLLEY

NOW YOU CAN HAVE HOLLEY PERFORMANCE WITH WEBER TUNEABILITY

By David Vizard

Here's a fully assembled carburetor employing a power plate in both the primary and secondary sides. The increased thickness of the Power Plate means fuel bowls are spaced farther apart, needing a longer fuel line between the two.

Holley is the standard in performance carbs. Even in applications like Grand National racing, where carburetion is limited to one fairly small carb, the power levels exceeding 600 hp are incredible. For wide-open throttle operation the Holley 4-barrel is unequalled. And basically, the only adjustable features on a Holley are idle mixture, float level, and main jet sizing.

Holley has introduced a more sophisticated series of carbs, called their "Mile-a-Dial" and "Quarter Mile-a-Dial" electronically adjusted carbs, but these systems only affect the main jet metering system. However, serious engine builders and tuners throughout the country have been seeking the services of specialty carb shops to change the other circuits like idle, intermediate, and power enrichment. As Holley is quick to point out, not all of these shops have the expertise to effect an improvement, and when you consider that most of these circuit changes are machined modifications, you could just possibly end up with a junk carb or one that will only work with one engine combination.

And then there is the question of driveability. Modified engines place different demands on carbs than stockers. High camshaft overlap causes pulsing in the intake tract at low speeds, and a low vacuum signal usually adds up to poor throttle response and driveability. Much of the custom work performed on Holleys is strictly to improve driveability in areas other than full-throttle operation.

Bob Noguez of Redline, importers of Weber carburetors, has spent considerable time dyno-tuning both domestic and imported carbs. Bob liked the way the Weber was totally and easily adjustable. He could calibrate all circuits by changing main jets, air correctors, emulsion tubes, and idle jets, all of which are quickly accessible. By designing a new metering plate, Bob was able to combine the best Holley and Weber had to offer: a Holley carb with Weber adjustability. Redline then teamed up with Interco, an East Coast Weber distributor, and finally Weber of Italy, to put his patented metering plate, the "Power Plate," into production.

The true function of the Redline Power Plate is probably best stated by Redline's Steve Murphy. "The Power Plate is not essentially a power producer in itself. It should be looked upon as a device which will quickly and simply allow the Holley user to extract full potential by accurately and simply calibrating all three of the most important phases of calibration: idle, transition, and full power. Having said that, though, we do expect users to see some gains in power due to the various circuit refinements."

HOT ROD substantiated Steve Murphy's statement through a dyno test of a hot street 355 Chevy small-block. While horespower gains throughout the rpm range were in the 2 to 4 hp magnitude, the Power Plate actually provided a 10-hp increase at 5000 rpm on a 280-hp engine

But more than the power increase, we learned just how easily and quickly the Holley can be "dialed in." It's possible to change a pair of main jets in 30 to 45 seconds. By pulling the main jet you expose three variables open to change, which would not be the case with a stock Holley. The main jet is plugged into the end of the emulsion tube. Changing an emulsion tube changes the shape of the fuel delivery curve to the engine. By selecting an emulsion tube with a different hole pattern, the mixture can be enriched or leaned at various points within the rpm range. On top of the emulsion tube is the air corrector jet. By changing to a larger one we can flatten the fuel delivery curve, or vice versa. We optimized an identical 600-cfm 1850 Holley the old fashioned way, and it took nearly seven hours to get where we were with

WEBERIZE YOUR HOLLEY

the Power Plate in 30 minutes. Not only that, but the changes made to the other Holley are permanent.

With the Redline Power Plate we also solved a very knotty calibration problem on a Monza daily driver. In this case it required only a Power Plate on the primary side of the carb, which would be the case for most street applications. Initially the Power Plate is being introduced for 4150 and 4160 series Holleys. Depending upon how it is received, the Power Plate is likely to be expanded to cover virtually all square and spread bore-type carburetors, electronic and mechanical alike. **HR**

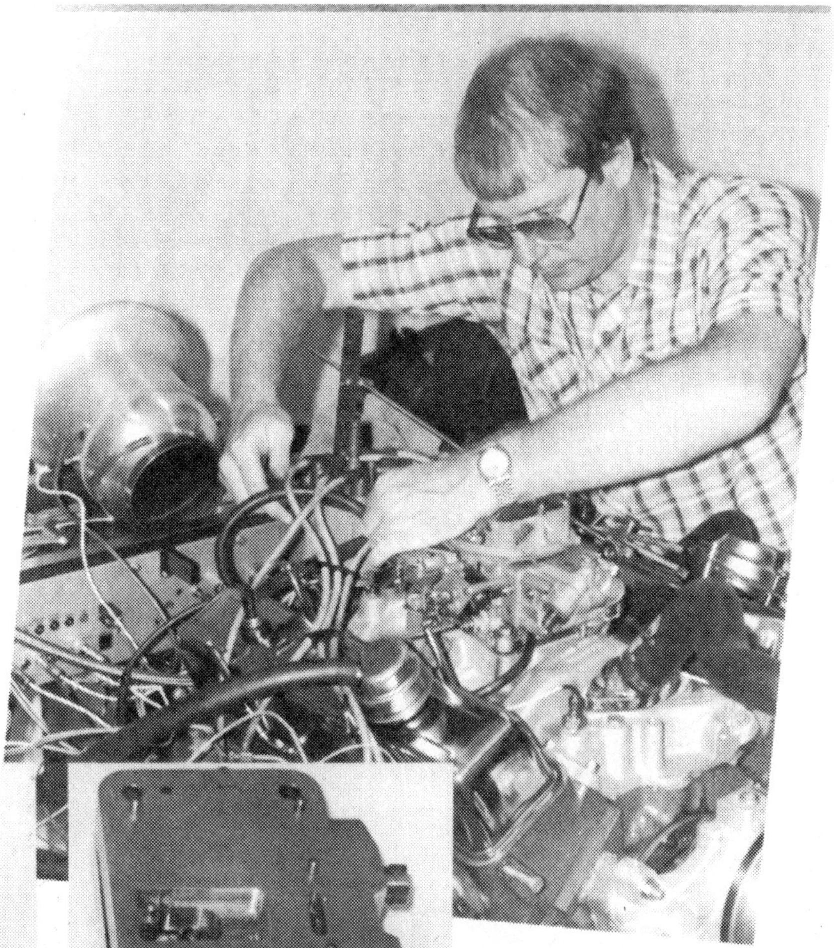

Here, our much-worked dyno mule motor undergoes a change from the stock Holley to the Weber metering blocks. Changeover time: 10 minutes maximum.

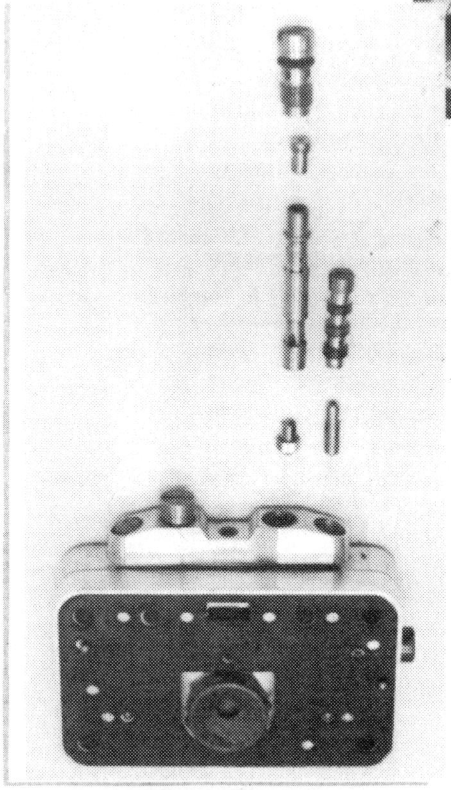

A view of the Power Plate with the main and idle system calibration components pulled out to show you what it contains. The idle system, on the right, consists of an idle jet and a combined threaded idle jet holder/emulsion tube. The longer column on the left has, at the bottom, a main jet, which plugs into the emulsion tube; and into the top of that, the air corrector is installed. Over the air corrector a threaded holder secures the assembly into the metering block.

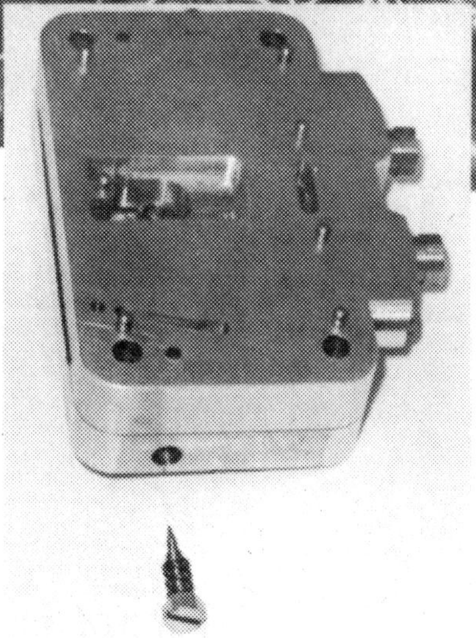

Idle mixture adjustment is achieved by conventional-tapered screw. On our test units it was found that idle mixture adjustment was very sensitive. A ⅛-turn made a relatively large change in the mixture.

The "Power Plate" kit comes with everything you will need to be up and running, including gaskets and long float bowl screws.

Here the power valve has been removed from the Power Plate. The power valve restriction channel (arrow) is, unlike a Holley, a removable jet so it can be sized to suit requirements.

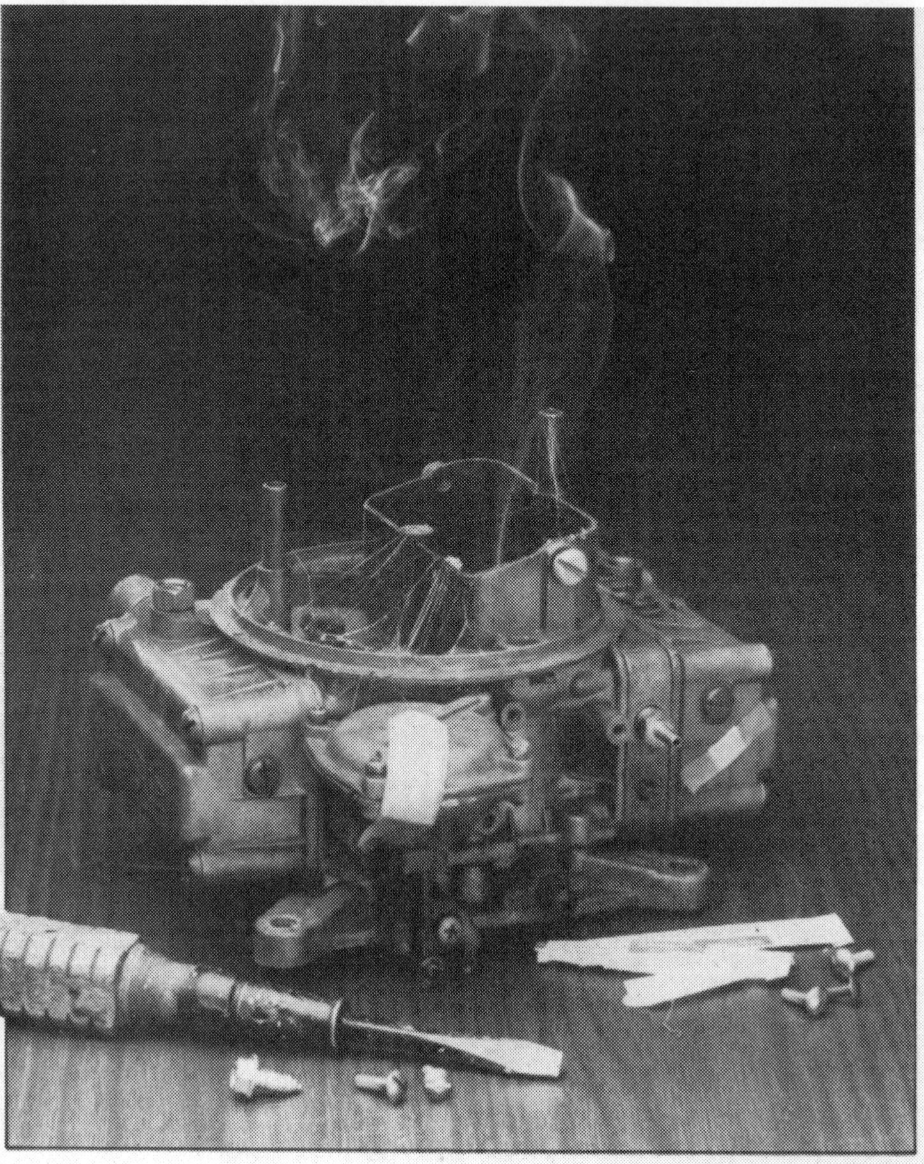

CARB GOOFS

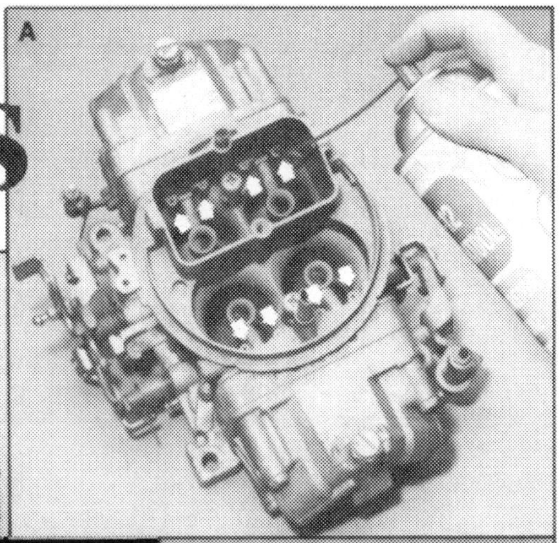

A

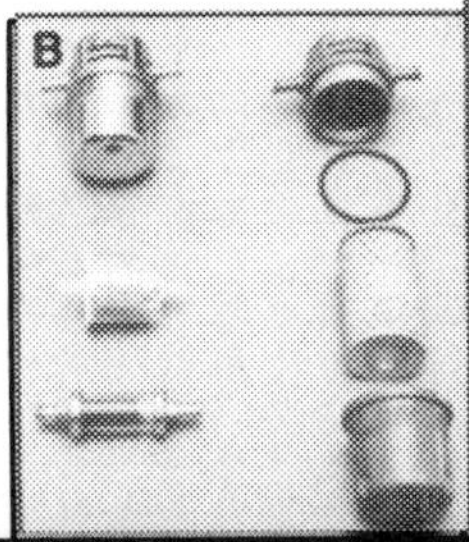

B

1/DIRT IN THE CARBURETOR

C

While our lead shot is admittedly a rather extreme example, the greatest enemy of a carburetor is dirt—internal or external. External contamination is caused by the everyday hazards of living in the real world, with its dust, debris, grease, and grime. The best defense is a good air cleaner. Surprisingly, Chuck says that lots of people don't use an air cleaner, or, if they do have one, never bother to replace a dirty element. Even with a clean element, the carb will eventually get dirty, so every so often it pays to give it a quick degrease. The air bleeds are especially sensitive to dirt, and should be periodically hit with spray-on carb cleaner (**A**).

Keeping the inside clean is just as important as keeping the outside in shape. Dirt and impurities in the fuel system can be prevented from plugging or damaging the carb by the installation of a good inline fuel filter (**B**). Cannister types offer the least flow restriction, and many have replaceable elements. See-through glass types allow easy observation of dirt buildup, indicating the proper time for replacement. Install inline filters on the pressure side of the fuel pump to minimize vapor lock tendencies. Many Holleys come stock with small sintered bronze fuel filters installed inside the fuel inlet fittings; while better than no filter at all, they're restrictive and ideally should be removed and replaced with one of the inline types.

Eventually, any delicate part wears out and requires rebuilding. A carburetor is no exception. Holley factory rebuild kits have the right high-quality parts needed to restore the carb to an as-new condition (**C**).

2/INCORRECT POWER VALVE SELECTION AND INSTALLATION

A

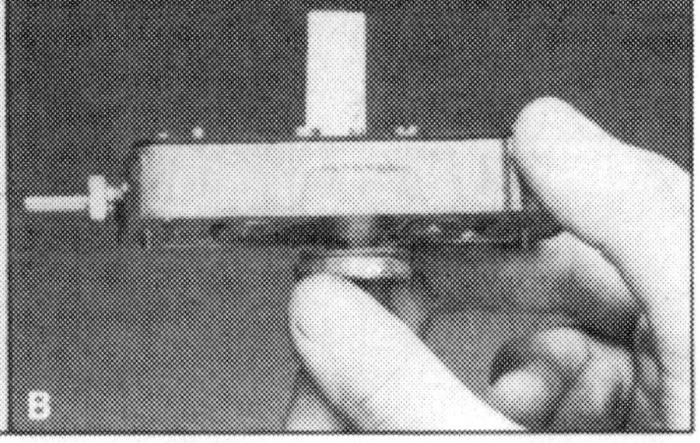

B

The Power Valve (PV) is probably the most misunderstood part in a Holley carb. Never remove the primary PV on a street car. The PV augments the fuel suppy to maintain correct metering under acceleration; if it's removed, the main jets must be richened to compensate, in turn causing an over-rich condition in the part-throttle cruise mode. A few large-cfm double-pumpers have secondary PV's. Make sure the secondary PV has the same opening point as the primary PV. Or, since the secondaries are calibrated for wide-open throttle (WOT) operation, the secondary PV can be removed and replaced with a PN 26-36 plug. You must jet up four to six sizes to compensate.

PV's are numbered according to the vacuum level below which they open. (An "85" valve opens below 8.5 inches Hg.) To prevent unnecessarily rich fuel metering under light loads, select a valve that opens at least 1 to 1½ inches Hg below the average manifold vacuum at idle on a stick-shift car (use a vacuum gauge).

On some cars equipped with automatic transmissions and tight torque converters, when the vehicle is placed in gear the vacuum drops to the point where the PV opens prematurely (below 1500-2000 rpm). Holley recommends that automatics be checked in gear.

PV's with rectangular "picture windows" offer less flow restriction than "stock" types with holes. Whichever design is used, to prevent a flooding or over-rich condition at idle it is important to match the correct PV gasket with the particular PV shoulder design in use (**A**). An 8R-1597 round gasket goes on valves with an internal shoulder; 8R-669 with three points is for external shoulder valves.

The valve and gasket are best installed by holding the metering block upside down and level with the eyes (**B**). This prevents the gasket from falling off the PV shoulder and causing a flooding condition.

3/WRONG METERING BLOCK GASKETS

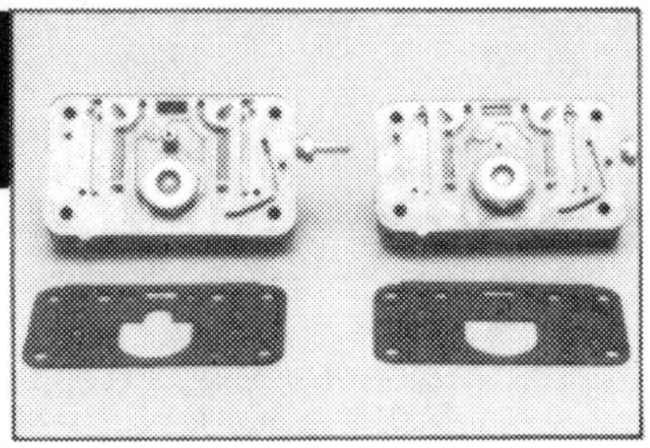

Metering block gaskets are another potential trouble source. Some non-Holley cork gaskets tend to shrink, distort, or flake. Always use Holley gaskets made from paper. It is necessary to correctly match the gasket style with the particular metering block in use. Metering blocks with an O-ring-sealed accelerator pump transfer tube (**left**) use gasket 108-31 that's cut to clear the tube. Any attempt to use this gasket with the metering block on the right will draw fuel from the accelerator pump system and supply a full-rich mixture to the engine. Instead, use gasket 108-29. Never attempt to cut or modify part 108-29 to work in place of 108-31, since small leftover particle residue will invariably be sucked into the transfer tube.

4/WRONG TYPE OR DESIGN NEEDLE-AND-SEAT

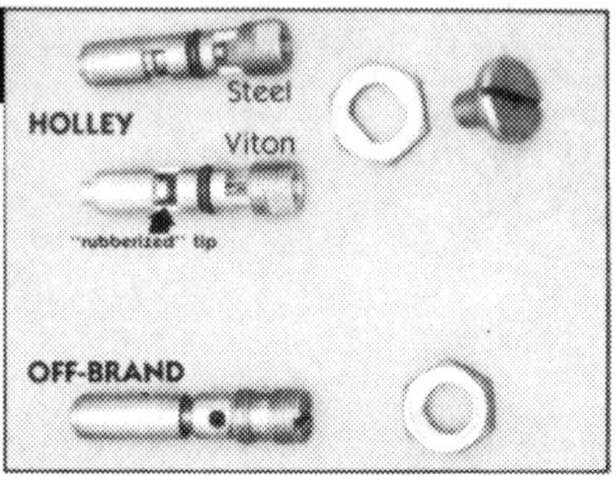

Many non-Holley carb rebuilding kits contain incorrectly designed needle-and-seats. The worst of these can be quickly identified by the recessed screw and nut adjustment (the brass nut is on top, instead of under, the screw). This setup tears up the float bowl because it has the wrong thread pitch, and usually can't hold fuel pressure, either. It pays to use only genuine Holley viton needle-and-seats on gasoline-fueled vehicles.

The .110-inch diameter (6-504) seat is the biggest offered with a viton needle; the latest versions utilize slots instead of small round holes, thus offering increased flow. While larger-orifice steel needles and seats are available, they are intended for use with exotic fuels only, and should be avoided on any gasoline-powered vehicle since they won't seal as well as the viton design.

5/IMPROPER VACUUM LINE CONNECTION

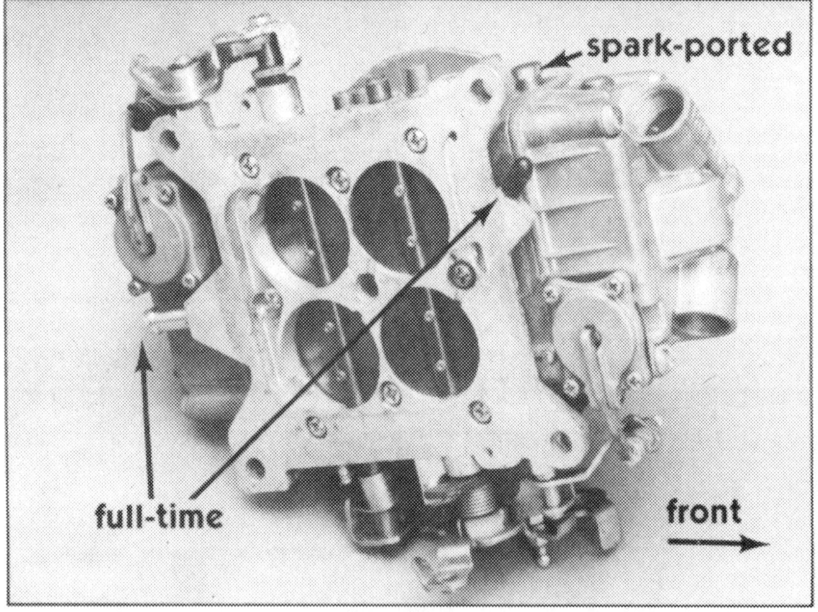

When it comes to correct vacuum line hookup, "any port in a storm" just won't cut it. Many times off-the-line hesitation is blamed unfairly on the carb, when actually the vacuum advance line has been attached to the wrong port on the carb. The correct vacuum port to use will vary depending on the engine and the type of distributor vacuum "can" in use. Any source below the carburetor throttle valves, including the carburetor baseplate and intake manifold, is known as "manifold vacuum" or "constant vacuum"; it is highest at idle and decreases as the throttle valves open. Any location above the throttle valves (such as the metering block port) is known as "spark-ported vacuum"; it is lowest at idle and increases with throttle opening. Early engines usually use manifold vacuum; late-models, spark-ported vacuum. Which is best for you depends on a number of factors (as discussed in "Talking Tech," December 1982, pp. 99-100). The photo shows you where to hook up the vacuum lines on a typical Holley carb.

6/MISADJUSTED FLOAT LEVEL

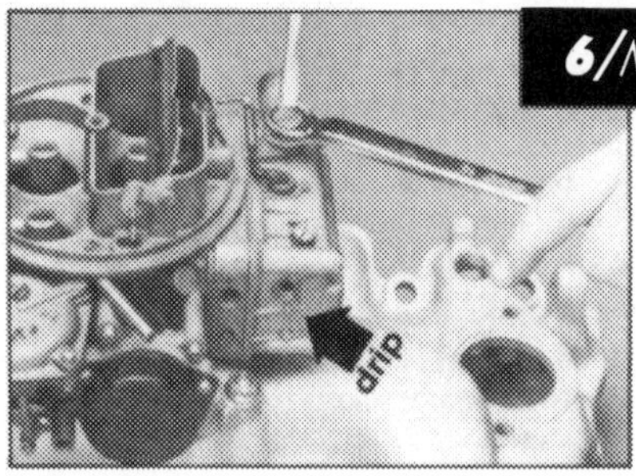

Many carburetor problems are traceable to incorrect float level adjustment. Too low a float level causes a lean mixture, resulting in late or hard starting, a flat spot in acceleration, or stumble. Too high a float level causes a rich mixture, resulting in the engine racing after starting, percolation, and even fuel pouring out the bowl vents.

Since most Holleys have externally adjustable floats, there's no excuse for living with the wrong float level. To adjust the float level on Holley needle-and-seats, first remove the sight plug in the primary float bowl. Loosen the set screw and turn the adjustment nut until fuel barely drips out the sight plug hole. (The engine should be running, or if you have an electric fuel pump, it should be turned on). Turning the adjustment nut clockwise raises the float level; counterclockwise lowers it. Once the primary side is adjusted, repeat the procedure on the secondary side.

7/PART NUMBERING SYSTEM CONFUSION

Holley's numbering system nomenclature is sometimes confusing. Power valves are numbered according to the vacuum level below which they open (**see 2**). Accelerator pump discharge nozzles are numbered according to their nozzle size (**see 12**). Main jets are numbered from 40 to 100, with 40 being the leanest and 100 the richest. The basic part number is 122-(jet size). Jet numbers don't correspond to any particular drill size, since they are rated by flow rather than area. Holley emissions carbs have "close limit" main jets which feature only 1.5-percent average flow variation within each jet size (instead of a standard jet's average 3-percent variation). Their finer tolerance could benefit the tuner in some applications.

CARB GOOFS

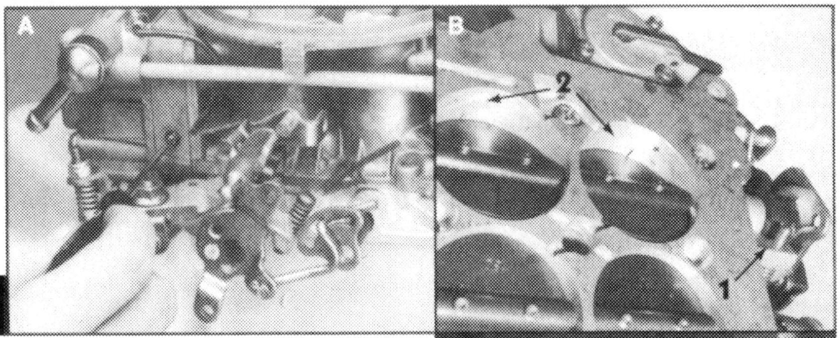

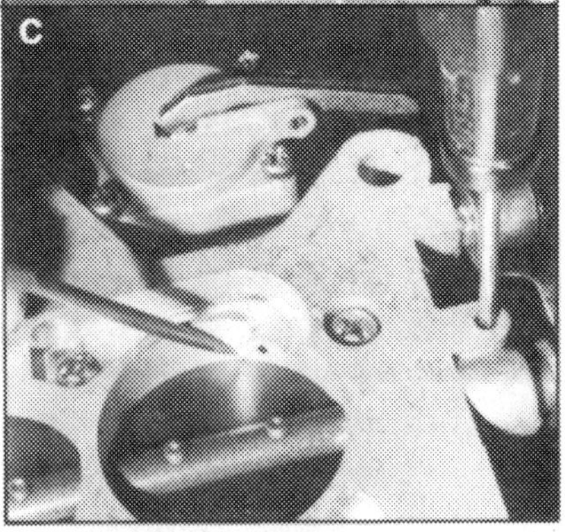

8/POOR IDLE QUALITY

Since the idle mixture screws are responsible for idle quality and even part-throttle cruise transition, their correct adjustment is vital for maintaining smooth driveability and avoiding a stumble. Ideally, the idle mixture screws, located on the primary metering block **(A)**, should be adjusted to obtain the highest manifold vacuum reading, with each screw approximately the same number of turns out from the "lightly bottomed" position. If a vacuum gauge is unavailable, back the screws out 1½ full turns; next, turn the screws clockwise until lean misfire occurs, finally backing out to best or smoothest idle. Some emissions Holleys have a "backwards" idle system, in which turning the mixture screws clockwise richens the mixture (these have special tags affixed near the idle screws). Such carbs usually have insufficient adjustability to work successfully with high-perf cams.

Manual transmission cars should be adjusted in neutral. Adjust automatics in Drive because of their large rpm drop. Fine-tune the idle speed with the curb idle screw. ***(arrow)*** If the car diesels after shutoff, add anti-diesel solenoid.

If, on carbs with conventional idle systems and high-lift cams, it is difficult to get the vehicle to idle properly, it's because the primary throttle blades are excessively open (via the curb idle speed screw, **(B/arrow 1)**, thereby exposing the primary transfer slot **(B/arrow 2)** and creating an over-rich condition. To correct this, adjust the secondary throttle screw so that the secondary throttle blades will just barely expose the secondary idle fuel transfer slot **(C)**. This will permit a reduction of the primary throttle opening so the idle screws again become functional.

9/IMPROPER ACCELERATOR PUMP ADJUSTMENT

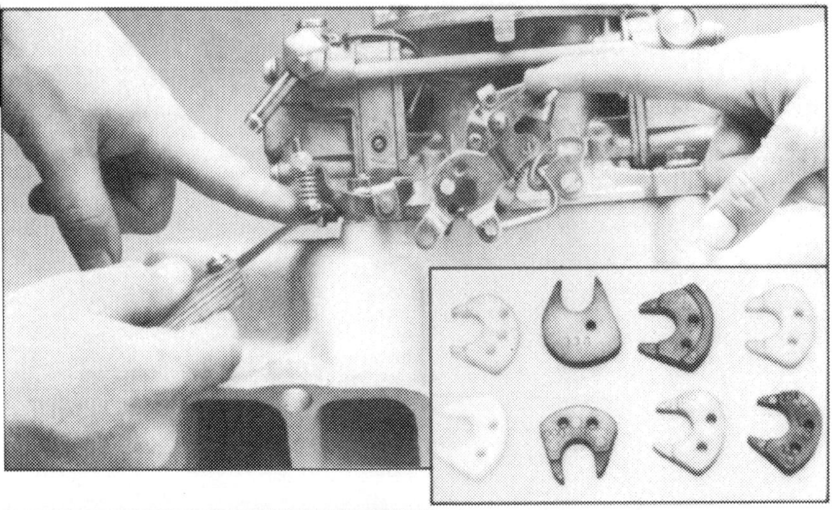

Everyone and his brother have their pet accelerator pump adjustment methods. Holley recommends the following procedure: open the throttle all the way (WOT). Push the pump arm lever down, and then adjust the pump over-ride spring to obtain .010-.015 clearance between the pump arm and lever. Unfortunately, you need three hands!

Pump cams usually have two installation positions. Hole No. 1 offers quicker response, but in most cases hole No. 2 supplies more capacity and a longer duration pump shot. Street cars rarely need to change cams.

10/WRONG AIR CLEANER FOR APPLICATION

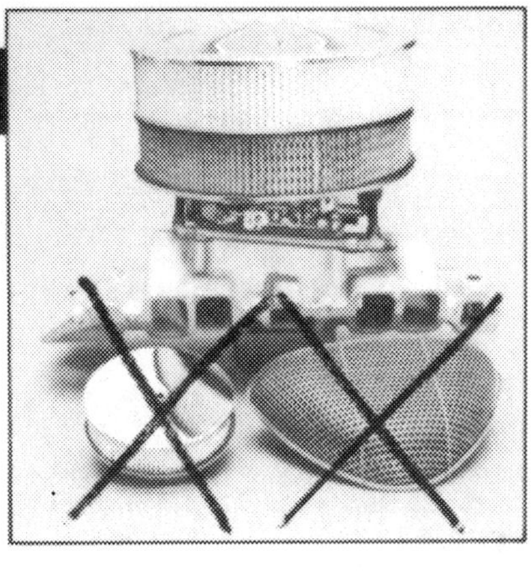

Air cleaner design is critical for obtaining max performance. It's one case where too big is way better than too small! In extreme cases, a rich mixture will result from a too-small element. The average 350-inch Chevy running a 600-cfm carb requires a 3-inch tall x 14-inch diameter air cleaner. (Alternatively, 9-inch diameter x 4 or 5 inches high will work satisfactorily.) Tests have shown that double-stacking air cleaner elements, if practical, lets the carb achieve its full flow potential.

A minimum ¾-inch clearance is required between the bowl vent and all parts of the air cleaner housing. Some air cleaners, especially the various flat foam styles, come too close, preventing proper equalization of the bowl.

11/FUEL PRESSURE TOO HIGH OR TOO LOW

Wrong fuel pressure is responsible for many carb woes. Ideally, pressure should range between 5 and 6 psi at all times. Pressures lower than 4 psi will result in fuel starvation. Pressures above 8 psi will blow the needle off the seat. A regulator is required to control excessive fuel pressure when using high-volume electric fuel pumps or the new Holley NASCAR mechanical pumps, but is not normally needed with standard high-performance mechanical pumps. The only way to make sure pressure on any pump is correct is to use a fuel pressure gauge, installed at or between the regulator and carb inlet (or between pump and inlet, if a regulator is not used).

12/OFF-THE-LINE STUMBLE

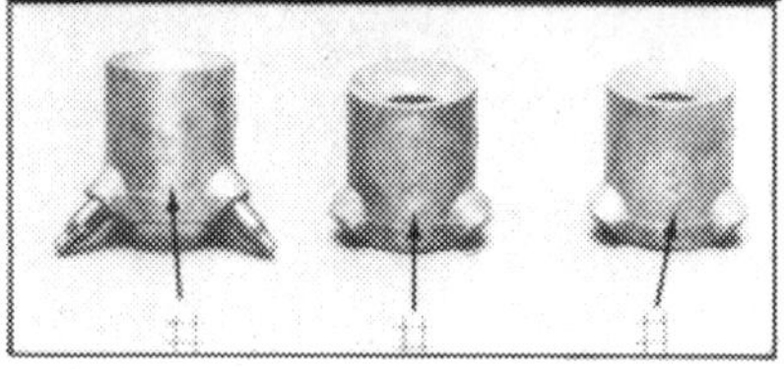

If, after proper adjustment of the idle mixture and throttle stops **(see 7)**, and the accelerator pump arm-to-lever clearance **(see 9)**, there is an instantaneous off-the-line stumble, and no black smoke comes out the tailpipe, the accelerator pump discharge nozzle (shooter) size should be increased. If the car stumbles and black smoke does come out the tailpipe, reduce the shooter size. Shooters are numbered according to nozzle size. One stamped "27" has .027-inch diameter nozzles. A variety of different styles are available. Chuck likes the long-tube version, since it directs the shot closer to the booster venturi's center, thus promoting better fuel atomization. The .040 and larger shooter sizes require a hollow hold-down screw to ensure adequate fuel delivery.

13/OVERCARBURETION

This has been touched on many times, but Holley reports that "too big a carburetor" is still one of the biggest problems they encounter. Assuming the average street car has a 350 engine operating to 6000 rpm, automatic trans with mild converter, near-stock cam, and mild rearend gear ratio, a 600-cfm carb, or even a spread-bore 4360 450-cfm carb, is plenty. This can be shown by the standard formula used to select carburetor size:

$$\textbf{carb cfm} = \textbf{VE}\left(\frac{\textbf{Engine cid x max rpm}}{\textbf{3456}}\right)$$

"VE" stands for "volumetric efficiency," or the percent of air actually taken into an engine, compared to the mass which the engine displacement would theoretically take in if there were no losses. Normally aspirated street engines rarely exceed 85 percent; most, in fact, are closer to 75 percent. But even assuming 85 percent VE (.85), our hypothetical 350 still only needs a 516-cfm carb!

In addition to carb size, secondary actuation method is also critical. On the street, vacuum secondary carbs are preferred over double-pumpers, since they operate by sensing engine demand, metering the right amount of air to maintain induction system velocity. Only if you have big gears, a blower, a wild cam, and/or a manual trans should a double-pumper be considered.

14/VACUUM SECONDARY BOGS ON OPENING

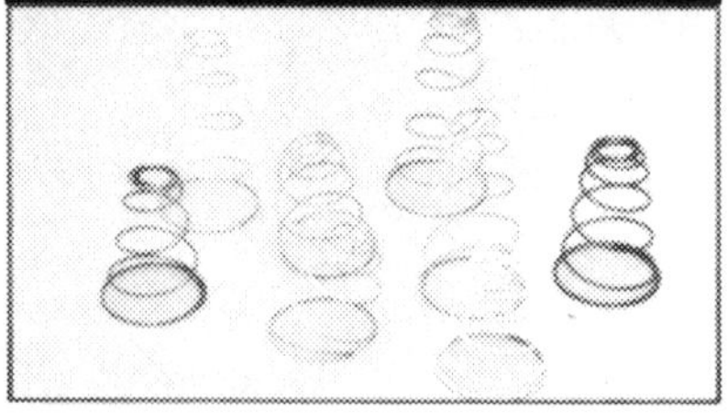

Many people end up (incorrectly) with double-pumpers because the vacuum secondaries "bog." But the vacuum secondaries have greater fuel economy potential, and they're easily tuned with secondary diaphragm spring kit 20-13. Bogging and hesitation are caused by the secondaries opening too quickly; go to progressively heavier springs from the kit to correct this. If the engine is sluggish in response to full throttle, try lighter springs. The proper springs should be selected before playing with accelerator pump shooter size **(see No. 12, above)**, so you can utilize as small a squirter as possible and enhance fuel economy potential.

15/OVERHEATING

Late-model smog motors, or any motors installed in tight engine compartments sometimes encounter vapor lock or fuel percolation due to high underhood ambient temperatures. Minimize vapor lock by routing fuel lines as far away from heat sources as practicable. Percolation is solved by using a phenolic insulator **(arrow)** between the carb and intake manifold. Chevy also offers an aluminum heat insulation plate, part No. 3969835. In any case, avoid stacking gaskets; you'll end up cracking the baseplate. **HR**

(Editor's note: Parts in photos supplied courtesy of Earl's Supply, Holley Replacement Parts Division, Bob Jennings Automotive, Service Center #2, and Snap-On Tools Corp.)

GOOD GOLLY MISS HOLLEY!

Street and Strip Tips for One of the Most Tuneable Carbs Going

Throughout the years, there have been loads of articles written covering all sorts of exotic modifications for the basic Model 4150/4160 Holley 4-barrels. However, for most street and street/strip use, you can use the "stock" parts to dial in virtually any carb. Since the Holley is among the most tuneable carbs going, there's a plethora of readily available, off-the-shelf factory parts.

To get the straight scoop, we visited one of the leading Holley carb "gurus"—Murray Jensen. Jensen, Edelbrock's former carb-building ace, and partner Bob Szabo, today run Fuel Curve Engineering, a small but growing shop (74-818 Joni Dr., Palm Desert, CA 92660, 619/340-2443). What he told us exploded a lot of the Holley tuning misconceptions we've all been laboring under these past years. So here's to jettisoning excess baggage....

I. SELECTION & INSPECTION

Which Carb? Selecting the right carb for the right job is one of the most important steps in achieving a happy relationship with your Holley. All of the world's greatest tuning tricks can only "band-aid" a mismatched combination.

The carb to use depends not on some theoretical whiz-kid selection process, but rather on a down-to-earth approach that consists of building a total combination that will do what you need to do—the first time. Engine size, camshaft, intake manifold, rpm range, vehicle weight, trans type, final drive ratio, clutch, and torque converter stall speed all factor into the carb selection process. For example, "with a manual trans, you can use more carb; a bigger gear, you can use more carb. On the other hand, with an automatic trans and stock converter, a smaller carb is better. Single-plane intakes usually have less bottom end than dual-planes, so on the street, a smaller carb would be required with a single-plane, all other factors being equal.

"As a general rule, assuming a 350-inch motor, use a 600 or 650 cfm carb with near-stock gears. If running a 4.11 gear, a 700 or 750 will put you in the ballpark. More radical engines may require an 800 or even an 850. The goal is to select a carb that supplies sufficient air to allow the engine to breathe and yet maintains sufficient velocity to generate enough kinetic energy in the induction system to fill the cylinder."

Holley 4-barrels come with both vacuum and mechanical secondaries. Vacuum secondary carbs are best suited for mild street applications and four-wheel drive hill climbers. They work well with standard torque converters, and are generally more forgiving than the double-pumpers since they work by sensing engine demand, metering the right amount of air to maintain induction system velocity. Vacuum secondary carbs are offered in 390, 600, 725, and 750 cfm flow rating; 600 cfm carbs intended for specific applications are by far the most plentiful. Lighter cars with hot cams, big gears, and extended rpm ranges should lean toward the more tuneable double-pumpers. Such cars require the extra fuel enrichment offered by the double-pumpers, which come in 50 cfm increments from 600 all the way to 850. There are also special application units with various "weird" cfm ratings, including an 830, and even a 350 and 375!

In addition to secondary actuation method, standard-flange Holley 4-barrel carbs can also be classified according to two basic model configurations—the Model 4150 and Model 4160. Model 4150s have a fuel metering block with conventional main jets on both the primary and secondary sides, while the cheaper 4160s have a rear metering plate which does not have removable jets or power valves. All the common double-pumpers are Model 4150 carbs, while the majority of vacuum-actuated carbs are 4160s. If you do not plan to do much recalibrating, 4160s are okay for street use. For serious performance, the 4150s are preferred because the replaceable jets allow easy tuneability. If desired, 4160s may be converted to the 4150 configuration by using the appropriate Holley conversion kit; this is probably cheaper in the long run than buying all the different metering plates.

There are also two different types of float bowls—center pivot and side pivot. Nearly all Model 4160s (with the exception of the 3310-2) have side-pivot bowls. On 4150s, it's dependent on application and list number. The common double-pumpers are all center-pivot. Center-pivot bowls are preferred for vehicles utilizing conventional (fore and aft) carb mounting that's subject to high G acceleration

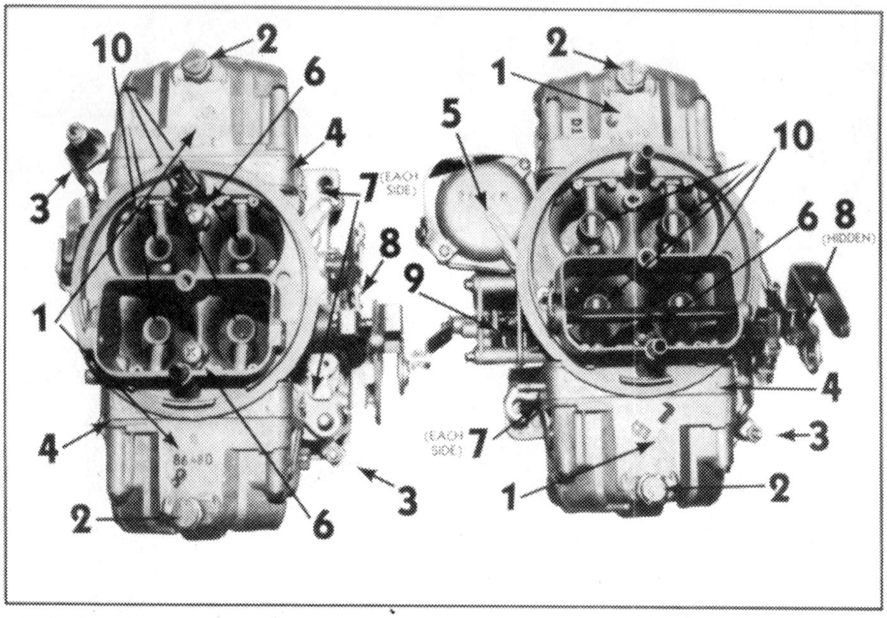

Here's the scorecard so you can tell the players. On the left, a late competition "double-pumper" with mechanical secondaries, rear mixture screws, and rear accelerator pump. On the right, a vacuum secondary carb with rear metering plate.

1. Float bowl
2. Needle-and-seat assembly
3. Accelerator pump arm and spring
4. Metering block containing jets and (usually) power valve
5. Secondary vacuum diaphragm and spring (under housing)
6. Accelerator pump shooter
7. Idle mixture adjusting screws
8. Curb idle screw
9. Fast idle screw (for adjusting choke rpm)
10. Booster venturis

Fuel Curve recommends adjusting the float level using a drill bit as a gauge. With the bowl inverted, place drill bit between bowl and float. On center pivot bowls, measurement is made above raised area **(arrow)**. *Floats should barely clear bit; loosen the set screw and turn the adjusting nut as necessary.*

Access on a Holley to jets and power valves is easy, thanks to the carb's modular construction. Merely remove the float bowl attaching screws, and there's the metering block containing jets and power valves. Bowl whistle **(arrow**, *part No. 26-40) prevents fuel slosh out bowl vent.*

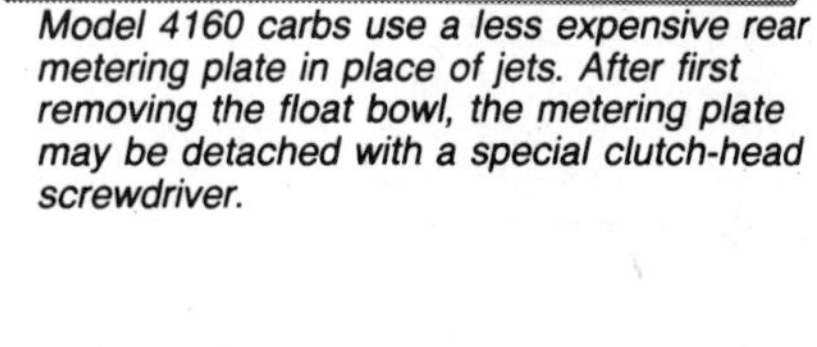

Model 4160 carbs use a less expensive rear metering plate in place of jets. After first removing the float bowl, the metering plate may be detached with a special clutch-head screwdriver.

forces, including drag cars and road racers. Besides superior fuel handling, the center-pivot bowls offer increased fuel capacity due to greater internal volume and separate fuel inlet systems.

Initial Inspection. After unpacking your new carb, Fuel Curve recommends its complete disassembly and inspection. You must make sure the carb was correctly assembled, and that no physical damage occurred in shipping. Inspect the accelerator pump system for correct operation. Check for any binding in the linkage, and that the linkage is correctly adjusted. Check and adjust the float setting after disassembly. Believe it or not, you will rarely get correct adjustment by using the sight plug located in the bowl side. With the engine running there's simply too much vibration to accurately gauge fuel level with that method! Instead, with the carb disassembled and the bowl inverted, use an appropriately-sized drill bit **(see photos)** to set the level. This method provides the correct float setting for 95 percent of all applications. Incidentally, there are two types of floats: brass and nitrophyl. Fuel Curve prefers brass floats, since their correct operation may easily be verified. By contrast, the nitrophyl float may slowly become saturated with fuel, leading to a "sink, not swim" situation. Nitrophyl floats additionally may be adversely effected by exotic fuels and fuel additives.

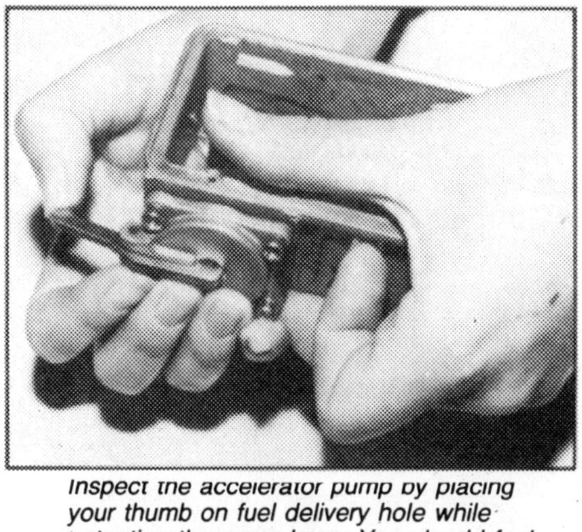

Inspect the accelerator pump by placing your thumb on fuel delivery hole while actuating the pump lever. You should feel pressure if pump diaphragm is okay. The pump should spring back into position when the lever is released. If it doesn't, someone forgot to install the spring.

It's important to use the right gasket with the right metering block. Paper gaskets are preferred over the cork type, which tend to shrink and distort. Brown paper gaskets are better in turn than black paper ones, since the former won't stick as much, simplifying in-pit metering changes. Metering blocks with an O-ring-sealed accelerator pump transfer tube ***(left)*** *use gasket 108-31 that's cut to clear the tube. Any attempt to use this gasket with block on right will draw fuel from the accelerator pump system and supply a full-rich mixture to the engine. Instead, use gasket 108-29. Power valve* ***(arrow)*** *is removed with a 1-inch wrench.*

Holley needle-and-seat assemblies, from left: Two-piece unit is used on low-performance carbs with non-adjustable float bowls. Adjustable seats with holes should be replaced with slotted type that offers less flow restriction; .110 viton size works best on the street. Finally, for exotic fuels there are steel units available, including Fuel Curve's own two-piece design featuring tuneable, replaceable needles.

II. TUNING

Needles and Seats. Our introduction to all the things you can play with on a Holley begins with the needle and seat. For street use, "you can't go wrong" with a .110-inch-diameter viton needle and seat, "which is large enough for nearly any gas-fueled engine." Although some stock and/or emissions carbs come with smaller needle-and-seats, replacement of the as-equipped assembly with the .110 seat will enhance top end flow with no sacrifice in part-throttle economy. Make sure you get the latest versions that utilize slots instead of small round holes for increased flow (the .110 viton seat number is 6-504).

The needle and seat assembly is prone to dirt contamination, and sometimes the O-ring seal may dry out and crack, resulting in fuel seepage around the outside of the seat. If you're experiencing problems, look down the carb at idle and see if fuel is running out of the booster or bowl vent; if it is, chances are it's either the needle and seat or an incorrectly adjusted or malfunctioning float.

Out-of-the-box double-pumper secondaries often don't open fully, because the secondary linkage rod is incorrectly adjusted ***(right)****. Bend linkage as shown* ***(left)****.*

Power Valves. Generally, primary power valves should not be plugged on street-driven vehicles. Since the power valve augments the fuel supply to maintain correct metering under hard acceleration, removing the power valve requires richening up the main jets. This in turn causes an overrich condition in the part-throttle cruise mode.

Power valves are numbered according to the vacuum level below which they open. An "85" valve thus opens below 8.5 inches Hg. To prevent unnecessarily rich fuel metering under light loads, select a valve that opens at least 1 to 1½ inches Hg below the average manifold vacuum level at idle.

Also available are two-stage power valves that open partially at one vacuum level and then fully at a lower vacuum level. While these valves may improve gas mileage on lightweight, mildly-cammed vehicles, they should be avoided on modified street engines or race cars, or in any carburetor that has power valve channel restrictions larger than .076 inch.

Holley power valves (PV) have been criticized for their alleged tendency to "blow out." Fuel Curve responds that PV failure is only symptomatic of a problem elsewhere—backfiring, excessively lean carburetor calibration, ignition system woes, etc.

Idle Mixture System. Idle mixture screws are located on the primary metering block, and are primarily responsible for curb idle quality and even part-throttle cruise transition. Preliminary idle adjustment is made by "lightly bottoming" the screws, then backing off 1½-full turns. Next, using a vacuum gauge, adjust the screws to obtain the highest possible vacuum reading, with each screw approximately the same number of turns out from the "lightly bottomed" position. If a vacuum gauge is unavailable, slowly turn the screws clockwise (in) until misfire occurs, then back out to best or smoothest idle. On most Holleys, turning the screws counterclockwise richens the mixture; however, on late emissions carbs the mixture screws control the amount of air (instead of fuel) permitted into the idle system. Here turning the screws *in* richens the mixture since the volume of air is decreased. Such carbs do not usually offer sufficient adjustability for high-performance applications.

Even with a conventional idle system, sometimes with single-plane intakes and high-lift cams, you may encounter difficulty in obtaining the desired idle quality. Due to the increased cam overlap, low manifold vacuum, and/or low induction system velocity, the throttle must be opened further to obtain adequate mix-

Conventional Holley power valves **(left** *and* **center)** *are numbered according to the vacuum level below which they open* **(arrow)**. *"Picture-window"-type high-performance valve offers less flow restriction than "stock" valve with holes, and does not affect fuel economy since the PVCR is still the limiting factor. Weird unit on right is two-stage type for fuel economy applications. Two different series of picture-window valves are available* **(see list below)**, *depending on your PVCR size. Be sure to use the right gasket with the right valve. 8R-1597 round gasket goes on valves with an internal shoulder* **(left)**; *8R-669 with points is for external shoulder valve.*

High-Performance "Window" Power Valve Part Numbers

Opening Point (inch Hg)	Under .090 PVCR	Over .090 PVCR
2.5	125-25	—
3.5	125-35	—
4.5	125-45	—
5.0	125-50	—
6.5	125-65	125-165
7.5	125-75	—
8.5	125-85	125-185
10.5	125-95	125-1005

Two-Stage Power Valve Part Numbers

Opening Points

1st Stage	2nd Stage	
12.5	5.5	125-206
10.5	5	125-207
10.5	5.5	125-208

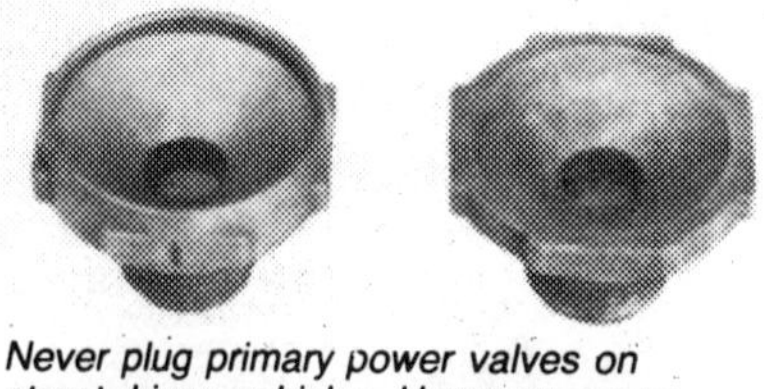

Never plug primary power valves on street-driven vehicles. However, some double-pumper carbs have secondary power valves. Fuel Curve feels since the secondaries are calibrated for WOT operation to begin with, there is no real need for such augmentation. If desired, you may jet up the secondaries six to eight sizes, and use a 26-36 plug to block the passage. To avoid interference with the main body screw boss, when used on the secondary side the plug must have its shoulder machined off **(right)**.

Here's a neat tip for the weekend warrior who races on Sunday and drives to market on Monday: Jet the carb for regular "corked-up" motoring, and compensate for WOT open-exhaust leanness by increasing the PVCR size **(arrows)**. *Initially, go up about .003-inch from the as-delivered hole size. Continue in .001 increments until the spark plugs look okay under open exhaust.*

Late emissions carbs have a reverse idle system: turning in the mixture screws richens the mixture. If on such a carb turning the screws "full in" fails to effect idle quality, the idle feed restriction in the metering block is too restrictive to flow enough fuel. Either get the right high-performance-oriented carb or see a specialist like Fuel Curve, since (because of the unconventional metering) you cannot merely swap other blocks on without extensive mods.

ture flow, but in so doing the increased throttle angle exposes the transfer slot. Idle mixture starts to flow through the transfer slot, causing the engine to run so rich that it no longer responds to the idle mixture screws. Correction of this problem consists of properly adjusting the throttle stops and, if necessary, drilling holes in the throttle blades **(see photos)**.

However, if you're planning the whole combo from scratch and wish to avoid the problem to begin with, or if the combo is so radical that you still have idle quality problems even after drilling holes in the blades, consider procuring the special double-pumpers (0-8156, 750 cfm; 0-8162, 850 cfm) built with rear secondary idle mixture screws or those with both rear idle mixture screws and annular discharge booster venturis for increased bottom end response (0-9379, 750 cfm; 0-9381, 830 cfm; 0-9380, 850 cfm). Such "four-hole idle systems" offer better fuel distribution and more precise and adjustable mixture control, even on the street. And, yes, if all else fails, they can be treated to the throttle plate drill—but on all four butterflies, of course. Fuel Curve can convert existing 4150s to the four-hole idle system.

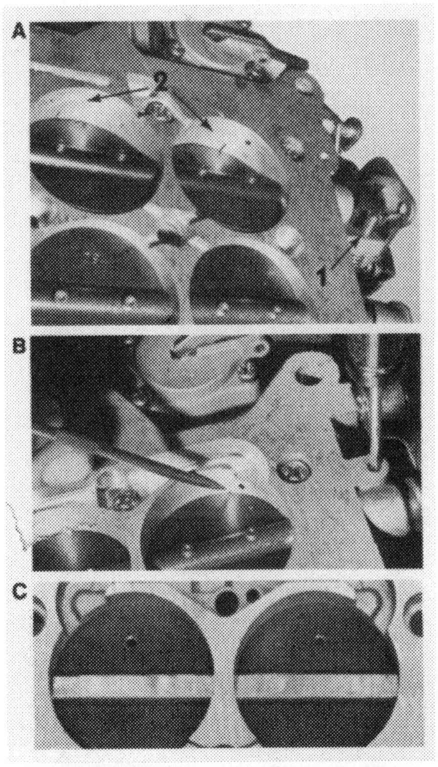

*If with high-lift cams turning the idle mixture screws has little or no effect, it's because the throttle was excessively open (via the curb idle speed screw, **A, arrow 1**), thereby exposing the primary transfer slot (**A, arrow 2**) and creating an overrich condition. To correct this, adjust the secondary throttle screw so the secondary throttle blades will just barely expose the secondary idle fuel transfer slot (**B**). This will permit a reduction of the primary throttle opening to the point where the idle screws should again become functional. If after performing this adjustment you still experience problems, get out the drill set and drill two .093-inch holes centered in the primary throttle blades (**C**). Since air will pass through the new holes, the throttle plate may now be closed to the point where idle adjustability should return. This mod is generally required with cams having over 270 degrees advertised duration, especially when combined with large plenum single-plane intakes like Edelbrock's Victor series.*

Accelerator Pump System.

Like his unconventional approach to float adjustment, Murray does not go by the book when it comes to pump travel adjustment either. Forget about the ".015-.020 feeler gauge at wide open throttle (WOT)" routine—it just isn't the ticket for achieving optimum pump efficiency. Ideally, the pump arm should move as soon as the accelerator pump cam moves. This is achieved by first installing the cam in the position offering the most capacity. For most cams, this is Position 2; an exception is the No. 330 cam, where Position 1 offers the best shot.

Next, adjust the pump spring—it should have some play so the pump lever won't get bent, yet be tight enough so the pump lever spring won't collapse under hard acceleration. If after proper adjustment the pump still lags, you must bend the pump arm as shown in the photos.

Accelerator pumps may be tuned through a trial-and-error effort by changing pump cams and shooter sizes. Usually, the as-delivered cam works best for most applications, and it merely becomes a matter of varying shooter (nozzle) size. If after proper adjustment of the idle mixture screws, throttle stops, and accelerator pump arm there is an instantaneous off-the-line stumble, and no black smoke comes out the tailpipe, shooter size should be increased. If the car stumbles and black smoke comes out the tailpipe, reduce the shooter size. If you end up having to increase shooter size beyond .035 on a 30cc pump, and the stumble still is not entirely eliminated, consider a different pump cam. On the street, this rarely is a problem, as most double-pumpers come through with the preferred No. 330 cam. Since this cam is very quick and also has three different positions, it solves most problems.

(continued)

*To obtain correct 30cc accelerator pump operation, adjust the override spring length from ½ to ⅝ inch as necessary (**shown**). On a 50cc pump, adjust to obtain ⅛ inch of visible thread.*

HOLLEY!

Some double-pumpers come with a large 50cc pump. These rarely need the adjustment arm bent, although the procedure is similar to the smaller cam arms. The as-delivered cam with the big pump is usually right on, but here again you can play with shooters as necessary. Be sure to check that there is adequate under-carb clearance for this longer-stroke pump. With some extremely large-plenum intakes (like the Smokey Ram), it may be necessary to install 50cc pump kit 20-11 up front.

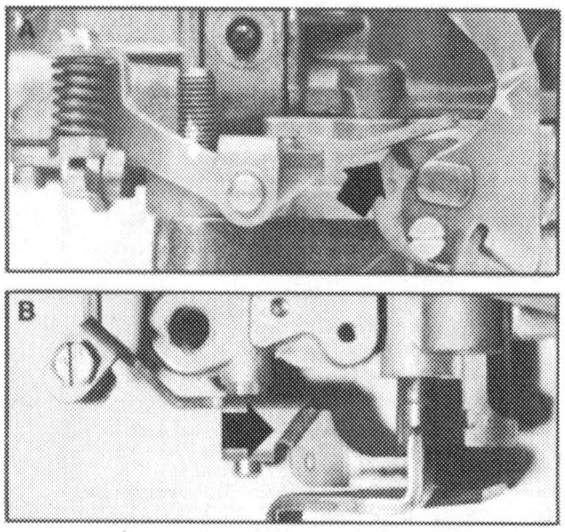

These photos show the proper cam-to-arm relationship necessary for optimum accelerator pump operation. In **A**, *throttle linkage was cut away so you can see the proper initial contact point between the arm and cam heel (black line). This is achieved by bending arm along line shown in* **B**.

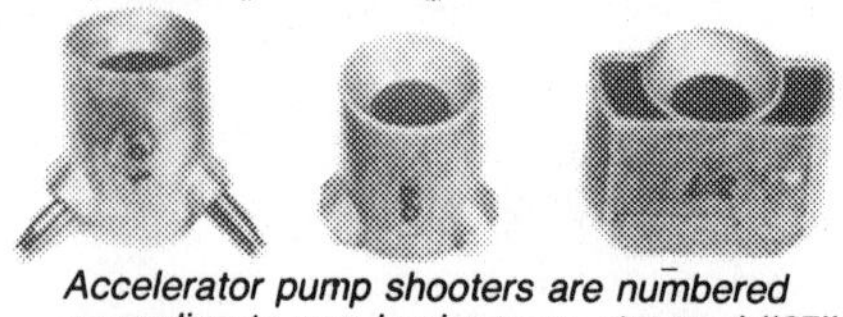

Accelerator pump shooters are numbered according to nozzle size—one stamped "27" would have .027-inch-diameter nozzles. They're available in a variety of shapes and sizes: Long-tube version **(left)** *directs the pump shot closer to the booster venturi center, but as Fuel Curve feels it may "siphon" or "pull" fuel under high air flow conditions, they prefer the standard nozzle* **(center)** *for most applications. The 4165/4175 nozzles use a special "anti-pullover" design* **(right)**; *this can be installed in 4150/4160 carbs if fuel siphoning still is a problem even after trying the standard nozzle.*

Pump Discharge Nozzle Part Numbers

Hole Size	Tube	Straight	Anti-Pullover
.025	121-25	121-125	121-225
.028	121-28	121-128	121-228
.031	121-31	121-131	121-231
.035	121-35	121-135	—
.037	121-37	121-137	121-237
.040	121-40	121-140	121-240
.042	121-42	121-142	121-242
.045	121-45	121-145	121-245
.047	—	121-147	121-247
.050	—	121-150	121-250
.052	—	121-152	121-252

Main Jets. The main metering system is composed of replaceable jets or (on Model 4160 secondaries) metering plates. Jets should never be drilled, since this may damage the jets' entry and exit features, causing the fuel to swirl. Since jets are rated by flow (not size), just because the jet was drilled one drill size larger does not necessarily mean that the jet flow is correspondingly increased. Rear metering plates may be drilled, but then you have started down the road of no return. It's far better to either procure extra plates, or convert the carb to 4150 configuration.

Usually, the "out-of-the-box" jetting is fairly close to correct. If you find that a jet richer than eight sizes over "stock" is required, there is something seriously wrong elsewhere. There may be insufficient delivery from the fuel pump, an air cleaner cover blocking the bowl vents, vacuum leaks, etc.

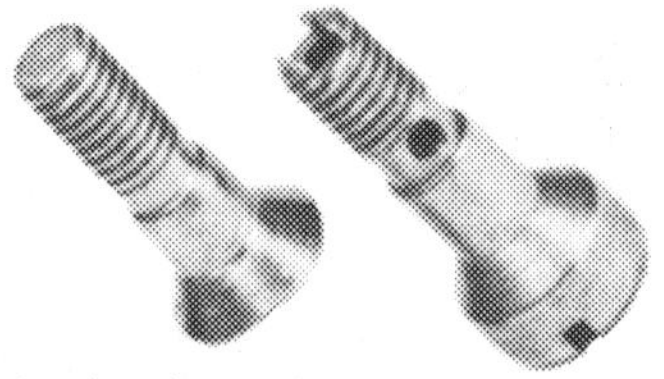

Shooter sizes larger than .035 inch require a hollow hold-down screw **(right)** *to ensure adequate fuel delivery.*

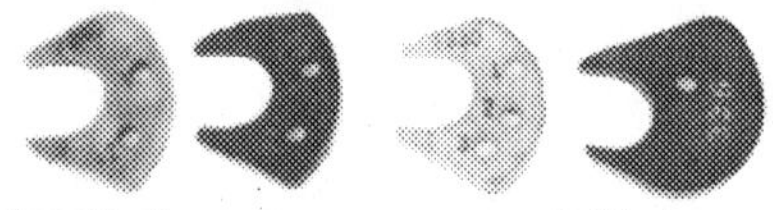

Accelerator pump cams are available in many profiles. Fuel Curve feels the 330 cam has the best opening rate and works well with 30cc pumps. Note the third adjustment hole. One hole 336 cam is strictly for 50cc "Reo" pumps and should not be used with the smaller pumps or you could blow the diaphragm. Part No. 20-12 gets you a cam assortment.

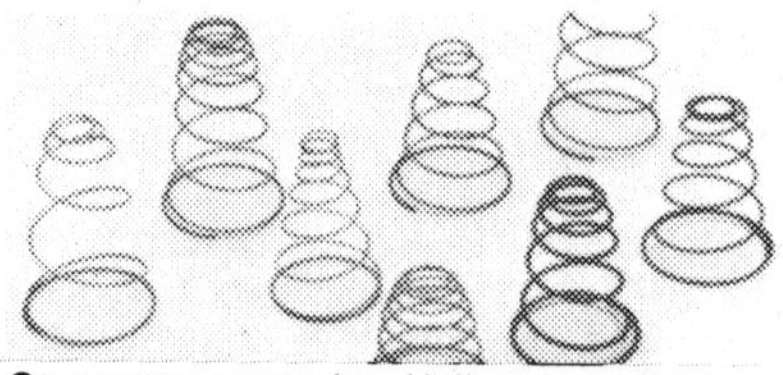

On vacuum-secondary Holleys, accelerator pump squirter size must also be coordinated with the secondary-vacuum diaphragm spring to prevent bogging and hesitation caused by the secondaries opening too quickly. Go to progressively heavier springs to correct this. If the engine is sluggish in response to full throttle, try lighter springs. The proper springs should be selected before varying accelerator pump shooter size, so you can utilize as small a squirter as possible.

Secondary Diaphragm Spring Kit 20-13

Color	Relative Load
White	Lightest
Yellow	Lighter
Yellow	Light
Purple	Med. Light
Plain (no color)	Medium
Brown	Med. Heavy
Black	Heavy

Because there is about a 3 percent flow range within each jet size, and about a 4.5 percent difference in flow between average jets in two consecutive sizes, going up or down only one jet size at a time may not seem to appreciably affect the mixture if the tolerance stack-up places the two jets' flow close to each other. Fuel Curve feels that jetting changes therefore should be made in two number increments. Some Holley emissions carbs have "close limit" main jets with only 1.5 percent average flow variation in jets of the same size.

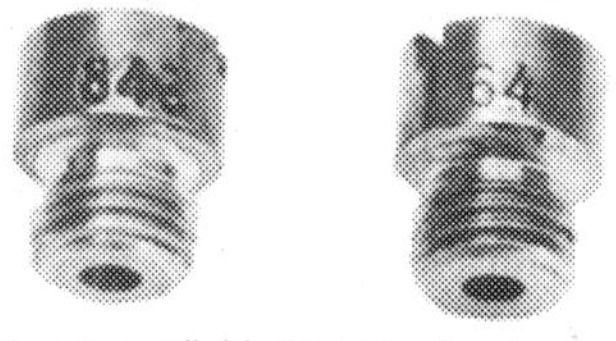

Jets are available in 1-number increments ranging from 40 to 100, with 100 being the richest. The basic part number is 122-(jet size). Holley emissions carbs have "close limit" main jets with reduced flow variation that could benefit the tuner in some applications—order basic part number 122-(jet number).

Chokes. It is important that the choke unloader adjustment be correct. With the throttle in the wide open position, the choke blade should open ⅛ to ¼ inch. This allows enough air for the engine to start if it is flooded and the choke is still closed. Use care to hook electric chokes up to a full 12-volt *non-resistor* circuit (not the coil!). As an initial setting, adjust the choke "at the center of the index." Some engines may require a leaner setting; very few need richening. A 45-267 Electro-Dyn heat sensor can be hooked to the engine's exhaust manifold to provide better operation with these chokes. Late "-2" series double-pumpers may be converted to electric choke operation with kit 45-224.

Improper operation of OEM divorced or non-electrical integral chokes can often be traced to plugged exhaust passages, especially on Chevys. The tuner often incorrectly responds to the condition by leaning the choke, only compounding the problem. Instead, bite the bullet and remove the intake to unplug the passages.

While the foregoing may seem quite involved, the Holley really is a basic, easily modified, super-responsive carb—that's why it's such a favorite among racers. With a little patience, and avoidance of the "King Kong" syndrome, you can dial that Holley right in for street and strip. **HR**

The "-2" double-pumpers may be converted to electric choke operation, but the rear accelerator pump cam boss must be ground for clearance.

HOLLEY TUNING TIPS

Basic Steps to Smooth, Responsive Performance from That New (or Used) Carb

By C.J. Baker

A great many articles have been written about special racing or performance modifications to Holley carbs, but articles covering the basic tuning steps for Holley carbs on street-driven vehicles are a real rarity. Consequently, we're going to review the common tuning procedures that will help you get the best driveability, performance, and economy from a new, or used, Holley carb.

Before beginning any carb adjustments, a few preliminary steps should be taken to ensure the carb is properly matched to the engine and that the fuel system is delivering an adequate supply of clean gasoline to the carb. Additionally, the induction system must be free of vacuum leaks and the carb needs to be free of any dirt or debris that may be restricting critical air and fuel passages. Unless all of these prerequisites are met, carb adjustments are meaningless since they'd only be covering up a more serious problem. So begin by checking fuel pressure. It should be between 4½ and 7 psi at all times.

Ensure cleanliness of the fuel entering the carb by installing an inline fuel filter between the fuel pump and the carb, if one isn't already in place. If your car already has an inline filter, check to make sure it is clean and not restricting fuel flow. Similarly, if you're starting with a used carb or one that has been in service for a while, verify that all of the fuel and air passages are clean. Start by running a wire down the idle air bleeds and high speed air bleeds while spraying a little carb or choke cleaner into them. Generally clean the outside of the carb with the cleaner, too. If the carb is really dirty, it should be disassembled and thoroughly cleaned with carb cleaner before beginning any tuning. This is the only way to guarantee that the carb is capable of functioning as intended.

Picking the right size carb for an engine usually means checking the application list in the Holley catalog, but if you or the local speed shop can't come up with such an application list, a good rule of thumb for selecting the *maximum* cfm (cubic feet per minute) rating that should be used on a street engine is:

$$\text{CFM} = \frac{\text{cubic inches} \times \text{max. rpm} \times .85}{3456}$$

When using this formula, be realistic about the maximum rpm of the engine. If you never rev the engine above 4800 rpm, don't plug 6000 rpm into the formula just because that's how high you'd like to see it rev; stick with the real figure. The constant of .85 represents the volumetric efficiency of the engine, and .85 (or 85 percent) is a pretty good ballpark figure for most street engines. Again, don't deceive yourself about this. It takes a pretty sophisticated racing engine to exceed 90 percent volumetric efficiency.

As an example, we'll compute the carb cfm requirements of a 350-cubic-inch engine with 5000 rpm capability.

$$\text{CFM} = \frac{350 \times 5000 \times .85}{3456} = \frac{1487500}{3456}$$
$$= 430 \text{ cfm}$$

Surprised? See, you didn't need that big 780-cfm carb after all. A carb more than 10 percent bigger than the actual needs of the engine will usually provide poorer throttle response and economy than one that does match the engine. In most cases, a carb of *exactly* the right size won't be offered, but at least stay close to the right figure. If your engine's cfm requirements fall right in the middle between two different cfm-rated carbs, choose the smaller carb for economy or the larger one for performance. If still in doubt, pick the smaller one. Forget

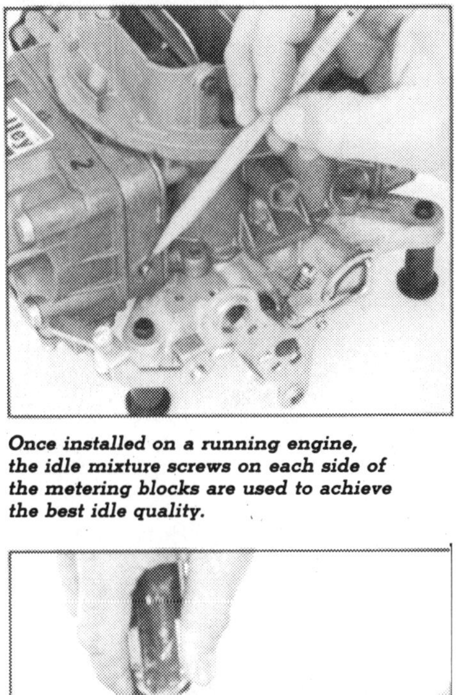

Once installed on a running engine, the idle mixture screws on each side of the metering blocks are used to achieve the best idle quality.

These high-speed and idle air bleeds (see arrows) can be cleared of dirt and debris with a fine wire and carburetor cleaner. Fine wire such as used on parts tags works well.

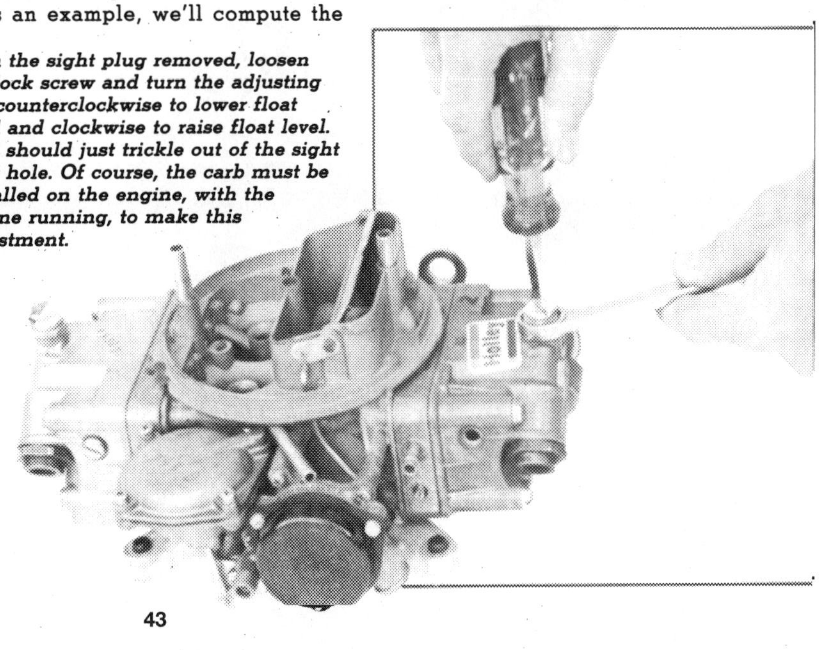

With the sight plug removed, loosen the lock screw and turn the adjusting nut counterclockwise to lower float level and clockwise to raise float level. Fuel should just trickle out of the sight plug hole. Of course, the carb must be installed on the engine, with the engine running, to make this adjustment.

about those big carbs the racers are using. Most of them are doing it wrong too, and besides, we're talking about a street engine.

Now that we know the right size carb, have established that it's clean and getting enough fuel, it must be mounted and checked for vacuum leaks. Be sure all hoses are connected and that any unused vacuum tubes are capped. In most cases, the use of a heat insulator-type gasket between the manifold and the carb will help stabilize the fuel temperature in the carb for consistent fuel metering. If everything checks up to this point, it's time to start tuning (all of these tuning steps should be done with the engine warmed up to normal operating temperature and the choke fully open).

The very first thing on a carburetor that must be set is the float level. The fuel level in the float bowls, as determined by the float level (and the fuel pressure and fuel specific gravity), affects the fuel metering of both the idle and main circuits within the carb. The float level is easily checked and adjusted on Holley carbs thanks to sight plugs and external adjustments on the float bowls. The fuel should just barely trickle out of the sight plug hole with the engine running at idle (see photo). This done, set the desired idle speed.

Idle mixture is the next item that should be adjusted. Idle mixture screws are located on the ends of the metering block(s). Using either a tachometer or a vacuum gauge, the idle mixture screws should be adjusted to obtain the highest possible vacuum reading or idle speed. Additionally, the screws should be adjusted approximately the same number of turns out from the "full in" position. Settings within 1/4 to 1/2 turn of each other are acceptable.

You may encounter a situation where the idle mixture screws don't seem to have any appreciable effect on the idle quality. This is usually caused by the primary throttle plates being too far open at the desired idle speed. You may have found it necessary to set the idle speed stop screw on the primaries at such an extensive opening just to get the correct idle speed. The cure for this situation is to adjust the *secondary* throttle screw (accessible from the bottom of the carb—see photo) 1/4 to 1/2 turn to provide more secondary air flow. This in turn will permit a reduction of the primary throttle opening at idle to the point where the idle mixture screws will become functional. A good rule of thumb when adjusting the secondary throttle stop is to position the throttle plates so that the bottoms of the fuel transfer slots are just barely exposed. Such a setting will provide a smooth transition into the secondary main circuit as the secondaries open.

If, after proper adjustment of the primary and secondary throttle stops, and the idle mixture screws, a road test reveals a bog or hesitation when the throttle is quickly opened, it may be necessary to increase the size of the accelerator pump squirter(s). Holley squirters are numbered corresponding to the size of the squirter nozzles. For example, a squirter stamped 27 indicates .027-inch diameter nozzles. Holley offers squirters in a variety of sizes, but you can also drill them with wire drills and a pin vise. Increase nozzle size one wire drill size at a time until the bog clears up. Although Holley also offers a number of different accelerator pump cams, it is rarely necessary to change these for street applications, especially on vacuum-secondary carburetors.

On vacuum-secondary Holleys, accelerator pump squirter size must also be coordinated with the secondary vacuum diaphragm spring to prevent bogging and hesitation. If the secondaries open too soon or too quickly, a bog will result. Using the next heavier spring can correct this. If the engine seems down on power or sluggish in response to full

The secondary throttle plate opening at idle can be adjusted by turning this screw on the bottom of the carb. The carb must be off the engine to make this adjustment. The idle fuel transfer slot (heavy arrow) should just barely be exposed when viewed from the bottom.

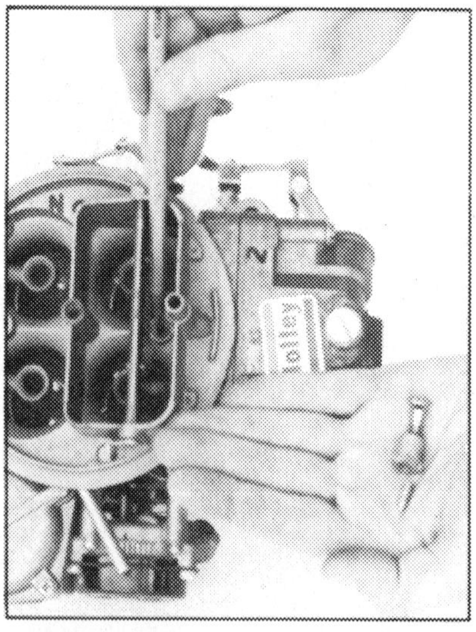

The accelerator pump squirter nozzle is held in place by one screw. Under the squirter is the small check-valve needle (shown in hand). If this check-valve needle falls out when removing the squirter, it must be reinstalled with the pointed end down.

The vacuum-secondary diaphragm spring is located under this cover. The entire secondary diaphragm assembly will have to be removed from the carb to get this cover off unless you drill a hole through the air cleaner flange to gain access to the back screw.

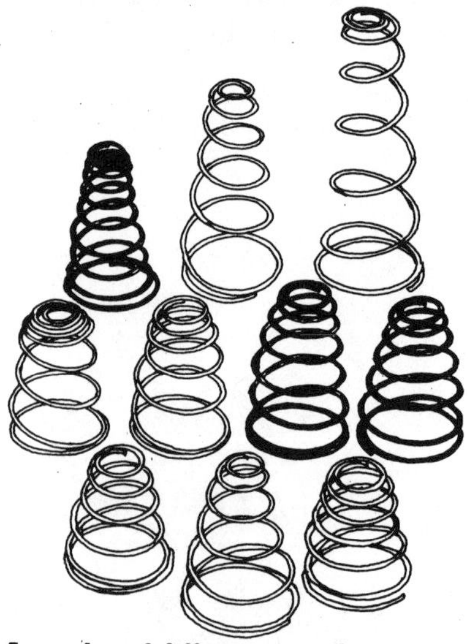

A number of different secondary diaphragm springs are available from Holley to tailor the opening point and opening rate of the secondary throttle plates (see text).

HOLLEY TUNING TIPS

throttle, try the next lighter spring. These springs are color-coded: white, yellow, red, pink, purple, green, orange, plain (no color), brown, and black, from the lightest to the heaviest, respectively. The proper spring should be selected before varying accelerator pump squirter nozzle size in an attempt to use as small a pump squirter as possible. Holley offers an assortment of the most commonly used springs under part No. 20-13.

The only remaining circuits in the carburetor that may need further tuning are the main circuits and power circuits. Main circuit metering is controlled by jets or metering plates. In the majority of cases, the "out-of-the-box" jetting is very close to perfect, but altitude considerations and engine configuration may require slight adjustment. Holley offers a high-altitude jetting kit (part No. 85R-6069 for 6619, 6909, 6919, 6979, 6989, and 7009 carbs, and part No. 85R-6068 for 7454, 7455, 7456, 7555, and 7556 carbs). In any case, jetting changes should always be done with numbered Holley jets or metering plates. Jets and plates should *never* be drilled. If you find it necessary to rejet the carb more than five numbers either way for the "out-of-the-box" jetting, there is probably something wrong, such as a vacuum leak, restricted passage, or inadequate fuel delivery system.

The power circuit is also important to the main circuit. Tailoring the power valve opening point (by manifold vacuum level) will provide smooth off-idle

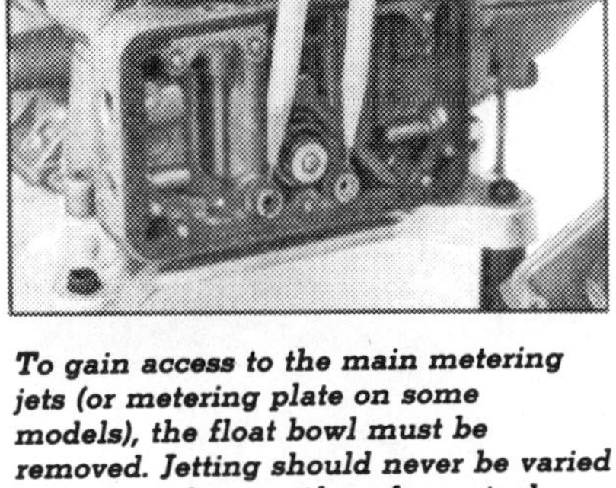

To gain access to the main metering jets (or metering plate on some models), the float bowl must be removed. Jetting should never be varied more than five numbers from stock.

acceleration and proper fuel metering under high-load situations. Power valves should *never* be plugged on street-driven vehicles since idle quality, driveability, and fuel economy will suffer horrendously. Power valves are numbered according to the vacuum level (in inches of Mercury, Hg.) below which they are open. For example, a power valve stamped 85 opens at 8.5 inches Hg. and stays open at all vacuum levels lower than 8.5 inches Hg. When open, the power valve allows extra fuel to flow into the main metering circuit.

To correctly match the power valve(s) to the application, select one(s) that opens *at least* 1 to 1½ inches Hg. below the manifold vacuum level that the engine maintains at idle in gear. This will prevent unnecessarily rich fuel metering under light loads. Fox example, an engine that idles at 8 inches Hg. should have power valves stamped 65 or lower. If the power valves that came in the carb are already stamped 1½ or more below the idle vacuum level, do not change them.

Holley also offers two-stage power valves that open partially at one vacuum level and then open fully at a lower vacuum level. Such two-stage power valves can improve gas mileage on some lightweight vehicles with very mild cam timing, but they should be avoided on modified street engines or in any carburetor that has power valve restrictions (the two holes in the metering block just inside the power valve gasket) larger than .076 inch.

The final setting is the choke adjustment. At room temperature on a cold engine, the choke blade should close so that it leaves an air gap of ⅛ to ¼-inch when the throttle is depressed. This is just a rule of thumb, and variations may be required depending on the climatic conditions in your area.

As a closing comment, the air cleaner you select can also affect how a carb performs. The air cleaner top should be at least ¾ inch above the bowl vents to prevent fuel siphoning and to allow the air to turn smoothly into the carb throats. Tuning a Holley carb is easy and straightforward if you just take the steps in the right order. Give it a try. You'll be surprised at the results. **HR**

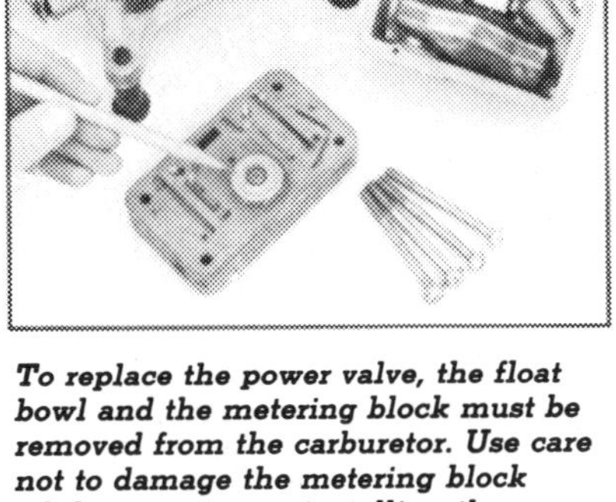

To replace the power valve, the float bowl and the metering block must be removed from the carburetor. Use care not to damage the metering block while removing or installing the power valve.

Holley Secondary Metering Plates

Many Holley high-performance carburetors use a secondary metering plate (part Number 34BP-2007) with drilled restrictions rather than removable secondary jets. When required, it is recommended that this secondary metering plate be changed rather than drilled. Listed below are the specifications for the Holley 34BP-2007 secondary metering plate. The diagram shows secondary idle feed hole—see hole "B" in chart, and secondary main metering restriction—see hole "A" in chart. Order by the suffix dash numbers following the part number. Example: 34BP-2007-14 will have .098 main metering restriction and .070 idle feed hole.

Main Hole "A"	Idle Hole "B"	Order By Part Number	Part Stamped
.052	.026	34BP-2007-7	7
.052	.029	34BP-2007-34	34
.055	.026	34BP-2007-3	3
.059	.026	34BP-2007-4	4
.059	.029	34BP-2007-32	32
.059	.035	34BP-2007-40	40
.063	.026	34BP-2007-5	5
.064	.028	34BP-2007-18	18
.064	.029	34BP-2007-30	30
.064	.031	34BP-2007-13	13
.064	.043	34BP-2007-33	33
.067	.026	34BP-2007-8	8
.067	.028	34BP-2007-23	23
.067	.029	34BP-2007-16	16
.067	.031	34BP-2007-9	9
.067	.035	34BP-2007-36	36
.070	.026	34BP-2007-6	6
.070	.028	34BP-2007-19	19
.070	.031	34BP-2007-20	20
.070	.053	34BP-2007-41	41
.071	.029	34BP-2007-35	35
.073	.029	34BP-2007-39	39
.073	.031	34BP-2007-37	37
.073	.040	34BP-2007-17	17
.076	.026	34BP-2007-10	10
.076	.028	34BP-2007-22	22
.076	.029	34BP-2007-43	43
.076	.031	34BP-2007-12	12
.076	.035	34BP-5113-3	3
.076	.040	34BP-2007-28	28
.078	.029	34BP-2007-38	38
.079	.031	34BP-2007-11	11
.079	.035	34BP-2007-24	24
.081	.029	34BP-2007-44	44
.081	.040	34BP-2007-21	21
.081	.052	34BP-2007-31	31
.081	.063	34BP-2007-29	29
.082	.031	34BP-2007-46	46
.086	.043	34BP-2007-25	25
.089	.031	34BP-2007-47	47
.089	.037	34BP-5113-5	5
.089	.040	34BP-2007-27	27
.089	.043	34BP-2007-26	26
.093	.040	34BP-5113-4	4
.094	.070	34BP-2007-15	15
.096	.031	34BP-2007-50	50
.096	.040	34BP-2007-45	45
.098	.070	34BP-2007-14	14
.113	.026	34BP-2007-42	42

Able to fine tune engines in all kinds of weather...

Adjustable for altitude and barometric pressure...

Calculating mileage, fuel consumption, elapsed time, and range...

Is it the engine of tomorrow?

Is it a mechanic in a can?

No, it's...

Super CARB!

By C.J. Baker

Wouldn't it be super to be able to turn a knob on the dash to select the best fuel mixture for performance? Or to dial-in the best mileage for road and weather conditions? Or vary the mixture to compensate for differences in fuel quality? Well, thanks to the folks at Holley, now you can do just that. What we're referring to is Holley's new line of electronically controlled carburetors, and you won't need X-ray vision to determine the immediate results of mixture changes if you also select the optional driving computer, which accurately assesses fuel flow in relation to vehicle speed and time.

Essentially, Holley is offering a manually adjustable electronic feedback carburetor system with which the driver can alter the air/fuel ratio, within prescribed limits, to suit vehicle operating conditions. This is a spin-off of the new generation of electronic feedback carbs that have been developed for use on many late-model stockers out of Detroit. Such systems employ a mixture control solenoid to vary metering according to an electronic signal generated by a non-adjustable computer in response to a series of environmental and operating sensors. The Holley adjustable retrofit system eliminates all of the automatic sensors and allows the driver to electronically vary the air/fuel mixture with direct control over the mixture control solenoid.

But how will the driver know when he's dialed-in the optimum setting? That's where the optional mileage-fuel flow-time computer comes into play. In addition to the electronically controlled carburetor, the top-of-the-line Holley kit incorporates a sophisticated, programmable computer to display instantaneous and average fuel mileage, speed, elapsed time, fuel flow, total fuel consumption, fuel remaining, and projected range before all fuel is consumed. It also contains a quartz clock. A touch of a button permits the display to be in either U.S. or metric units.

Such a computer is a natural outgrowth of advanced electronics applied to the automobile, making this kit a revolutionary advance in retrofit carburetion and analytic displays for the enthusiast. Last month we acquainted you with adjustable ignition controls, and now these Holley kits offer variable fuel metering, as well.

To evaluate Holley's new system, we installed a prototype kit on an '81 Dodge Charger 2.2. Systems are currently available for retrofitting '79 Chevettes and '78-'80 Ommi and Horizon models to replace the stock, non-electronic, Model 6500 staged 2-barrel carburetors, but our prototype unit was designed as a replacement for a vehicle that already had a stock electronic feedback 2-barrel carb on it, thus overriding the vehicle's automatic feedback system. Kits will be offered for Model 2300 and 4160 2 and 4-barrel carbs, with multiple kit applications on 4150, 4165, and 4175 models. In other words, kits will soon be available for virtually all of Holley's most popular street carbs. These kits will be sold with completely modified carburetors, or with just the nec-

essary parts to update existing carbs, depending on the customer's needs.

The accompanying photos outline the installation, which took the better part of a day to complete. Once installed and calibrated, the computer proved to be a fascinating tool for evaluating vehicle efficiency and driving techniques. Such data as fuel remaining in the tank can quickly save the embarrassment of running out of fuel, and when displaying mileage, it becomes an exciting test of driving skills to achieve the highest possible fuel economy. As for the adjustability of fuel metering, the adjustment range is limited to ±9 percent, thus preventing the possibility of burning holes in pistons due to excessive leaning or washing the cylinder walls with overrich mixtures. Theoretically, mileage gains up to 9 percent might be possible under extreme conditions, but we were only able to achieve gains in the order of 4-5 percent on our already frugal Charger. Best results were obtained by leaning the mixture in city driving where numerous stoplights and prolonged idling are encountered.

Will these "Super Carbs" make a car faster than a speeding bullet? More powerful than a locomotive? Able to achieve tall mileage numbers in a single bound? No. No. Maybe. But call them what you will, these variable feedback systems are an evolutionary step beyond the purely mechanical carbs of days gone by. **HR**

Shown installed in the fuel line on our Dodge Charger 2.2, the fuel sensor measures flow down to .01 gallon. Also visible in this picture is the vacuum control switch.

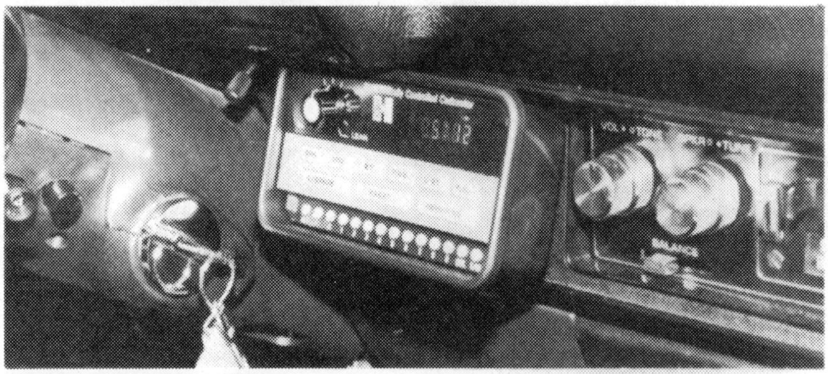

ABOVE— *A position was selected to provide good visibility yet keep the computer out of direct sunlight. The base was blind-riveted to the dash and a ¾-inch hole drilled for the wiring harness.*
Once mounted, the lower face plate of the computer snaps off to provide access to the programming and calibration buttons. Once calibrated, the face plate is installed to prevent accidental touching of these controls. Incidentally, depressing the button marked "6" also provides display of fuel flow in gallons (or liters) per hour.

BELOW— *The modified float bowl and metering block which are the necessary parts required to upgrade existing Model 2300, 4160, 4150, 4165, and 4175 2 and 4-barrel carbs. These components will be part of a kit to convert these carbs to electronically controlled configurations, either with or without the diagnostic computer, as the customer desires.*

To measure distance accurately, special magnets are attached to the driveshaft (or front half-shaft as on this front-wheel-drive installation) and a sensor mounted adjacent to the magnets. Safety wiring ensures that the magnets stay in place on the rotating shaft.

LEFT— *Complete system contains necessary hardware to facilitate the installation, which along with calibration, takes the better part of a day. Normal retail selling price for this system is just under $300.*
The basic components of the Holley electronic feedback carburetor system include a replacement carburetor with an electric mixture control solenoid, a fuel flow sensor, a manifold vacuum control switch, and the programmable computer and display panel.

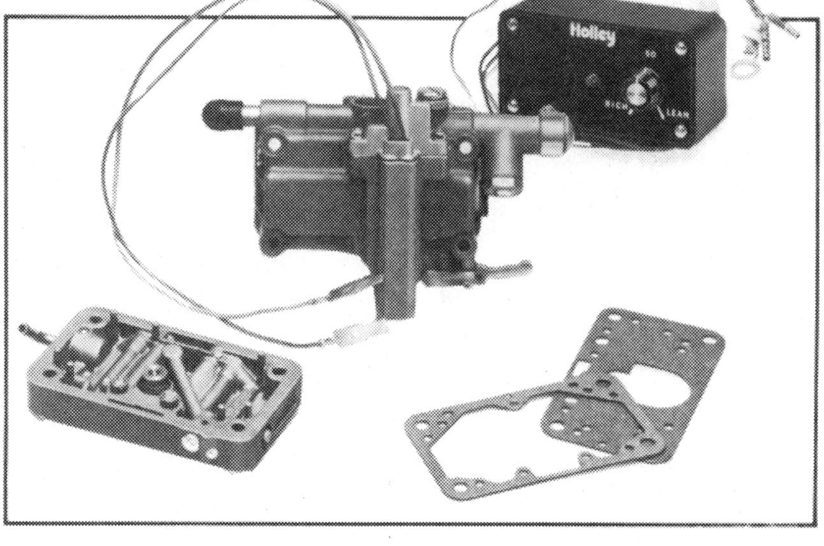

Alcohol

Two New Holley Performance Carburetors Specially Calibrated for Alcohol

By C.J. Baker

FUEL ALTERED

Converting to alcohol as a fuel for a racing engine used to be difficult, requiring extensive carburetor rework or a switch to fuel injection. The reason for this is that far more fuel flow is required when using alcohol (roughly twice as much) than when gasoline is used. Many carburetors simply cannot be modified for such a large increase in fuel flow, and even those that can require extensive reworking of all fuel passages and circuits, including the idle circuit, to function properly. Such modification requires equipment, skills, and a knowledge of complex carburetion principles that many racers do not possess, yet the racing associations are encouraging a switch to alcohol, and additional power *can* be realized from most engines when they are converted from gas to alcohol. Consequently, many racers would like to switch to alcohol to gain a competitive edge, but they don't know how to go about it. To help solve this dilemma, Holley has introduced two new carburetors specially modified and calibrated for methanol.

To appreciate the complications of such a switch, it is necessary to understand the peculiarities of alcohol. There are two basic types of alcohol commonly used as an automotive fuel—ethyl alcohol (ethanol), which is derived from grain, and methyl alcohol (methanol), which is a wood by-product. Ethanol has a higher heat content than methanol, and in comparison to gasoline, about 1.6 to 1.7 times as much ethanol must be used to obtain a proper air/fuel mixture.

Methanol requires flow rates 2.2 to 2.3 times greater than gasoline due to its lower heat content, making it the less desirable of the two versions, yet methanol is the commonly used variety since it is more readily available than ethanol in most areas. Consequently, Holley's new carbs are calibrated for methanol, although ethanol could be used by rejetting the main metering circuits and leaning the idle mixture.

Because of the fuel flow limitations, Holley's alcohol 4-barrel, the O-8156X, is limited to an air flow of 750 cfm (cubic-feet per minute). The fuel passages simply cannot be enlarged any further to supply higher air flow rates. The O-8156X is a dual-inlet with center-piv-

The main discharge passage on the alcohol Holley carbs is enlarged to handle the necessary fuel flow. The large REO accelerator pump(s) are also used, with enlarged squirter nozzles and hollow-drilled nozzle screws. The 4-barrel model uses REO pumps on both the primary and secondary sides.

FUEL ALTERED continued

ot float bowls, mechanical-secondary, double-pumper carburetor. If air flow rates greater than 750 cfm are desired, Holley recommends using two smaller carbs, such as their second alcohol model, the O-4412X 2-barrel, rated at 500 cfm. To complement the fuel flow requirements of these carburetors, *everything* else in the fuel system must be capable of delivering an adequate supply of alcohol, from the fuel tank pickup(s) to the fuel line (minimum ½-inch unrestricted inside diameter), fuel filter (a large canister-type is recommended), fuel pump(s), and pressure regulator(s).

Aside from the fuel flow requirements imposed by the use of methanol, other problems complicate the conversion. Methanol is a super solvent—it will knock everything loose in an existing fuel system, such as scale and rust, so frequent fuel filter replacement is a must. In fact, methanol cleans so well that after a period of time, the inside of the fuel tank and fuel passages will take on a polished appearance.

Methanol is also a corrosive, attacking the base metal of the carburetor body, bowls, and metering components. Consequently, all alcohol should be drained from the vehicle (including the carburetor) at the end of each event. Storage of alcohol is also a hassle since methanol readily assimilates moisture from the air, contaminating the fuel. Water contamination must be kept below one-half percent to avoid problems.

There are other maintenance problems with alcohol. Alcohol tends to extract the "plasticizers" from rubber-type compounds, causing a general hardening of the materials, which results in splitting, cracking, and a loss of sealing. This means that metal fuel lines must be used instead of rubber-type hoses. If a flexible line must be used, a true braided steel fuel line, such as that used in aircraft, should be selected. Such braided lines don't have rubber liners, so there is no problem. Also, all the carburetor gaskets, O-rings, and seals, along with fuel pump and regulator diaphragms need to be periodically replaced. The Holley electric fuel pump service kits are No. 12-806 for the check valve (pressure relief) assembly, and No. 12-808 for the lower housing. The regulator service kit is No. 12-807.

Alcohols also cause problems with the viton-tipped inlet needle assemblies in most carburetors, so Holley uses their No. 18BP-109AS steel inlet needle assemblies in their new alcohol carbs. Brass floats are also used exclusively since methanol attacks nitrophyl floats, causing them to swell, increase in weight, and soften.

These new carburetors solve many of the problems confronting the racer who wants to switch over to alcohol, yet tuning and adjustment is identical to other Holley carbs, using standard Holley jets. However, racers will have to learn a few new tricks, too. Alcohol doesn't color plugs like gasoline does, and peak combustion temperatures are lower. Alcohol also burns slower than gasoline, so ignition timing needs revision for peak performance. Like so many things these days, it's a new ball game, but these Holley carbs are the first step toward making your car fuel altered! **HR**

All fuel passages in the metering block are enlarged, including the main metering wells and the angled top transfer channel. The idle fuel bleeds are enlarged and the power valve channel restrictions are opened to supply extra fuel. Since the power valves are an important part of the fuel delivery circuit on these alcohol carbs, they should never be plugged.

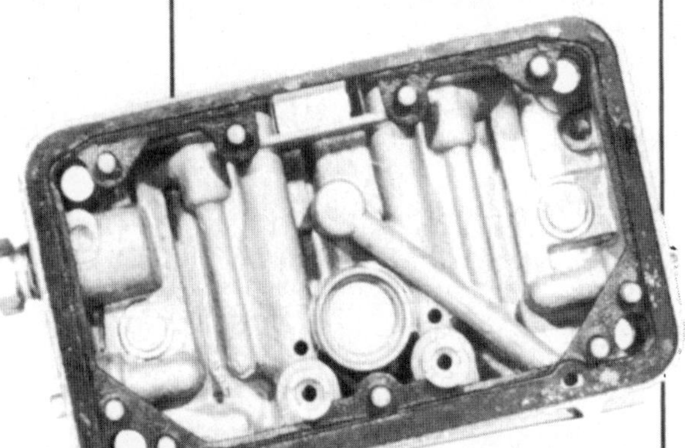

On the bowl side of the metering block, precisely calibrated holes are drilled just above the jets to provide a constant fuel flow into the main metering wells. Tuning adjustments are then made by varying the main jet size, just as on gasoline carburetors.

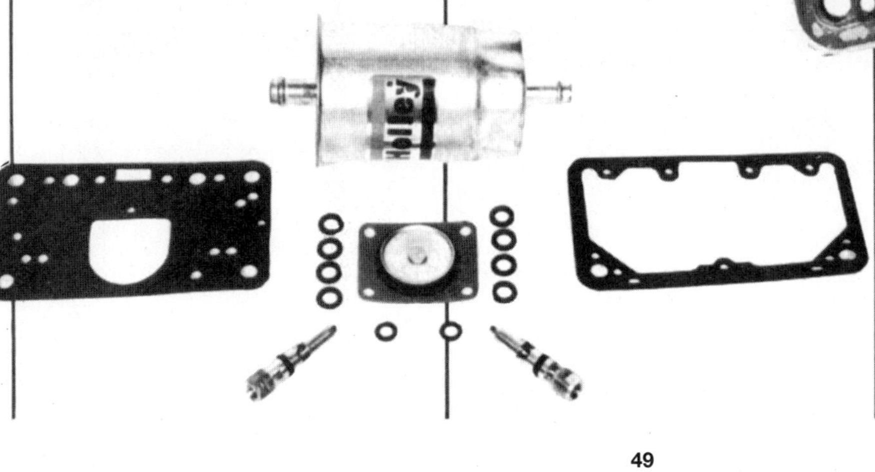

These rubberized carburetor components require periodic replacement to prevent hardening, splitting, and cracking. The fuel filter (top) also needs frequent replacement due to the cleaning action of alcohol in the fuel system. The large .130-inch-diameter steel inlet needle assemblies (foreground) must be used with alcohol to provide flow without swelling.

Jet Set

By C.J. Baker

This New Kit Offers Dual Jetting Versatility to Holley Carburetors

Wouldn't it be nice if you could select different carburetor jetting right from the driver's seat to match the driving conditions at hand? Well, if your vehicle is equipped with any one of Holley's numerous modular 2- or 4-barrel carbs (those that use a separate float blowl and metering block held in place by four bowl screws), there's an innovative new kit on the market that permits remote switching between two previously selected sets of jets. On 4-barrel carbs, this kit is fitted to the primary side of the carburetor only, since it is expected that secondary jetting matching the richer of the two sets of jets in the primary circuit will be retained for performance and added fuel enrichment under high-load conditions when the secondaries come into play.

For example, let's assume you have a high-performance street machine

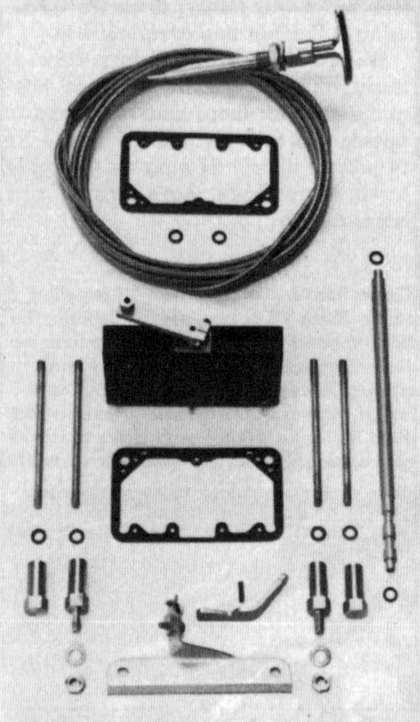

A typical Motr Trim kit contains all the parts needed to modify specific Holley modular carburetors, including gaskets, a new accelerator pump arm, control cable, brackets and even an extended fuel transfer tube for those models that use a single fuel inlet. Anti-slosh tubes, close-tolerance Holley jets and special jet tools are also offered separately.

Although it is assembled when you receive it, this disassembled Motr Trim module shows the precision pieces and workmanship that go into this top-quality kit. The module shown here hasn't been anodized yet, and it is also an early version that isn't drilled for slosh tubes or increased fuel circulation, as have all new production modules (see other photos).

Special bowl studs and nuts replace the original bowl screws. Provisions for extending the accelerator pump arm and bowl vent control arm (if used) are included in the kit, although the pump arm has not yet been installed in this photo.

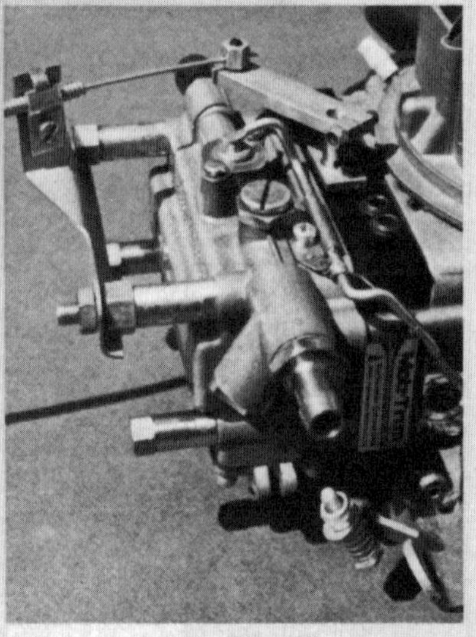

Fully assembled, the Motr Trim kit, which is used only on the primary side of the carburetor, gives the driver jetting control from the passenger compartment. No alterations to the basic carburetor that would prevent it from being returned to stock at a later date, such as when the car is traded or sold, are required.

Jet Set

that seems to run strongest with No. 80 jets all the way around, but while you're street cruising, that performance mixture tends to foul plugs while simultaneously killing gas mileage. Now, with a flick of the wrist, you could switch over to No. 76 jets on the primary side of that big Holley double-pumper to keep the plugs clean and improve the mileage.

Another situation could be that you live at one altitude but frequently drive to an area at a different altitude. Again, you could select the proper jetting for each location and switch over while en route. Still another application would be for the drag racer who likes to fine tune for maximum power and consistency as temperature and air density change.

In concept, this kit, called the Motr Trim carburetor conversion and manufactured by Vondee Performance Enterprises, Inc. (23136 Mariposa Ave., Torrance, CA 90502), is simply a block containing two sets of jets, fuel passages and a selector valve to permit switching from one set of jets to the other. The concept was the easy part—the hard part was designing it to work with all Holley modular carbs and to function smoothly without disrupting traditional Holley throttle response and driveability; but these goals have been achieved, and the patent rights have been applied for.

The racer net price on the Motr Trim kit is right at $140; but with the cost of gasoline and spark plugs these days, this could be a real bargain for the owners of dual-purpose performance machines. It's a high-quality, precision piece of equipment that modifies only the main metering circuit. The idle circuits, transition circuits, power valve circuit and accelerator pump circuits remain unchanged. Even the float levels remain the same, although correct float level and idle adjustment should be verified at the time of installation.

We installed a Motr Trim kit on the 600-cfm double-pumper on our 4.1-liter Buick V6-equipped '80 Skylark test car. Total installation time was less than an hour. The only complications that might conceivably arise involve air cleaner clearance for the jet selector valve and control cable atop the Motr Trim module if your air cleaner drops down around the perimeter of the carb. We were using the new Holley 9-inch open-element air cleaner, which caused no problems. Also, on dual fuel inlet carbs, the Motr Trim kit positions the bowls about ¾-inch farther apart; so depending on the fuel line kit used, one line may need to be lengthened slightly. We did encounter this problem, but it was minor, since we were using a flexible line arrangement.

We found the kit to perform flawlessly, allowing the use of No. 68 jets at the strip for those mid-14-second passes and then switching back to No. 64 jets for nearly 24 mpg on the drive home. In our book, that's having your cake and eating it too! **HR**

With the optional jet extraction tool, jets can be changed in the module in a matter of seconds without draining the float bowl, adding further versatility for the racer or performance enthusiast.

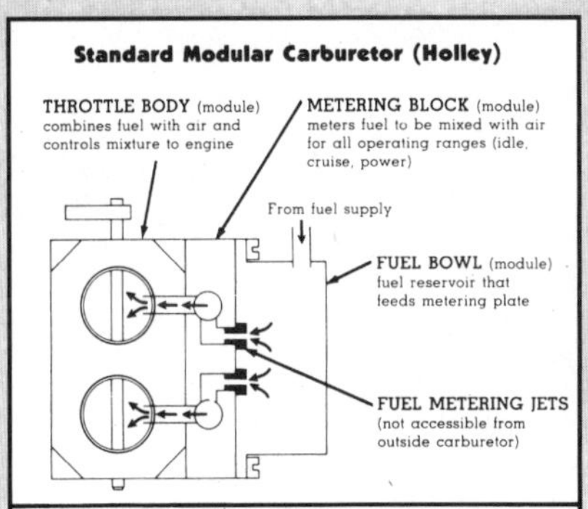

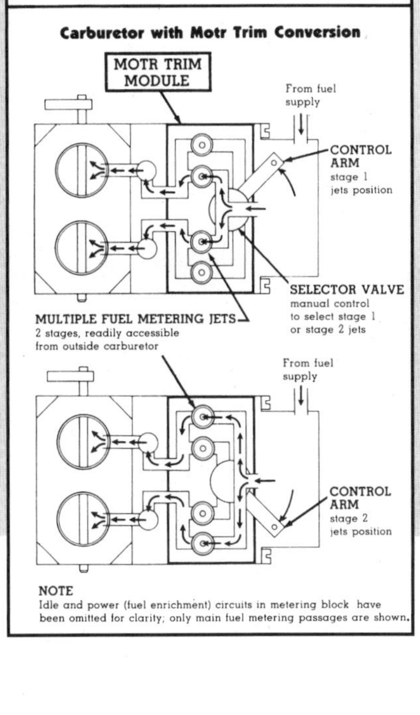

This is the production configuration of the Motr Trim module, shown here fitted with the optional anti-slosh tubes which are highly recommended for high-performance vehicles that experience rapid acceleration, braking, cornering or attitude changes.

Shown here is the Motr Trim kit installed on the Buick V6 in our Project Skylark. The kit performed flawlessly, offering maximum power jetting for performance and lean cruising jetting for economy without any loss of response or driveability, but we did need the anti-slosh tubes to make the module compatible with this vehicle's capabilities.

1 *Begin by completely disassembling the carburetor. Remove both float bowls, the throttle plate and the secondary metering plate, if so equipped.*

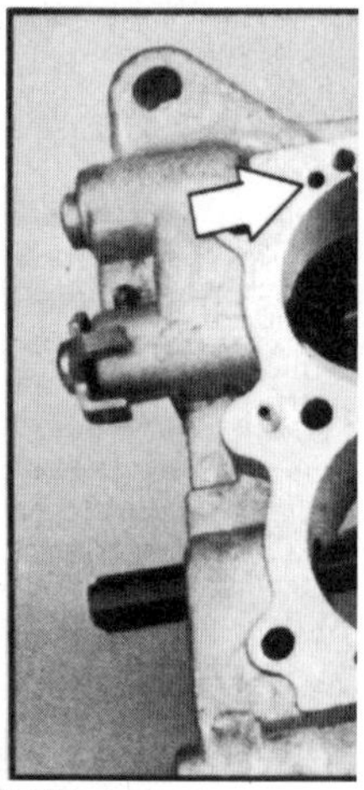

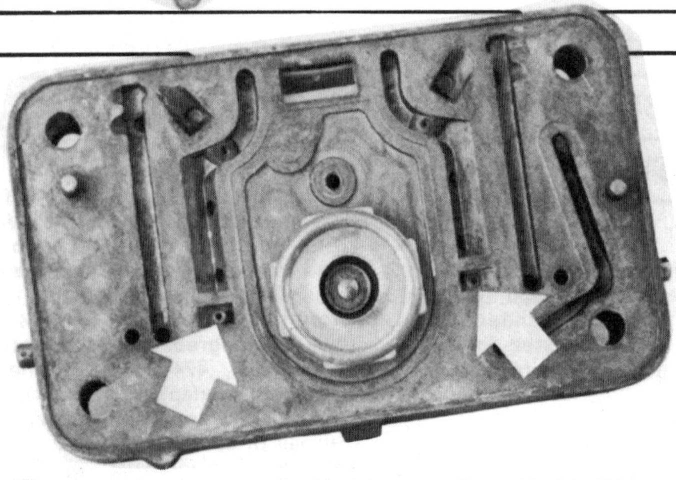

2 *Although the primary metering block from virtually any Model 4150 or 4160 Holley carb can be used, you should find one that has exposed idle fuel restrictions (arrows). Some primary metering blocks do not have these idle fuel restrictions exposed. For our modification, these idle fuel restrictions should be approximately .040-inch in diameter. Some metering blocks will require enlargement of these restrictions to this diameter.*

4 *Drill out the idle air bleeds (arrows) to the same size as those on the primary side of the carb. This is usually in the neighborhood of .070-inch.*

3 *The secondary metering plate, or metering block, must be replaced with a primary metering block identical to the one on the carburetor, which has idle adjustment screws in it. Using a No.31 drill, work through the two holes indicated on the main carb body until the drill breaks through the first time. Do not drill any further. These holes will now permit the flow of idle fuel through the two corresponding holes indicated on the metering block.*

INSTANT ACTION!

HOW TO ADD AN ADJUSTABLE SECONDARY IDLE CIRCUIT TO HOLLEY PERFORMANCE CARBS FOR LIGHTNING REFLEXES AT THE DRAGS

By C.J. Baker

5 *To complete the modification, two holes must be drilled in the secondary throttle bores corresponding to the two fuel passages indicated by the arrows.*

6 *These holes (arrows) must be drilled downstream from the throttle plate when it is in the closed position. The distance below the throttle plate should be equal to that of the idle discharge holes on the primary side of the carb. The size of the holes isn't critical, although they must be at least as large as those on the primary side. You may have to use a drill larger than the primary diameter simply to avoid breakage.*

7 *Drilling these holes is tricky. Take your time and be very careful to avoid breaking the drill. To get clearance for the drill, these holes will have to be drilled at an angle, as indicated in this photo.*

Drag racers will tinker with anything in the interest of performance, and carburetors seem to be high on the list. In most cases, the effort doesn't really make much of an improvement, but occasionally racers come up with a modification that is so effective that Holley will incorporate it into their line. Such is the case with the adjustable secondary idle circuit.

Several years ago racers began adding adjustable secondary idle circuits to the performance-oriented versions of the Model 4150 and 4160 Holley carbs. This modification vastly improves the throttle response and "stageability" of drag cars so equipped by increasing the idle fuel flow and priming the secondary fuel metering circuits for a faster secondary response time. It works so well, in fact, that Holley added six carburetors to their 4150 series, incorporating the adjustable secondary idle circuit. They are the No. O-8156 750-cfm, the No. O-8162 850-cfm, the No. O-8804 830-cfm, and three other versions—Nos. O-9379, O-9380 and O-9381—which have the same respective airflow ratings as the first three, but also incorporate Holley's new annular discharge booster venturi. This is all well and good if you're in the market for a new carb, but if you already own one of the countless thousands of mechanical secondary 4150 or 4160 series performance carbs, we've got the lowdown on how you can update your carb for this improved response.

For complete details, we contacted Joe Sherman, who, along with Cyril James, modified the O-4224 660-cfm carbs atop the tunnel-ram manifold on his '77 Nova (HOT ROD, December '79 issue) for the adjustable secondary idle circuits. Joe indicated that after the modification had been made, the engine didn't lose as many rpm when the automatic transmission was pulled into gear, the car was easier to stage, and the throttle response was vastly improved to the point where starting-line traction became marginal because of the instant power output.

Essentially, the modification amounts to simply making the secondary side of the carburetor identical to the primary side by substituting a primary metering block for the existing secondary metering block or metering plate. The primary metering block is used because it has the idle adjustment screws. Obtaining a suitable primary metering block is the biggest hassle of this modification, since Holley doesn't sell primary metering blocks separately. The simplest approach is to buy a used 4150 or 4160 series carb at a swap meet for around $5 to $10 just to get the metering block. (Save the rest of the used carb for spare parts.) The accompanying photos detail the simple drilling that must be done to make the secondary idle circuit functional.

Tuning the carburetor after this modification has been made will be quite critical as the idle adjustment screws become very sensitive. A good starting place is to back the idle adjustment screws out ½ to ⅝ of a turn from the full in position. Secondary jetting usually needs to be enriched somewhat too. Again, a good starting place would be to increase secondary jet *area* by about 10 percent.

With this modification completed, the stock-sized accelerator pump squirter nozzle size should be adequate. This mod should be especially beneficial for cars with automatic transmissions that need good low-end power to launch the car. It's simple and it's cheap. **HR**

By John Reed

AIRED OUT

A LOOK INTO THE BASICS OF MODIFIED NASCAR CARBURETORS. INTERESTINGLY, SOME IDEAS ARE THE SAME, REGARDLESS OF YOUR TYPE OF RACING

Whenever the National Association for Stock Car Auto Racing chooses to curb speeds in its Grand National division—and this happens frequently—it usually looks at carburetors. Approximately 80 percent of NASCAR's speed-cutting rules in the last five years have involved carburetor restrictions. Future rules are sure to be similar.

Carburetor builders like me find this extremely discouraging—the toughest part of the game to live with. Without questioning NASCAR's motives or methods, the hard fact remains that the amount of time and development work apt to be wiped out by a single rule change is staggering. We live with it, and will continue to, because, unfortunately, carburetor rules make sense. Any racing engine is nothing more than a glorified air pump. It puts out horsepower in direct proportion to the amount of air passing through it—or not passing through it. Reducing carburetor efficiency and capacity is a devastatingly effective way of reducing speeds.

NASCAR's 1978 carb rules aren't as highly restrictive as a few seasons back—then no polishing or milling of any sort was tolerated. But they're restrictive enough. One four-barrel carb is maximum, and the size is limited. Any Holley 4150-series carb with 1 11/16-inch throttle bores can be used, or any 4500-series carb can be used if restrictor rings are installed to reduce the two-inch throttle bores to 1 11/16 inches. The main exception to this rule involves the Chevrolet Laguna-bodied cars, which are further restricted to a 4150-series carb with a maximum *venturi* diameter of 1 1/16 inches, as opposed to a venturi diameter of 1⅜ inches on most other 4150s being used. Consequently, flow is restricted to about 600 cfm, which effectively kills the Laguna from competition.

In respect to the 358-cubic-inch Chevrolet V8 (the dominant engine in '78 Grand National competition), we at Reed Cams prefer the Holley Model R4779-2, rated at 750 cfm. We could select a larger capacity carburetor; but merely using a larger one, while producing more air capacity, wouldn't provide the really dense charge of high-velocity air we like to ram into the cylinders. Velocity is more important than total flow, provided you have enough basic flow to supply the engine. Flow beyond those basic needs is useless.

The following modifications refer primarily to the R4779-2 carburetor. Holley has done an excellent job of providing five basic circuits in this carb. When properly balanced, no additional circuits are needed.

AIR FLOW

The first thing we do with a stock Holley is to uncrate it and place it on our air flow bench and, later, the dy-

ABOUT THE AUTHOR

As a sixth-grader at Winona Park Elementary School, John Reed built (as part of a class project) a flathead engine complete with moving parts and cam timing so realistic that if it had been made of metal instead of wood, it would have run.

Today that pioneer engine is part of the bric-a-brac in the office of John Reed's 27,000-square-foot Reed Cams facility in Decatur, Georgia. At 37, Reed has polished and expanded his mechanical touch, and he is now the president of a corporation with three-quarters of a million dollars worth of speed-producing equipment. Next to furthering the development of the internal combustion engine, Reed's second passion is the collection and restoration of vintage airplanes.

A cam grinder active in carburetor building? Reed admits it may seem unusual. "Why does a cam manufacturer get involved in this?" he ponders. "I can't answer that. He just does. If he's interested in his engine parts doing a good job, he doesn't want some customer saying, 'Well, that guy's camshaft is good, but it can't perform because everything else he builds or sells is lousy.' Major components in an engine have to work together, and I might as well build as many as I can myself."

PHOTOGRAPHY: C.J. BAKER

namometer. However, the air flow bench is the key. Of the approximately 14½ hours of time spent on each carburetor, the greatest portion is spent on the air bench. We work hard equalizing the vacuum pulled through all air system bleed passages, all main well exit ports in the end plates, all jets, all booster main nozzles and idle circuits. Well balanced and equalized vacuum pulls on the flow bench assure that the fuel metering is controllable and accurate for the engine's air mixture requirements.

Next we totally disassemble the carburetor; radically increasing the dry air flow capacity and velocity is our first interest. Later, of course, we become extremely serious about increasing the fuel flow and metering it in proportion to the new higher air flow figures.

As far as modifications are concerned, the first thing that goes is the choke horn. We mill it away. By hand grinding, we recontour and reradius all entries to the velocity stacks (carburetor throats). The size of each venturi is not increased, but nevertheless we pick up a great deal of air flow simply by cleaning up the entries to each individual venturi. The standard kind of roll configuration used by Holley is flattened out a great deal. We turn it into a 22-degree angle, which on this carburetor is perfect. Conversely, the 10-degree deposit angle at the bottom of the venturi isn't recontoured at all, just highly polished. Air flow tests have convinced us that it's impossible to improve this angle.

The next area of concern is the base plate, and specifically the butterflies. Stock base plates have fully rounded ⅜-inch shafts, with the butterflies attached by two screws. Screw heads disturb air flow. Off they come, and in their places we heliarc-weld the butterflies to the shafts, with a very fine (but tough) stainless steel weld. We also thin the shafts and radius their vertical edges. All this allows us to increase air flow by *no less than 22 percent.*

That's quite a gain, but the interesting thing is that when we reassemble the base plate to the carburetor main body and recheck the air flow, the gain increases even more, often to 30 percent. This is a phenomenon of air flow. It is better to have a pre-established vertical column of air (in this case, the main body) above the hole, because air tumbling around corners actually reduces the effective diameter of the hole, choking it. The vertical column allows air to ram through without turbulence and choking.

To assure ourselves that all holes (air passages) have been drilled properly and filed and flattened as we like, we check and recheck everything many times. That's basically all we do with raw air flow. There are other metering passages in the main body, but we confine ourselves to cleaning them up. We don't enlarge them. The exceptions are the air bleeds, but I'll get to them momentarily.

FUEL FLOW

Gasoline stored in the fuel bowls is forced into the main jets, then goes through metering plates before dumping into the center section and getting blasted out through boosters into the air stream. The fuel bowl includes a fuel cutoff system—the needle—that, in conjunction with the float, adjusts the fuel level in the bowl; an accelerator pump, the diaphragm of which is located in the bowl; a vent for air, so there's no vacuum in the bowl; and a brass float.

We replace the brass float with a phenolic unit made of resin or plastic material that can be reshaped. Reshaping the float is important. Because side forces exceeding 1.5 G are exerted on cars at certain of NASCAR's flat-cornered short tracks, fuel frequently gets forced to the right side of the bowl at a 45-degree angle. So steep an angle can easily uncover a main jet and pull pure air instead of fuel into the main well. To eliminate that danger, we cut off one end of the phenolic float to a 45-degree angle, then use epoxy substance on the freshly cut end, preventing its somewhat porous surface from allowing fuel into the float and making it sink. With the end of the float corresponding to the 45-degree angle of the fuel in the corners, we've tricked the float into staying down, admitting extra fuel to keep the main jets covered, yet returning to the proper bowl level on the straightaways.

The accelerator pump physically pumps fuel through the squirters directly into the venturis on this 750-cfm carb. However, we don't want the engine flooded with too much sudden fuel, so we leave the stock accelerator pump, a 20cc-capacity unit, and change only the diameter of the squirter nozzles and the lift of the actuating cams which control the volume and rate of fuel discharge.

(continued overleaf)

Dramatic air flow gains are made by cleaning up the throttle plate of the carb. Note the comparison of the stock and modified throttle plates above. The throttle shafts are slabbed, the throttle butterfly screws removed and the butterflies heliarc-welded to the shafts to create the minimum obstruction to air flow. These mods alone are responsible for a 22-percent flow increase, and that figure climbs even higher when the carb is assembled, as outlined in the text.

AIRED OUT

Other than through the idle system, the primary path that fuel follows to exit the float bowls is through the main jets into the main wells. There are two main wells in each end plate. Each main well is fed by one main jet (along with other passages) and serves as a central fuel collection point to supply each venturi. Consequently, the main well is the granddaddy of them all—the main fuel sup-

Modification of the main body begins with the removal of the choke horn, which is milled away. Then a high-speed cutter is used to smooth up and recontour the entry into each carb throat, achieving an optimum 22-degree entry angle. Fine-grit abrasive hardrolls are used to finish-polish the main body. The booster nozzles are cleaned up, but not recontoured, except at the very bottom to make it function as a more effective venturi.

ply to the carburetor. Parts of the main well are in the end plate, and part is in the center section of the carburetor. We enlarge everything, delicately drilling straight through the main well. We eliminate any steps or humps in the main well that might restrict flow, enlarging it to a consistent .172-inch diameter.

Solid fuel in the main well can't properly atomize in the venturis until it's mixed with air. That's why Holley puts in so many nice little air bleeds, forcing air into the main well and the idle circuit, which also feeds off the bottom of the main well. Probably the most critical work we do involves increasing the size of these air bleeds in proportion to the new higher air and fuel flow figures. We are looking for one vacuum draw figure for the idle positions and another for the open position of the butterflies. These numbers are not the same, because the vacuum in the venturis varies continuously with the throttle position.

We leave the idle system's air bleed diameters alone; the others we increase in direct proportion to the increased vacuum we've achieved at the jets—about five percent. The only time the carburetor is operating on the idle system is when the car is running slowly on the pace lap, or under the yellow (slow down) caution situation. The dangerous combination of very light throttle pressure, the short cam timing of a Grand National car, and the increased air we're drawing through the carburetor could lead to mixture leaning, and possible galling of cylinders and pistons. Therefore, to be on the safe side, we enlarge the idle restrictor jets (fuel bleeds).

The last thing that happens to the fuel/air mixture in the main well system before it is dissipated in the engine is that it is blasted through the main fuel nozzle in the venturi booster. The booster is actually an added vacuum pump—pull is generated by the venturi and the booster itself. We use the stock Holley booster, but its internal bore is recontoured very slightly at the bottom. This creates a rolled effect. Now the booster looks like a small venturi and obviously acts like one by speeding up air and increasing fuel dispersion into that air. High velocity always equals low pressure, and this aids the booster in supplying well-atomized fuel to the engine.

OVERVIEW

We're never truly finished with carburetor modifications. Testing and retesting continue on our flow bench and dyno. We pull the same vacuum against the carburetor on the bench as the manifold pulls against it on a running engine at wide-open throttle. Also, some part-throttle open tests are conducted to assure the carburetor's balanced condition to engine response. We want to arrive at a flat brake specific fuel curve through the engine's anticipated working rpm range.

Our dyno, with its laminar air flow measurement in cubic feet per minute, enables us to monitor all air flow requirements of the engine running at all speeds, as well as manifold vacuum and accurate fuel brake specific curves. Such measurements taken on a dyno are enormous aids in "dialing in" a carburetor for a specific race track.

What Reed Cams does with manifolds, carburetor adapters, air cleaners and cylinder heads—all vital parts of the induction system affecting not only power but that all-important factor in circle-track racing called "response" time—is another story for another time. What you have just read is the high state of the art of carburetor modification as Reed Cams knows it today—before NASCAR makes changes and we have to scrap it and begin learning something new. **HR**

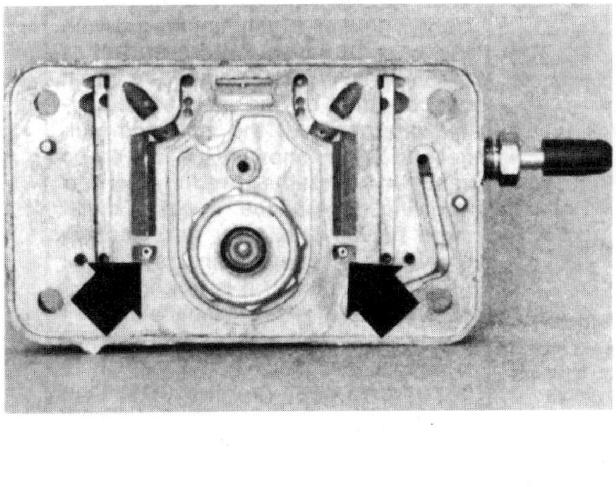

The bottoms of the venturis in the main body are polished, but not recontoured. The standard brass floats are removed from the float bowls and replaced with phenolic floats (available from Holley). To compensate for fuel slosh in the corners, one end of the float is trimmed at a 45-degree angle. To prevent a lean condition during low-speed operation on the idle circuit, the idle fuel bleeds (jets) are enlarged slightly.

Reed assures precise fuel and air metering from each carburetor they build by checking each and every one on both the flow bench and the dyno. Total engine air flow (demand) is accurately measured on the running dyno engine, through the use of this massive laminar air flow device. Fuel consumption is carefully monitored, as is exhaust temperature, to check distribution.

DOUBLE BARRELED

YOU DON'T HAVE TO THINK SMALL WHEN YOU CONSIDER THE POSSIBILITY OF A RACING TWO-BARREL

By C.J. Baker

The evolution of sophisticated two-barrel racing carbs began in the mid-'60s with the AHRA formula Stock classification system which provided distinction between cars running multiple carburetors (Formula 1), a single four-barrel (Formula 2) or a single two-barrel (Formula 3). This same classification system is also being utilized by IHRA. At first, the Formula 3 racers relied on the Rochester 2GC and Holley R4112-AAS 500-cfm carbs. However, the rules also permitted any modification to the carburetor, provided the carburetor started life as a two-barrel and not a four-barrel. Consequently, aftermarket suppliers began supplying some truly incredible modified Holley two-barrels for Formula 3 racing. The first of these suppliers was Braswell Carburetion in Tucson, Arizona. In fact, Braswell's annular discharge of the early 500-cfm two-barrel was actually the forerunner of the Holley R6425-AAA 650-cfm annular discharge two-barrel released in '72. Those early Braswell reworks boasted an air flow of 710 cfm (rated at 3-inch Hg. manifold vacuum) and were responsible for setting most of the Formula 3 records at that time.

With the introduction of the new Holley 650-cfm carb, Braswell and another firm, C&S Specialty in St. Charles, Missouri, began offering reworked versions of this larger carb. Both companies achieved flow rates

If these don't look like your average two-barrel carbs, it's because they're not. Both the C&S unit (top) and the Braswell version shown here are rated at nearly 1000 cfm!

of over 800 cfm. Basically, they reshaped the venturi entry and discharge angles, enlarged the throttle plates, reworked the fuel metering passages and added special velocity stacks. These modified carbs showed a .1- to .2-second improvement over the stock Holley 650 two-barrel. And again, the records fell. Now both firms have released still larger versions of their reworked Holley two-barrels.

C&S has enlarged the venturi diameter from 1.550 inches to 1.650 inches, with correspondingly larger fuel metering passages and a tremendous fuel handling capability. This is in conjunction with their proven 2-inch throttle plates and unique cast-resin velocity stacks. To employ the 2-inch throttle plates, a ¾-inch cast-resin spacer plate is inserted between the main body and the throttle body to provide a smooth transition.

Braswell has taken a different approach, retaining their proven 1.530-inch venturi diameter and their smoothly radiused stainless steel velocity stacks, but inserting a ¾-inch aluminum spacer below the throttle body, which is machined to provide a smooth transition into a single oblong throttle plate similar to that used on the secondary side of the Holley three-barrel. And of course the fuel passages have been enlarged accordingly.

Both manufacturers' carburetors

now offer flow rates approaching 1000-cfm! If you translate that into flow at 1½ inches Hg. of manifold vacuum at which four-barrel carbs are rated, the flow is well over 700 cfm, so the performance potential of these huge two-barrels is nearly equivalent to that of a large four-barrel on all but very large displacement engines. Increases of another .2-second and 2 mph have already been recorded. But there's another plus, and it's a big one. These big two-barrels achieve their incredible flow by careful contouring of the venturis and matching them with velocity stacks to achieve high velocity flow. Similar development is rarely attempted on four-barrels, since they can achieve the needed flow with the extra two-barrels. Therefore, this relatively high velocity flow through the

DOUBLE BARRELED

(continued on page 130)

These photos show the extent of modification on the Holley 650 annular discharge 2-bbl to achieve nearly 1000-cfm flow capacity. The C&S carb (molded velocity stacks and twin throttle plates) has the venturi diameter enlarged to 1.65 inches, with correspondingly big fuel passages. The Braswell carb (metal velocity stacks and single throttle plate) uses 1.53-inch venturis but enhances flow by enlarging the signal area immediately under the venturis with the single throttle opening. Note too the use of screw-in air bleeds for both the idle and high-speed air bleeds to enable precise fuel curve tuning. Both carbs use ¾-inch spacer plates between main and throttle body.

THEY'RE UNSINKABLE, MOLLY BROWN

In the September '76 issue of HOT ROD, we told you about the Fast Jet Kit for Holley carburetors equipped with center-pivot float bowls. This kit is manufactured by C&S Specialty (85 Susan Drive, St. Charles, Missouri 63301), the same company that modifies one of the big two-barrels in the accompanying article. Basically, the kit reduces the time required to change jets, while simultaneously reducing the fire hazard. However, there was one possible hang-up. As originally sold, access for the jet tool required dimpling the brass float. All too often, such dimpling would result in a slight leak which would eventually sink the float, causing flooding. The cure is incredibly simple. Holley offers a closed-cell Nitro Phyl replacement float (part No. 16BP-451A), so clearance notches can be easily filed. Now all of the new C&S Fast Jet Kits include these floats. Holley suggests that the surface of the filed portion of the float be sealed with a *thin* coat of two-part epoxy cement, but chances are that you can skip this step without ever encountering any problems unless the float is frequently exposed to very hot fuel. **HR**

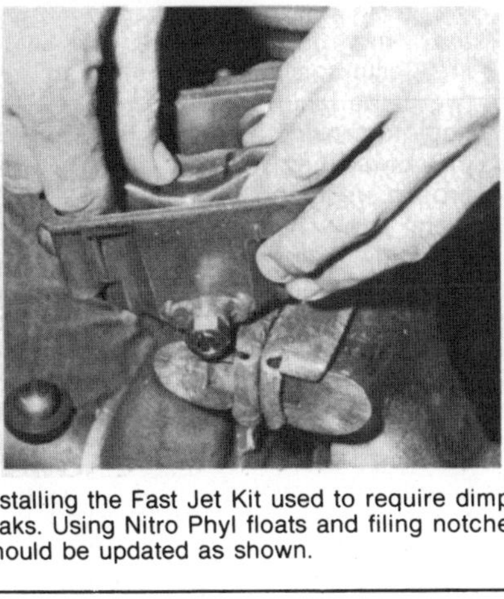

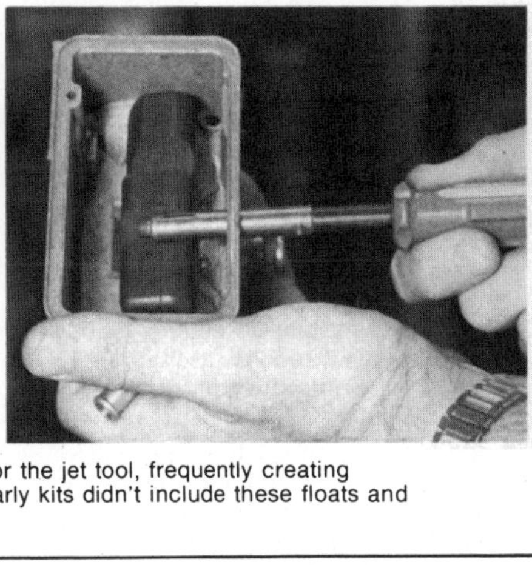

Installing the Fast Jet Kit used to require dimpling the brass float to provide clearance for the jet tool, frequently creating leaks. Using Nitro Phyl floats and filing notches in them positively solves the problem. Early kits didn't include these floats and should be updated as shown.

OPTIMUM PERFORMANCE FROM A HOLLEY CARBURETOR IN A STREET APPLICATION IS JUST A MATTER OF . . .

FINE TUNING

By C.J. Baker

A couple of months ago we presented an in-depth feature discussing the relative merits (or lack of them) of various off-road modifications currently being performed on Holley carburetors, and the favorable response to that article has been overwhelming. But what about the "on-street" use of Holley carbs? Why not offer a similar tech feature to help solve the problems frequently encountered with the street use of Holley carbs? Why not, indeed! And to make such a feature as authoritative as possible, we again sought firsthand information direct from the manufacturer by contacting one of Holley's best-known tech reps, Chuck Gulledge.

Based on his extensive experience in correcting fuel system woes for numerous participants at street rod, street machine and van events across the country, Chuck has developed an orderly procedure for correcting such common complaints as poor idle quality, bogging, lack of power, poor economy, leaks and stalling. We've summarized his procedures and corrective measures. By the time you've finished reading this article, you may be surprised to discover, as we were, that very few of the so-called "carburetion" problems are caused by the carburetor. More often, it is the remainder of the fuel system, lack of maintenance or proper filtering, or the installation that has precipitated the malfunction. Read on. You'll see what we mean.

"Dirt and other foreign material in the fuel probably cause more carburetion problems than any other single thing."

In the vast majority of cases, enthusiasts aren't dealing with new cars, so it cannot be assumed that the gas tank or fuel lines are clean and corrosion-free. Instead, the hot rodder's car is usually several years old or, in some cases, it has been fabricated from scratch. In either case, the probability of dirt, rust, fiberglass particles, metal filings or welding slag and scale in the gas tank or fuel lines is quite high. And of course there's no guarantee about the cleanliness of the fuel coming out of the nozzle at the filling station either. Therefore, the installation of a good fuel filter just ahead of the carburetor is essential. Most Holley carbs come with an internal fuel filter positioned just behind the fuel inlet. Many previous magazine articles have recommended the removal of such filters to prevent a possible fuel restriction in the event that these relatively small filters should become partially or completely plugged. However, this is not a good idea unless a high-quality in-line filter has been installed ahead of the carburetor fuel inlet, and even then the internal filters provide good insurance for a street-driven vehicle. If the in-line filter is doing its job, the internal filters should remain clean, and in the event that the in-line filter should become plugged in the middle of nowhere, it could then be by-passed temporarily, since the fuel reaching the critical small metering orifices within the carburetor is still filtered.

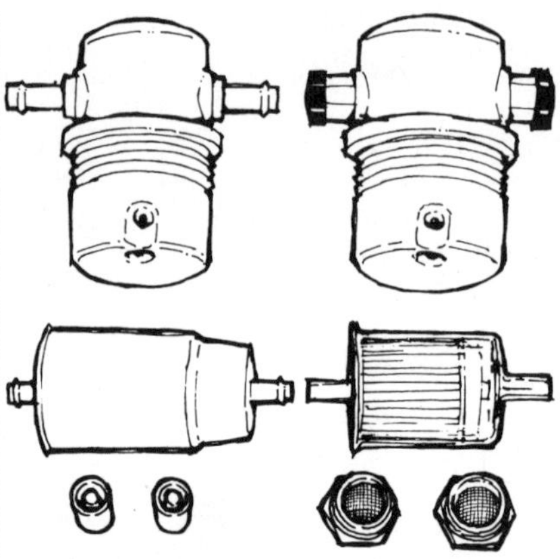

Shown from top to bottom are in-line filters that use replaceable filter cartridges, disposable in-line filters and the common sintered bronze carburetor filters. Also shown are some special fuel inlet nut strainers that Holley developed for NASCAR.

Some street rodders prefer not to run filters because of space problems or appearance considerations, but this can be a costly mistake. A piece of grid could easily cause an inlet needle to stick open, resulting in fuel overflow from the bowl vents and a probable fire on a running engine. The fact that filters do get dirty and require periodic replacement is testimonial to the fact that dirt and contaminants *do* find their way into gasoline. So if the car in question is exhibiting symptoms indicative of restricted fuel flow, such as a lack of high-speed power, lean surging, popping back through the carburetor or outright fuel starvation, one of the first things to check or replace is the fuel filter(s).

"Any filter, regardless of brand, style or size, is preferable to no filter at all."

Another potential problem involving fuel contamination deals with the

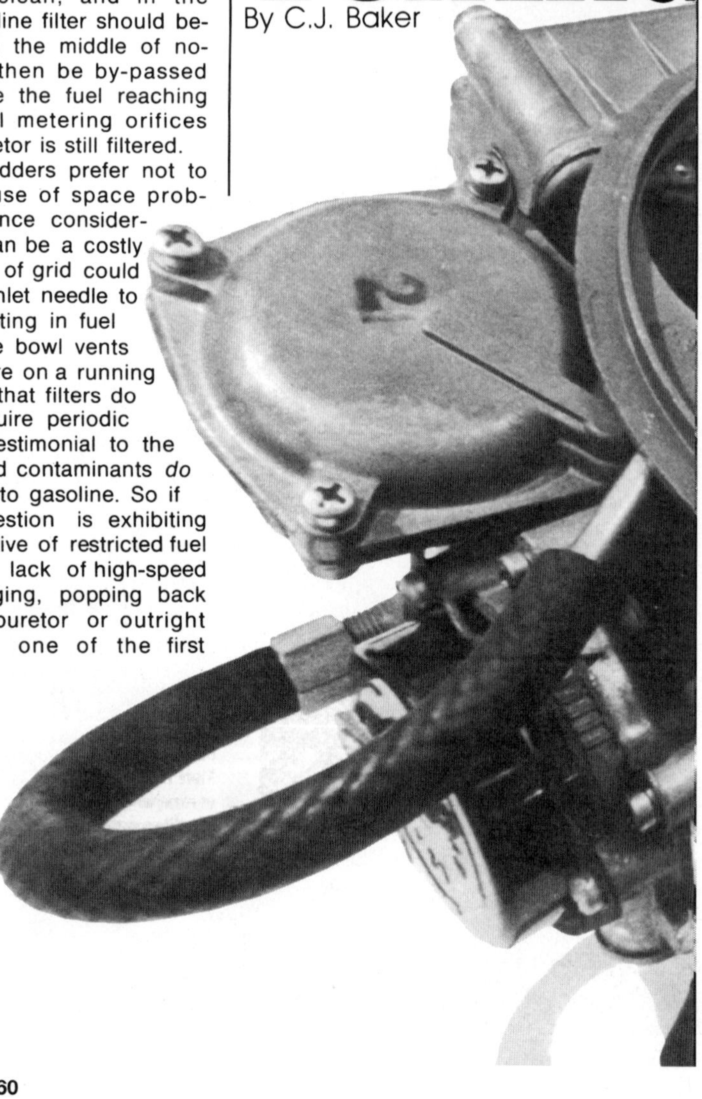

possibility of water or other substances in the gasoline. This is quite likely in humid climates on older cars where the gas tank is vented to atmosphere, especially if the vehicle sits outdoors for extended periods. Storage also presents the likelihood of stale or stagnant fuel. Gasoline does not have unlimited storage life. In time, the complex gasoline molecules begin to deteriorate and the light aromatic compounds tend to evaporate away. The result is that the fuel's octane rating, ignitability and even its flammability are reduced, which could cause uneven combustion and a rough running engine—if it'll run at all.

"If there's any doubt about the fitness of the fuel, it should be drained from the tank and flushed from the fuel lines and replaced with fresh gasoline."

But there are other properties of gasoline that can manifest themselves as apparent carburetion problems when in fact the carburetor is working properly. Most notable is the Reid Vapor Pressure (RVP). This is the fuel's tendency to evaporate, changing from a liquid to a vapor in relation to changes in temperature and pressure. Unfortunately, this is another property of gasoline that has become extremely inconsistent and unpredictable in recent years. Consequently, many enthusiasts are now faced with severe fuel percolation on high-performance engines that produce high underhood temperatures. The problem becomes critically apparent when a hot engine is restarted after sitting for five to 30 minutes, especially if the driver stands on the throttle while the fuel that became superheated in the lines immediately adjacent to the hot engine is still entering the carburetor.

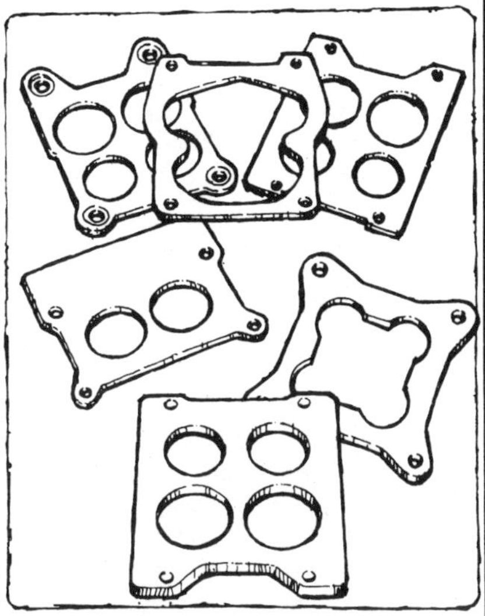

Heat insulators are available in different shapes and thicknesses for most two- and four-barrel carburetors.

"Heat insulators should be used whenever hood clearance permits."

To reduce the probability of fuel percolation and hot fuel *flash* into the vapor state, heat insulator gaskets that go between the carburetor and the intake manifold are available for most carburetors. The fuel line should also be routed to keep it as cool as possible. In really severe cases, a fresh-air duct can also be run from the grille to vent cool air

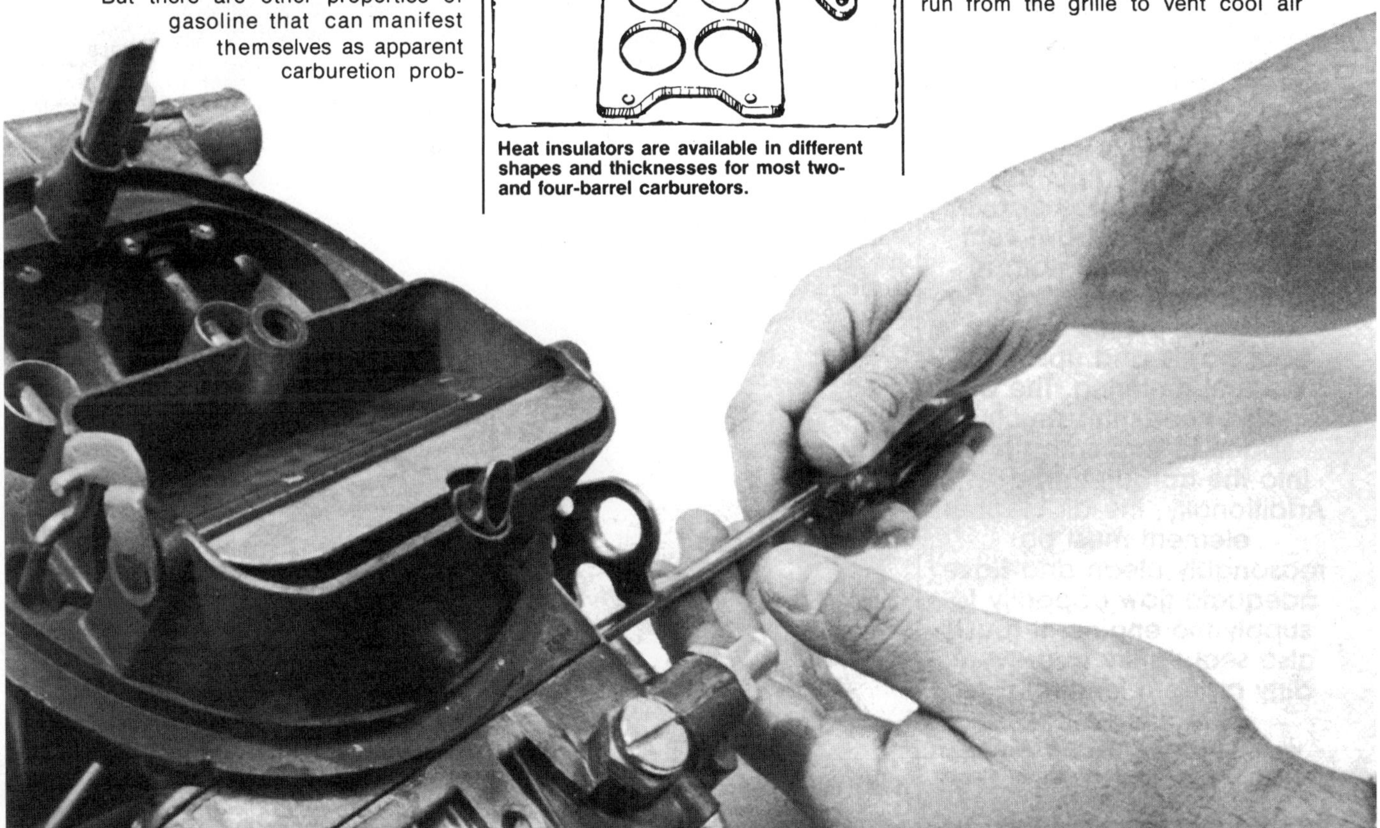

FINE TUNING
(CONTINUED)

directly on the carburetor body. A second cool-air duct to supply cool air to the air cleaner also helps reduce inlet air temperatures.

"Adequate fuel pressure and volume are essential to proper carburetor operation."

Naturally, a good fuel pump and unrestricted fuel lines of a size capable of supplying the engine's needs must be used. Verifying these conditions is another initial inspection step that should be checked before assuming a problem exists within the carburetor.

With the fuel delivery system checked for proper operation, correct installation of the carburetor must be checked. We've already mentioned heat insulator gaskets, but proper linkage attachment and adjustment, choke adjustment and adequate air cleaner clearance must be verified.

"It is imperative to have at least ¾-inch of internal air cleaner clearance above the tops of the bowl vent tubes to prevent normal air flow from disrupting the reference pressure in the float bowls and upsetting the fuel metering. The air also needs this much space to turn smoothly into the carburetor bores. Additionally, the air cleaner element must be reasonably clean and have adequate flow capacity to supply the engine. It must also seal tightly to prevent dirty air from leaking past the element."

While inspecting the installation of the carburetor, make a careful check for vacuum leaks. Such leaks are frequently the cause of poor engine performance that is blamed on the carburetor. Look for disconnected, loose or cracked vacuum lines or tube plugs. Check too for leaking intake manifold gaskets.

If everything has checked out up to this point and the problem still exists, *and* you're sure that some other system, such as the ignition or valvetrain, isn't malfunctioning, it is time to begin working on the carburetor.

"The very first thing on a carburetor that must always be checked is the float level."

The fuel level in the float bowls, as determined by the float level (and the fuel pressure and fuel specific gravity), affects the fuel metering of both the idle and main circuits within the carburetor. Fortunately, the float level is easily checked and adjusted on Holley carbs, thanks to the bowl sight plugs and external adjustment. The fuel should just barely trickle out of the sight plug hole with the engine running at idle. Of course the float level must be reset whenever a new inlet needle assembly is installed. And incidentally, don't be lured into installing oversized inlet needle assemblies. The standard Holley .110-inch Viton needle assembly is more than adequate for any street application. It is also the most trouble-free.

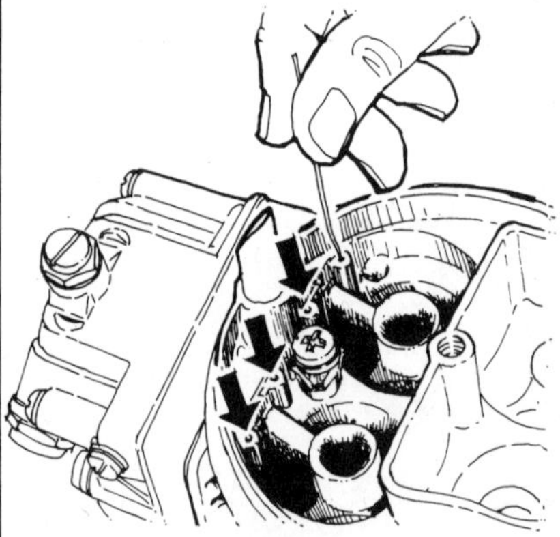

These high-speed and idle air bleeds (see arrows) can be cleared of dirt and debris with a fine wire and carburetor cleaner. The fine wire such as used on parts tags works great.

Still other inspections should be made with the carb on the engine. Verify that the idle and high-speed air bleeds, located adjacent to the bowl vent tube(s), are not plugged or restricted. These tiny orifices are especially susceptible to plugging with dirt or fuel residue. A quick squirt of carburetor cleaner and probing with a fine wire is usually sufficient to clear these passages. Disassembly of the carburetor is generally required to permit further inspection, cleaning and adjustment to resolve more serious problems.

"Whenever a carburetor is disassembled, it should be thoroughly inspected for dirt or restrictions in the metering orifices, discharge ports and fuel transfer slots."

This last item, the fuel transfer slots, is especially important, since proper fuel flow from these slots is required for a clean transition from the idle circuit to the main circuit. Many accelerator pump circuits have been erroneously reworked to cover up a bog that occurs because the transfer slots were plugged or the throttle plate stops incorrectly adjusted to get proper action from the transfer slots. As with the previous check of idle and high-speed bleeds, a fine wire and some carb cleaner should be adequate to purge these passages.

Another item to check while the carb is apart is the flatness of the main body metering block mating surfaces. This is rarely a problem on new carbs, but on units that have been in service for a while, verify that these surfaces haven't warped or distorted more than .010-inch. A light dressing with a fine flat file is all that's required to restore it to an acceptable condition, but don't overdo it. Now we're ready to begin tuning the carburetor to the application.

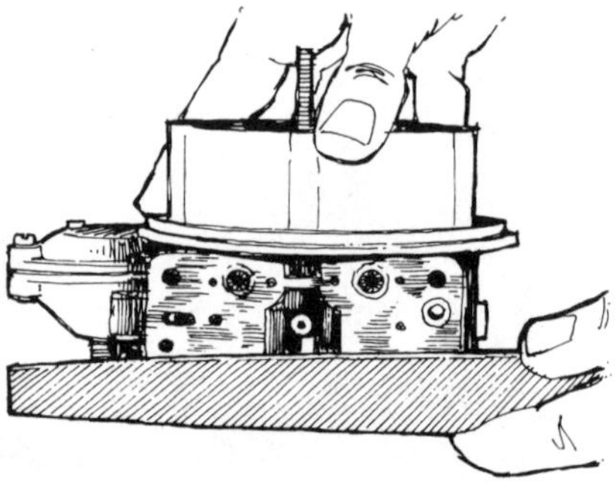

A fine-toothed flat file can be used to restore the metering block surface of the main body to acceptable flatness.

The idle circuit is especially critical on street-driven vehicles. You may have encountered a situation where the idle mixture adjustment screws didn't seem to have any appreciable effect on the idle quality. This is usually caused by the primary throttle plates being too far open at idle. You may have found it necessary to set the idle stop screw on the primaries at such an excessive opening just to get the correct idle speed. However, the cure for the situation is to adjust

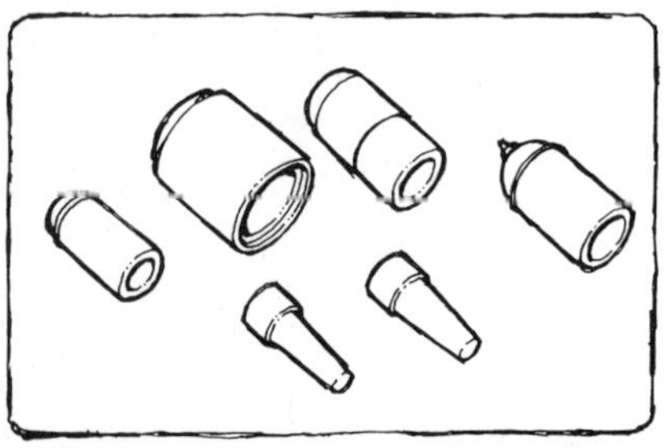

Holley offers assortments of rubber tube plugs to seal off vacuum tubes that may not be used in a particular application. Plastic hose plugs are also available.

the *secondary* throttle screw (accessible from the bottom of the carb—see photo) ¼- to ½-turn to provide more secondary idle air flow. This in turn will permit a reduction of the primary throttle opening at idle to the point where the idle mixture screws will once again become functional. It is very seldom necessary to resort to racing modifications on the idle circuits of street-driven carburetors, such as drilling idle air or fuel bleeds, installing restricting wires in the idle fuel metering orificies or drilling idle air holes in the throttle plates.

"A good rule of thumb when adjusting the secondary throttle stop is to position the throttle plates so that the bottoms of the fuel transfer slots are just barely exposed. This setting will also provide a smooth transition into the secondary main circuit."

If, after proper adjustment of the primary and secondary throttle stops, a bog or hesitation occurs when the throttle is opened quickly, it may be necessary to increase the size of the accelerator pump squirter(s). Holley squirters are numbered corresponding to the size of the squirter nozzle. For example, a squirter stamped 27 indicates .027-inch-diameter nozzles. Holley offers squirters in a variety of sizes, but you can also drill them with wire drills and a pin vise. Increase nozzle size *one* wire drill size at a time until the bog clears up.

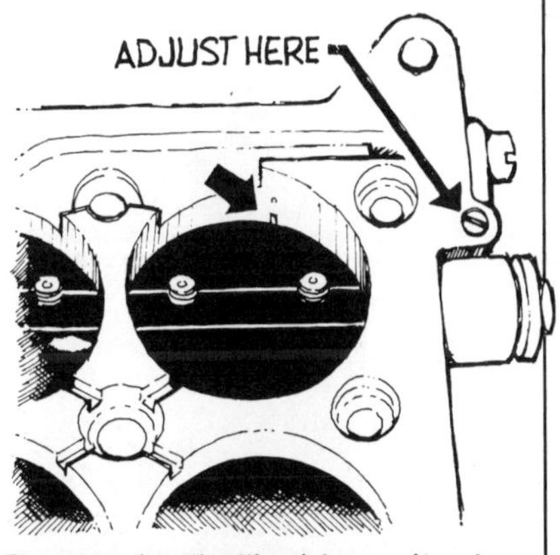

The secondary throttle plate opening at idle can be adjusted by turning this concealed stop screw. The idle fuel transfer slot (heavy arrow) should just barely be exposed when viewed from the bottom.

On vacuum-secondary Holleys, accelerator pump squirter size must also be coordinated with the secondary vacuum diaphragm spring to prevent bogging and hesitation. If the secondaries open too soon or too quickly, a bog will result. Using the next heavier spring can correct this. If the engine seems down on power or sluggish in response to full throttle, try the next lighter spring. These springs are color-coded: white, yellow, red, pink, purple, green, orange, plain (no color), brown and black, from the lightest to the heaviest, respectively.

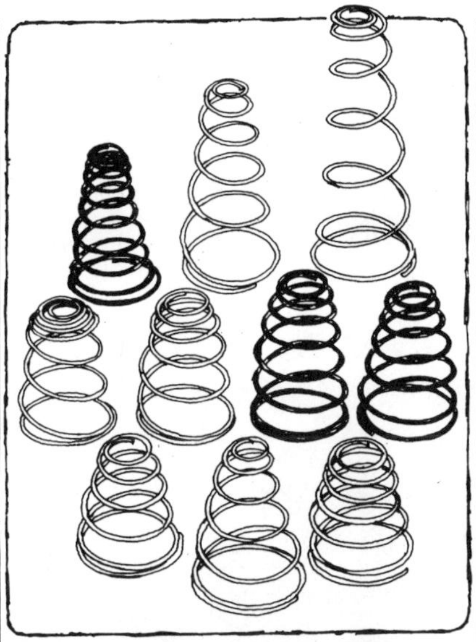
Vacuum secondary diaphragm springs come in an assortment of lengths and stiffness to tailor the secondary opening point and the opening rate. See text for details.

"The proper spring should be selected before varying nozzle size in an attempt to use as small an accelerator pump squirter as possible. Holley offers an assortment of the most commonly used springs under part No. 85BP-3185."

The only remaining circuits in the carburetor that may need further tuning are the main circuits and the power circuits. Main circuit metering is controlled by jets or metering plates. In the majority of cases, the "out-of-the-box" jetting is very close to perfect, but altitude considerations and engine configuration may require slight adjustment. Holley even offers a high-altitude jetting kit (part No. 85R-4707) for the 6619, 6909, 6919, 6979, 6989 and 7009 RV carbs. In any case, jetting changes should always be done with numbered Holley jets or metering plates. Jets should *never* be drilled. If you find it necessary to rejet the carb more than five numbers either way from the "out-of-the-box" jetting, there is probably something wrong, such as a vacuum leak, restricted passage or inadequate fuel delivery system.

The power circuit is also important to the main circuit. Tailoring the power valve opening point (by manifold vacuum level) will provide smooth off-idle acceleration and proper fuel metering under high-load situations. Power valves are numbered according to the vacuum level (in inches of Mercury, Hg.) below which they are open. For example, a power valve stamped 85 opens at 8.5 inches Hg. and stays open at all vacuum levels lower than 8.5 inches Hg.

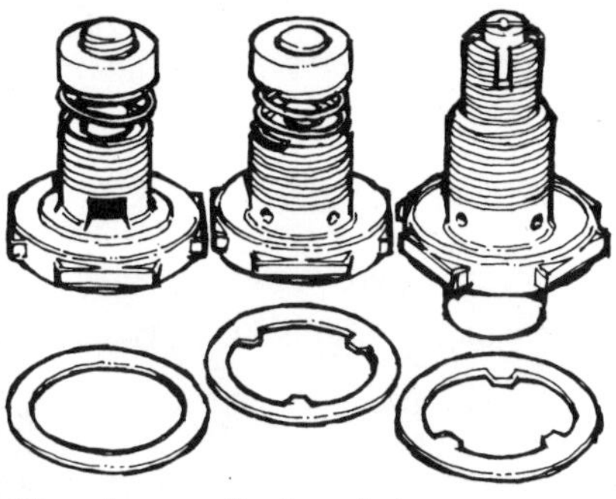
Shown here are the bay window power valve (left), the four-hole single-stage power valve (a six-hole version is also offered) and the two-stage power valve (right). Note the gasket shape that must be used with each design. The bay window version is preferable in the majority of applications.

To correctly match the power valve(s) to the application, select one(s) that open *at least* 1 to 1½ inches Hg. below the manifold vacuum level that the engine maintains at idle *in gear.* This will prevent unnecessarily rich fuel metering under light loads. For example, if the engine idles at 8 inches Hg., a power valve rating of 65 should be selected. If the power valve(s) that came in the carburetor from the factory is already 1½ inches Hg. or more below the idle vacuum level, there is no need to install power valves with higher ratings to get within 1½ inches Hg.—just stay with the stock rating. Holley also offers two-stage power valves that open partially at one vacuum level and then open fully at a lower vacuum level. Such two-stage power valves can improve gas mileage on some lightweight vehicles with very mild cam timing, but they should be avoided on modified street machines or in any carburetor that has power valve restrictions larger than .076-inch.

These tips for fine tuning and maintaining Holley carbs on street-driven vehicles should be all that's needed to straighten out those problems normally encountered by enthusiasts, because as we pointed out in the beginning, most "carburetor" problems have little or nothing to do with the actual carb. But just in case you do run into something more complex, Holley also sells an excellent technical manual (part No. 36-73) to bail you out. **HR**

A LOT OF SO-CALLED EXPERTS ARE CHARGING A LOT OF MONEY TO REWORK HOLLEY CARBS FOR COMPETITION USE, BUT DO YOU REALLY GET WHAT YOU PAY FOR?

THE TRUTH ABOUT... HOLLEY CARB MODS

By C.J. Baker

It's no secret that even the competition-designed versions of Holley carburetors can be reworked, cleaned up and modified from their "out-of-the-box" condition to achieve increases in performance. The racers know this, and so does Holley. The engineers at Holley continually evaluate those things that are being done to their carburetors to determine if those mods are truly effective. If they are advantageous, and if such improvements are feasible from a production standpoint, they are often incorporated into revisions of existing models. And even if those modifications are prohibitive on a production basis, Holley still passes the information along in Tech Clinics, periodic service bulletins and their Holley Carburetors & Manifolds manual (part No. 36-73) which describes system operation, disassembly, reassembly, specifications and modification of various models.

But the story doesn't stop there. Lately Holley has been seeing an increasing number of their carburetors that have been altered by various

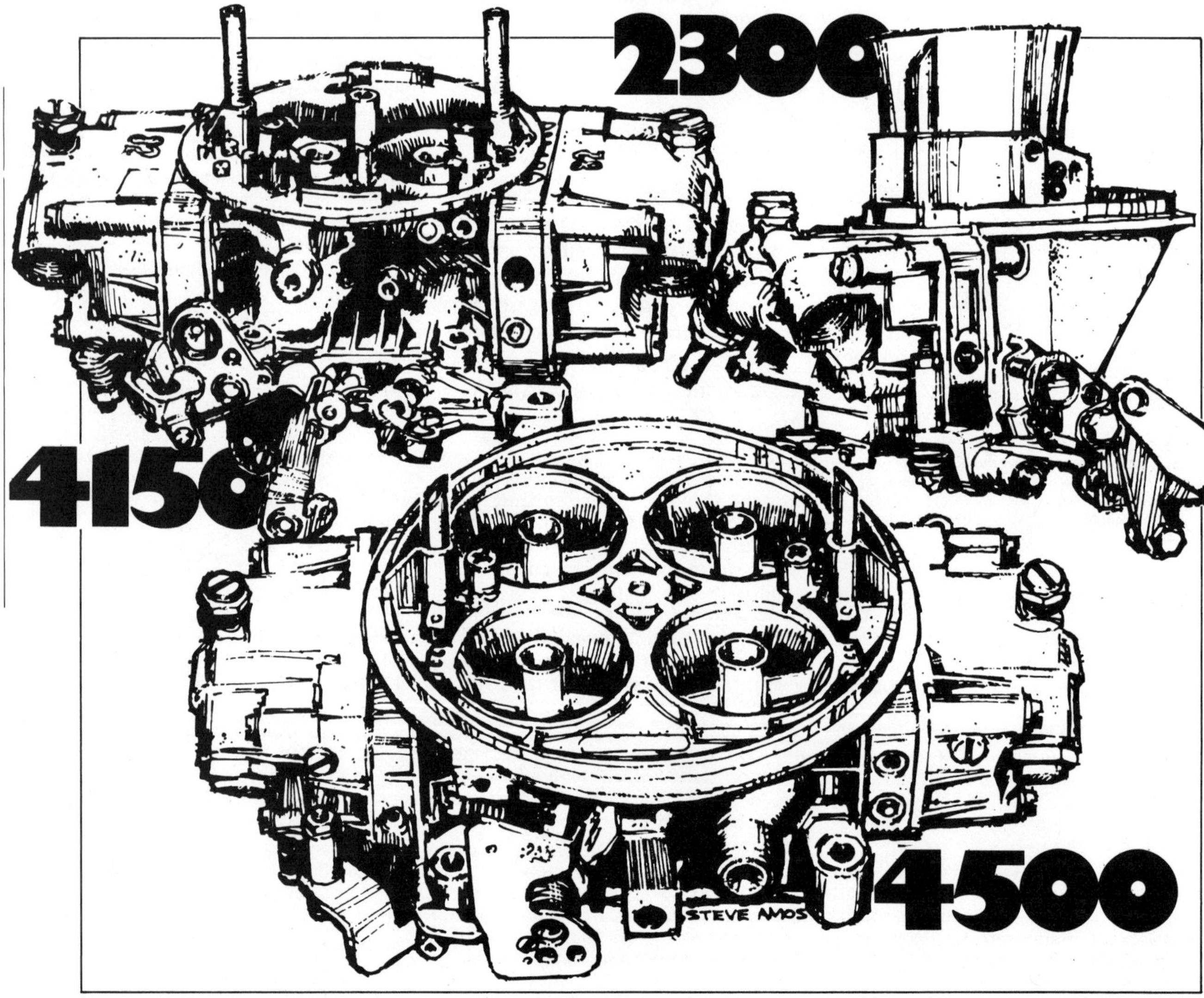

CARB MODS

people who have made a very lucrative business out of such modification. Unfortunately, sometimes the modifications being performed, while costly, are of questionable value in terms of performance gained and quality of workmanship. Some of these mods, however, do work quite well. When there is a problem, it stems from the fact that those modifications are un-engineered and each carburetor is rarely tested to assure consistent results. Many of the aftermarket modifiers are not presently equipped to wet-flow carburetors to simulate the actual flow of both air *and* fuel. Instead, they measure only air flow and pressure drops, if they check anything at all. But fuel flow is just as critical. One firm, Braswell Carburetion, is currently constructing a wet-flow bench, but as this is written, it is still not in operation.

Because obtaining the ultimate in performance from a carburetor is so important to many racers, Holley has taken a firm stand on current modifications—what's good, what's bad and what's only eyewash, regardless of cost. They have begun pointing out the real value of what's being done, and more important, they are offering recommendations on how to do those modifications yourself, or to obtain equivalent results by different or cheaper methods. Holley has made this information available to HOT ROD, and we have condensed and summarized it for you here. Remember, modifying carburetors requires careful planning and a certain level of technical proficiency. Don't attempt to perform a modification until you understand its precise purpose, and whether it is correct for your application. Above all, proceed slowly and make changes in small increments, checking your progress as you go to see if what you're doing really helps.

THE 4150 SERIES

The Model 4150 Holley carb was introduced in 1957. This series includes all of their standard-flange, double-pumper performance carburetors in use today. These carburetors are currently being modified by numerous people, with the tab running between $185 and $525, plus the carburetor! Any way you look at it, that's a lot of money. Various tricks are employed by each modifier. In most cases the tricks are of *some* benefit, but in many cases Holley feels that equivalent gains can be achieved by cheaper and easier methods, or you can make the same modifications yourself. Here's what's being done.

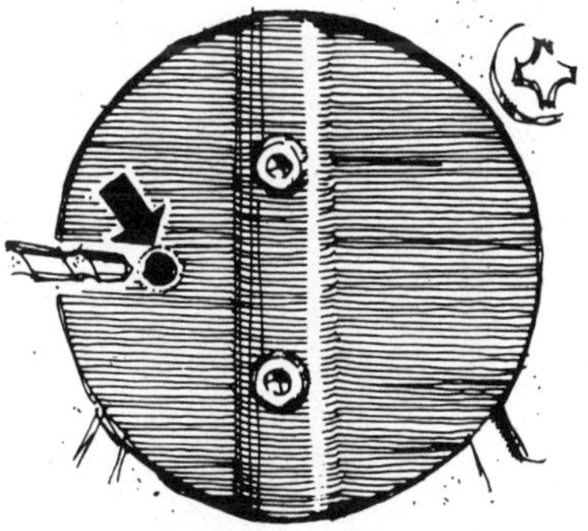

Idle quality can be improved on radically cammed engines by drilling a 3/32-inch hole in each throttle plate, as shown. The hole should be on the same side of the throttle shaft as the idle discharge port, meaning the upstream side.

The idle circuit usually gets considerable attention, in an attempt to provide finer tuning capability, prevent loading and sharpen throttle response. One way of doing this is to install a primary metering block on the secondary side, along with machining the required idle passages and extending the secondary transfer slots. This modification does permit greater flexibility in fine-tuning the carb, but in many cases, it's not really necessary. Another approach is to

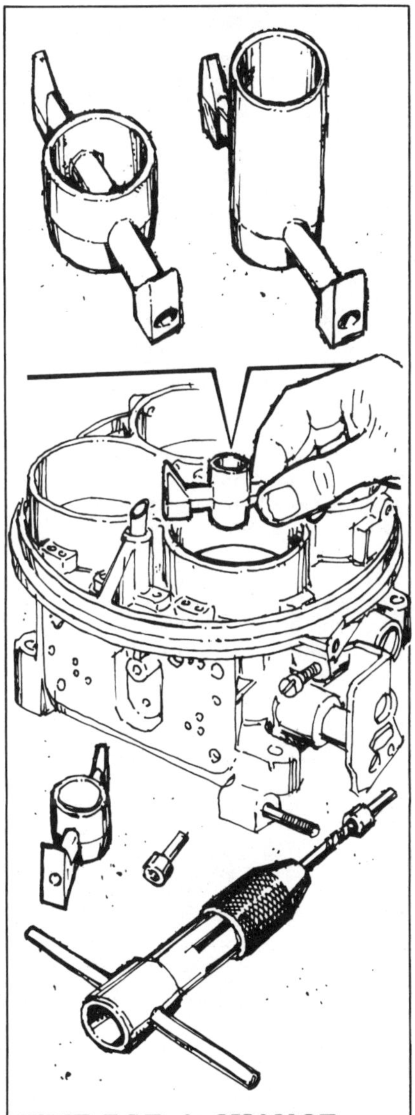

TIME FOR A CHANGE

Changing the booster venturis, as recommended on certain models and applications of 4500-series carburetors, isn't particularly difficult, but it does require a stud extractor and replacement fuel passage brass bushings (part No. 14R-826). The booster venturi is held in place by the brass bushing. Using the extractor, simultaneously twist and pull the bushing from the main body. The booster can then be removed. A few gentle taps with a wooden dowel from below may be required to free the booster from its alignment slots. To reinstall a booster, simply tap into the alignment slots until the fuel passage hole in the booster lines up with the hole in the main body, and then tap the new bushing into place with a plastic hammer. **HR**

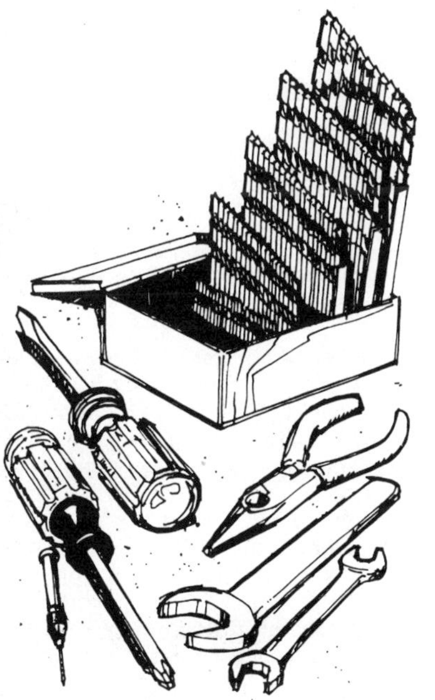

TOOL SCHOOL

Working on and modifying carburetors requires specific tools such as those shown here for all but the most complex alterations. At a minimum, you should have a complete set of wire drills, a pin vise, needle-nose pliers, screwdrivers, 1-inch, ⅜- and 7/16-inch open-end wrenches. Also useful for cleanup work is a set of cheap jeweler's files for removing casting flashing. **HR** drill holes approximately 3/32-inch in diameter in each of the throttle plates to permit extra air flow under low-idle vacuum conditions with high-overlap camshafts. This modification works well, and you can do it easily with a drill motor. It is also being incorporated on some production Holley carbs. In still other cases, the idle air bleeds are removed and the passages tapped for replaceable bleeds

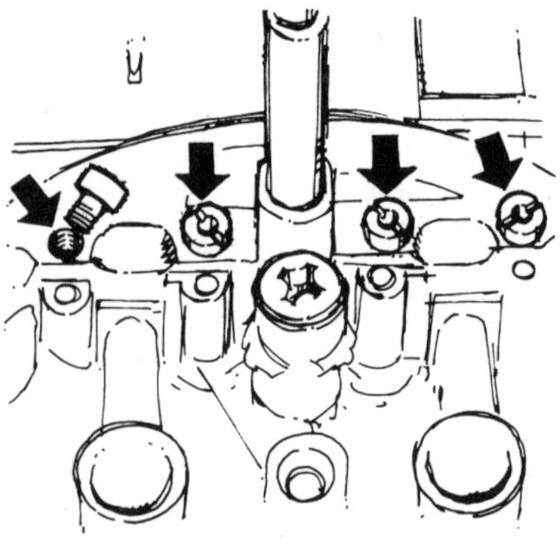

Replaceable idle- and high-speed air bleeds can be used to vary calibration of these circuits, but Holley recommends varying the fuel restrictions for these circuits rather than the air bleeds.

in various sizes. Others simply drill or resize the existing air bleeds, but Holley feels this is a poor way to alter calibration, since it also affects the metering signal. Idle mixture is better controlled by sizing the idle

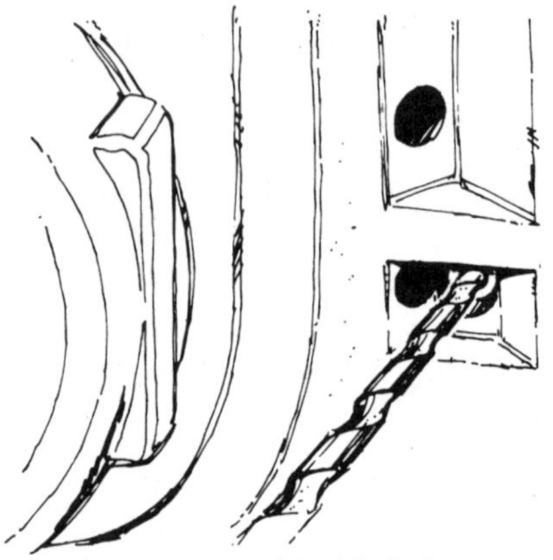

Changing the size of the idle fuel restriction bushing can bring the idle adjustment screws into an operational range. Enlarging the bushing richens the idle mixture. To lean the idle mixture, small pieces of wire may have to be inserted in the orifice, as shown in another illustration.

fuel restriction, if required, to bring the idle mixture adjustment screws into an operational range.

Idle quality is also affected on many competition engines by the power valves (if the carburetor is so equipped), due to the oscillations in the manifold when a high-overlap cam is used. Some modifiers block the vacuum passage to the power valve(s), effectively opening the power valve(s) at all times. This is a definite mistake, frequently causing loading during staging and immediately after burnouts. A power valve supplements fuel flow from the main jets during very low vacuum (high engine demand) conditions and prevents overrich mixtures during high vacuum (low demand) conditions. Consequently, power valves should be used whenever possible, although they must be correctly matched to the normal idle vacuum of the engine. The power valve should open 1- to 1½-inch Hg. below the observed vacuum level of the engine at idle or while staged, whichever is lower.

Most modified 4150-series carburetors have the choke shroud (air horn) milled off. This is generally a desirable modification, if the machined edges are radiused. This aids air en-

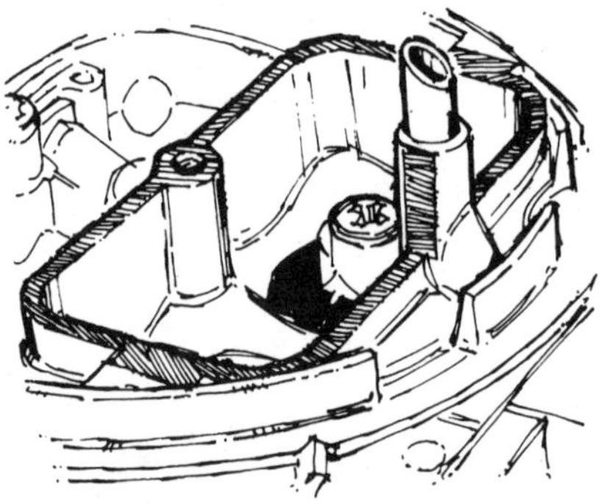

Milling off the air horn generally increases the total air flow of the carb and aids air entry to the venturis. Do not mill off the boss around the vent tube. Smooth and radius the edges after machining.

try to the carburetor and increases air flow slightly, especially on larger venturi carbs. Still others polish the venturi. Any polishing beyond the simple removal of casting flashing is generally a waste of time. Other approaches to improving air flow include slabbing the throttle shafts and screws and/or the installation of larger bore throttle plates with the venturi blended in accordingly. Tests indicate that such procedures, when properly executed, do increase air flow appreciably. Some modified 850-cfm carbs showed flow rates approaching 1000 cfm, but it must be remembered that increased flow to produce top-end power usually reduces the metering signal at lower air flows, often creating lean spots or "holes" in the fuel flow curve, which can result in an overall performance loss. Size alone does not guarantee better performance.

Because leaning and flat spots are a problem, one modifier installs the metering blocks from either the 6214 or 6464 Dominator carburetors incorporating not only the idle circuits on

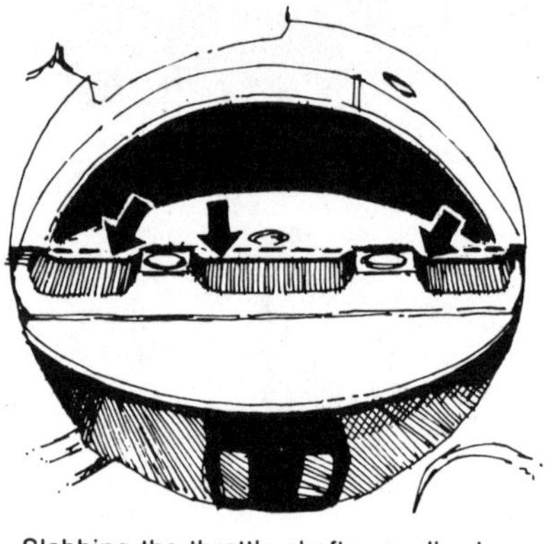

Slabbing the throttle shafts usually shows a slight improvement in air flow, but unless you really need that extra 5 cfm or so, it usually isn't worth the trouble. There's more to gain by installing larger throttle plates in most cases, often as much as 40 cfm.

both primary and secondary sides, as previously discussed, but also intermediate circuits to smooth the transition into the main metering circuit. Corresponding passages and air bleeds are also added. Such changes require wet-flow testing of each and every unit to assure consistent and accurate results, which does not currently seem to be the

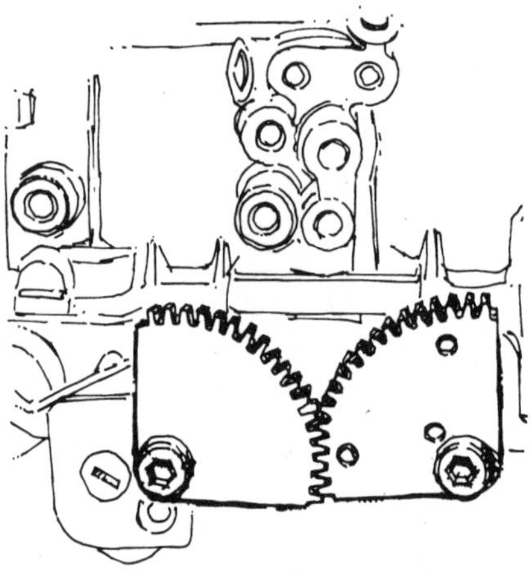

Mechanical secondary kits that employ gear drives are definitely not recommended, as they may cause premature wear, bending or binding of the throttle shafts.

situation. A better solution here is a revision of the accelerator pump circuit to achieve the same results. For example, on the 850-cfm double-pumper, Holley recommends the installation of a 50cc pump on the primary side. The secondary side already has a 50cc pump. The accelerator pump shooters are then opened to .042-inch on the primary and .047-

CARB MODS

inch on the secondary sides. This approach is far easier and cheaper than adding an intermediate system, and the results are comparable in most cases.

In concluding the modification of 4150-series carbs, several other things are being done which are either of no benefit or detrimental to carburetor performance. These mods include: chamfering of all main body

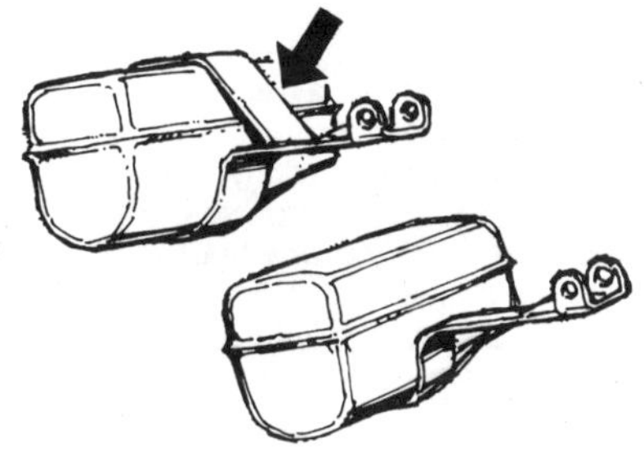

Strengthening struts on brass floats are totally unnecessary. There's no way the fuel forces within a carb can bend one, so all the strut does is add undesirable weight to the float, making it less responsive to fuel movement.

passages, installation of 1:1 gear-type linkage, adding strengthening struts to brass floats, cutting down Nitrophyl composition floats, and even opening of the power valve vacuum chamber to add extra float bowl

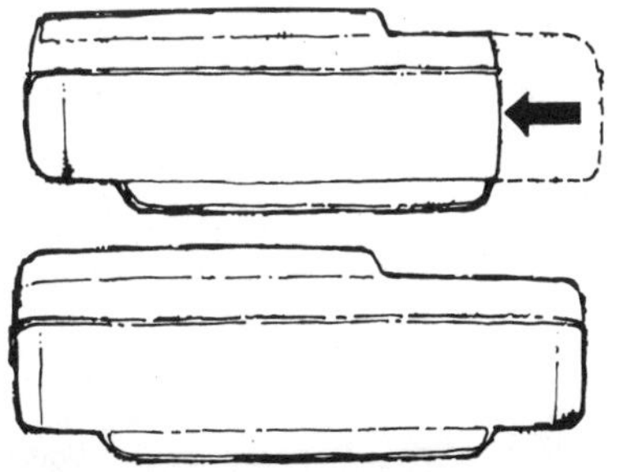

Another detrimental modification is cutting down the Nitrophyl composition floats. This trick can work in special oval-track and some drag applications where specific fuel slosh conditions may exist, but usually it does nothing more than reduce the float.

volume on those carburetors where power valves aren't used. Holley has not seen any benefit from these modifications.

THE 2300 SERIES

The 2300 series includes the 500- and 650-cfm performance two-barrel carburetors used in IHRA and AHRA class racing. Two of the companies modifying these carbs are C&S Specialty and Braswell Carburetion, at a cost of about $200 to $265, plus the cost of the carburetor. Considering the extent of the modification, including installation of larger annular discharge venturis and larger throttle plates along with a complete reworking of the fuel circuitry, and the improvement in flow and performance that is achieved, these carburetors are among the best values in modified carburetion currently available—if you run in such a class. Both units seem to work very well as delivered, although Holley field technicians report a reduced metering signal at lower air flows on some applications with the C&S unit, which can present difficult calibration problems. In such cases, the problem can generally be remedied by using .045-inch accelerator pump shooters, a 4.5-inch-Hg. power valve, and drilling the main jets to .144-inch, as No. 100 jets (.128-inch diameter) seem to be too lean.

Editor's Note: *Personal experience with this carburetor on large-displacement engines with radical cam timing have produced the above conditions. An alternate fix that provides greater tuning flexibility is to enlarge the power valve restrictions to .113-inch. Then conventional Holley numbered jets in the No. 95 to No. 100 range can be used to achieve the optimum air/fuel ratio and tune for varying altitude and atmospheric conditions. Such changes have already been incorporated on the latest versions of the C&S carbs for big-inch applications.*

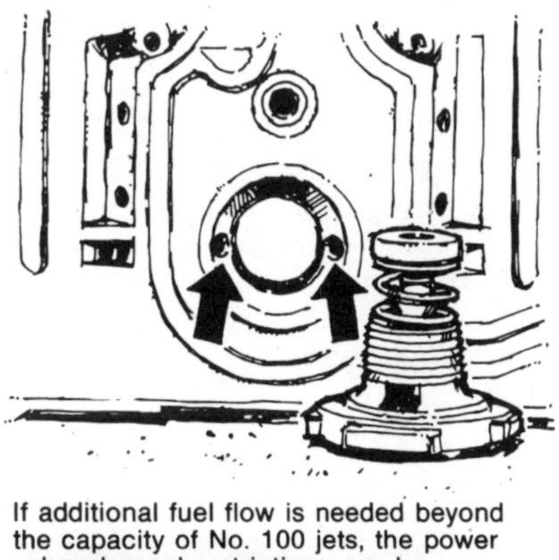

If additional fuel flow is needed beyond the capacity of No. 100 jets, the power valve channel restrictions can be enlarged. Make enlargements in small increments until the desired jetting is achieved. Also make sure the fuel passages can handle the total flow of the main jet and the power valve (see Area sidebar).

THE 4500 SERIES

The 4500 series includes all of the wide-spaced-flange, large-bore-venturi Dominator carburetors. Modification prices for these units range from $185 to over $700, plus the cost of the carburetor. The most expensive mods include enlargement of the throttle bores from 2- to 2⅛-inch for a flow capacity in excess of 1300 cfm. Such modification is extensive, requiring remachining of the venturis, movement of fuel passages, changes of the booster venturis and a general cleanup of the unit. As with similar modification to the 4150-series and 2300-series carbs, increased total air flow decreases low flow metering signals, which are already marginal on the 4500-series carbs. And since it is doubtful that the manifolds and cylinder heads on most Econorails, for which this modification is intended, can really use the extra top-end flow, the decrease in low-end performance may actually hurt overall performance.

Lesser modifications of Dominator carburetors generally includes little more than removal of the air horns, replacement of the booster venturis, enlargement of certain fuel passages and a general cleanup of the unit, since most 4500-series carbs already have idle circuitry on both primary and secondary sides. Additionally, the Dominator is offered in a variety of configurations, with and without progressive linkage and with and without intermediate circuits. In other words, optimum performance from a 4500 carb amounts to little more than selecting the right model and performing simple do-it-yourself mods on it. Consequently, Holley has made the following recommendations for specific applications.

For Econorails or single Dominator applications, use either the R-7320 (1150-cfm) or R-8082 (1050-cfm) version with 2.5-inch-Hg. power valves on both primary and secondary sides

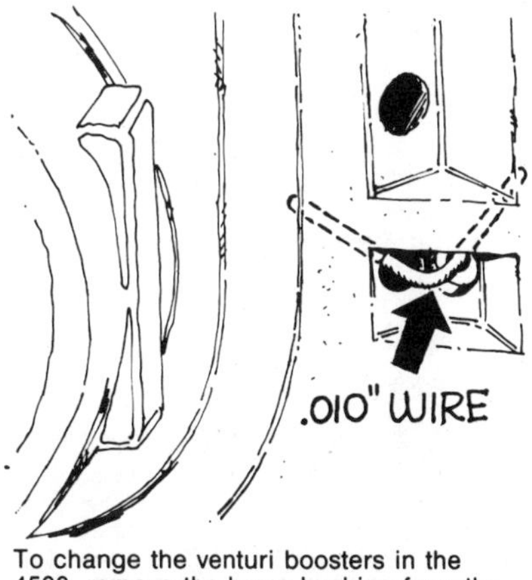

To change the venturi boosters in the 4500, remove the brass bushing from the main body with a stud extractor. Where recommended, the shorter 21R-718 boosters should be installed. These shorter boosters are more efficient than the long booster used on some 4500 models.

without modification to the power valve channel restrictions. Drill the accelerator pump shooters to .042-

(continued on page 96)

HOLLEY CARB MODS

(continued from page 50)

inch and use the 85BP-5229 hollow shooter screws. Begin with No. 88 jets all the way around and fine-tune jetting from there. The idle system should also be leaned by inserting a .010-inch wire, ½-inch long, in each idle fuel restriction. Adjust the idle so that all four adjustment screws are backed out the same amount, usually starting with them out 1½ turns. Both the primary and secondary throttles should also be opened the same amount at the desired idle speed. And don't bother milling off the air horns in this application.

On all applications where two 4500s are used, the air horns should be milled off and the entry to each venturi smoothed and radiused.

Chrysler Hemi cars using the Weiand "Hole Shot" manifold should use R-6214 carbs, 50cc accelerator pumps with .031- and .035-inch shooters on the primary and secondary sides, respectively. Install the shorter 21R-718 booster venturis, use stock jetting, and adjust the idle screws evenly.

On Ford 351C engines or Weiand-equipped Chevrolet 396 and 402 big-block engines with dual 4500s, use the R-6464 carbs. Only the air horns need be milled off here, since stock jetting, accelerator pumps, shooters and the idle circuitry are all acceptable as is.

Big-block Chevrolet engines of 427 cubic inches or more with the Weiand manifold should use the R-6214 carbs with machined air horns, the 21R-718 booster venturis, 50cc pumps, .035-inch shooters and jetting somewhere between No. 88 and No. 95. Begin with No. 90 or No. 92 jets and tune from there. Additionally, the intermediate discharge passages need to be restricted to .046-inch with brass plugs, and the idle air bleeds should be enlarged two wire drill sizes.

Beyond the specific modifications listed here, Holley carbs need little more than a careful cleanup of casting flashing in the venturis and booster venturis, along with a close general inspection to see that all passages were correctly drilled at the factory. Now that you have the facts, those of you who are budget-minded might consider doing it all yourself. **HR**

SIZING UP THE FLOW

Although it isn't completely accurate in terms of wall friction and boundary flow, a close approximation of fuel flow through a passage can be achieved using cross-sectional areas. The area of a circle is calculated from the formula $A = \pi r^2$, where A is the area, π is 3.1416 and r is the radius (half the diameter) of the circle. For example, a No. 100 jet has a diameter of .128-inch. Therefore, its radius is half of that, or .064-inch. Putting that into our formula (thank Heaven for pocket calculators), $A = 3.1416 \times .064^2$. This comes out to .012868 square inches.

Now let's suppose we want to calculate how big the fuel passages must be in the metering block and main body of a carburetor to adequately handle all of the fuel passed by a No. 100 jet and a power valve restriction of .113-inch. We already know from our previous example that the area of the jet is .012868 square inches. Performing similar calculations on a .113-inch orifice yields an area of .010029 square inches. Adding these two areas together gives a total area of .022897 square inches. Therefore, every part of the fuel passage must have a cross-sectional area at least that big to pass the fuel. Working backward in our formula, we can find the radius of a circle with that much area from $r = \sqrt{\frac{A}{\pi}}$. Again plugging this into our pocket calculator, $r = \sqrt{\frac{.022897}{3.1416}}$. It all comes out to $r = .08537$-inch. That means the diameter of the fuel passage must be .171-inch.

This same procedure can be used for determining just how much power valve restrictions should be enlarged or for determining accelerator pump passage restrictions, etc. It all seems a bit complicated at first glance, but it's really a snap with any pocket calculator that has a square root function on it. **HR**

BY C. J. BAKER

under 60sec.

AT LAST, A SIMPLE WAY TO EXTERNALLY CHANGE HOLLEY JETS

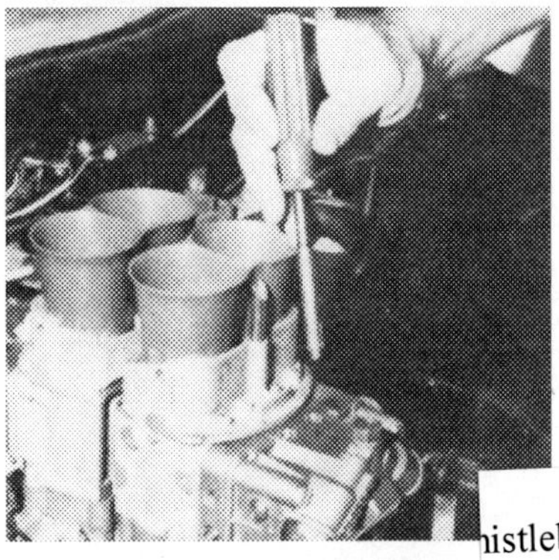

Jet changing becomes a snap, even o
prototype developed by C&S and C.J. I
for Formula I IHRA/AHRA drag ra
which features twin C&S two-barrel c

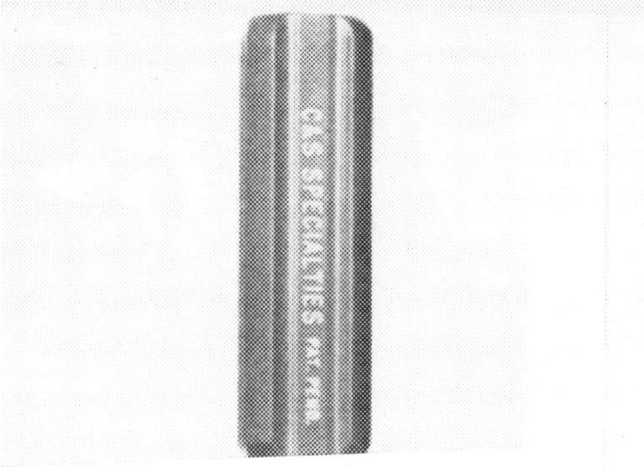

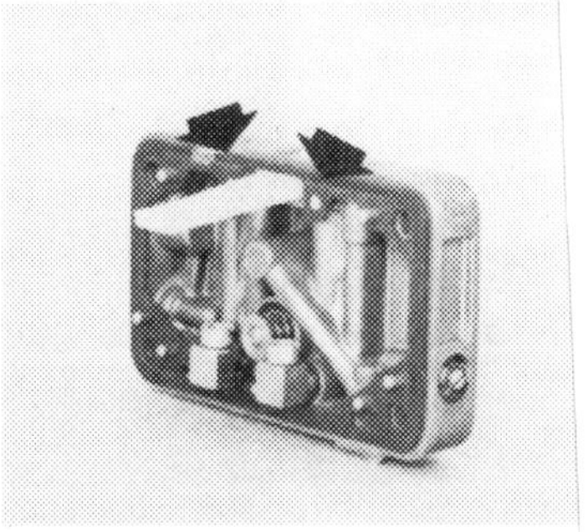

Jet fittings screw into the metering bloc
place of existing jets. Two of the ga
pins must be filed or sawed down to ga
height for tool access clearance.

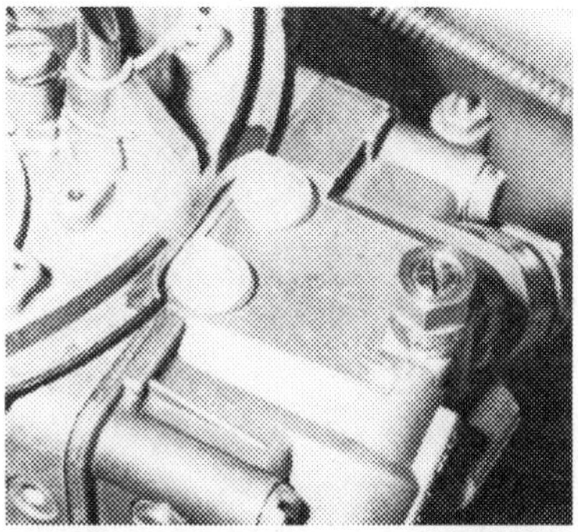

The larger air cleaner flange on Holley 4500 Dominator carbs must be notched with a file, as shown, for adequate jet tool and access plug clearance.

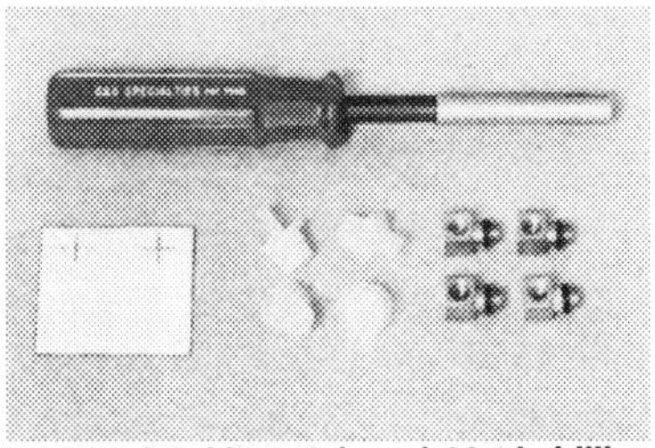

The complete kit contains a jet tool, drilling template, access plugs, brass jet fittings and detailed instructions. Racer net is $22.50. Installation is simple.

Occasionally a product com
along that works so well y
wonder why somebody dic
think of it long before. Such a pro
uct is the C&S Fast Jet Kit
center-pivot float bowl-type Holl
carburetors.

There have been other kits d
signed to permit external jetti
changes, but they required extensi
carburetor reworking. The beauty
the C&S kit (85 Susan Drive, S
Charles, Missouri 63301) is that it r
tains stock Holley jets. Additional
the kit is easy to install, requirin
nothing more than drilling two hol
in the top of the float bowl(s), as l
cated by a supplied template, screv
ing two special O-ringed brass fi
tings into the metering block(s), in
place of the existing jets, and dimpling the float(s) slightly. And in case you're wondering whether the reloca-

...

than 60 seconds on both ends of a four-barrel, using the special jet removal/installation tool supplied with the kit. In fact, the jets *could* even be changed while the engine is running! Gone is the need for frequent bowl gasket replacement, and gone too is fire hazard of draining the float bowls on a hot engine.

This kit is almost certain to become a "must" for any serious racer who is interested in achieving optimum performance for the existing temperature, humidity and air density conditions. Even the street/strip enthusiast will find it useful for quickly changing from his economy/emission street jetting to an all-out performance jetting for weekend competition. ■■

histleblower: $80B Budget to oppress Targeted

Sneaking into Google Street View ... with your sneak...

EMI Seeks to Bar #EFF From Cloud-Music Case | Epic...

Appeals Court: Pirate Bay Admins Still Guilty, Now...

Golden Shield Music Turns Web Censorship Into Art ...

The Real Story Behind NASA's Resurrected Space Pla...

On The Financial Horizon, Germany Is Rising | via ...

A New World Disorder: Obama puts up American landm...

U.S. Government

KITCHEN TABLE CARB REBUILD

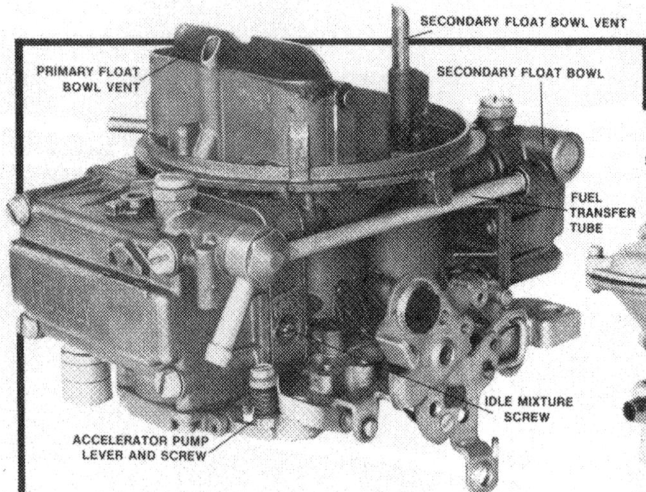

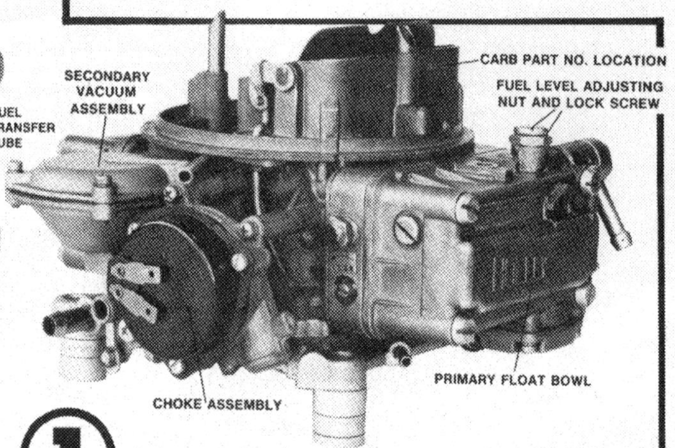

1

The exploded view of the Holley provided with the repair kit, combined with these external shots, will help should you misplace a part or two. However, as the following photos will illustrate, any mistakes can be avoided by confining your repairs to only one section at a time. With only minor differences, these external photos apply to the majority of 4150 and 4160 Holley carbs.

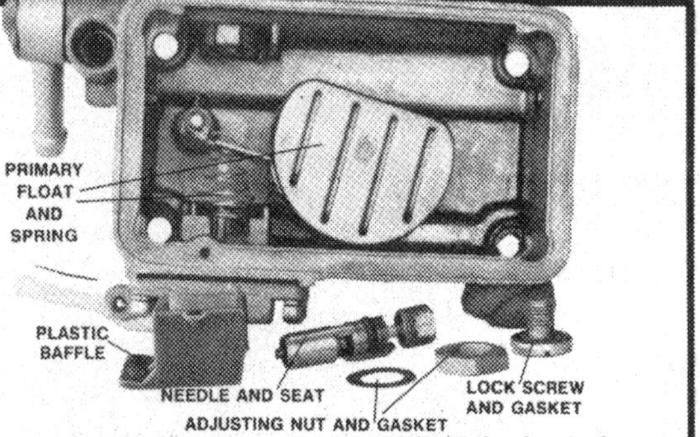

2

Remove the primary float bowl from the front of the carb and, from it, remove the components shown in the photo. Inspect float for pinholes and replace if any are found. Soak bowl in cleaner, rinse in thinner and return float and spring to bowl.

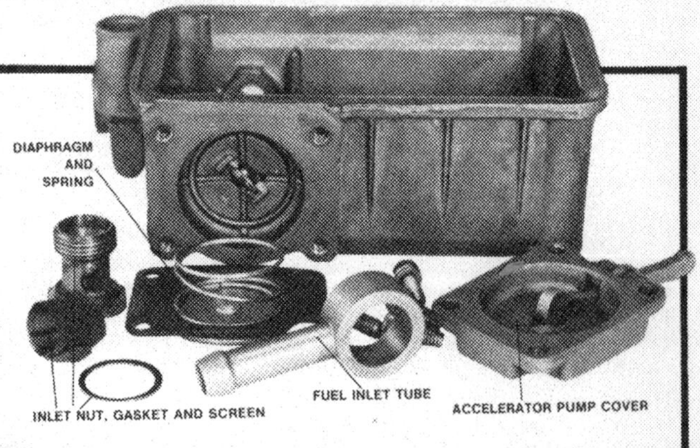

3

Remove the parts shown in photo, clean, and return, using new diaphragm (convex side out). Install new gasket on sight plug, new O-ring on needle/seat, and install in bowl with nut, screw, baffle. Repeat for secondary.

ANY FOUR-BARREL REQUIRES A PERIODIC REBUILD, AND A QUALIFIED SHOP IS AS NEAR AS YOUR KITCHEN— JUST SIT DOWN AND DIG IN

By Jon Jay

The kitchen table, usually nothing fancier than chrome tubing and formica top, has become a significant part of the American home. It's always been much more than just a place to eat corn flakes. Traditionally, it's a place for poker games, building models, doing homework and those dreaded tax forms. For the person without a driveway or a garage, it's also a great place to rebuild a sick carb, so move over, cereal; make room for the Holley.

Before you begin, there are a few precautions that must be taken. Remember to drain all fuel from the float bowls before taking the carburetor into the house. Gasoline on the rugs will not promote family peace. The job is going to require a carburetor cleaner containing chlorinated hydrocarbons, and prolonged exposure to its fumes is not recommended. Food, plants, pets and the human body are all affected, and for that reason, all carb soaking should be done outside. The carburetor should be disassembled one section at a time, soaked in the cleaning solution, and then rinsed in paint thinner or water before any parts are replaced. The cleaning agents are highly flammable, and to use them near a stove or other source of heat would be asking for trouble. It's also suggested that gloves be worn to protect your hands against the harsh chemicals. Carburetor cleaner is nasty stuff but it sure does its intended job.

For this project, we've chosen one of the more popular examples of Holley's new breed of smaller four-barrels—the 6619-1. For years, the majority of street machines breathed through large double-pump, dual-feed Holleys. Most of these cars were overcarbureted, and it took nothing less than a gas crisis to make this a reality. As fuel prices rose, the big four-barrels were ultimately shelved for smaller, more practical alternatives, and with a '74 introduction, the 6619-1 filled the bill nicely. At 600 cfm, the single-feed, single-accelerator pump carburetor is the perfect size for most V8s. Throttle response is smooth and crisp, economy is good, and emissions are easily brought into line.

The No. 6619-1 is Holley's identification (or list number, as Holley calls it) for this specific four-barrel carburetor, and it is just one of many small-cfm carbs included in the 4150 and 4160 model series. This series has been around for a number of years, and all the carbs within this particular family are very similar. Most parts are interchangeable;

5

Position new gaskets on the metering block. Outside gasket will go on backward, so be careful. Both the float bowls and block are complete, but keep clean until the time comes for installation. Using a clutch-head screwdriver, remove the secondary metering body and plate. The gasket, plate, plate gasket and metering body (in that order) are to be installed later; in the meantime, don't lose those screws.

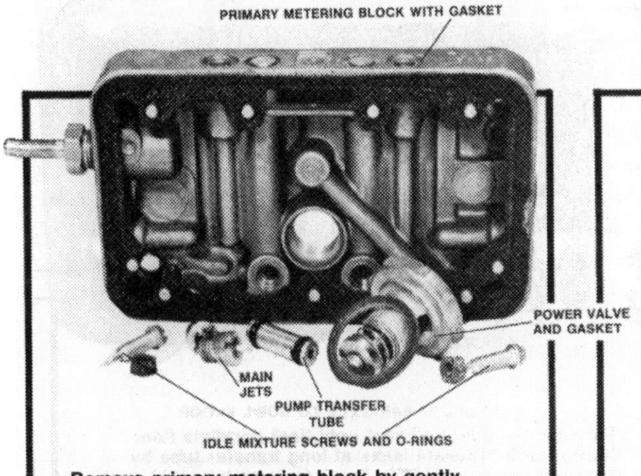

Remove primary metering block by gently prying between air horn and block. Clean block and remove parts shown in photo. Main jets are same size, so don't worry about mixing them. Install new power valve and gasket and other parts, using new O-rings and gaskets. Return mixture screws to same holes, snug down, back out 1½ turns each.

4

Separate the throttle body from the main body, after noting the position of the six screws. Remove plastic choke cover and soak both bodies in cleaner. When clean, install new gasket, assemble and install choke cover.

6

consequently, the rebuilding procedure described here applies to many more part numbers than just the 6619-1. The parts number list of 4150/4160 Holleys is much too long to include here, so compare the outward appearance of your carb to the photos in figure 1. If they appear to be the same, with only minor differences (for example, some carbs in this series have a secondary metering block like the primary, complete with power valve and jets), then the following how-to also applies to you.

Every carburetor requires some periodic repairs, and after two years of service, the 6619-1 may be exhibiting symptoms of wear for some of its owners. If the engine has recently been tuned but it continues to idle roughly, hesitate during acceleration or use excessive amounts of fuel, then the carburetor may be the problem. Fuel leakage around the carb, flooding, eye-burning fumes of unburned gas at the tailpipe and insensitivity to fuel/air adjustments are further symptoms of a tired carburetor.

To many, the carburetor is a complex, mysterious instrument containing many little components, and for those reasons, many people are afraid to tamper with it. Despite its complexities, the Holley four-barrel is actually quite simple to rebuild (probably the simplest of all four-barrels, and it has the great advantage of external float bowl adjustment) if the project is handled systematically. That is, don't be overeager with its disassembly. Taking something apart is always easier than putting it together, so don't get carried away. A professional carburetor rebuilder completely disassembles a carburetor, throws all the pieces into a basket, and within minutes, the unit is back together and operating smoothly. Of course that kind of expertise comes with lots of experience. For the beginner, such an optimistic approach would yield nothing more than a basket full of unfamiliar parts. The safest way to rebuild a carburetor is to work on one section at a time. As you will see, the first step is to remove the float bowl. The bowl is cleaned and then each part is removed—one piece at a time—cleaned or replaced and then returned to its place in the bowl assembly. When the bowl has been serviced, the next section (the metering block) is removed, cleaned and serviced. Doing the job in this manner may take longer, but it ensures that there will be no mistakes. Cleaning should always precede disassembly because it's so easy to forget what went where while parts are soaking.

The entire project requires a clutch-head screwdriver (about $3) to remove the secondary metering body (not necessary for carbs with secondary metering blocks) and just a few basic hand tools: a wide- and a thin-blade screwdriver, a small- and a medium-point Phillips screwdriver, a 5/16-inch open-end wrench, 7/16-inch, ⅝-inch, 11/16-inch and ½-inch (for the carb-to-manifold bolts) box-end wrenches. If you own these tools, or if you can borrow them, then your monetary investment is relatively small. You will need to buy a carburetor repair kit for a Holley 6619-1 (about $8 retail), a 1-gallon can of paint thinner ($1) and a 1-gallon can of carburetor cleaner (about $5). The repair kit includes a complete gasket set, needles and seats, a new power valve, an accelerator pump diaphragm and all the required O-rings. Instructions are also included, but they usually aren't too helpful. The same kit is used for many 4150/4160 models; consequently, you will end up with some extra gaskets, so don't panic. For a Holley carb, the best kit is made by Holley and sold by most speed shops. There are several other kits on the market, but try to stick with the popular name brands, as they are usually the best.

Carburetor cleaner is available at all auto supply stores. The 1-gallon can is a good investment, because it's usually large enough to accommodate the entire carburetor for thorough soaking, and it often includes a wire-mesh parts basket. Cleanliness is critically important to carburetor operation, so choose a good cleaner.

The beginner should allow a full Saturday for the job, and a total investment of approximately $17, for a $30-$40 saving and that feeling of satisfaction from doing the job himself . . . right on his kitchen table. ■■

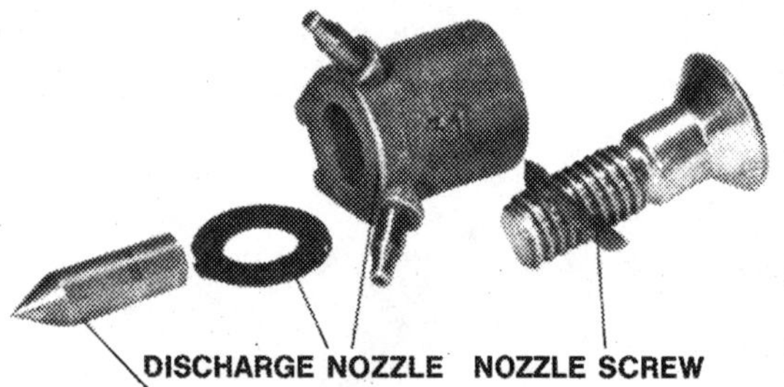

7

If you disassemble the secondary vacuum assembly for cleaning, you will find a small check ball inside. It's easy to lose, so be careful. Looking down through the top of the carb, there is a Phillips screw directly below the choke plate. It secures the accelerator pump discharge nozzle, two gaskets and a small needle. Clean, replace gaskets, and install with needle point down and nozzles facing toward center of carb. Install the metering block and secondary metering body, plate and gaskets, and tighten screws in a diagonal sequence.

Turn each needle and seat until float parallels floor (top) of bowl. Prevent leaks at long transfer tube by coating O-rings with STP. Small transfer tube will seat when bowl is tightened. With car running, adjust floats till fuel appears at lower threads of sight plug holes.

8

big mouth

"People who buy subcompacts after they've had larger cars usually wind up hating them," Bill Mitchell said resolutely. "After they've had 'em for a few weeks, they find out how slow they really are and it makes them want to sell the car." As the operator of Motion Minicar, Inc. (Baldwin, Long Island), Mitchell is privy to such confessions and maintains a livelihood by them, increasing the respiratory capacity of these pocket street bombs from Audis to Z-cars.

The most frequent visitations come from the owners of V6 Capris. The 2600cc and 2800cc engines have something of an advantage over the rest of the field in that the engines are considerably larger in displacement. But even that offers little consolation. The owner of our subject Capri had already put a kicker to his car's deep breathing with headers and a free-flow exhaust system, but he felt the 5200 series Autolite/Weber two-barrel (the same carburetor found on 2000cc Pinto four-cylinders) was just too restrictive. He was right. They don't even work well on the smaller engines, so on his car it's almost like having a rag stuffed in the venturi.

Mitchell already knew the culprit and had the situation corrected. The plan is as simple as bolting on an R-4643-1 350-cfm Holley dual-throat carburetor. And as simple as making a 1-inch-thick aluminum plate to adapt it, and modifying a GM throttle hold-down. All stuff you can do or have done, *or* forget all that and buy a kit from Minicar. What it'll do is make your V6 pull two lengths farther out of the tollbooth on any of those guys who gave you trouble before

Adapter plate made by Bill Mitchell at Motion Minicar is a one-inch-thick aluminum unit. The four mounting holes are countersunk and use Allen head capscrews so the plate will flush-fit on the manifold.

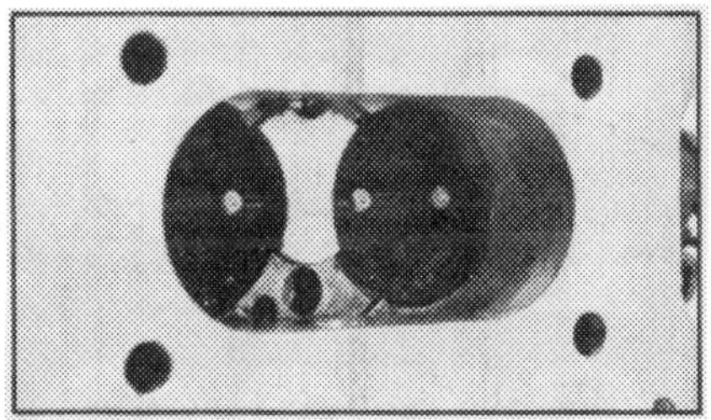

Close-up view of the adapter plate reveals that only one oblong hole is used instead of two, one for each throttle butterfly. The one large hole needs to be only slightly larger than the throttle butterflies to prevent binding and sticking on hot engines, allow plenty of room for flow to develop.

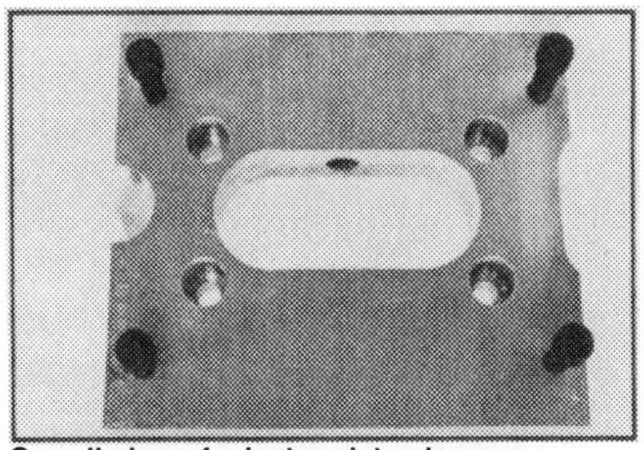

Overall view of adapter plate shows notches for linkage clearance, main bore hole, countersunk mounting holes and four holes to mount the new, larger carb on top.

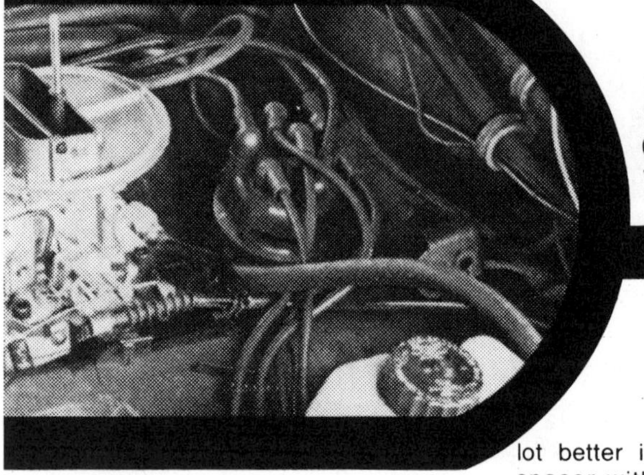

A LARGER, MORE EFFICIENT CARBURETOR IS WAITING FOR YOUR V6 MUSTANG OR CAPRI

and still retain the same gas mileage as you did with the tiny pot-metal fixture you had before.

The carburetor is a Holley replacement for a V8 Ford, chosen because it has a choke and provision for a choke stove for the exhaust manifold. If you're in a bind or there's a big-block Chevy tri-power upside down in the grit of your garage, well, the center carb from it will work, too.

Big Mouth will sit on the manifold a lot better if you have an aluminum spacer with four countersunk 5/16 Allen-head screws and an opening that approximates the baseplate of the Holley. It doesn't have to be exact, just a little larger than the throttle bore itself. Tap a ¼-inch hole in the back of it and thread it so you can fit a 90-degree elbow and a ¼-inch nipple. Then you'll need a small-block Nova or Camaro throttle cable tie-down, part No. 6261814, and an ordinary flat washer. This is so you can make an anchor for the Capri cable and hook it to the carb base at the same time. Tack-weld the washer in the slot so that you have an "O". Grind off the slag and spray-paint it black.

When you bolt it back together, the thing will fire and run, if you remembered to attach the fuel line, but the jets will be too large. The Holley comes with No. 61s, but even headers will require a cut to 59; a stock exhaust system would need something at least two or three sizes smaller and perhaps an accelerator pump adjustment. The choke heat stove can be cobbled from a parts store kit, and your Big Mouth will have crisp response, start instantly, and be like Maalox for your minicar nausea. ■■

Small-block Chevrolet throttle cable tie-down, part No. 6261814, is shown on left. On right is the same part with a flat washer welded in place to guide the Capri or Mustang or Pinto throttle cable and anchor it at the same time.

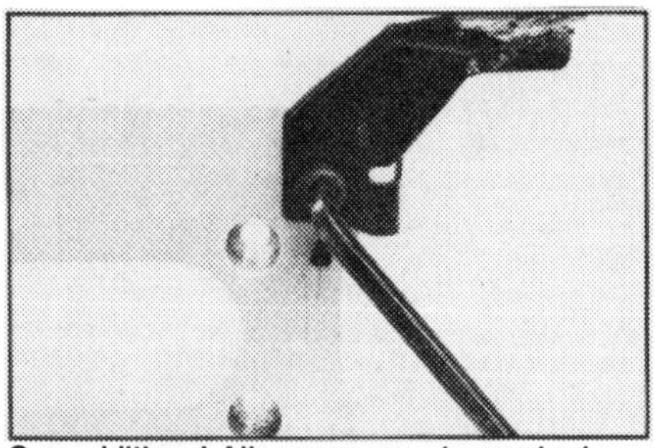

One additional Allen capscrew is required to bolt the new cable guide to the adapter plate on the left rear corner before the whole assembly goes on the engine.

The bare carb, Holley's R-4643-1 AAS, is a 350-cfm unit that has all the linkage and accessories needed.

Bob Thompson Reveals His Inner Knowledge of Tunnel-Ram Holleys ■ Norman Mayersohn

Any newcomer to Modified Production-style racing has a slew of baffling problems to solve: the innumerable possibilities of car/weight/engine combinations; the complexities of choosing a drivetrain that will be efficient as well as reliable; and sorting out all of the possible sources for major engine components, such as heads, pistons and camshafts. But there is one thing that does seem to remain constant: The induction system for the MP and Gas classes almost always consists of two Holley four-barrel carburetors on plenum-type tunnel-ram manifolds.

LORD OF THE BUTTERFLIES

Bob Thompson has quite a few years experience setting up these induction systems to work on both his Indy-winning big-block Corvette and his current 600-hp 302-cubic-inch D/MP combination. We managed to nail the elusive Thompson down one Saturday morning during a break in his post-Supernationals dyno and drag strip test session, and he agreed to let us in on the basics of tunnel-ram carb preparation.

We expected a Jenkins-like black magic approach; Thompson rather surprised us by revealing that the out-of-the-box No. 4224 Holley tunnel-ram carbs will operate just fine with a bit of attention to two details: preventing fuel from sloshing out of the bowl vents and into the venturis under acceleration, and fixing the idle circuit to prevent an over-rich condition. Both fixes aim to improve driveability on the strip as well as in the pits.

Assuming that we are preparing the carbs for an all-out drag strip application, the first step is to strip the carbs down to the main bodies and have a competent machinist whack off the air horn which surrounds the two primary barrels of each carb (it is no longer needed to house the choke assembly). This allows all four holes of the Holley to "see" the same air flow coming down the venturis without any restriction. This is also a good time to drop by the speed shop and order a metering block conversion kit, Holley No. 85R-3548.

While you're waiting for all of that to happen, it would be worthwhile to look at the entire in-car fuel system to ensure that the carbs will get adequate fuel. Thompson uses a ¾-inch-diameter pickup which sits in a lowered reservoir in his tank. The pickup has a screen over the end of it, and the ¾-inch line runs all the way to the red Holley electric pump. From the pump forward, Bob has a single piece of ⅝-inch-i.d. Aeroquip line running to the engine compartment, where it is split by a Y-fitting. Two 5/16-inch fuel lines run to the carbs, each one having a Holley regulator set to 7 psi in the line. The linkage mounts should be solid for both reliability and good pedal feel for the driver.

With the carbs back in your eager little mitts, it's time to get down to the serious work. Replace the stock needle and seats with the Viton-tip .110-inch set, Holley No. 18BP-135AS. Thompson swears by the Viton parts because they are not leak-prone, even if some grit manages to get caught in a vital spot.

Next, you want to yank out the stock bowl vent tubes with a pair of pliers and tap the remaining holes to 5/16-18 thread size; no drilling is needed. Block the vents permanently with an appropriate length bolt, using Loctite to ensure that it won't vibrate its way into the venturi. Since the fuel bowl still needs a vent to atmospheric pressure to operate, an external vent must be fitted. Grind down the end of the last rib at the forward end (front of the car) of the float bowl and drill a hole large enough for a ⅛-inch pipe tap approximately ½-inch from the edge of the bowl. Tap the hole two or three threads deep and thread a 45-degree angle fitting into the hole, making sure to use Teflon tape to prevent leaks. You don't want the fitting nipple too far into the float bowl, because it may interfere with the float.

The vent will probably work best pointed to the rear of the car, but this depends on hood scoop design. Some scoops may have enough air flow across the nipple tip to actually siphon gas out of the vent at high speed. In any case, this vent fix should eliminate all of the car's tendencies to load up after a hard burnout, since it will eliminate the problem of gas sloshing down into the venturis. The vent baffle or whistle can be discarded and the rivet replaced, since the stock vent is now blocked.

The factory-stock 660-cfm carbs come from Holley with removable jets on the primary side and a metering plate in the secondaries. Thompson converts the secondaries to screw-in jets with the 85R-3548 kit, which includes the new metering block, longer transfer tube (fuel pipe) and screws, and all the necessary gaskets. This gets installed as is, except for replacing the stock rounded float with a rectangular brass part, No. 16R231A. There is no provision for a power valve, so you need not worry about that; all that needs be done is to trim off the unneeded brass tabs from the new float, leaving only those that hold the closing spring in place.

There are a number of ways to lean down the idle circuits on Holleys, but if you've tried any of the wire restriction tricks, you know that

With stock air horn machined off, stock bowl vents are pliered out. Hole is tapped for 5/16-18 thread, and Loctited bolt is run down into orifice to block it.

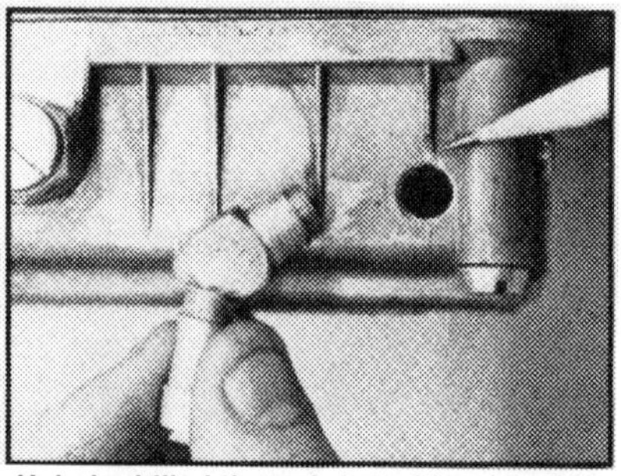

Hole is drilled through top of bowl in this location, tapped with ⅛-inch pipe tap to accept a 45-degree angle fitting with threads wrapped by Teflon tape.

Photo of typical Thompson 660 Holley reveals the chopped air horn, block stock vent tube, new vent on top of bowl at left, metering block and long transfer tube in place (85R-3548 Holley kit) and recontoured secondary opening ramp on linkage.

it's hard to really make a worthwhile difference. Bob Thompson feels that the only right method is to use a true orifice rather than a makeshift restrictor, and to have that orifice as near the point where the fuel enters the venturi (down near the baseplate) as possible. To this end, he uses a .150-inch-diameter lead shot to fill the two idle *fuel* bleed holes in the main body of the carburetor. Using a hammer and needle-nose pliers, he forces the shot into the outer two holes at the primary end of the main body and cuts off the excess flashing with a razor blade. Drill the plugged hole to .038-to-.042-inch, using a drill bit in a hand-held pin vise, stopping when you feel the bit break through the other side of the shot, and the idle restriction is complete. There is never a need to fool with the idle air bleed holes at the top of the carb, since this is a racing application. The lead shot balls can be obtained from any hunting store or gun shop.

With the carb stripped to its barest essentials, Thompson recommends some clean-up work. There will be a small casting flash left at the top of the venturis by the manufacturing process; this can be easily removed with a scraper or sharp knife. Flip the carburetor body over and you will notice that there is considerable mismatch between the bottom of the venturi and the baseplate. Thompson fixes this with a 60-degree valve seat cutter, grinding just enough to make a nice matchup at the mating surface between the body and baseplate. Use a gradual chamfer and don't allow the venturi to get any bigger than the i.d. of the throttle bore. Don't grind, enlarge or hone the rest of the venturi at all. It'll get you nowhere, slowly.

As for the rest of the carb body,

Not every single area and subassembly of the 660 Holley needs rebopping. Thompson prefers stock booster assemblies, older accelerator pump nozzles unmodified and stock 50cc pumpers.

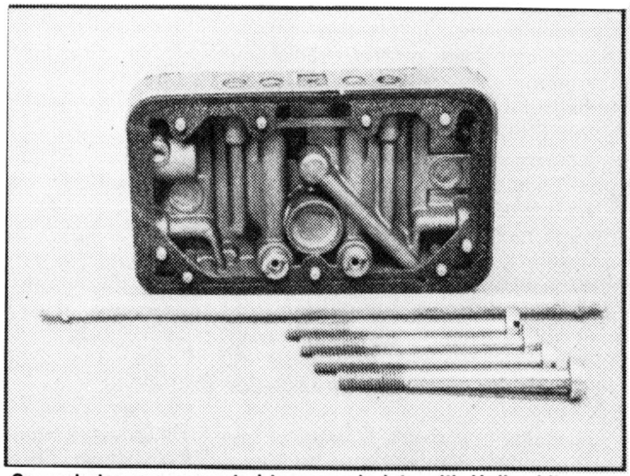

Secondaries are converted to screw-in jets with Holley kit 85R-3548, containing jet block, jets, longer screws, longer fuel transfer tube and gaskets. Jets end up eight square.

BUTTERFLIES

Bob leaves things much as they come from Holley. There are no booster modifications, and he uses a stock, old-style accelerator pump nozzle with brass outlet tubes. Thompson does not drill or modify the nozzles in any way. The 4224s come with the 50cc accelerator pumps to begin with, and this is adequate for any tunnel-ram application just the way it comes out of the box.

Moving to the external workings of the 660s, there is some worthwhile finagling to be done on the carb linkage. On the accelerator pump arm lever, Bob feels that you should always attempt to have the same amount of pump spring extending from its housing—7/16-inch is a good figure to use. The spring bolt should just touch the pump arm with zero clearance, but no preload, so as to save all of the movement for the starting line. At the throttle linkage arm, the accelerator pump cam should be a maroon piece and should be moved to the No. 2 position to delay pump action.

Thompson has also found an advantage to delaying the secondary opening in that it is now stored for a maximum shot of air/fuel mix as the car leaves the starting line. The secondary opening ramp on the linkage is recontoured to get this effect, and can be duplicated with a hand grinder from the accompanying photo. He also cuts the linkage return loop off at this time for manifold and distributor clearance. The accelerator pump and secondary delay work not only aid in giving the extra boost at the starting line and in first gear, but they also allow the car to run cleaner on the return road and in the pits. Spark plugs will still tend to be black at the front of the motor after some idling, but Bob says this is normal and means nothing.

A good number of the faster MP cars are now using baseplates from the 850-cfm carbs on their 660s. Bob has made this modification to his carbs, but he says this is not the sort of thing that brings an instant e.t. improvement. It is far from a bolt-on job—it's mostly an attempt to slow down the velocity of the air flowing through the carb to help keep metering within the limits of the 660. With small-block power levels hovering at 2 hp per cubic inch, Bob feels that they are reaching a saturation point and that 750- and 850-cfm carbs will come into use on tunnel-rams soon anyway.

Experience with the familiar yellow Corvette has taught Bob that the best way to prevent float settings from changing constantly is to *not* drain the carb bowls after putting the race car back on its transporter—that just leaves them to hang free and rattle on the tow home. Float levels should be set to ⅛-inch from the top of the bowl, adjusted with the bowl upside down (off the carburetor, naturally) and not with the sight window. Initial setup calls for jetting somewhere between 72 and 75, depending on displacement and hood scoop design, using "square" jetting throughout the carbs. Idle bleed screws usually work just fine when turned all the way in and backed out ½-turn. Bob idles his 302 at 1500 rpm, so you know that there's no low-speed sloppiness in the carbs at all.

A lot of supposed competition carburetors have come and gone since the 660 Holley came into general use, but the dependable and responsive 4224 continues to be by far the most popular tunnel-ram carburetor. While its days may be numbered for Pro Stockers or big-blocks, for most applications a pair of 4224 four-barrels is still by far the best way to get 1320 cfm for any 1320-foot application. ■■

A closer look at the recontoured secondary opening ramp on the linkage shows that it has been trimmed of the linkage return loop for manifold, ignition clearance.

All float bowls get Viton .110 needle and seat assemblies, Holley 16R231A rectangular brass float assemblies trimmed of extra tabs. Floats are set ⅛-inch from top of bowl.

Main bodies are flogged to lean down idle circuit. Two .150-inch lead shots are forced into idle fuel bleed holes, bored to .038-.042 with pin vise and cleaned of all flashing in area.

This photo shows the basic requirements of the grinding job needed to delay secondary opening for clean running and best launch. Shot was taken before loop was trimmed off arm.

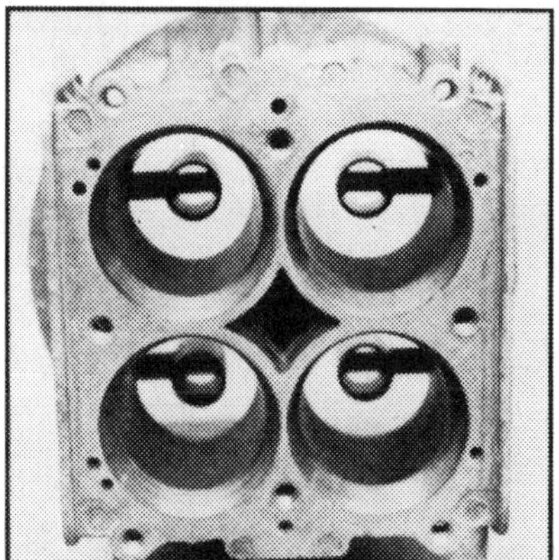

Mismatch between venturi and baseplate is fixed with a 60-degree valve seat cutter, grinding just enough to match up the two orifices. Venturi shouldn't be larger than throttle-bore i.d.

Whether it's performance, emissions or economy, they've got a carb for you!

Holly's new 4500 and other stories

By Terry Cook ■ There was a time when a carburetor manufacturer who was involved with the aftermarket had only to concern himself with developing hardware which was designed to make cars go faster. Today there is an increased responsibility to concern oneself with things like improved emissions and increased fuel economy. In light of these new objectives, the people at Holley have been hard at work redeveloping some existing hardware "families" to meet the needs for low emissions and improved fuel economy, as well as maintaining their commitment to the high-performance end of things. Out of all this labor, three different types of carburetors have emerged, and they have three different uses.

The first of these new redesigned carburetors we will concern ourselves with is the revised edition of the famed 4500 performance carb, the No. R7320. The unit is a large four-barrel with a flow rating of 1150 cfm (cubic feet per minute) and is designed for use on the drag strip in the Pro Stock, Modified Production, Gas Dragster classes where dual quads are permitted, as well as in the NHRA Econorail classes and AHRA Super Stock applications where the rules restrict the induction system to a single four-barrel carb. Those of you who are hard-core carburetor fans will remember that the 4500 was introduced in '69, and all of the 4500s were rated at 1050 cfm, with the exception of the few 1150-cfm pieces which were made for the Ford 302 Trans-Am program.

In the past five years, drag engine induction science has changed, and the trend seems to be going toward the IR (individual runner) design. In general, the plenum sizes have been shrinking, creating a bigger demand for fuel delivery in instances such as the mini-plenum-equipped big-blocks. Holley has responded to this need for a bigger carb by redesigning the 4500, but they simultaneously decided to incorporate everything good the Holley field tech reps had accumulated over the past year of flogging the major NHRA drag events with their technical aid participation program. These reps have dissected hundred of carbs which have been "tricked up" by the nation's most renowned carb magicians. In some cases, they found it rather disturbing, because some racers had been charged as much as $300 per carb for the trickery, and the rework consisted of nothing more than a half-hour of labor. As a result, the Holley engineers decided to include everything that worked in the internal redesign of the 4500, so that when

LEFT—A box-stock 7320/4500 on the left is compared to a modified 7320 with its air horn properly removed. Air bleed location should not be altered since they need to be in turbulent air.

Holly's new 4500

someone paid $300 list price for a new 4500, he wouldn't have to lay out that much again to have some guy smoke it over. Incidentally, while the list price of the 4500 is about $300, if you shop around, you can buy the new units at racer's net somewhere in the $200-to-$220 range, or so it is said. The only

Modal Summary

Mode Type	*HC*	*CO*	*NOX*
Idle	2.928	76.81	.240
Accel	8.470	146.89	5.527
Decel	2.973	64.29	.452
Cruise	11.467	201.02	6.226

1974 302-c.i.d. Maverick Stock Manifold Stock Carburetor (2-bbl Motorcraft)

Mass Emissions (Grams/Mile)

	HC	*CO*	*NOX*
	3.045	66.58	1.768

1974 Federal Standard

	(3.4)	(39.3)	(3.0)

Fuel Economy (Miles/Gallon) 13.63*

**This represents city driving*

Modal Summary

Mode Type	*HC*	*CO*	*NOX*
Idle	2.731	23.23	.246
Accel	8.442	57.81	10.482
Decel	3.150	23.88	.623
Cruise	11.741	146.19	11.478

1974 302-c.i.d. Maverick Holley R6989 RV Carb Edelbrock Torker 289 Manifold

Mass Emissions (Grams/Mile)

	HC	*CO*	*NOX*
	2.497	28.22	2.715
	(3.4)	(39.3)	(3.3)

1974 Federal Standard

Fuel Economy (Miles/Gallon) 15.05*

**This represents city driving*

thing the Holley guys didn't do to the carb to make it work better was to remove the air horn. They left it on for several reasons. First, the tooling would have been so expensive that each new 4500 would have had to cost about $25 more, whereas you can take it to a ma-

ABOVE—Choke-side view of new 4175, list No. 6926, the new spread-bore carb with vacuum-operated secondaries and a "divorced" choke assembly too.

chine shop, along with the directions supplied with the carb, and have the air horn cut off for about $10. The air horn was also left on because some cars with certain hood configurations seem to run better with the air horn left on. One modification which the Holley engineers purposely did not perform was to smooth out the bores with epoxy, which seems to be a popular trick. They didn't do this, because the air bleeds require a turbulent area to function properly.

Some of the internal improvements to the 4500 include increased fuel passage size to make the carb more responsive to jet changes, the addition of different boosters to make the main system start sooner to improve the low-end and mid-range fuel metering, and the improvement of the idle system, which now features a wide range of adjustment. Another improvement was the addition of big accelerator pumps on both sides to get even the heaviest car launched properly. The carburetor has been tailored to big-block applications, but a Trick Kit is available from Holley for those who wish to use the carb on small-block engines.

One modern approach the Holley engineers added to the 4500 was to make it a "systems" carburetor when they redesigned it. This simply means that they made it easier to change parts in the carb, depending upon the different applications. One example of the systems approach is that since a power valve is desirable for single quad uses but not needed in dual four-barrel applications, machining was performed so that you simply remove a blank plug and screw in a power valve if you need it. Previously you couldn't put a power valve in the 4500s, but now you can do it in a minute or two, thus permitting the metering to be tailored to allow

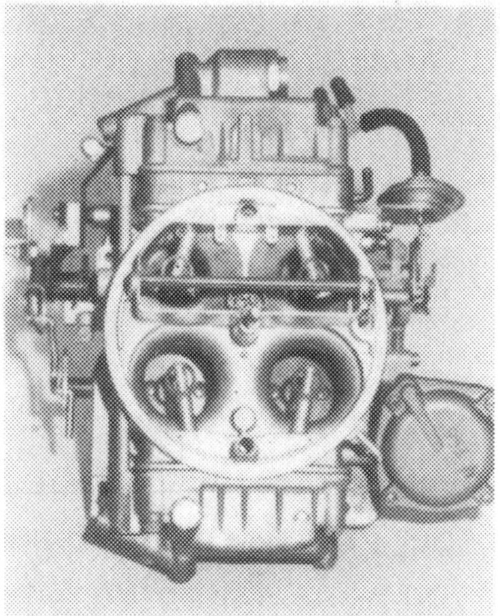

ABOVE—Top view of 4175 carb. This and the 4165 are both rated at 650 cfm, offer a good balance of emissions and economy and performance spread-bore style.

richer settings. The intermediate systems which were included in the old 4500s have been eliminated, because nobody is using the deep plenum manifolds any more.

The final change for the new 4500 series was the use of staged linkage which allows the primaries to open 45 degrees sooner than the secondaries to facilitate easier staging. A kit is available for those who are fed up with getting ripped off by the magic carb rebuilders. Out of the box it should perform as well as the high-dollar rebuilds.

The second series or family of carbs the guys at Holley have come up with is an emissions-oriented unit. Right off.

Continued

Holly's new 4500

that sounds misleading, as if all of Holley's other aftermarket carbs are "dirty" emissions-wise. Nothing could be further from the truth, as the management of Holley is strict in their feeling that the Holley hardware should be designed to operate at least 20 to 25 percent *under* the current emissions requirements spelled out by the federal government. And, thankfully, Holley has the know-how and wherewithal to do it. The engineers admit that the secret to designing and building clean carbs lies in providing excellent distribution at *all* rpm.

Once this level of emissions performance has been attained, then the engineers start working on other characteristics of the carburetor, such as fuel economy or performance. Remember, now, we're speaking of carburetors intended for street use here. In light of the time, money and engineering brain-power Holley has expended upon emissions, it's only natural that they have arrived at a good, low-emissions carburetor which you can bolt onto your existing manifold. The unit we're pointing up to here is Holley's new pair of spread-bore carbs, the 4175 and 4165. Remember that the term "spread-bore" refers to the fact that the carb has small primaries and large secondary bores. The 4175 is equipped with vacuum-controlled secondaries, while the 4165 has mechanically operated secondaries. The engineers have found that the vacuum-operated secondaries offer a more substantial improvement in fuel economy, so lean toward the 4175 as a result. Both the 4175 and 4165 are rated at 650 cfm.

Holley is working on a raft of these carburetors, and by April they should have them available as a replacement

ABOVE—R-7320 comes with plugs for power valves, but 25R595A65 valves and No. 86 jets convert it for one-carb racing.

for the Rochester Quadrajet in the following applications: 1973-'74 Buick 350- and 455-cubic-inch engines, 1973-'74 Chevrolet 350- and 454-cubic-inch engines, and 1973 Chevrolet light-duty truck equipped with the 350-cubic-inch engine. In addition, the carb serves as a replacement for the Carter Thermo-Quad in the following cases: 1973-'74 Chrysler 340- and 400-cubic-inch engines, the 1973-'74 Chrysler 440-inch (A-120) standard engines, as well as the 1973-'74 Chrysler 440-inch (A-134) performance engines. In all instances mentioned, the 4175 and 4165 will bolt right up to the existing manifold, and everything needed to make the swap is provided in the box with the carb. Remember, these carburetors have excellent emissions characteristics, good performance and good economy, and can be bolted directly to a number of OEM intake manifolds.

The third family of new Holley carbs we will discuss is a special street fuel economy four-barrel, rated at 600 cfm and designed to be used with aftermarket intake manifolds. Here we are talking about a square carburetor (all four bores are the same diameter, 1 9/16 inches) which the Holley people like to refer to as an RV (recreational vehicle) carb. The fuel economy improvement of this unit is cited in the accompanying data tables, showing a gain of about 1½ miles per gallon in this instance. The unit has a suggested list price of about $129, but you can shop around and find one for close to $100 new at some speed shops. The carb is equipped with

ABOVE—Photo shows how to cut air horn and radius entries. Cut should be level with air cleaner flanges.

an electric choke in order to facilitate proper choke hookup and operation. The choke is ignition-activated from a 12-volt source, such as the field connection in the alternator. This is a vacuum-operated secondary carb and has a number of different models available, each with a fuel delivery curve tailored to the applications indicated. While there are a host of different model numbers, it is essentially the same unit. The applications are as follows: For General Motors—Use No. 6619-1 for 1972 and prior engines or No. 6979 for 1973 and later with EGR* For Ford Motor Co.—Use No. 6619 for 1972 and prior engines or No. 6989 for 1973 and later with EGR For Chrysler Corp.—Use No. 6909 for 1972 and prior engines or No. 7009 for 1973 only, with EGR (*EGR = Exhaust Gas Recirculation)

So there you have it, a wrap-up of the latest hardware from Holley. Whether you're into drag racing, low emissions or just improved fuel economy, it appears that the people from Warren, Michigan have a piece of equipment to suit your needs. ■■

There's more to tuning a Holley carb than just changing jets

POWER VALVE PARADE

Text & photos by C.J. Baker ■ One of the least understood metering systems in Holley carburetors is the power enrichment circuit and the power valve. What does a power valve do? Does it regulate fuel flow? Can it be used to tune an engine for peak performance? These are some of the questions that we are frequently asked by phone, by mail and at the tracks across the country. To clear up these questions, we'll outline the function of the power valve, the different power valves that are available and how they can be used for best performance.

The power valve, or power enrichment circuit, is designed to supplement the main metering circuit of the carburetor under high-load conditions by supplying additional fuel to increase the flow of the main circuit. Under normal cruising conditions, where the engine is propelling the vehicle down the road at a fairly constant speed, the main jets meter the fuel into the main circuit for delivery to the air stream. Main jet size is selected to meter fuel for an air/fuel ratio in the range of 16:1 to 13.5:1, depending on application, for optimum driveability, economy and emission characteristics under these conditions. However, when the throttle is depressed for additional engine output, either for acceleration or to overcome a heavier load, such as a hill, the air/fuel ratio needs to be temporarily enriched to approximately 12:1 to provide optimum power until the load is overcome and normal cruising conditions are resumed. The power valve provides this temporary enrichment by essentially opening two additional metered passages (you could think of them as two additional fixed-size jets) to the main circuit. Operation of the power valve is controlled by manifold vacuum. Manifold vacuum draws against the diaphragm in the power valve and closes the power valve against its preset spring tension. When the engine load increases, manifold vacuum drops and the power valve spring forces the power valve open to admit more fuel.

From the above description you can readily determine two things: First, when the power valve functions is dependent on the manifold vacuum level at which the power valve is set to open. And second, the power valve is simply an on-off valve. It does *not* regulate the amount of fuel that flows through the power enrichment circuit. That is controlled by the two metered passages, or "power valve channel restrictions," as they are usually called, located immediately behind the power valve seat in the metering block (see photo).

Now that we know what a power valve does and how it does it, we'll look a little more closely at power valve differences. Holley has equipped its carburetors with power valves of four distinct designs, excluding the unique two-stage power valves, which we'll cover later. These four designs include the four-hole, six-hole, "small picture window" and "large picture window" models (see photo). The only real difference in these designs is the amount of fuel they are capable of passing. The four-hole and six-hole versions are capable of flowing enough fuel to supply two power valve channel restrictions up to .067-inch diameter. If your carb has bigger channel restrictions than this, the four- and six-hole power valves will actually choke off proper flow. And incidentally, Holley is in the process of phasing out these four- and six-hole power valves in favor of the "picture window" models. The small picture window power valve can supply channel restrictions up to .095-inch, and the large picture window version can handle up to .120-inch. The only trick here is to be sure you have a big enough power valve to feed your carb's channel restrictions. The best choice is usually the small picture window model, since only a very few special race-oriented carburetors have channel restrictions over .095-inch. Additionally, the small picture window model is available in nine different vacuum settings, whereas the large picture window power valve is offered in only three.

This brings us to the next area of concern: vacuum settings. Each power valve is stamped with a two-digit number. As previously indicated, this number has nothing to do with power valve flow. Instead, it is the vacuum level below which the power valve is open. For instance, if it is stamped "85," that power valve will be open at all vacuum levels below 8.5 inches Hg. This is the same scale of measurement used on most manifold vacuum gauges. For street applications, the goal is to select a power valve with a vacuum setting below the manifold vacuum normally generated by the engine in the cruise mode. If the power valve vacuum setting is too high, the power valve will be open during cruising conditions, creating an unnecessary and wasteful "full rich" air/fuel ratio mixture which will decrease gas mileage, increase emissions and shorten spark plug and piston ring life.

For racing applications, the vacuum setting must be low enough to keep the power valve closed while driving around in the pits and while staged waiting for the green, to prevent the engine from loading up and bogging. When the throttle is nailed, manifold vacuum falls to nearly zero and the power valve then will fully open.

Two areas of concern when determining the vacuum setting include elevation and whether the vehicle normally maintains much manifold vacuum. Late-model emission engines and engines with a lot of camshaft duration and overlap do not usually carry much manifold vacuum, so power valves with a low setting must be used. Additionally, Holley recommends that the power valve opening be delayed by 1.5 to 2.0 inches Hg for each 3000-foot increase in elevation. For example, if the correct power valve for your engine has a setting of 8.5 inches Hg (stamped 85) and you move to a location 3000 feet higher, the power valve should be changed to one with a setting of 6.5 inches Hg (stamped 65) to maintain the same relative opening point. This is because engine manifold vacuum decreases as altitude increases.

With the conventional power valves covered, we'll look at Holley's latest offering, the two-stage power valve. You will recall that we said that the conventional power valve is simply an on-off mechanism. You can readily appreciate that if engine loading is increased (manifold vacuum decreased) just enough to open the power valve, the mixture immediately goes to "full rich," when something in between "full rich" and normal cruising might have been adequate to overcome the load. Thus the idea of a two-stage power valve that opens partially when manifold vacuum falls to a certain point and a second stage that fully opens the power valve if manifold vacuum continues to drop to a second predetermined level.

Holley engineering tests have shown that such a two-stage power valve can restore some driveability to vehicles equipped with exhaust gas recirculation (EGR) and also increase fuel economy in certain class vehicles. However, such staged power valves do *not* show improvements in *all* vehicles, nor is any guarantee made or implied by Holley regarding mileage or emissions. Holley has published the following guidelines as to what vehicles and carburetors are *likely* to show improvements through the use of two-stage power valves.

1. The two-stage power valve works most effectively in vehicles with a weight-per-cubic-inch ratio of 14:1 or greater.
2. Two-stage power valves should not be used in any vehicle which sees occasional drag strip use, or in any carburetors with power valve channel restrictions greater than .060-inch (this includes the R-3310 and the conventional double-pumper line, which require high-capacity power valves). Excessive leanness could result from the limited-flow capacity of the two-stage valve.
3. Greatest fuel economy increases will be seen on relatively heavy vehicles used in rolling or mountainous terrain, or in stop and go suburban driving (this includes vehicles pulling trailers).
4. Two-stage valves will probably yield the greatest economy increases in recreational vehicles equipped with the R-6619 and R-6909 carburetors, as well as some model 4165-equipped full-size sedans and station wagons.

Holley produces these two-stage power valves in two different versions. The first carries part No. 25BP-475A-12, which is designed for most applications. Stage 1 opens at 12 inches Hg and Stage 2 opens at 6 inches Hg. The second version, part No. 25BP-475A-25, is designed for vehicles that will see frequent high-altitude use above 4000 feet. Its opening points are delayed to 10 and 5 inches Hg, respectively.

One last word of caution. Holley makes two different power valve gaskets, and it is essential that the correct gasket be used with the power valve being installed. The 8R-669 gasket, which has three tabs on its inside diameter, must be used with all four- and six-hole and two-stage power valves. The 8R-1597 gasket, which has no tabs, must be used with all picture window power valves. ■■

The small hole at the tips of the pen is one of the two power valve channel restrictions. The power valve must be capable of supplying adequate flow for these restrictions.

Shown from left to right are the four-hole, six-hole, small picture window and large picture window power valves.

Shown at the right is a two-stage power valve which has the same flow limitations as the four- and six-hole power valves on the left.

here's two ya

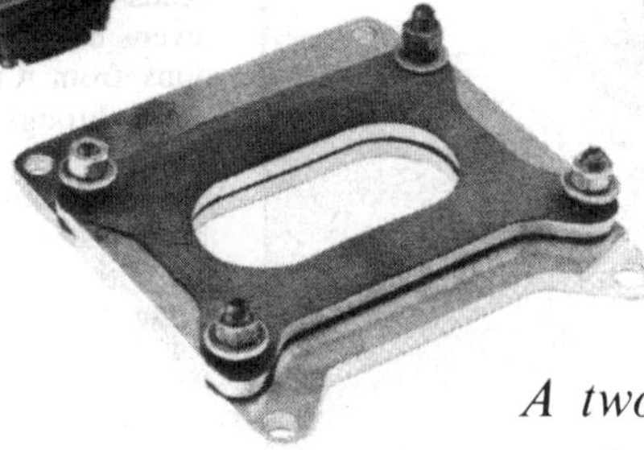

A two-barrel carb on a drag race car! But why?

Text & Photos by C.J. Baker ■ It all started back in the middle '60s when the American Hot Rod Association decided that stock-bodied cars should not be lumped together solely on the basis of engine size and weight at drag races. Instead, the AHRA felt that separate categories should exist that were based on how the vehicle was equipped or modified; i.e., automatic transmission versus manual transmission, OEM manifolds versus aftermarket manifolds, and the list went on and on.

They called these categorical, or equipment, distinctions *formulas.* By the time they were done, there were more classes and formulas than there were cars at an average drag race. In fact, there got to be as many as 10 to 15 formulas within specific classes.

Obviously the system was unwieldy and virtually unworkable, but one good thing did emerge out of it. One of the equipment distinctions was carburetor type; i.e., multiple carburetors versus single four-barrel setups versus single two-barrel carburetion. And eventually the entire formula system was streamlined down to a manageable three formulas within the Formula Stock, Formula Sports and Hot Rod classifications, with the three basic formulas based solely on carburetion. Formula 1 is for cars equipped with multiple carburetion, factory supercharged or fuel injection. Formula 2 is for single-four-barrel-equipped vehicles, and Formula 3 is restricted to single-two-barrel cars. This is the system currently in use today in both the AHRA and the IHRA, International Hot Rod Association, which has adopted similar class rules.

Actually racers are free to run any one of the three formulas, as there are no restrictions against swapping carburetion setups, and this has brought about a curious phenomenon. Normally most people think of stock-bodied drag race cars as being equipped with multiple carbs, or at least a giant four-barrel, but within the formula classes there is a surprisingly large percentage (perhaps as high as 40 percent) of Formula 3 two-barrel racers.

Because this approach seems contrary to the fundamentals of racing—to go as fast as possible—we decided to investigate this phenomenon, query the racers as to why they have chosen the Formula 3 setup and to take a look at the two-barrel carburetors that are currently being used. We began by visiting the first major race on the '74 AHRA Grand American schedule, the AHRA Winternationals at Bee Line Raceway in Phoenix, Arizona.

The general feeling among the Formula 3 racers is that the two-barrel carburetor offers ease of adjustment, simplified jet changing, less trouble and greater consistency than similar four-barrel setups. Royce Fowler of Independence, Missouri, who campaigned his highly successful 3-GA '69 Pontiac GTO during the '73 season, pointed out that tuning time is at a minimum during national events and that the reduced time required to rejet and adjust a two-barrel is definitely beneficial.

The Griffen & Hodge team from Phoenix, Arizona, who race a 3-DH Camaro, had quite another reason for choosing a two-barrel. They felt that the Formula 3 approach was the only *practical* way to be competitive, in light of the hard Formula 1 and Formula 2 national records in their class. And perhaps this is the real underlying issue behind the two-barrel phenomenon. Vance Brady, the AHRA Technical Director, suggested that some good-running AHRA racers have chosen to switch to Formula 3 to avoid the possibility of having to face a high-dollar NHRA Super Stock car (NHRA does not have a comparable two-barrel category) which might be entered in an AHRA event during class eliminations. This assures the AHRA two-barrel car of making it into Super Stock eliminations, where the cars compete on a more equitable handicap start and break-out system. Still other two-barrel racers scoff at this reasoning and claim that it's simply fun to go fast with a two-barrel and that it's cheaper too.

Regardless of where you turn, the justifications for selecting the two-barrel setup are many and varied. However, once the decision to run a two-barrel has been made, the selection of the most efficient two-barrel carburetor becomes critical. The rules state that all carburetor modifications are legal, with one exception. A four-barrel carb may not be sawed in half to make a giant two-barrel in Formula Stock or Formula Sports, although this procedure is legal in the Hot Rod classification. One enterprising racer removed the center divider between the primary bores on a Holley three-barrel, claiming that the modified carburetor was then a two-barrel. Although it is not specifically prohibited in the rule book, Vance Brady informs us that such a modification is definitely *not* legal.

In any event, the carburetor that is used almost exclusively, although some are highly modified, is the Holley two-barrel. Holley introduced its well-known R4112AAS 500-cfm carb in the late '60s. It was an immediate success with the Formula 3 racers, since it represented the largest two-barrel available. However, in 1969, a modified version of the 500-cfm two-barrel suddenly emerged from Tucson, Arizona. It was the brainchild of J.D. Braswell and his son Dave. The Braswells enlarged the size of the carburetor throats and replaced the normal fuel discharge tube and venturi booster rings with a unique annular fuel discharge system. This meant that fuel entered the air stream from a ring of small discharge ports around the perimeter of the venturi. The result was a carburetor that flowed over 600-cfm and offered fantastic response. As with any good idea, it was an immediate hit with the racers.

Then, in 1972, Holley released its own version of the annular-discharge two-barrel, R6425AAA, rated at 650-cfm (see Two-Barreled Dynamite, Mar. '73 HRM). Actually the Holley 650-cfm carb and the Braswell modified 500-cfm Holley offer very comparable performance, and these are the two carburetors which are found on almost all Formula 3 race cars today.

But we began to wonder what might happen if the Braswells were to work their magic on the Holley 650-cfm version. We contacted Dave Braswell at his firm, POW Engineering, 2901 North El Burrito Avenue, Tucson, Arizona 85705, and got some good news. Such a carburetor was already under development. In fact, there was even a prototype that we could test. In a comparative evaluation of this new Braswell two-barrel against a Holley 650-cfm carb on our own record-holding 3-GA Chevelle (featured elsewhere in this issue) at Irwindale Raceway, the Braswell showed a consistent .13-second and 2-mph improvement. Dave calculates the flow rate of his modified two-barrel at over 700 cfm. And more good news. By the time you read this, Braswell's new "monster" two-barrel should be available to the racers.

So the next time you go the drags, don't laugh at the guy with the two-barrel carb. Unless you've got a really good-running four-barrel car, he just might blow your doors off. Here's two ya! ■■

BELOW—Although buried beneath hood scoop cool air seals, velocity stacks and air flow improvers, this Holley 650-cfm two-barrel is the workhorse carb on C.J.'s 3-GA Chevelle.

BELOW—Equal in popularity and performance to the Holley 650 carb is the Braswell modified 500-cfm Holley two-barrel, shown here on Ray Montoya's 3-FA '71 Dodge 383.

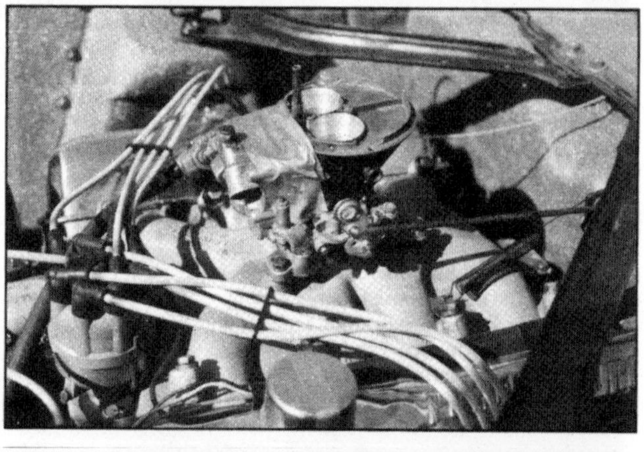

ABOVE—And this is the latest and fastest of the Braswell two-barrels. Dave Braswell, shown adjusting float level, claims the flow rate is over 700 cfm on this one. Performance improvements substantiate the claim (see text). ABOVE RIGHT—Since all carb modifications are legal, Jon Kaase of Westlake, Ohio, fabricated his own annular-discharge two-barrel from a Holley 500-cfm model. Jon claims this carb is worth .3-second over any other two-barrel on his 3-B Mustang. RIGHT—In the Hot Rod class it's legal to cut a four-barrel in half to make a big two-barrel, and that's just what you're looking at on Sammy Crissman's small-block Camaro. This carb is half of a 4500 Holley Dominator model. Jerry Gigot performed the dissection.

CARBURETORS WITH ESP

If what you're looking for is Emissions Street Performance from your car, van, pickup or motor home, these Holleys have it

By Jim McCraw ■ It's been a long time coming, but the people at Holley Carburetor Company have released a whole new line of carbs with their own brand of ESP: emissions street performance. These new carbs get better readings on pollutant detection equipment; they provide more power and a higher level of drivability and response than previous street models; and the new family is available in a wide range of models and cfm ratings. They're intended for street use on cars, RVs, pickups and towing rigs for most car and truck manufacturers.

Members of the new ESP family have been recalibrated and re-engineered to be as efficient and as "universal" as possible, and the two best examples of this are the new 650 and 750 double-pumper models, the 6708 and 6709. Both bodies use 1½-inch primary and 1¾-inch secondary throttle bores, with different venturi sizes accounting for cfm ratings. But both models have a whole list of modifications: at least two additional vacuum ports for evaporative canister purging and power brake boosters; a kickdown lever arm for use on Ford cars with automatics; evaporative emission provision on the primary side for Chrysler products; a universal throttle lever arm; manual choke provision; race-type fuel bowls; split accelerator pump sizing with a 30cc pump on the primary side and a 50cc "Reo" pump on the secondaries; higher capacity, single-stage power valves that open at the same vacuum signal levels as earlier models but flow more; universal solenoid bracket for late-model cars using electric throttle positioners; cast-in boosters, integral with the main body, that avoid "pullover" or dumping of extra fuel due to stray vacuum signals around the booster area; an idle transfer system; revised secondary throttle linkage with fewer pieces; an idle mixture adjustment feature that controls *air bleed,* rather than *fuel feed* (you back the screw out to adjust idle mixture rather than turning it in, and these models have a tag attached to remind the user of the reversed rotation). The smaller capacity 6708 is recommended for 327-400-cubic-inch engines without EGR, and the 750-cfm 6709 is recommended for 340-455-cubic-inch engines, also without EGR systems. These models, like the others we'll be talking about, have been listed in the latest Holley literature but have not, in fact, been available through speed shop channels until just recently.

Probably the second most popular of the new ESP carbs from Holley will be the "universal" 780-cfm vacuum-secondary model called the 7010. The 7010 carries almost all of the same fea-

ABOVE—A brand-new entry into the ever-expanding line of Holley carbs for street applications is the 7010, a "universal" vacuum-secondary carb with an improved diaphragm, center-inlet bowls, electric choke provision and a gang of features designed to make it more drivable and more economical, even with a 780-cfm rating. BELOW—For Chrysler RV and street cars of 318–413 inches, the 6909 600-cfm has similar internal improvements, too.

tures as the double-pumper mechanical-linkage models, except that the secondary opening is triggered by an improved vacuum diaphragm and spring setup. It has center-inlet bowls, electric choke operation, and is recommended by Holley for use on 350-455-cubic-inch engines without factory EGR systems.

The largest sub-group within the new ESP family consists of three variations on the "old" 6619 replacement carb for RV equipment that have caught on like wildfire. The new 600-cfm 6619 model has an improved quick-off choke assembly; transfer tubes from the bowls, fitted with O-rings for improved high-temperature operation; universal throttle linkage and solenoid bracket for throttle stops; and basic calibration changes for improved performance and drivability. The new 6619 is targeted for '70–'72 Chevrolet 307, 350 and 400 small-block engines in RVs, pickups and towing rigs without EGR. Its companion model is the 6979, for '73–'74 350 and 454 Chevrolet engines *with* EGR systems, and is also a 600-cfm.

The next member of the four-digit clan that's new is the 6909, and like the others in this sub-group, it is a 600-cfm unit with $1\frac{9}{16}$-inch throttle bores all around, with vacuum secondaries. It's calibrated for '70–'72 Chrysler RV wedge engines from 318 to 413 cubic inches and has the special Chrysler evaporative emission valve on the primary bowl, universal throttle linkage and solenoid bracket, electric choke, a reworked choke assembly for quick warm-up and a kickdown lever attached. The 6909 uses side-pivot float attachment and a single-inlet fuel bowl, and is not intended for EGR applications. The companion unit for this series is the 7009, which is calibrated for 340-through-440 Chrysler wedges of '73–'74 using exhaust gas recirculation systems.

The Ford-oriented member of the family is the 6919, for use on non-EGR '70–'72 RV engines from 302 to 390 cubic inches. It differs slightly from the others in that it has a special two-stage power valve, along with the other improvements. The EGR model, for '73–'74 302 engines with EGR, is the 6989. All members of this particular family use metering blocks on the primary sides and metering plates on the secondaries.

So one of the largest re-engineering and research programs in Holley's recent history has culminated in this line of emissions street performance carburetors with an eye to clean-running, power-producing, smooth street performance in everything from ponycars to motor homes; and as new models are added for specific applications, we'll bring you the new numbers and specs. ■■

TOP—The 6708 on the left and the 4777 on the right are the street and track models of the same carb, with the 6708 having "universal" hardware. ABOVE—4165 spread-bore, 6708 double-pumper and 4781 race carb.

ABOVE—The Holley 6619 was originally designed for 350 Chevrolet street car and RV engines, but has now been improved for wider use and has spawned a whole new series of carbs for late models with and without the EGR feature.

QUICK-CHANGE CARB KITS

A whole new line of kits from Holley enables the shadetree mechanic to remove, clean and completely recalibrate any popular model for increased performance

By Jim McCraw ■ If there is one single area of high-performance tuning and modification that is generally regarded as a branch of alchemy or black arts, it is probably carburetor rebuilding and recalibration. The carburetor of a modern automobile engine has nearly as many moving parts as the rest of the engine, and probably more internal passages than the block and heads have for oil and water. Many enthusiasts who are involved in building a street machine or race car simply unbolt the carburetor when the time comes, take it to the neighborhood expert, along with some money, and go back a few days later to pick up their reworked piece, which may or may not yield better performance than when it was stock. But that doesn't have to be the case anymore. Holley Carburetor Company recently released a whole new series of "Trick Kits" for their two-, three- and four-barrel carburetor families, and these kits go far beyond anything previously available in offering the enthusiast more performance from his existing carburetion system. While some of the components in the kits have been available through Holley and its dealers, the Trick Kits combine a standard rebuilding kit with the pieces necessary for high-performance modifications *and* a set of instructions and tips.

In the kit shown in the photograph, there are 60 individual pieces, and in many cases, there will be parts left over after a complete rebuild and modification flog, since the kits are somewhat "universal" by design and cover many different members of the Holley carburetor family under the following list numbers: 2300, 3160, 4150, 4160, 4165 and 4500. If you are familiar with the family, you can see that every one of the most popular carburetors is covered.

The instructions supplied with the Trick Kits are long, extremely well-detailed, even humorous in some instances; altogether some of the best and clearest instructions we've come across on the subject of carburetion. The kit user is instructed to remove, disassemble, clean and reassemble the carb with the provided pieces, reinstall it, check its operation in stock form, and *then* begin the step-by-step modifications. Pieces included with our kit, other than stock gaskets and washers, included two 110 Viton needle-and-seat assemblies, four different diaphragm springs for adjusting secondary opening, two "65" power valves, three accelerator pump cams, three different "shooter assemblies," secondary jet slosh tubes, accelerator pump diaphragm, plastic "whistle" assembly for the primary vent tube, and a brass vent assembly for secondaries. These pieces, the stock gasket and washer sets, and the all-important instructions can get even the rankest amateur off to a better understanding of his Holley carb, what makes it tick, and how to make it tick to quicker times. ■■

TWO-BARREL PLACEMENT SCIENCE

If someone showed you how to gain as much as 40 hp by moving your carburetor about an inch, would you do it?

DYNAMOMETER COMPARISON OF CARBURETOR PLACEMENT

Engine: 302 Chevy (Performance Modified)
Manifold: Edelbrock Tarantula
Carburetor: 500-cfm Holley Two-Barrel

RPM	Horsepower with center carburetor placement	Horsepower with forward carburetor placement	Net Gain
5000	255	270	+15
5500	275	298	+23
6000	273	313	+40
6500	290	312	+22
7000	298	305	+ 7
7500	279	284	+ 5
8000	247	278	+31

Source: Edelbrock Equipment Company

Text & photos by C.J. Baker ■ It's not uncommon to walk through the pits at AHRA and IHRA drag races and find numerous stockers equipped with two-barrel carbs. At first this seems to be a contradiction. Why would a racer put a seemingly restrictive two-barrel on a fully prepared race engine? The answer is two-fold. First, both AHRA and IHRA offer special Formula 3 classes for two-barrel carbureted engines, and second, the current Holley 500-cfm and 650-cfm two-barrel carbs aren't as restrictive as you might think (see "Double Barreled Dynamite," March '73 HRM). These carbs work very well, even on highly modified engines.

Since the Formula 3 rules permit the use of aftermarket manifolds, and because the manifold manufacturers don't offer specific two-barrel performance mani-

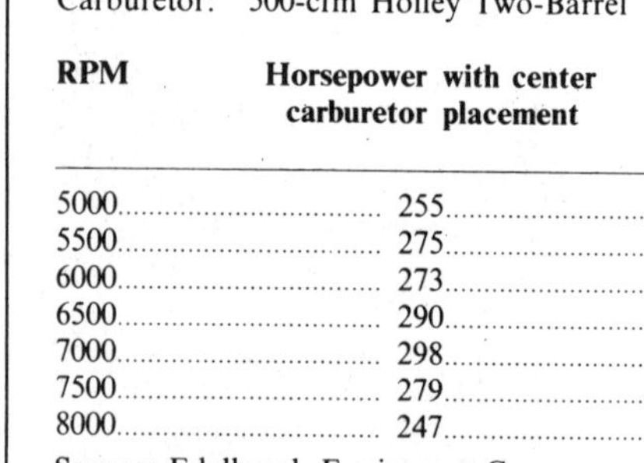

Edelbrock Torker & Tarantula-like 2-bbl positioned in same location 4-bbl would occupy. Eelco makes a forward-placement adapter plate of this type.

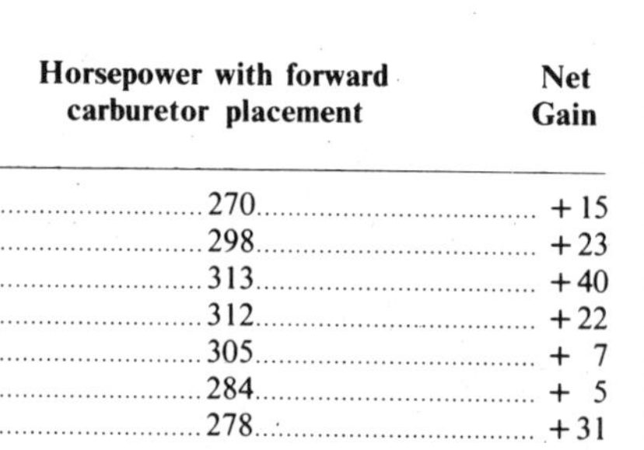

Offenhauser Port-O-Sonic works well with carb positioned just slightly forward of plenum centerline. Offy makes adapter to locate the carb.

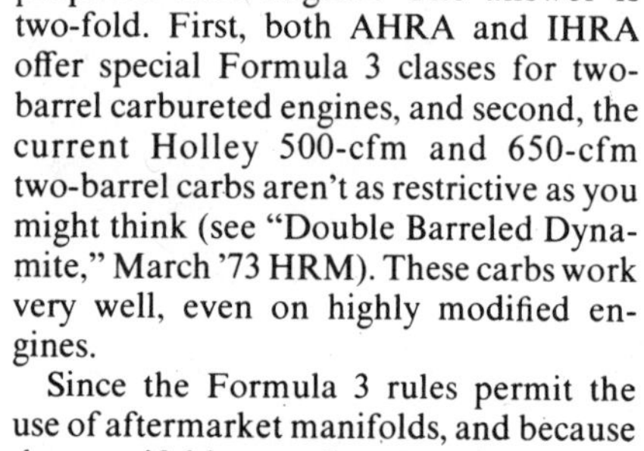

Weiand X-terminator manifold seems to favor placement of carb in exact middle of plenum opening. Eelco also makes an adapter for this placement.

folds, most competitors use a single-plane four-barrel manifold, such as the Edelbrock Torker or Tarantula, the Offenhauser Port-O-Sonic or the Weiand X-terminator, with a two-barrel adapter. At first glance, the obvious place to locate the two-barrel carb would be directly over the middle of the manifold plenum opening, but that may or may not be correct (see chart on Edelbrock dyno tests). In some cases, depending on manifold type or brand and the degree of engine modification, moving the carburetor forward can result in fantastic power gains on both street-driven and strip-only cars. Here's why.

On street-driven cars, the manifold sees a lot of partial-throttle operation. During partial-throttle operation, the throttle plates of the carburetor are at an angle which tends to divert the carburetor flow in a direction behind the actual placement of the carb. This rearward flow will cause the rear cylinders to run rich, while the front cylinders will be starved if the carburetor is centrally located over the plenum on some single-plane manifolds. Relocating the carburetor toward the front of the plenum lets the flow effectively enter the middle of the plenum chamber and correct distribution can be restored. Since the amount of forward relocation varies with different manifolds and different engines, slot the bolt holes in the adapter and experiment until optimum performance is obtained.

Now that we've covered partial-throttle operation, let's consider wide-open-throttle (WOT) flow, such as encountered with a drag strip car. Here again it would seem like central placement of the carb would be ideal for equal WOT distribution, since the throttle plates aren't diverting the flow. And in fact, on some manifolds, such as the Weiand X-terminator, central placement seems to be the answer. But because of differences in runner shape, runner angles and plenum configuration, another problem, reversion, can upset WOT flow from a centrally mounted carburetor. This is especially true with the Edelbrock single-plane manifolds, in which the reversion phenomenon seems to be at its greatest in the middle of the plenum chamber. Simply relocating the carburetor in front of this reversion concentration allows the carburetor to flow a greater volume of air and fuel without disruption.

Shown here are photographs of each of the three major manufacturers' single-plane manifolds, with a Holley performance two-barrel mounted in its recommended position. But it must be emphasized that these locations are only suggested starting points and that adjustments in location should be tried to optimize the performance of your particular combination. And incidentally, these same principles frequently have an appreciable effect on four-barrel carbs too. So if you're looking for a little more power (aren't we all?), give it a try. You might be pleasantly surprised. ■■

DOUBLE BARRELED DYNAMITE

KA-BOOM!

Holley's got a new 650-cfm two-barrel. Here's how it compares to a couple of Holley's other popular performance carbs and a stock Quadrajet

Text & photos by C.J. Baker ■ When we first heard about Holley's 650-cfm annular-discharge-venturi two-barrel carburetor, the R-6425AAA, the comment was made, "Why that thing must be almost as good as a four-barrel." And so the idea for this article was born. If the annular-venturi two-barrel is almost equal to a four-barrel, let's compare it to a popular-sized performance four-barrel, such as the R-4780AAA 800-cfm double-pumper, and see just how closely it performs. Let's also try an R-4412AAS 500-cfm two-barrel to see just how much better this new unit really is.

Then the idea really began to snowball. Since we had decided to use a nearly stock 325-hp 396 Chevelle for the tests, it would be necessary to replace the stock Quadrajet carb and spread-bore manifold with one that's compatible with the two- and four-barrel Holley flange patterns. Conveniently, Edelbrock had just completed the development of their new Torker manifold for oval-port big-block Chevys. So why not test the performance gain offered by the Torker manifold, as well as that of the carburetors, over the stock setup? The possibilities were becoming endless. Then, at the last minute, Holley came out with still another carb, the R-6619AAA 600-cfm four-barrel, which is designed for good emission characteristics. Since the emission thing has become such an important part of the automotive scene, why not test this new four-barrel too?

So, loaded with the Torker manifold and an armload of carburetors, we set out for the strip. The '68 Chevelle was selected as our test car for several reasons. First of all, it had an automatic transmission which, left in Drive, would permit consistent runs with exactly the same shift points (5000 rpm) being used each time. In this way we would be sure that differences in the elapsed times would reflect the performance of the carburetors and manifold rather than the driver. Second, the 325-hp 396 was big enough to use the amount of air that the carburetors were capable of flowing, with the single exception of the 800-cfm unit. Additionally, this engine, in combination with a 3.31:1 rearend ratio, did not result in a car that was fast enough to have tire slippage problems from run to run. And third, the car was available for the tests.

Naturally the first runs were made using the stock cast-iron manifold and the Quadrajet carb, producing a best e.t. of 14.58 seconds at 98.18 mph. During these runs we noticed a slight amount of rear wheel hop when leaving the line. Although this condition wasn't severe enough to detract from the car's performance, we did feel that the condition would worsen with the anticipated power gains of the new carburetors and manifold. Consequently, we

ABOVE— Stock cast-iron Q' jet manifold was replaced with Edelbrock's latest oval-port Torker to accept the Holley carbs.

contacted Jim Thomas at Genuine Suspension in Costa Mesa, California, for a set of his No-Hopp traction bars, which eliminated the problem. The e.t. remained the same after the installation of the bars.

Next we installed the Torker manifold, using an adapter to mate the Q'jet carb to the standard flange bolt pattern of the manifold. Jim McFarland, the guiding hand behind the development of the Edelbrock Torker, cautioned us that the single-plane Torker manifold might not show any improvement when used in conjunction with a spread-bore carb, such as the Quadrajet, due to the inherent distribution imbalance caused by the huge secondaries and small primaries. But even with the distribution imbalance, the new combination cranked out a 14.47 at 98.38 mph, more than a tenth faster than the stock manifold. Later testing by Edelbrock has shown that enrichening the primaries approximately eight to ten percent and leaning the secondaries by a similar amount restores mixture distribution at wide-open throttle for drag strip applications, but these mods also destroy the streetability of the Q'jet.

Now that we had baseline test figures for the stock setup and the new manifold, we were ready to test the Holley carbs. To mount the two-barrel carbs on the Torker manifold, it was necessary to make an adapter plate. Although there are aftermarket adapters available, they are one inch thick, which adds to the effective size of the manifold plenum chamber. This additional plenum volume hurts low-speed throttle response and upsets some of the designed flow characteristics of the

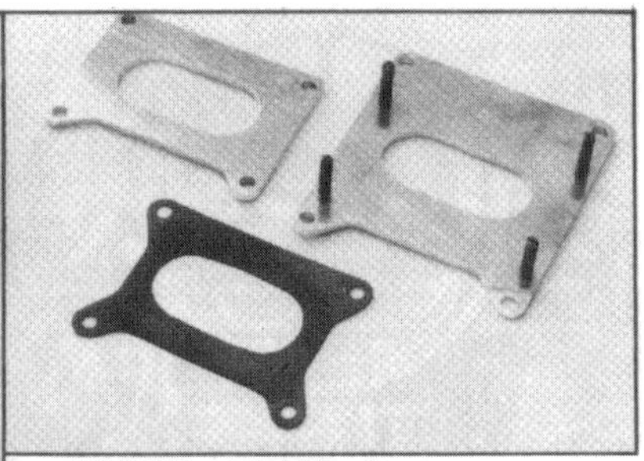

ABOVE—A 2-bbl adapter plate was made from two pieces of ¼-inch aluminum plate. Gaskets were used as patterns.

manifold. The one-inch addition in carburetor height would also have created hood clearance problems on our test car.

We made our adapter, using carburetor flange gaskets as a pattern. It was necessary to cut two ¼-inch aluminum plates, as shown, to provide clearance for the 50cc accelerator pump while centrally locating the two-barrel carb over the manifold opening. Even using two plates, it was still necessary to grind away a portion of the left front bolt tab and the head of the bolt, along with notching the accelerator pump arm, as shown, to get adequate

(Continued)

ABOVE—These carbs, ranging from the stock Q' jet on the left to the 500- and 650-cfm 2-bbls in the center and the 800-cfm 4-bbl on the right, were tested on the Torker manifold with the appropriate adapters and the AC open-element air cleaner. LEFT—The 50cc Reo accelerator pump and arm on the 2-bbl carbs caused clearance problems on the adapter plate. The problem was cured by filing away half the thickness of the bolt tab, half the thickness of the bolt head and by notching the pump arm (see detail shot below). Sender unit in the fuel line (center foreground) monitors fuel pressure.

clearance. This resulted in a finished adapter plate that was only ½-inch thick, with a minimum increase in plenum volume.

With the adapter plate completed and the 500-cfm two-barrel in place, we made another series of runs. The result was a 14.59 at 96.69, only .01-second slower than the stock setup. Of course the mph was down, due to restricted top end breathing, but this gives an indication of the fantastic throttle response and torque increase, as the car left the line and in the mid-range.

Now for the super two-barrel. This new 650-cfm annular-discharge-venturi carb is unlike anything Holley has previously offered. The annular-discharge feature means that rather than introducing the fuel into the middle of each throat via a small tube and booster rings in the conventional manner, this carb has a series of small holes drilled around the inner circumference of the venturi through which fuel is introduced into the airstream. Although an annular-discharge venturi is considerably more expensive to manufacture, there are two distinct advantages: First, there are no tubes or boosters cluttering up the throats to restrict the airflow; and second, at partial throttle opening the high-speed airflow is close to the throat walls where the fuel discharge ports are located, permitting precise fuel metering at less than peak stream velocities. To further increase the airflow rate, the throttle bores and venturis are 1/16-inch larger in diameter than those of the 500-cfm two-barrel.

We bolted on the 650-cfm carb and fired the engine. Throttle response was every bit as sharp as it had been with the smaller 500-cfm two-barrel, so we were sure the increased flow capacity would result in a performance jump. The clocks confirmed our predictions with a 14.39 at 97.71 mph. That's two tenths and more than one mile per hour faster than the 500-cfm two-barrel. Equally impressive was the feel of the car. It left the line well and pulled strongly all the way through the quarter-mile.

Back to the pits again to bolt on the 800-cfm four-barrel. With the double-pumper in place, we returned to the line. The car initially left hard, but about fifty feet out of the gate the rate of acceleration briefly slackened and then recovered to pull all the way through the lights, turning a 14.24 at 98.88 mph. This kind of performance with a mid-range flat spot is predictable when a mechanical-secondary carburetor, especially one as large as the 800-cfm unit, is used on a near-stock engine with a standard-stall-speed torque convertor (see "How to Select the Right Carb" elsewhere in this issue). The interesting fact was that the popular performance four-barrel ran only .15-second and 1.17 mph faster than the 650-cfm two-barrel. That's not really much faster, considering the differences in the carburetors.

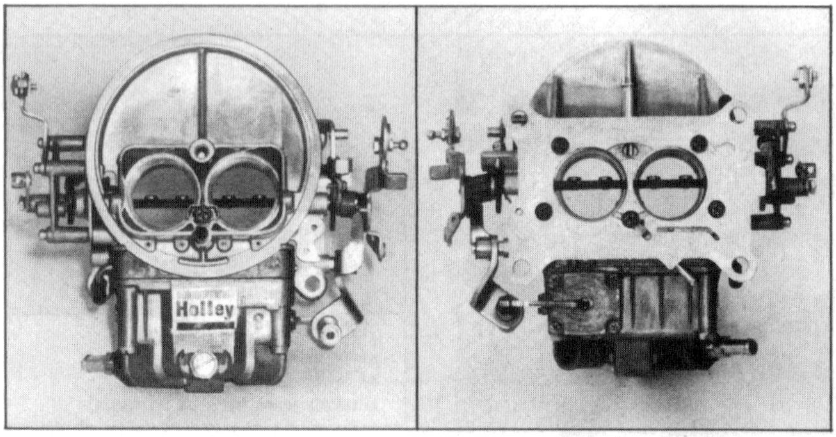

ABOVE—Down-the-throat shots show uncluttered, high-flow, annular-discharge venturis. BELOW—R-6619AAA emissions 4-bbl was fast as well as super-clean.

Our last test was the 600-cfm vacuum secondary emission four-barrel. Frankly, we hadn't expected much from a super-clean emission carb, but we were in for a surprise. With the 600-cfm unit in place, the car recorded its best e.t. and mph of the entire test session: 14.21/99.35. If anything, the car left harder than it had with the two-barrel carbs, and of course there was no flat spot, because the vacuum secondaries didn't open until the engine could use them. But the real eye-opener was the miles per hour. There was no fall-off in the upper rpm ranges, indicating that the 600-cfm flow rating was indeed accurate and that the fuel metering was consistent across the entire operating range.

In conclusion, our tests showed that the new 650-cfm two-barrel is indeed very close in performance potential to a comparable performance four-barrel. The operation of this two-barrel would be even more impressive on a small-block V8 of the 350-cubic-inch variety. So whatever your preference, two-barrel or four-barrel, either the new 650-cfm two-shooter or the 600-cfm super-clean four-barrel will make a dynamite carb for most engines. ■■

PERFORMANCE COMPARISON (325-hp/396-c.i. '68 Chevelle)

Manifold/Carburetor	E.T./MPH
Stock/Quadrajet	14.58/98.18
Torker/Quadrajet	14.47/98.38
Torker/500-cfm 2-bbl	14.59/96.69
Torker/650-cfm 2-bbl	14.39/97.71
Torker/800-cfm 4-bbl	14.24/98.88
Torker/600-cfm 4-bbl	14.21/99.35

CARBURETOR SPECIFICATIONS

CARB	BARRELS	CFM	PRIMARY THROTTLE BORE DIA. (INCHES)	PRIMARY VENTURI DIA. (INCHES)	SECONDARY THROTTLE BORE DIA. (INCHES)	SECONDARY VENTURI DIA. (INCHES)	SECONDARY OPERATION
Rochester Quadrajet	4	750	1 3/8	1 5/32	2 1/4	1 13/16	Mechanical (Air Valve)
Holley R-4412AAS	2	500	1 11/16	1 3/8	–	–	–
Holley R-6425AAA	2	650	1 3/4	1 7/16	–	–	–
Holley R-4780AAA	4	800	1 11/16	1 3/8	1 11/16	1 7/16	Mechanical
Holley R-6619AAA	4	600	1 9/16	1 1/4	1 9/16	1 5/16	Vacuum

CARB MAGIC

The skinny on what to do when running No. 4334 660-cfm carbs on a Modified Production engine

Text & photos by John Dianna ■ Admittedly, there is no magic involved in preparing a set of carbs for Modified Production racing, not even for World Champion Paul Blevins. Paul does, however, perform certain modifications that he feels add to his car's performance. In addition to his noted G/MP Chevy Nomad (11.3 e.t.s), Paul has been running a '66 Corvette. With the car weighing in at 2950 pounds and powered by a carbureted 327 engine, he has shocked the Modified racing fraternity with impressive 10.19 clockings. More recently he backed up this performance by winning the '72 NHRA World Finals, and in the process set a new D/A National Record by turning 10.55 at the 3000-foot-elevation Amarillo strip. This 10.55 e.t. factors out to an impressive 10.23 (the old record stood at 10.47). Paul's other accomplishments include winning Modified Eliminator at the '71 Grandnationals, '71 Supernationals and the '72 Indy Nationals. He has also been the NHRA Division I Points Champion for three consecutive years. Not a bad record, to say the least.

Because Paul is one of the most respected Modified racers, he gets a horde of questions put to him concerning his various combinations and modifications to the car. Probably the most sought-after information is Paul's carburetor recommendations. As a starter, Paul likes using the Holley No. 4224 660-cfm carbs for most engine displacements. He completely disassembles the carbs, and the first item to get attention is the throttle lever cam that actuates the mechanical secondaries. The secondary pickup slot (on the throttle lever) is filed so that the secondary throttle shaft isn't opened until the primaries are nearly half open. This helps eliminate engine stumble while driving through the staging lanes and while staging the car. In its original form, the secondary cam slot is about ½-inch long. The length of

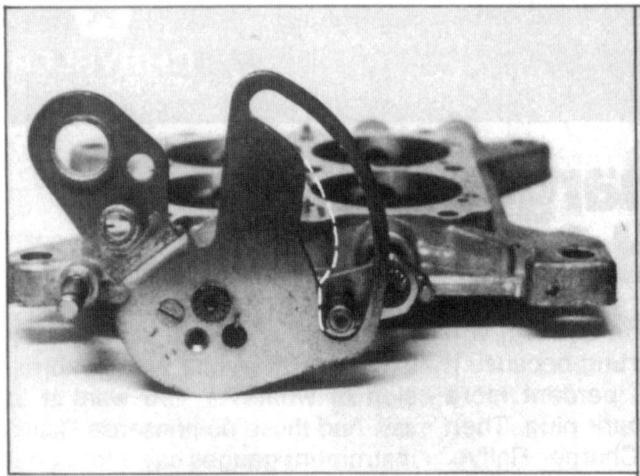

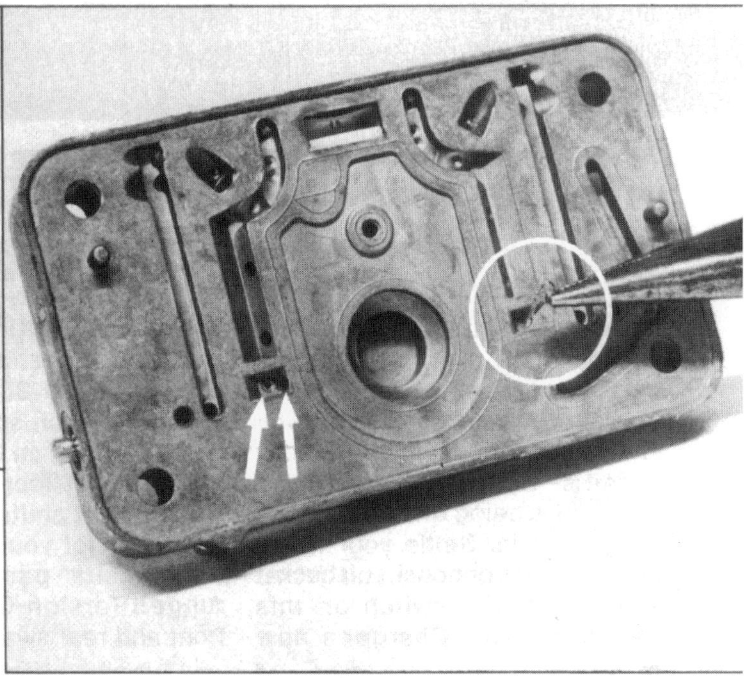

ABOVE—To delay the secondary throttle plate opening, the pickup slot in the throttle lever requires some grinding. The dotted white line is the stock configuration. Notice the amount of material removed. RIGHT—To lean down the idle circuit, .018-inch-diameter wire is inserted in the idle feed restriction bushings (see arrows). The wire should be bent in a V shape (see example in circle).

this slot can be increased to about ⅝-inch for most cars, and even farther for those that see the clutch fly at around 8500 rpm. The slot can be lengthened simply by scribing a line along the bottom of the slot and filing by hand. Depending on the length of the new slot, the straight bottom line may need to be curved to further retard secondary throttle plate opening (see photo).

Although the above throttle lever modification greatly helps the drivability of the carbs, they still should be leaned down in the idle circuit. There are a few ways to go about this. The primary air bleeds can be enlarged, the idle transfer channels in the main body can be reduced in size, and the idle feed restriction bushings in the jet block can be reduced. Because brass carburetor bushings are not readily available to the buying public, the easiest way to lean down the idle system is to reduce the diameter of the existing idle feed restriction bushings (in the primary jet plate) with .018-inch-diameter wire. Paul cuts ½-inch lengths of wire and bends them in a V shape. He then inserts a wire in each of the two restriction bushings. Actually, one end of the wire extends in the bushing and the other into the main well channel. The V shape of the wire and the gasket will keep the wire in place. To lean the idle circuit even further (if a particular application required it), the idle air bleeds can be increased in size by hand drilling. They are already quite large, but can be increased a drill size or two. These mods should help eliminate the loading up and stumbling caused by an overrich idle circuit.

In an effort to equalize air flow between the primaries and secondaries, Paul mills off the air horns at the tops of the carburetors. Care should be taken here, as the material is thin and it is quite possible to break off the air horn lower than you have intentions of milling it. Small careful cuts will get the job done. Once the air horn has been removed, the sharp corners left by the cutter should be rounded with sandpaper or a rotary grinder. Paul also examines all the core joints around

(Continued)

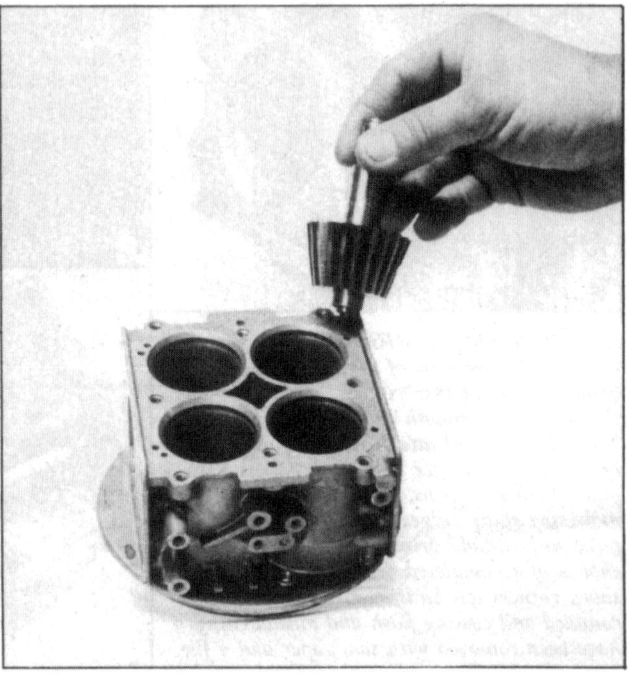

ABOVE—Examples of carb bodies with and without the air horn removed. The removal of the air horn helps equalize air flow between the primaries and secondaries. LEFT—Arrows show the primary and secondary air bleeds. Enlargement of these holes will also lean the idle circuit. Be careful not to get carried away; a slight increase in size will affect the idle mixture. RIGHT—Throttle bores can be blended using valve seat cutter and hand grinder.

CARB MAGIC

the venturis and smooths any mismatching. At the bottom of the venturis there are ridges. These rings (or ridges) can be removed with a 75-degree valve seat cutting tool. The cutter will remove the ridges, but the new bottom angle requires hand blending toward the center of the venturis. If this modification is attempted, be sure not to cut the venturi diameters larger than the 1 11/16-inch base plate holes.

As far as fuel delivery is concerned, the stock .101-inch needles and seats should be replaced with the .110-inch Holley units. Float levels can be adjusted using number drills as we have previously discussed ("Racer by Request," March '72 HRM), or you can make sure the floats are as high as possible by following Paul's recommendations. He mounts the carbs on the side of the manifold (hanging off the side of the manifold flange rather than directly over the plenum) and turns on his electric pump. He backs off the float level until gas just starts leaking out of the idle circuit (this occurs just before the fuel leaks out of the main metering tube). Once the carb starts to leak, Paul shuts off the electric pump, lowers the float level ever so slightly and reduces the volume of fuel in the primary bowls by actuating the accelerator pump a few times. (Secondary fuel level can be lowered by removing the sight plug on the side of the bowl.) He then turns on the electric pump again to make sure the carbs don't leak from too high a fuel level. At this point, the floats are adjusted as high as possible. For an extra margin of safety, it would be wise to lower the levels another 1/16-inch so that you don't experience spillover on hard burnouts.

The accelerator pump delivery is important on a Modified car. In stock form, the carbs come with the pump cam mounted in the No. 1 position. With the cam in this position, the pump squirt is nearly all-in as soon as the secondaries are cracked. To delay or extend the delivery, the cam should be reversed and mounted in the No. 2 position. This will give you a good starting point with the No. 25 discharge nozzle. However, if you require longer delivery, the cam, spring tension and nozzle size will need to be tailored to your individual situation.

A few final tips include grinding the ends of the transfer tubes (tube that mounts between primary and secondary fuel bowls) to ease O-ring installation. Make sure all main body, jet plate, metering plate and cover, and fuel bowl mating surfaces are clean and free of nicks. All gasket sealing surfaces should be in good condition or leaks will occur. Speaking of gaskets, Paul recommends making your own plenum-to-base gaskets out of soft asbestos material (similar to intake gasket material), as this will allow additional crush and aid sealing. Paul also recommends using Permatex Hy-Tak spray sealant on the plenum and the base, and sandwiches these aluminum isolator plates between two gaskets. Paul mounts his carbs sideways, but he doesn't feel it is absolutely necessary for a Modified car. The sideways mounting position helps eliminate unloading the jets on severe acceleration and also allows the installation of a secondary jet block in place of the metering plate. With the carbs in this position, jet changes can be completed without removing the carbs from the manifold. When mounting the carbs in this position, be sure the fuel inlets are at the rear to avoid the chance of the engine loading up from spillover as the floats get slammed shut during braking. Also, the secondary floats are not closed on hard acceleration, as is the case when the carbs are mounted conventionally. The stock No. 76 jets and No. 12 metering plate are recommended for a ballpark jet combination. If jetting changes are necessary, leaner jets will most likely be needed.

Minimum fuel pressure should be 7 pounds at idle and not lower than 5½ or 6 pounds through the lights. Other than making sure you make an adequate throttle linkage (one that won't flex or break), that's about the skinny on Blevins' carb modifications. As you can see, Paul performs no magic; just basic racer sense is all that's used. So if your carbs need a little help, try the Blevins remedy. It sure works for him. ■■

ABOVE & ABOVE RIGHT—This is a before and after view of the throttle bore blending. Notice the "rings" are removed and that there is a smooth transition into the bores. Pay careful attention not to make the bore diameters larger than those in the base plate. If this happens, you will have to remove any sharp ridges in the top of the base plate with a hand grinder. RIGHT—This shot is of a completed 4224 Holley carb. All sharp corners left by the mill have been rounded and casting flash and mismatching have been removed with sandpaper and a file.

How to JET YOUR HOLLEY

Making jet changes can increase performance as well as economy

The most important step in rejetting a carb is being able to tell when you've reached the maximum richness mixture for a given engine. As you probably know, the carburetor manufacturers calibrate their OEM carbs for all-around driving performance and do not tailor individual carbs for specific areas. High-performance aftermarket carbs are calibrated quite differently from the OEM carburetors. Most aftermarket carbs won't require more than a step or two change in jet size (providing you don't live in an area such as "mile-high" Denver, Colorado) for optimum street performance as long as the carb selection meets the requirements of the engine. A common mistake among the street set is to over-carburate (use too large a carburetor) or over-richen (use too large a jet) an engine, causing low-speed flooding, poor throttle response, lower fuel economy and excessive emissions.

Once you've decided to change the jets in your Holley, check the unit for its jet numbers. To do this, remove the carb from the engine (the jets can be easily replaced by leaving the carb on the engine, but care must be taken when removing the float bowls, as fuel will run down on the intake manifold and possibly cause an engine fire) and turn it upside down to drain the gas from the float bowls. Remove the float bowls to expose the jets (the jets are located at the bottom of the metering blocks). Some Holley carbs do not have replaceable secondary jets; instead they have a secondary metering plate. To change the mixture ratio, the plate must either be replaced or drilled-out. The number stamped on the jets is not the actual orifice diameter (drill size); it's a reference number (see chart). When making a jet change, make the jumps small, notating engine performance and spark plug condition. Do this until you hurt engine performance; then go back to the closest jet size where you realized a good power increase and throttle response.

Holley jets are made with a precision-tapered orifice. Care should be taken not to disturb this taper. Be sure to use a screwdriver with a large enough blade so you do not damage the jets.

Once the jets have been replaced, inspect the bowl gaskets and replace if necessary. The single-inlet Holley carbs feed the secondary bowl via a transfer tube. This tube has a tendency to leak if not installed properly. When reassembling the carb, place the two rubber O-rings at the tips of the tube. Insert the tube in one mounted bowl, and when you are installing the second, the O-rings will be compressed into place. When starting the engine, be sure to check for gas leaks around the transfer tube, fuel-line inlets and bowl-to-metering-block parting lines. If you've never had a carb apart, don't be bashful—it's really a simple job. ■■

BELOW LEFT—Changing jets in a Holley carb is a simple task. It requires no special skill or talent, and only the bare necessities in tools. One important step to remember, after you have installed the float bowl and set float level, is to check the needle and seat for gas leakage. Simply shut the motor off and look down the venturis. BELOW RIGHT—The idle mixture is set with a screwdriver with the engine idling.

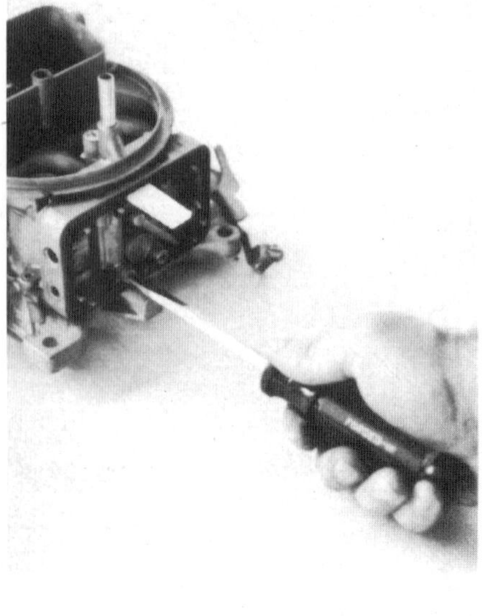

HOLLEY MAIN METERING JETS

JET NO.	DRILL SIZE	JET NO.	DRILL SIZE
45	.045	74	.081
47	.047	75	.082
48	.048	76	.084
49	.048	77	.086
50	.049	78	.089
51	.050	79	.091
52	.052	80	.093
53	.052	81	.093
54	.053	82	.093
55	.054	83	.094
56	.055	84	.099
57	.056	85	.100
58	.057	86	.101
59	.058	87	.103
60	.060	88	.104
61	.060	89	.104
62	.061	90	.104
63	.062	91	.105
64	.064	92	.105
65	.065	93	.105
66	.066	94	.108
67	.068	95	.118
68	.069	96	.118
69	.070	97	.125
70	.073	98	.125
71	.076	99	.125
72	.079	100	.128
73	.079		

• Carburetion is known as one of the keys to performance for the automotive enthusiast. Additional flow capacity, which means more cubic feet per minute (cfm) of fuel and air mixture, is usually necessary on engines where horsepower outputs are greater than stock. Original-equipment carbs are frequently a compromise between easy starting, driveability and performance with the original engine configuration. When higher performance is desired and engine modifications, such as adding a high-lift cam, a free-breathing aluminum manifold or a set of headers, are performed, a larger carb is almost a necessity to allow the other components to work to their maximum.

However, like most other modifications, changes in carburetion must be done with discretion or a loss in performance can result. Many times someone will choose a carb that is too large for the application and then wonder why the engine bogs or is unresponsive. It's a safe bet to say that the biggest problem facing the prospective buyer of a new carb is what size (cfm rating) to buy. This is a very difficult question to answer.

There are many variables that enter into this decision. These include engine size, valve size, cylinder head porting, intake manifold, camshaft, headers and, most important, intended usage. Let's face it: No one wants a "won't-run-below-4500-rpm engine" for the street! Unfortunately for the buyer, the range of engine/carb configurations is almost limitless. For example, a 300-inch V8 in a relatively mild state of tune can easily be overcarbureted by a mechanical-linkage 850-cfm carb, but an all-out 300-inch Trans-Am engine can accept over 2000 cfm!

The answer to this dilemma is not an easy one. Your best bet is to try to learn from the experience of others who have engines similar to yours. Manifold manufacturers are frequently a source of considerable help as a result of their dyno programs. Also, both Holley and Carter have long lists of recommendations for each manifold and engine in their catalogs, which can be seen at your specialty equipment dealer.

Regardless of the carb chosen, it cannot function properly if the remainder of the engine is not in satisfactory condition. The ignition system must be operating without fault and capable of providing adequate spark throughout the rpm range. Cylinder compression must also be high and uniform. No carburetor can function properly if it does not receive sufficient fuel. Some high-performance engine installations need only the addition of a high-capacity mechanical pump, while others demand the addition of an electric pump, or two. A fuel pressure gauge can tell you if you have sufficient fuel-feed capacity. (Check to see that the pressure doesn't fall below 4-5 psi under the high-load and -rpm conditions that would be experienced on the strip.)

In some cases you can bolt on a new carb and have good performance, but more likely than not, the new carburetor will have to be tailored to your engine through changes in jets, accelerator pump settings, vacuum diaphragms, metering rods, etc. Like all good things, carb modifications are not always easy, but they are well worth the trouble for the resulting improvements in power and driving smoothness.

BUYER'S GUIDE: CARTER and HOLLEY CARBURETORS

Increasing your engine's breathability is one of the easiest paths to performance. With the wide range of carbs available, there is sure to be a high-performance model to fit your exact needs

CARTER CARBURETORS The Carter Competition Series of carburetors was specially developed for both high-performance street and strip applications. Included in the series are five AFB four-barrel carbs, which can be used either singly or in pairs, two Thermo • Quad carbs and a special VW carb. All of these carbs have Carter's "metering rod" system of jetting. With this system, a rod with a step in its diameter is positioned in the main jet. Under high-load conditions, the rod is raised, which positions the smaller-diameter section of the tip in the jet, allowing more fuel to pass. Under light-load conditions, high engine vacuum causes the rod to be lowered in the jet and the corresponding larger-diameter section of the rod yields less fuel flow and greater economy.

The metering characteristics of the Carters allow easy adaptation to your engine by changing main metering jets and/or metering rods. Additionally, larger-diameter needle-and-seat assemblies are available, as are other competition-oriented accessories, such as mechanical and electrical fuel pumps, high-capacity filters and gasket kits.

DUNE BUGGY DEUCE

This specially devolped VW carb was designed for both on-road and off-road use. Since it is a two-barrel configuration, an accessory two-bore VW performance manifold must be added at the same time. There are many of these manifolds readily available. Metering changes are made in the same way as they are on the Deuce's bigger brothers; that is, both metering rods and jets are the same as used on the AFBs and Thermo • Quads, so these parts are readily available. Any necessary rod and jet chnges

Carter's Dune Buggy Deuce is designed to fit on VW accessory manifolds.

can be made without removing the carb from the manifold. Other features of the carb include an automatic electric choke which will operate on either six or 12 volts, double booster venturis for fast throttle response, a spark vacuum hookup, and a spring-loaded needle-and-seat assembly for rough terrain.

AFB

The AFBs have been one of the rodder's friends for many years. They were very popular with the Detroit manufacturers back in the pre-emissions-standards days when the additional complexities of the anti-smog measures were not required. The

The AFB can be used singly or in pairs. It was standard equipment on the Hemis.

AFBs are especially well-suited to dual-quad applications because of their small external dimensions, and most of the dual setups from Motown used these units.

The AFBs are available in three cfm ratings: 500, 625 and 750. In dual configuration, the cfm ratings are 1000, 1250, 1375 and 1500. All the competition series AFBs have a $5\frac{1}{8}$-inch-diameter air cleaner ring, a large-capacity accelerator pump, eight mounting holes for narrow or wide mounting patterns and various vacuum taps for PCV and spark pickups. The AFBs are the top of the line for Carter in the standard-bore configuration.

THERMO • QUAD

Thermo • Quad first saw service on the '71 MoPar 340 engines. In the past year, they have grown in popularity and are a desired and suitable replacement for other "spread-bore" carbs, such as the Rochester Quadrajet. The Thermo • Quad can be identified by its black plastic body, which is made of a nearly indestructible phenolic resin. The plastic keeps the fuel as much as 20° cooler than a metal body; this helps both per-

The Thermo • Quad has a simple three-piece construction with a phenolic resin body.

formance and evaporative emissions control. The carburetor is available in two sizes, 850 and 1000 cfm.

Both of the TOs incorporate moderate-sized primary venturis which permit rapid throttle response and good economy around town. The secondaries are a huge $2\frac{1}{4}$ inches in diameter, which really lets the air flow when you get on it. The difference in the cfm ratings is due to a difference in the size of the primary venturis. The secondaries do not use a venturi. Rather there is a spring-loaded flapper valve over the secondary portion of the carb which looks like a choke. Upon demand by the engine and throttle position, the door opens and acts as a venturi. This action permits an unobstructed flow for the air under maximum power conditions and also prevents a bogging or overcarbureted condition to occur when the throttle is suddenly opened.

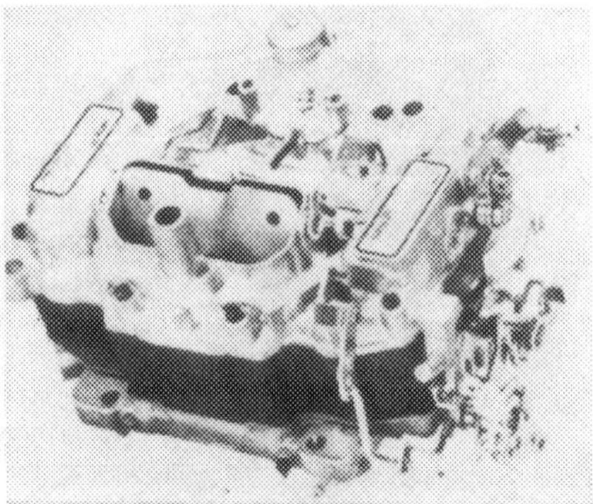

Cutaway of the Thermo • Quad showing the major components and operating parts.

The top section of the carb, known as the air horn, can be removed with the unit still bolted to the manifold. This permits easy changes of rods and jets. Early models had press-in jets, while the latest versions use screw-in types. Another feature of the carb is that the floats are hinged at the rear, which promotes easy opening on acceleration and when the engine has its highest demand for fuel, and easy closing on stopping, which helps eliminate flooding.

The Thermo • Quad, as does the Rochester

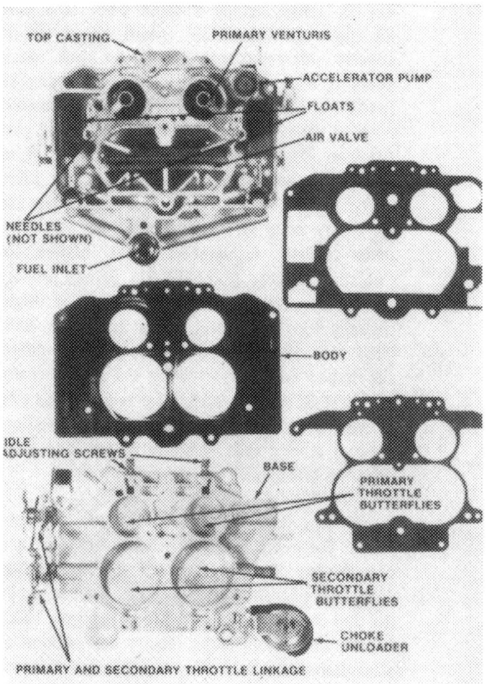

The Thermo • Quad is a spread-bore carb available in 850- and 1000-cfm ratings.

Quadrajet and the Holley 4165, requires a manifold which has the correct spread-bore spacing.

HOLLEY CARBURETORS Holley has been in the performance carburetor field for many years and, as a result, has a carb for almost any application. They have cfm ratings from 200, on their smallest VW carb, to 2300, for a pair of I.R. 4500s. The "modular" design of the high-performance Holleys has permitted the building of a large family of carburetors with the same basic metering system. As a result, almost all of their carbs can benefit from the same tips and trick pieces, a trait which has made them very popular.

The accompanying "blown-apart" diagram is of an R-3310AAS 780-cfm carb, which is typical of all Holley four-barrels. In the center is the main carb body. It may have either mechanical or vacuum-operated secondaries. The metering block is the "brains" of the carb. The primary block contains the main jets, the idle feed holes, the idle mixture screws, the main metering restriction and the power valve (which permits an extra flow of fuel under heavy loads). All com-

(Continued on following page)

BUYER'S GUIDE: CARBURETORS

monly performed changes, on such things as jets, are undertaken on the metering blocks. On the carb ends are the fuel bowls, which can be either the race bowls (pictured) or the "side-hung" bowls, which are adequate for most street applications. The race bowls have a larger needle-and-seat assembly and allow the float to pivot in the center of the bowl, which in turn prevents running lean or flooding during hard cornering or under race conditions.

Holley offers a number of kits which are most helpful in tailoring a specific carb to your engine. These include a center-pivot race bowl kit (parts No. 85R-3539), which allows you to convert side-hung types to center-pivot specs. There is also a secondary diaphragm spring kit (parts No. 85BP-3185) which contains an assortment of springs which will permit you to get your (vacuum-actuated) secondaries open as fast as possible without bogging. Main jet slosh tubes (parts No. 14BP-505) should be installed over the secondary main jets to prevent fuel starvation under severe acceleration. Vent baffles, or "whistles," (parts No. 59BP-91) permit primary fuel bowl venting under hard acceleration. High-capacity 50cc accelerator pumps (parts No. 85R-3354) are available, and an assortment of different-shaped cams for actuating the pumps (parts No. 85BP-3186) is available for tailoring the rate of accelerator pump fuel delivery. In addition, there is a very wide selection of jets and main metering blocks with various-sized orifices.

VW TWO-BARRELS

VW fans have their choice of three different carbs. There are two "Bug Spray" models and one "Buggy Spray" version. The 200-cfm Bug Spray is designed for 1300-1500cc V-Dubs, while the 300-cfm version is suited for the bigger 1500-1600cc mills. Both carbs require the addition of a two-barrel manifold. They are designed to adapt easily to the existing throttle linkage, and their electric chokes are for either six or 12 volts.

The 300-cfm "Buggy Spray" is a relatively new carb from Holley. It incorporates design features which permit proper operation in high angularity conditions as would occur when you take your Baja Bug or buggy out into the boonies. This one needs an air scoop for adequate clearance on standard VWs.

500-CFM TWO-BARREL

An easy way to give your two-barrel-equipped engine a new lease on life is to add one of these R-4412AAS carbs. Very popular in AHRA two-barrel class racing, this carb comes equipped with a center-pivoted float, hand choke fittings and the high-capacity 50cc accelerator pump. It will "bolt onto" Ford and Chevy two-barrel manifolds, as well as many others, with, at most, a simple adapter.

FOUR-BARRELS

Holley offers numerous four-barrel carburetors in the standard-bore configuration. They range from a brand-new 390-cfm unit for Pinto and Vega applications (R-6299AAA) to the 850-cfm units which have become the standard for performance. Generally, for a given cfm rating, there is both a vacuum-operated secondary and a mechanically operated secondary version. The vacuum secondaries permit the incorporation of a larger carburetor for street use without experiencing bogging when the throttle is suddenly opened. "Double-pumper" carbs incorporate mechanically operated secondaries and have two accelerator pumps, versus the single pump on the vacuum-triggered carbs. The double-pumpers are desirable for their fast response under racing conditions. The accelerator pump for the primary side of the carb works as it does on the vacuum carbs; that is, as the throttles open, fuel is squirted into the bores to richen up the suddenly increased flow of air. The double pumpers, unlike the vacuum carbs, have a second accelerator pump system which operates with the secondary throttles. (The vacuum-actuated secondaries compensate for the lack of the second pump.)

Holley's Bug Spray carb is available in 200- and 300-cfm ratings for on-road use.

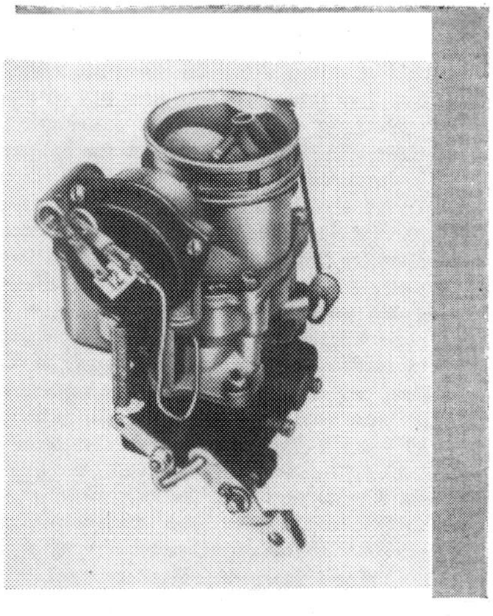

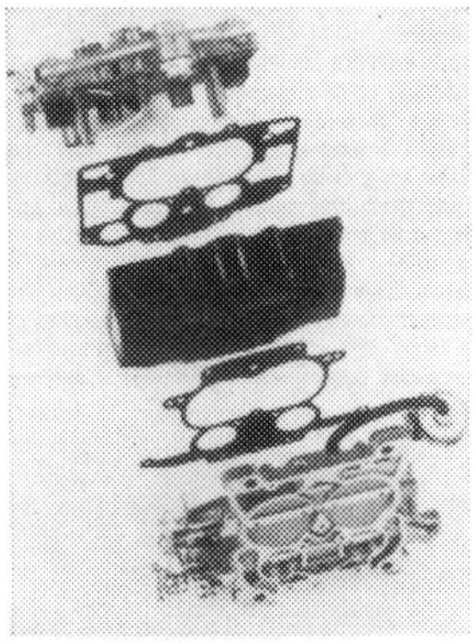

The 3310 is a very popular street carb, can be used on many different engines.

The Holley racing two-barrel has center-hung bowls, removable main jets and a hand-operated choke. Flows 500 cfm.

Twin accelerator pumps (arrows) are the significant feature of Double-Pumpers.

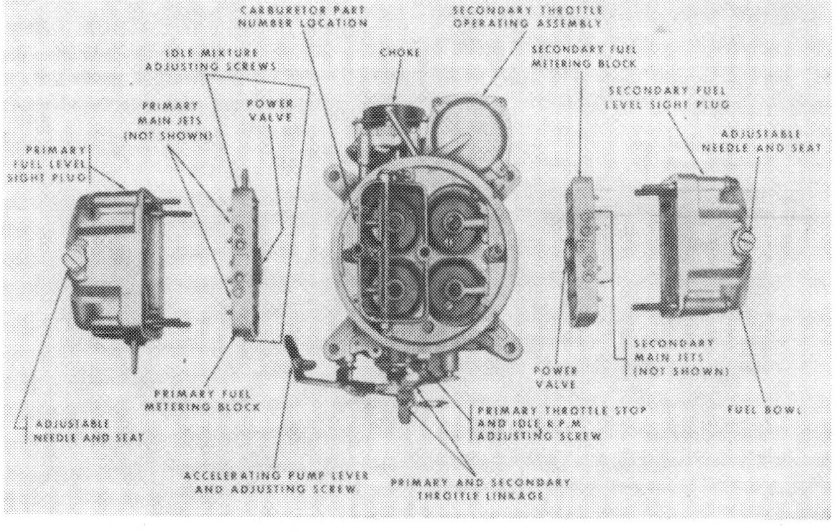

Cutaway view of the R-3310-AAS Holley is representative of all the Holley carbs. This model flows 780 cfm, has an automatic choke and vacuum-operated secondaries.

THE 4165 SPREAD-BORE

Holley introduced the 4165 carb in the winter of 1970 as a bolt-on replacement for Chevy/Rochester carbs. The 650-cfm spread-bore carb is suitable for a replacement on all Chevys through 454s; if you want even more performance, there are several 800-cfm versions, each tailored for a given application. Late news from Holley indicates that selected Pontiac, Buick and Olds engines have been included in the "bolt-on" list for the 4165.

4500

This is the "Big Mama" of the carburetor world. The 4500 has two-inch throttle bores and requires a special manifold. It will flow 1150 cfm, which makes it practical only for all-out racing engines or "moderately hot" big-blocks. The standard 4500 carries parts No. R-4575AAA. The I.R. 4500 is designed for individual runner systems, as used on the Pro Stocks. This one (R-621AAA) has both the primary and secondary throttles opening simultaneously.

TWO FOUR-BARRELS

Recommended dual four-barrel setups from Holley range from a pair of 600-cfm units yielding 1200 cfm to the 2300-cfm Pro Stock individual runner system. The latter uses a pair of the 4500s. Of special interest here is the R-6109AAA 750-cfm double-pumper which has been specially calibrated for use in pairs on "tunnel ram" manifolds.

THREE TWO-BARRELS

For you six-pack fans, the hot setup from the potent 440 MoPars is an R-4782AAA center carb and R-4783AAA outboard carbs. The carbs in this set all use center-hung race bowls and work well with progressive mechanical linkage. ■ ■

The R-6299-AAA is a new 390-cfm four-barrel for Pinto and Vega applications.

Chrome ram tubes are an accessory for the big 1150-cfm Holley Dominator.

CARTER SPECIFICATIONS

CARB NO.	CFM	VENTURI DIA. PRI.	VENTURI DIA. SEC.	THROTTLE-BORE DIAMETER PRI.	THROTTLE-BORE DIAMETER SEC.	CHOKE	NAME
DUNE BUGGY DEUCE							
CS 6123	255	1 1/16	—	1 5/16	—	Elect.	Deuce
AFB CARBURETORS							
CS 4758	500	1 3/16	1 1/4	1 7/16	1 11/16	Man.	AFB
CS 4761	500	1 3/16	1 1/4	1 7/16	1 11/16	None	AFB
CS 4759	625	1 3/16	1 9/16	1 7/16	1 11/16	Man.	AFB
CS 4760	750	1 7/16	1 9/16	1 11/16	1 11/16	Man.	AFB
CS 4762	750	1 7/16	1 9/16	1 11/16	1 11/16	None	AFB
THERMO QUAD CARBURETORS							
CS 4846	850	1 3/16	2 1/4	1 1/2	2 1/4	Man.	Thermo Quad
CS 4847	1000	1 3/8	2 1/4	1 1/2	2 1/4	Man.	Thermo Quad
AFB DUAL QUAD COMBINATIONS							
CS 4761 (front)	1000	1 3/16	1 1/4	1 7/16	1 11/16	None	AFB
CS 4758 (rear)		1 3/16	1 1/4	1 7/16	1 11/16	Man.	AFB
CS 4762 (front)	1250	1 7/16	1 9/16	1 11/16	1 11/16	None	AFB
CS 4758 (rear)		1 3/16	1 1/4	1 7/16	1 11/16	Man.	AFB
CS 4762 (front)	1375	1 7/16	1 9/16	1 11/16	1 11/16	None	AFB
CS 4759 (rear)		1 3/16	1 9/16	1 7/16	1 11/16	Man.	AFB
CS 4762 (front)	1500	1 7/16	1 9/16	1 11/16	1 11/16	None	AFB
CS 4760 (rear)		1 7/16	1 9/16	1 11/16	1 11/16	Man.	AFB

HOLLEY SPECIFICATIONS

CARB NO.	CFM	VENTURI DIA. PRI.	VENTURI DIA. SEC.	THROTTLE-BORE DIAMETER PRI.	THROTTLE-BORE DIAMETER SEC.	CHOKE	NAME
VW CARBS							
R-6244-AAA	200	1 5/16	—	1 7/16	—	Elect.	Bug Spray
R-4691-1AAS	300	1 5/32	—	1 7/16	—	Elect.	Bug Spray
R-6113-AAA	300	1 5/32	—	1 7/16	—	None	Buggy Spray
TWO-BARREL							
R-4412-AAS	500	1 3/8	—	1 11/16	—	Man.	Little Pro Stock
SINGLE FOUR-BARREL							
R-2818-1AAS	600	1 1/4	1 5/16	1 9/16	1 9/16	Auto.	Vacuum
R-4776-AAA	600	1 1/4	1 5/16	1 9/16	1 9/16	Man.	Double-Pumper
R-4777-AAA	650	1 1/4	1 5/16	1 11/16	1 11/16	Man.	Double-Pumper
R-4778-AAA	700	1 5/16	1 3/8	1 11/16	1 11/16	Man.	Double-Pumper
R-4118-AAS	715	1 5/16	1 3/8	1 11/16	1 11/16	Auto.	Vacuum
R-4779-AAA	750	1 3/8	1 3/8	1 11/16	1 11/16	Man.	Double-Pumper
R-3310-AAS	780	1 3/8	1 7/16	1 11/16	1 11/16	Auto.	Vacuum
R-4780-AAA	800	1 3/8	1 7/16	1 11/16	1 11/16	Man.	Double-Pumper
R-4781-AAA	850	1 9/16	1 9/16	1 11/16	1 11/16	Man.	Double-Pumper
SPREAD-BORE							
R-6210-AAA	600	1 5/32	1 3/8	1 3/8	2	Auto.	Spread-Bore
R-6212-AAA	800	1 5/32	1 23/32	1 3/8	2	Man.	Spread-Bore
R-6213-AAA	800	1 5/32	1 23/32	1 3/8	2	Auto.	Spread-Bore
4500							
R-4575-AAA	1150	1 11/16	1 11/16	2	2	Man.	Dominator
R-6214-AAA	1150	1 11/16	1 11/16	2	2	Man.	Individual-Runner
DUAL FOUR-BARREL							
R-1850-AAS	600 & 600	1 1/4	1 5/16	1 9/16	1 9/16	Auto.	Vacuum
R-4224-AAS	600 & 660	1 1/4	1 5/16	1 11/16	1 11/16	None	Center-Pump
R-3977-AAS	730 & 730	1 5/16	1 3/8	1 11/16	1 11/16	None	Ram-Log
R-6109-AAA	750 & 750	1 3/8	1 3/8	1 11/16	1 11/16	None	Double-Pumper
R-4543-AAS	850 & 850	1 9/16	1 9/16	1 3/4	1 3/4	None	Mechanical Sec.

MONS

HOLLEY'S FOUR-HOLE

by Steve Kelly

It has gone over 190 on a stocker

It was called everything from "trumpet mouth" to "mystery carburetor" at its debut at Daytona in February, but it was simply Holley's new 4500 series carburetor. The new 4-bbl helped win the race, coming in with Lee Roy Yarbrough on his first-place Ford, and it made the race a 100 percent Holley affair. No other carbs were used there. But the new long and low fuel passer never did get a name. There'll be a name soon, the result of a contest now being conducted by Holley. First prize is the only 4500 series carburetor to be released to the public this year. All other over-the-counter procuring will have to wait until January of 1970. That's how rare this new unit is, although factory race car use will be seen throughout 1969.

A short description of the carburetor takes in the design parameters that were outlined at the outset of its design project. These are: A. Four barrels arranged in normal configuration. B. 1.687-inch outlet diameters. C. Maximum practical airflow. D. Flange-to-flange height not to exceed 3.25 inches. E. No restriction on length and width.

The project was instituted by Ford, in conjunction with Holley, for their 1969 racing program, which also took in the new 429 Ford engine. That engine was supposed to have made its debut at Riverside, and there was a heavy "win Riverside" pitch all through 1968. Well, the engine and the carburetor didn't make Riverside, and only the carb got to Daytona legally. The 429 was used at Atlanta in March. New rules for 1969 allowed the use of only a single four-barrel carburetor on Grand National cars, rather than a pair of them as previously used. The objective in designing the new four-holer was to equal the performance of a dual setup with the single one as near as possible. Combined with new intake manifolding and some plain and fancy thinking, the die-hard 427 Ford tunnel-port engines even saw an increase in power at Daytona this year, with less carburetion. The new carburetor can be used in a dual four-barrel setup, as was done on the Bud Moore-built Mustang at Daytona's Trans-Am race in February. It's flexible and highly effective as a racing fuel inducer.

The center body is cast aluminum, and air passages are fully machined. Two types of center sections are used, one having no choke facilities, and the other with a manual choke assembly on top. The airhornless and unchoked model does look strange, lending itself to the "trumpet mouth" tag attached to it. The uppermost point on this one is 3.25 inches from the bottom flange point, just as specified in the design layout. The street model is 5.50 inches from lower flange to choke plate projection (choke open). The air cleaner flange diameter is 7.25 inches. Finding an air cleaner for this job will be somewhat easier than finding a manifold for it to fit, which means it's a good thing for manifold builders that the 4500 won't be out for at least six months; they can use the time to make up new intakes.

Bore spacing is healthy. Each of the two-inch throttle bores is separated by 2.75 inches, measured center to center, in any direction — forward, backward, or sideways. That caused some manifold redesign for the 427's, which was handled without difficulty by Ford metalworkers. The low height and large flow were the areas of concern, and these were the first orders of business accomplished for the project. While bore measurements are two inches normally, they have to be reduced to $1^{11}/_{16}$ by restrictor rings fitted in the bottom of the bores for legal NASCAR use. Throttle plates are therefore high in the air passages to allow fitting the restrictors. And when the restrictors are in place, the carburetor loses its venturi effect. Upper bore measurement is 2 inches, and center throat (venturi) diameter is $1^{11}/_{16}$ inches, the same as the outlet diameter with NASCAR rings installed. This makes it almost impossible to get a decent idle but does provide a Weber-like fuel delivery at high engine speed.

Fuel flow potential here is 1300 cfm. Presently the carburetor is flowing 960 cfm on stock cars, and 1100 cfm on dual-four

TER

won the first race it ran, and will deliver more gas than anything short of a service station

TOP — Flip side of 4500 Holley shows restrictor rings fitted for NASCAR use. Throttle plates are set high in bore. Actuating linkage is between right and left sides, thoroughly sealing assembly from grit and abuse when carb is in place. ABOVE — This is competition model without choke plate and air horn. Top flange diameter is 7¼ inches. LEFT — Striking illustration of difference between conventional 4-bbl carb bore-spacing and 4500 is seen here. Plexiglass overlay is size of standard shape. Manifold has been increased to 4500 base plate. BOTTOM — Here's the "street" model bolted to a 390 Mustang. Choke is manual, cable-operated. Carb is low, but over a foot long. Center section is aluminum.

carbed Trans-Am machines. This would make its application to a street-operated engine somewhat less than practical, unless that engine was a real heart-throbber.

Every facet of the new carburetor is super-duty. Throttle rods are a full half-inch in diameter, and the throttle plates are .090-inch-thick aluminum. The rods are slotted to accept the plates, and also slabbed down on each side of the plate to lessen air-flow restriction caused by rod projection into the air stream. The shafts are offset .030-inch to close under vacuum loading. Primary and secondary openings are the reverse of each other, which is conventional. The rods rest in sintered bronze bearings within the carburetor body, and shaft ends are sealed in the carb body. Throttle actuating levers are stamped from .156-inch (5/32) thick steel. All operating levers and rods are necessarily tough to prevent any bending or breaking under hard use. There have been more than a few instances of a driver pushing too hard on the throttle and breaking the linkage.

Throttle opening is by cam and roller method, with the mechanical linkage entirely buried in the center of the carburetor, between right- and left-side bores. This reduces the amount of grime that can deposit on the linkage and also puts the cam and roller movement in a central operating position rather than off to one extreme side. The new position reduces strain slightly. The throttles have a total movement of 80 degrees, from 40 degrees ahead of vertical at idle to 40 degrees aft of vertical at wide open. The secondaries are engaged at 30 degrees of primary staging or movement. Primary and secondary openings are simultaneous from that point.

The accelerating pumps are the cam-operated diaphragm-type, discharging into all four-barrels such as on the new Centra-Flow Holley carbs. The main discharge nozzles are hung over the air sections on tubular bars. The nozzles are machined brass boosters, soldered to the small tubes after assembly. The suspension tubes also serve as horizontal discharge channels, supplying fuel to the boosters.

Fuel floats are side-inlet, center-pivot-float race bowls. They're the same as those found in Holley Kit 85R-2937. They've been proven good, and there's no sense tampering with success.

No part of the carburetor has been left unnoticed and untouched when super-duty is considered. It's a true racing carburetor, as durable as it is high-performing. That's already been proven by the racers.

If you happen to spot one of these models around the race tracks, don't get too upset about not being able to get one of your own. While the 4500's won't be released through dealer outlet channels, they will be sold and delivered through the factory groups as original equipment during 1969. It hardly sounds like "original equipment," we know, but it sure does satisfy the "original" description. It's a very original piece of equipment. ■ ■

Those carbs don't always work to perfection right out of the box. A few bits and pieces and a little know-how will turn the trick

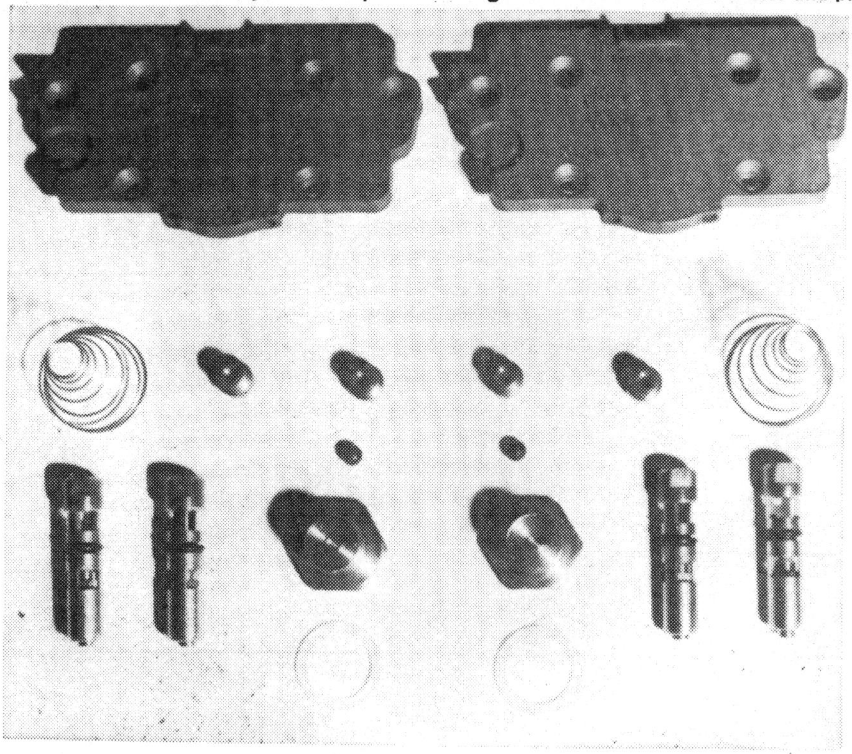

HOLLEY FIXER

Text and photos by Lee Kelley

Having trouble making those Holleys work on your new Tunnel Ram or Ram Log manifold? Many Edelbrock customers were complaining of below-par performances from their new carburetor-manifold combinations, so Vic Edelbrock took to the field (the drag strip) to determine the cause for all the commotion. What he found will be of interest to all performance enthusiasts who use Holleys, particularly those who use Tunnel Ram or Ram Log manifolds.

One of the first things Vic noticed was that most of the cars equipped with Holleys seemed to be "running out of gas" under hard acceleration. After some lengthy testing, Vic found that the float levels in the Holleys were not set high enough. He found that the old method of setting the floats (removing the inspection plug and setting the gas level just to the bottom of the inspection hole) was not adequate for performance engines. The reason? At idle, with high fuel pump pressure, the needles are jiggling off their seats (seeping) just a little and when the plugs are removed, fuel will spurt out, even though the float levels aren't high enough. If the floats are lowered, fuel volume in the bowls will be decreased and fuel will slosh away from the jets under acceleration. Fuel pump pressure was also found to be a culprit. Most competitors were running without a fuel pressure gauge, so they didn't know that their fuel system was inadequate. Under full throttle, some of the pump pressures tested were falling to as low as one pound. Vic compiled his findings, and here's the scoop on how to set up the Holleys and the related fuel system.

The carburetor that we'll use for an example is the 1850 (1-14) rated 600 cfm. Edelbrock uses two of these on his Tunnel Ram for engines of 327 cubic inches and under. The first step, once the carburetor is off the manifold, is to remove the primary bowl. Next, remove the power valve and plug the hole. Edelbrock uses a special plug with a gasket to assure that the two holes of the valve are closed. Remove the stock primary jets and replace them with Holley #75 jets. Use Holley jets only; they are flow-rated jets and should not be drilled to a larger size. The #75 jets will put the hot rodder in the ball park as far as jetting is concerned with the 1850 carbs on the Tunnel Ram. Slightly richer or leaner mixtures may be used for fine tuning.

Loosen the screw in the top of the bowl and remove the needle and seat. Replace the stock .097 needle and seat with Holley's vitron-tipped .110 version. The .097 has small holes in the body compared to the .110's "window" to let in more fuel. Turn the primary bowl upside down and adjust the float level until a #10 drill just barely passes between the top of the float and the top of the bowl. Secure the float in this position, and reassemble the primary bowl.

The next step involves the secondary bowl. Remove it, as well as the secondary plate, being careful not to damage the gasket. Drill the two top air bleed holes to .055-inch. This has nothing to do with the final jet mixture but it eliminates any bogging at low rpm with manual secondaries by richening up the mixture when the butterflies are just barely cracked. Next, drill the secondary plate (this carburetor has no secondary jets) from its stock .067- to .076-inch. When drilling the secondary plate, use a small hand drill; an electric drill will more than likely produce a larger hole than is desired because it is awkward to handle. Loosen the screw on the top of the secondary bowl, then pull out the needle and seat. Here again, replace the stock .097 with a .110 Holley. Set the float level for the secondary bowl following the same procedure outlined for the primary bowl, only this time use a #2 drill as a gauge. If anything, set float levels a little high. Reassemble the secondary bowl.

The final stage of modification involves the vacuum assembly. Remove the choke and vacuum assembly and disassemble the vacuum housing. Remove and discard the ball from the secondary diaphragm and replace the secondary spring with Holley spring No. 38-R-476 (or you can shorten the stock spring until the secondaries are fully open at 5000 rpm). Drill out the shot visible on the side of the carburetor with a #84 drill (.111) and then drill the hole from the primary venturi to the secondary diaphragm from .040- to .081-inch. Once the hole is drilled, plug the outside hole with a new piece of lead shot. Reassemble the vacuum housing and replace it and the choke. Do not leave the choke off the carburetor and do not remove the butterflies. Plug the holes that were used for the automatic choke in the carburetor and the vacuum for the ignition, and the Holley is ready to run.

Vic has put together a kit that will supply everything that is needed to complete these modifications. New secondary plates are provided so drilling the old ones is not necessary. Vic also passed along a few other tips in the carburetor-fuel system department. Run a fuel pressure gauge to have an idea what the fuel system is doing at all times. Maintain no less than 4½ pounds pressure through the traps; that may mean that the fuel pressure at idle will be as high as 7 pounds (pressure higher than 7 pounds at idle is not recommended). Use two electric fuel pumps in parallel, mounted below the pickup in the fuel tank. Mount these pumps as close to the fuel tank as possible. Gas volume is the key to success at high engine rpm. As we stated earlier, Vic found much of this information through testing on his Tunnel Ram, but Holley users everywhere can benefit from most of this information. ■■

A

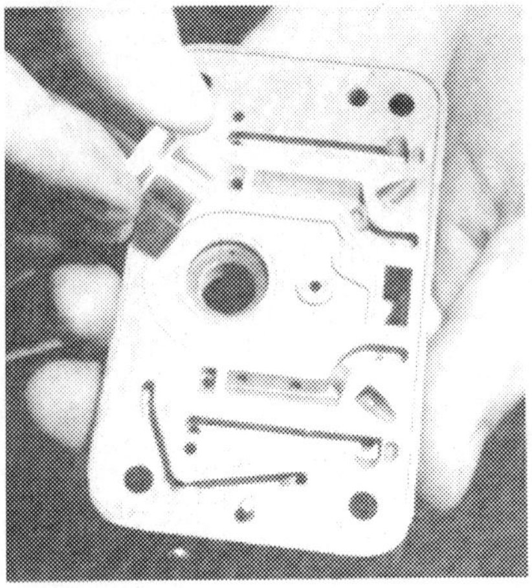

B

C

D

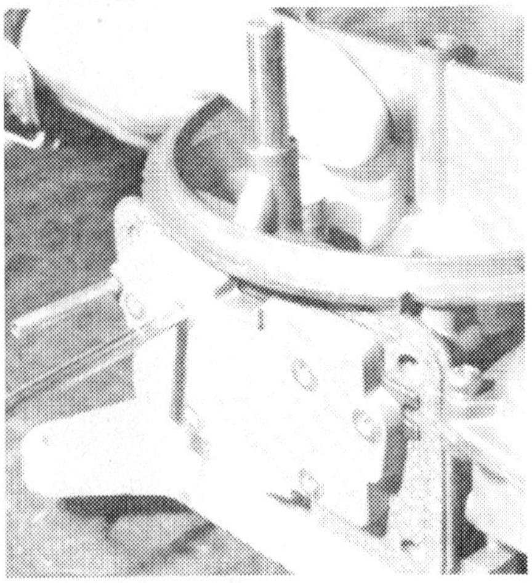

E

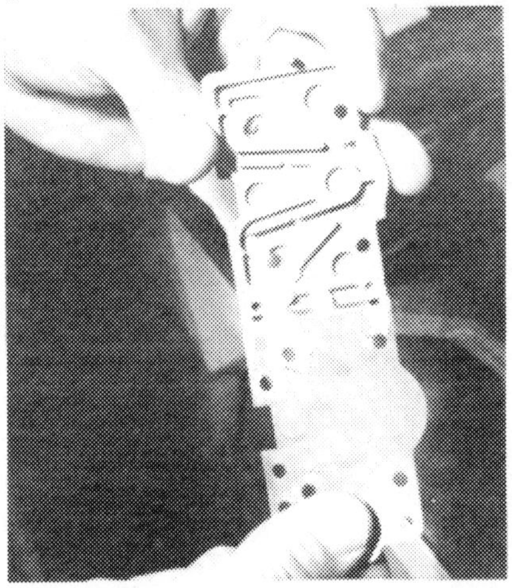

F

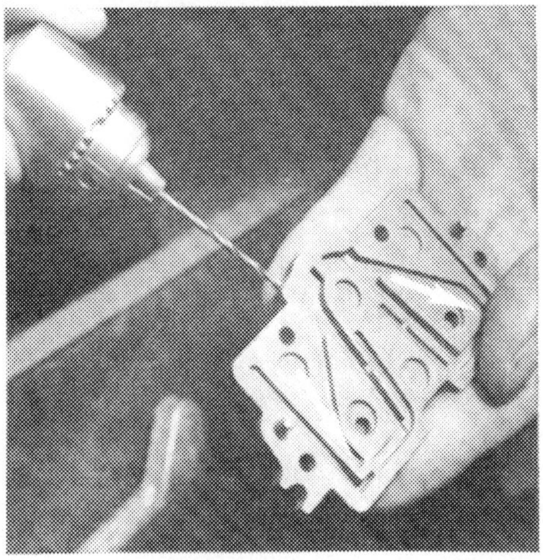

G

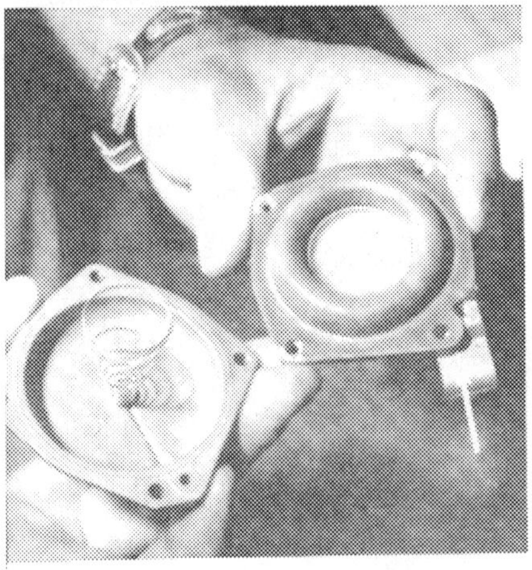

H

I

A. First step in the Holley modification is to remove the power valve and plug the hole with a special plug and gasket. Make sure the two small holes are sealed. B. Loosen the screw, remove and replace stock needle and seat with vitron-tipped .110 version. C. Turn the primary bowl upside down and adjust the float level until a #10 drill just barely passes between the top of the float and the top of the bowl. Secure the float in this position. D. Remove the secondary plate, being careful not to damage the gasket. E. If the gasket is not carefully removed, it will be destroyed. New gaskets should be checked closely to make sure that they are the proper ones for the carburetor. F. Use a small hand drill to enlarge the secondary plate from its stock .067-inch to .076-inch. Drill the top two air bleed holes (arrows) to .055-inch. G. After removing the choke and vacuum housings, remove and discard the ball from the secondary diaphragm (arrow) and replace the secondary spring with Holley spring No. 38-R-476 (or shorten the stock spring until the secondaries are fully open at 5000 rpm). H. Drill the hole from the primary venturi to the secondary diaphragm from .040- to .081-inch. Once the hole is drilled, plug the outside hole with a new piece of lead shot. I. Plug the holes (arrows) that were used for the automatic choke and the vacuum for the ignition. Edelbrock's new kit (opposite page) provides all components.

HOLLEY HINTS

Text and photos by John Thawley

These tips will bring those Holley four-barrels right up to snuff — and keep them there

A pair of Holley 3977's mounted on a big Edelbrock manifold represents a solid chunk of horsepower if, and only if, the two secondary vacuum advance inlets are connected. Without this hose engine runs poorly; investment is lost.

Several of the Holley four-barrel carburetors are among the most popular pieces of induction equipment on the performance market today. One of the reasons for the popularity stems from the fact the larger Holleys are easily modified.

First off, experimentation in jetting in most of the big Holleys may proceed at a rapid pace because of the ease of change. All of the primary metering blocks are removed by four large screws. In turn, the jets are screwed into the metering block. Jet size is stamped on the jet. On Holleys such as the 2818 and the 3310, secondary jetting is also handled by jets screwed into a metering block. On these carbs, changing jets on the secondary side is as easy as the job on the primary side. On a number of models though, secondary jetting is through a metering body with drilled restrictions rather than removable jets. When required, it is recommended that this secondary metering body be changed rather than attempting to enlarge the openings by drilling. If drilling is attempted, a hand drill must be used, since any power will rout too much metal from the tiny orifice. Enlarge only the two holes at the bottom of the secondary metering body when experimenting in this area. Again, the drilling is not recommended since the secondary metering body can be purchased in a wide variety of metering restriction sizes.

Bobby Meeks, who handles much of the Edelbrock experimentation work, urges that a tuner seek to maintain as close a relationship as possible between primary and secondary jet sizes. For instance, if an engine appears to be running too lean at the top end, try going to two jet sizes richer on the primary *and* secondary instead of moving four or five jets richer on the secondary side only.

Meeks also points out that staggered jet sizes in small-block Chevrolet installations are commonplace due to intake manifold flow characteristics. For example, a small-block Chevy running a Holley 3310 may have a 78 and 74 in the primary side and a 76 and 78 in the secondary metering block. Again, a close relationship of jet sizes between the two sides is important.

There are a number of necessary odds and ends which may mean the difference between a win or a loss on a run. The power valve on the Holley 3916 should have six holes drilled around the stem. If you're the proud owner of one of these big jugs, check this, since a number of carbs were inadvertently released with a four-hole power valve installed. The "65" stamped on the edge of the valve refers to the inches of vacuum required to open the valve. But Holley assumes you know where to insert the decimal point, thus, the 65 power valve opens at 6.5 inches of vacuum. (Boy, 65 inches would just about suck the air cleaner in!)

And speaking of air cleaners, if you think well of your engine, run a cleaner on the street, but make sure there is at least ⅜-inch clearance from any piece of the cleaner to the top of the pitot tubes. Holley makes a number of different lengths and several have the top end cut at an angle to gain extra clearance.

To change a big Holley from a vacuum operated secondary to a mechanical operation, Vic Edelbrock sells a very inexpensive little package which takes care of the job in short order. Also, the vacuum inlet must be plugged in the primary metering block when the distributor is changed from vacuum to mechanical advance.

On the dual carb installations on the big-block Ford and Chevy using the Holley 3977, a length of hose must be used to connect the two secondary vacuum advance inlets.

Some of the Super Stock racers have found that gasoline is climbing the back wall of the float bowl due to forward thrust of the car during the initial part of a hard run. The result is a pair of jets left without gasoline until the fluid can shift and reach the jets. Holley now makes a small tube which presses into the jet and protrudes into the bowl in order to gather fuel from a more central position in the bowl. The small tube is a press fit into all the Holley jets. Holley part number is 14R505.

One last word. Most top tuners will make only one change at a time when attempting to gain "just a little more" from an engine. In other words, they won't change jet size, advance timing and go two heat ranges colder on the plugs. Move slowly in the tuning steps you take, and most likely that big ol' Holley will help move you a little faster. ■■

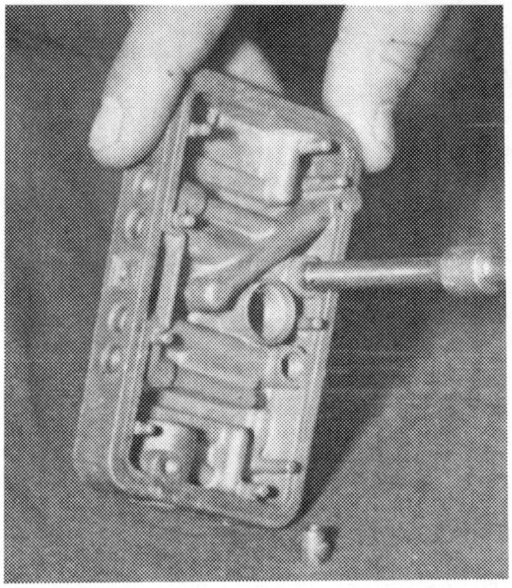

ABOVE – Jet changes come easily with screw-in jets. Size is stamped on edge. Metering block cannot be installed wrong. ABOVE RIGHT – Power valve does not regulate air/fuel mixture. Don't start drilling holes, but do check to see if correct valve is in carb. Plunger should be free, easy to move. RIGHT – Vacuum inlet on the edge of a metering block should be plugged if not connected to distributor, wipers, etc.

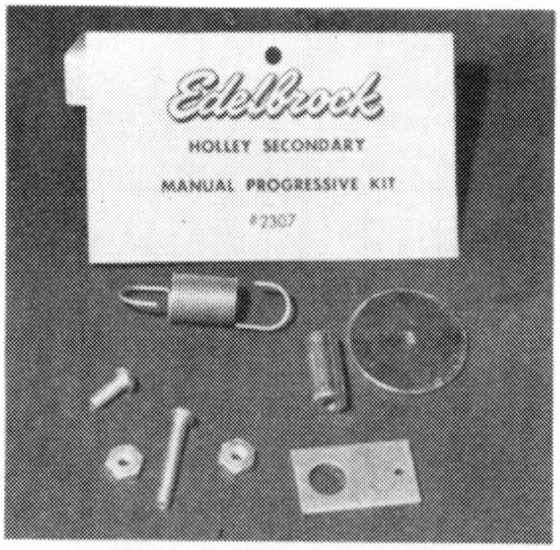

LEFT – Kit, such as Edelbrock's, takes care of opening secondary side of carb manually. Unit may be returned to vacuum operation later. BELOW LEFT – On models featuring drilled holes for secondary jetting, the two holes in bottom of metering body may be enlarged. Carefully! Best bet is to keep several blocks in various jet sizes for experimentation. Make certain that both pitot tubes have at least ⅜-inch clearance between any part of air cleaner.

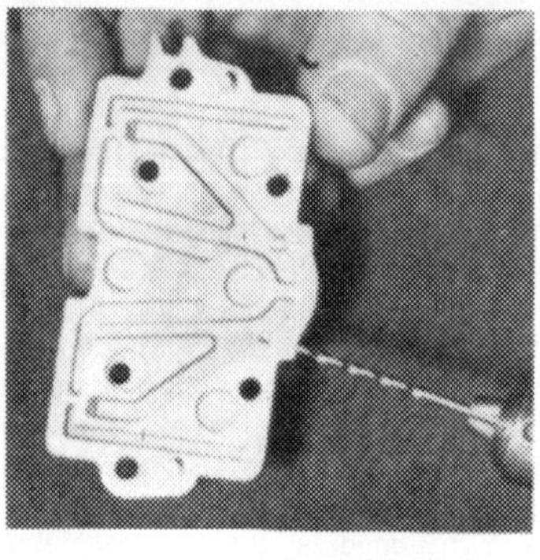

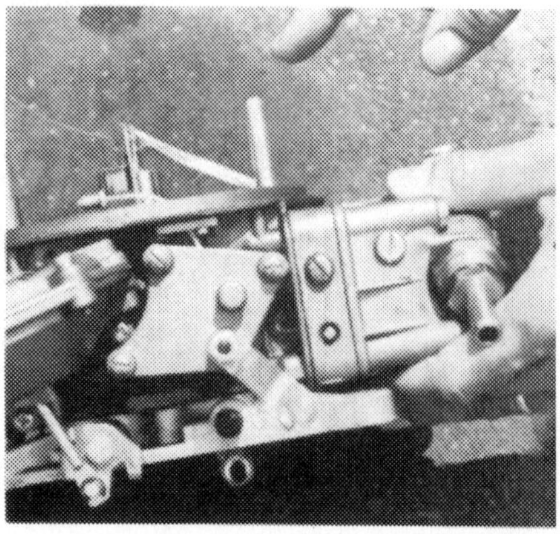

HOLLEY HOLES

Some Holley carburetors have a secondary metering body rather than the removable jets on the secondary side. How do you know what size the holes are? Don't, eh? Now you do. Give thanks to Ed Hecht of Holley Carburetor, who tipped us off on how to figure it all out. Take carb in hand and proceed thusly.

The secondary metering body has the part number embossed, upside down, just below the top rectangular opening on the front of the plate. The embossed number you're looking for is 34R-2007B. Just below this number, you'll see an identifying number stamped in the middle of the body. You guessed it. This number is also upside down.

The secondary metering body has two sets of holes in it. The chart which follows gives sizes for both sets of holes. The bottom jet is called the secondary metering restriction, or jet. The sizes of these jets are identified in the chart under the column heading of *Hole A*. The other set of holes is located on the back side of the plate near the top. These are called the secondary idle feed holes. The sizes of these jets will be found under *Hole B*. All of the hole restrictions are measured to a plus or minus .001.

Almost all of the Holley secondary metering bodies are covered in the chart. A few of the special Chrysler high-performance carbs are not covered by the chart, but the secondary metering body on this carb has a different basic part number. This means check part numbers before you start assuming the chart applies to the carb mounted atop your "go-toy."

Part Number	Hole "A"	Hole "B"	Stamped
34R-2007-3	#54 Dr.-.055	#71 Dr.-.026	3
34R-2007-4	#53 Dr.-.059	#71 Dr.-.026	4
34R-2007-5	#52 Dr.-.063	#71 Dr.-.026	5
34R-2007-6	#50 Dr.-.070	#71 Dr.-.026	6
34R-2007-7	#55 Dr.-.052	#71 Dr.-.026	7
34R-2007-8	#51 Dr.-.067	#71 Dr.-.026	8
34R-2007-9	#51 Dr.-.067	#68 Dr.-.031	9
34R-2007-10	#48 Dr.-.076	#71 Dr.-.026	10
34R-2007-11	#47 Dr.-.0785	#68 Dr.-.031	11
34R-2007-12	#48 Dr.-.076	#68 Dr.-.031	12
34R-2007-13	#52 Dr.-.0635	#68 Dr.-.031	13
34R-2007-14	#40 Dr.-.098	#50 Dr.-.070	14
34R-2007-15	#42 Dr.-.0935	#50 Dr.-.070	15
34R-2007-16	#51 Dr.-.067	#69 Dr.-.029	16
34R-2007-17	#49 Dr.-.073	#60 Dr.-.040	17
34R-2007-18	#52 Dr.-.0635	#70 Dr.-.028	18
34R-2007-19	#50 Dr.-.070	#70 Dr.-.028	19
34R-2007-20	#50 Dr.-.070	#68 Dr.-.031	20
34R-2007-21	#46 Dr.-.081	#60 Dr.-.040	21
34R-2007-22	#48 Dr.-.076	#70 Dr.-.028	22
34R-2007-23	#51 Dr.-.067	#70 Dr.-.028	23
34R-2007-24	#47 Dr.-.0785	#65 Dr.-.035	24
34R-2007-25	#44 Dr.-.086	#57 Dr.-.043	25

A SECONDARY CHANCE

It may be hard to believe but a new Holley 3-barrel can improve the mileage right along with adding a batch of horses

by Jim McFarland

Viewed from either side, Holley's new 3-bbl (3916A) presents engine with very healthy 6.62 square inches of venturi area. Notice the fact that a degree of modification was necessary to provide operating clearance for the carb's secondary butterfly. Bolt pattern is same as the conventional Holley 4-bbl design. Carb, as available from Edelbrock, comes with modified vacuum advance mechanism and other internal alterations. New intake manifold is forthcoming if you decide not to modify the plenum divider of the manifold you already have. It is recommended 3-bbl use special type air cleaner.

After a time, a fella sorta becomes accustomed to seeing some of the non-stock garnishments that turn up on this performance buggy or that. Consequently, it was with no particular note that we eyed the pair of stranded wires leading from beneath the hood and into the passenger compartment of this "you've seen one you've seen 'em all" '65 'Vette parked in the Edelbrock lot. "Say, that's your 'Vette isn't it, Vic?" Nobody else has a grin like that. "Course." And now another look at the red wires. "Where do those lead and what are they hooked to under the hood?" Up comes the hood. You're looking at the same giant of Holley we spotted. It seems Vic and the Holley Corporation had been street-testing a new cut of Holley 4-bolt, this one a 3-bbl that dyno-checked an increase (over a 4-bbl) of 33 hp at a crank speed of 5000 rpm when attached to an altered Edelbrock C-4B high-riser and nestled between the heads of the same 327-incher that formed the basis for the exhaustive program presented in HOT ROD of September '66. Not bad for gains resulting in the swap from a 3310 Holley 4-bbl to the new 3916A 3-bbl. A 3-bbl? Well, study the photos. Notice also that a certain amount of metalwork was required to facilitate secondary butterfly operation above the plenum's center divider.

The two red wires? Nothing more than a simple (but accurate) method to check secondary butterfly opening. An end of one wire was attached to the secondary arm, passed inside the driver's headquarters to one side of a 12-volt bulb in a red-lensed socket. The other side of the bulb carried an end from the second wire attached to a 12-volt source and on to a contact at the base of the carburetor (just below and behind the secondary throttle arm). As a function of manifold vacuum, the secondary half of the carb would open, bringing the stationary and mobile contacts together, lighting the bulb inside the car and indicating operation of the back half of the juice dispenser. Simple as ACB, or sumthin'. The gadget proved to be just the ticket for establishing changes required to make the 3916A compatible with all-round driving.

Did it help acceleration? At the time of this writing, the only data available was a comparison of 1st gear performance (3310 Holley vs. 3916A Holley) from 20-60 mph. Even in view of the fact this may leave something to be desired in terms of quarter-mile evaluation, the following results will provide you with sufficient evidence to see the suds this carb can supply: 20-60 mph required 4.4 seconds with the 3310 and only 3.6 seconds with the 3916A! Again, this was a 1st-gear-only test.

The 3916A operates quite well with vacuum-controlled secondary opening, and although the secondary spring is left stock (in the modifications that Vic makes), the third barrel is fully open by 4200 rpm. Other internal changes include a switch from the stock 3916A end floats to the center-pivot Chevy types, a swap to the Chevy-type throttle arm, and some diggin' in the jet department that amounts to final jetting of #80 primaries and secondary plates drilled to .086-inch. A special air cleaner is being designed by Hellings (Vic claims use of a stock cleaner will kill the engine) and is available in the same package that Edelbrock now has on the market. Incidentally, the new carb can be obtained only from the house of Edelbrock.

Specific dyno figures (in terms of corrected hp increases) ran as follows: 31 @ 4500, 33 @ 5000, 33 @ 5500, 25 @ 6500 and 20 @ 7000.

Almost forgot the fuel economy slant. Mileage was up; a little. So reflect on the situation. Better yet, slap a stamp on the nearest postcard and find out what sort of coinage it would take to put the complete manifold/carb package on your doorstep. It's a staunch performance-booster, even if you don't get the two little red wires. ■■

Curing Holley HICCUPS

SIMPLE TUNE-UPS FOR THE 600-CFM HOLLEY 1850

By Jeff Smith

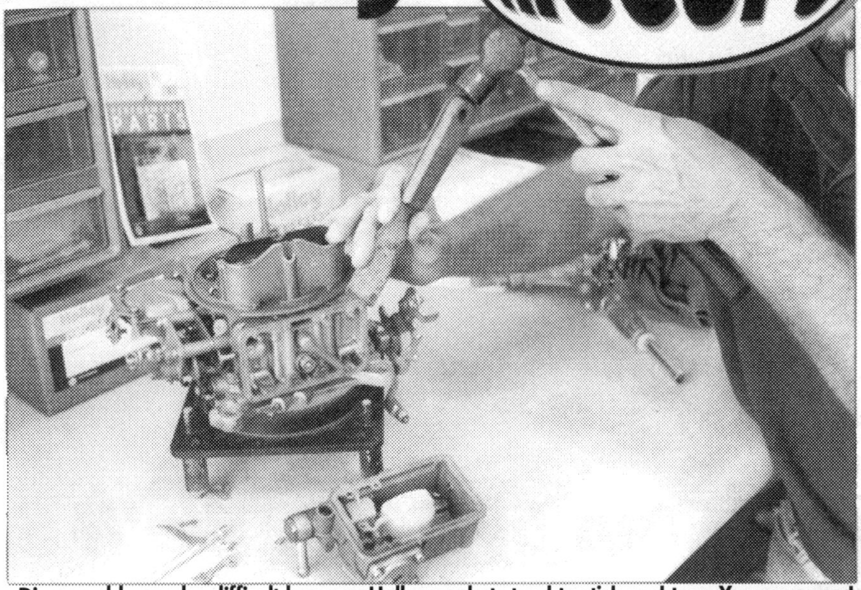

Disassembly can be difficult because Holley gaskets tend to stick and tear. You may need a gasket scraper and a small hammer to get it apart. Unbolt the primary and secondary float bowls, the diaphragm housing, and the primary metering block and baseplate.

The most popular street carburetor in the hot rod world is probably the 600-cfm Holley vacuum secondary carburetor, otherwise known as the 0-1850. They are popular for their wide range of application and low price. Mail-order speed shops sell a new 1850 for about $160, but used ones go for $25 to $50 at swap meets, and literally millions of them are out there.

According to Sean Murphy of Jones Performance Fuel Systems, driveability problems arise when hot rodders bolt this carburetor to the wrong engine. Murphy says, "I've seen 1850s bolted to anything from 350hp 289s to monster-cammed 454 Chevelles. Neither of these is a good idea."

Because the 1850 is designed with a somewhat lean idle circuit, driveability problems occur when it is used on an engine with a big camshaft and low manifold vacuum because the idle circuit cannot supply sufficient fuel to run at light throttle.

According to Murphy, the best application for this carb a 300- to 400ci engine equipped with stock or mild camshaft timing and 16 to 20 inches of idle manifold vacuum. Ideally, this engine would be in a 3000-pound or heavier car equipped with an automatic transmission and a 3.30:1 or numerically lower gear set. The problem with low manifold vacuum in the 1850 is it requires more idle throttle blade opening, which exposes most of the idle transfer slot to feed sufficient fuel to the engine to keep it running at idle. This often creates an overly rich idle mixture, which leads directly to a lean flat-spot just off-idle. This lean condition often leads to a backfire.

We'll examine the basic rebuild procedure for the 1850 as well as some quick fixes to cure the carb's hiccups. If you have an engine equipped with a big cam and low manifold idle vacuum, the 1850 can be reworked to deliver more

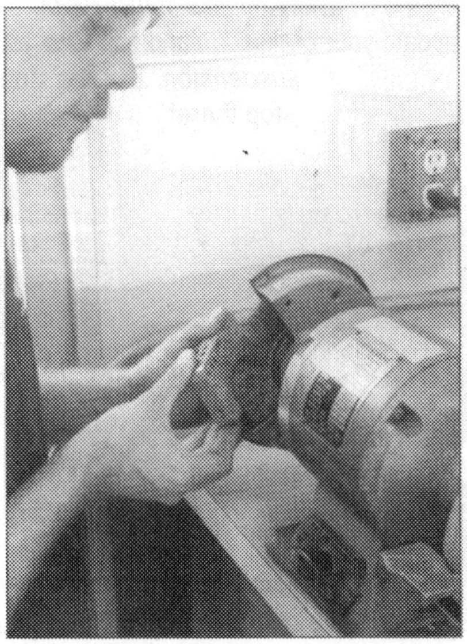

Removing the gaskets is a tedious job. Use a wire wheel to remove all of the gasket from all sealing surfaces to avoid a potential vacuum leak or a fuel leak.

The 1850 uses a secondary metering plate instead of a metering block with replaceable jets. Use a clutch-head screwdriver to remove the countersunk screws in the plate. If you don't have the special tool, you can make a crude one from a Phillips screwdriver.

Clean all passages with some sort of carb cleaner, then use compressed air to blow them out. This is especially important around the idle circuit passages since the idle feed restriction is 0.026 inch in diameter and easily plugged with dirt or gasket material.

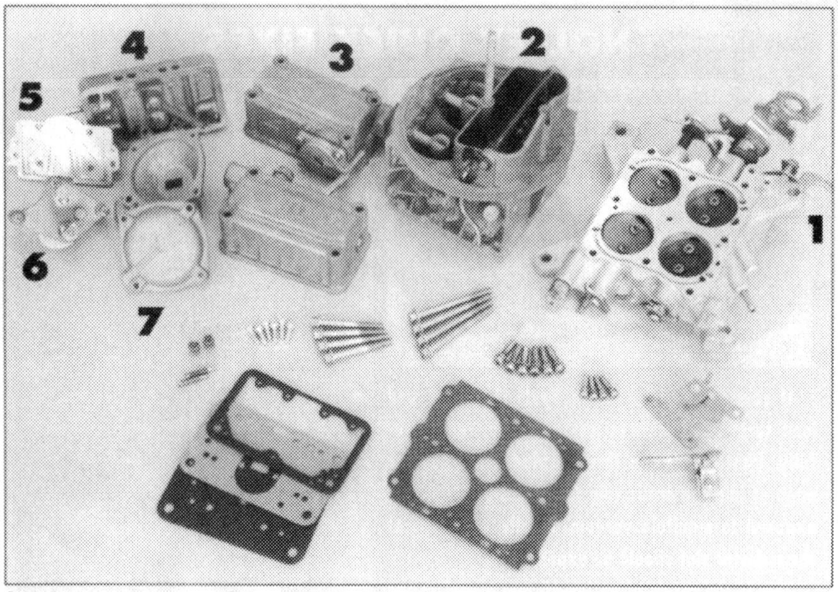

The main components of the 1850 include (1) the baseplate, (2) the main body, (3) the primary and secondary float bowls, (4) the primary metering block, (5) the secondary metering plate, (6) the manual choke mechanism, and (7) the vacuum secondary diaphragm housing.

To reassemble, fit the baseplate to the main body of the carburetor, making sure the gasket is oriented properly.

Remember to include the small steel ball (*arrow 1*) in the diaphragm housing location (*arrow 2*). This protects the fragile rubber diaphragm (*arrow 3*) from backfire damage.

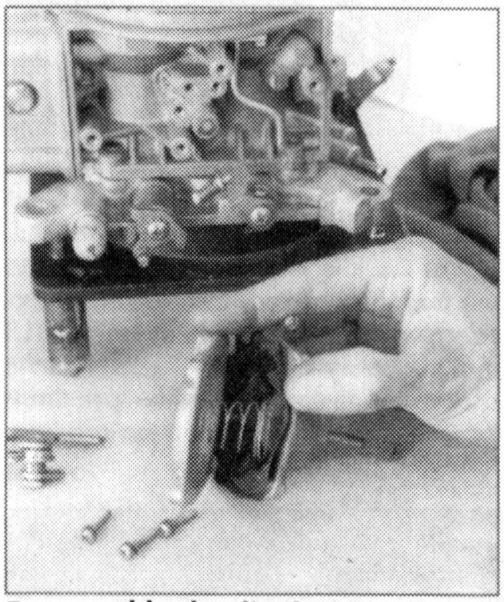

To assemble the diaphragm housing, place the spring in the cover, insert the steel ball in the lower housing, then carefully align the bolt holes of the cover with those in the rubber diaphragm. Be careful not to tear the rubber with the bolts.

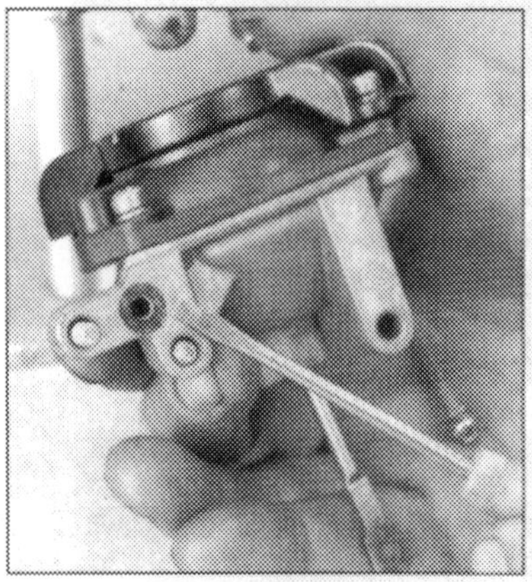

Be sure to include the small cork seal on the vacuum diaphragm housing where indicated.

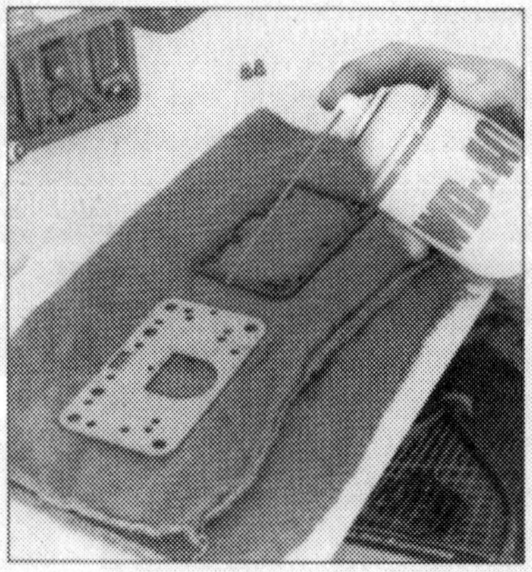

To prevent the gaskets from sticking the next time they are removed, apply some WD-40.

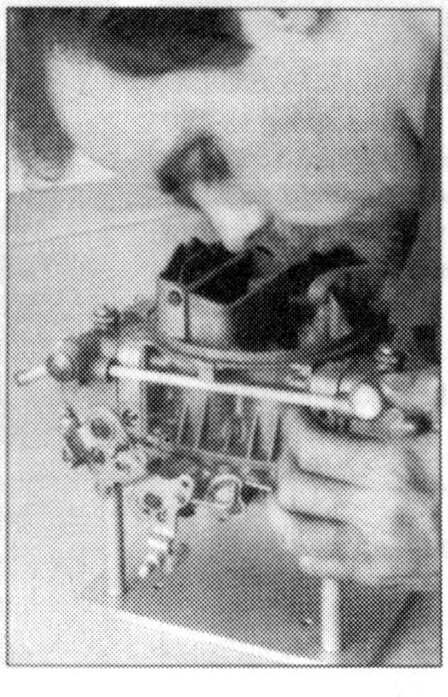

Take care when assembling the fuel-transfer tube between the primary and secondary float bowls. Place the O-rings on each end of the tube and insert the tube in the primary float bowl first. Slide the secondary float bowl over the O-ring, making sure you don't tear it. A dab of spit or oil will help to slide the O-ring in place.

fuel at idle, but it requires modifications best left to professional carburetor shops. (If further modification is required, consider investing in a carburetor with an inherently richer idle circuit, such as the Holley 3310.)

Let's tour the internals of this ubiquitous carb. Even if you've never rebuilt a carburetor before, the process is fairly simple. We'd recommend the help of a friend already familiar with Holley products. **HR**

Adjust the idle mixture screws 1.5 turns out from fully seated, then adjust the accelerator pump to provide at least 0.015 inch of clearance between the bolt and the accelerator pump arm at wide-open throttle. This prevents binding in the accelerator-pump cam assembly.

Install the carb, set the idle speed to 850 rpm, and tune the idle mixture screws to create the highest idle manifold vacuum. If you don't have a vacuum gauge, adjust the idle mixture screws to create the highest idle speed.

HOLLEY ACCESSORY PART NUMBERS

COMPONENT	PART NUMBER
Metering Block Kit	34-6
Quick-Change Secondary Kit	20-59
Secondary Spring Kit	20-13
Electric Choke Kit	45-223

HOLLEY QUICK FIXES

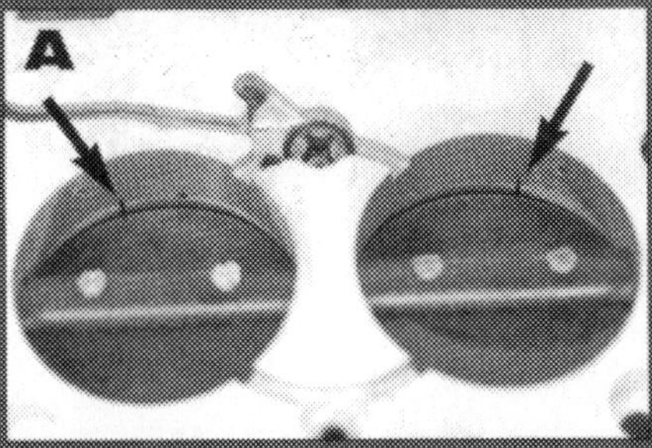

If your Holley idles rough, fouls the plugs, and lean-surges or backfires just past idle, read carefully. Photo A: If the throttle blades are too far open at idle the entire idle transfer slot (*arrows*) is exposed to manifold vacuum. Ideally, only half of the transfer slot should be exposed.

This is the result of an insufficient signal to the rather lean idle circuit on the 1850. To fix it, open the secondary throttle blades as shown in photo B. This allows more air into the engine at idle and allows you to close the primary throttle blades. This also reduces the signal to the idle transfer slot and makes the idle mixture screws more sensitive to adjustment.

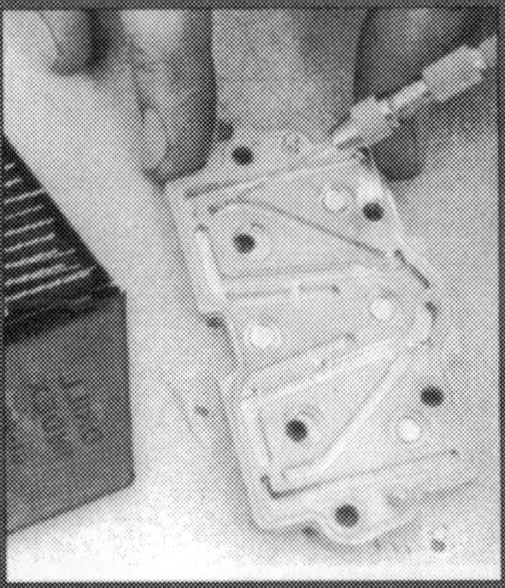

The 1850 often suffers from a slight hesitation just as the secondaries begin to open because the secondary fuel-transfer slot circuit is too lean. This seems to occur on cars equipped with high numerical gears and a loose converter or with a manual trans. Use a small, hand-held drill (often called a pin vise) to drill out the restriction as shown in the photo. Increase the drill-bit size in 0.002-inch increments until the hesitation disappears. Be sure the hesitation you're experiencing is not due to a too-light secondary spring that initiates the secondaries too soon. According to Murphy, the easiest way to cure the secondary bog is to install the Holley 1850 secondary metering block kit.

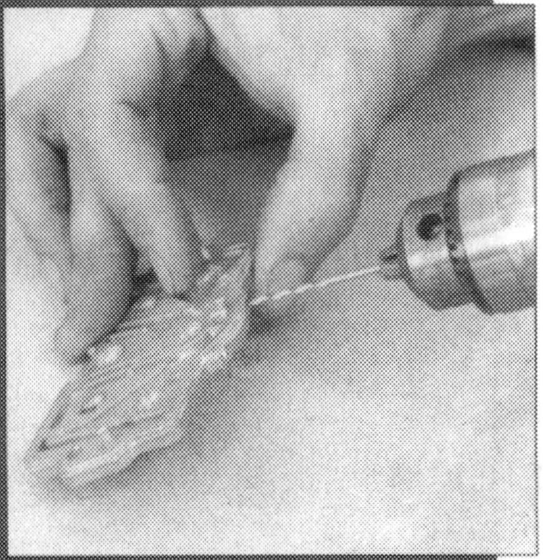

All 1850 Holleys have a secondary metering plate with fixed jets. You can convert the carb to a secondary metering block, but if you're on a budget, drill the secondary plate "jets" to richen the secondary mixture. The stock jetting is equivalent to a No. 65 jet (0.065-inch hole). According to Murphy, most engines respond to a 0.073-inch hole, which is roughly equivalent to a No. 70 jet.

SOURCE

Jones Performance Fuel Systems
Dept. HR12
17491 Apex Cir.
Huntington Beach, CA 92647
714/848-5500

EASY TECH

HOLLEY TUNE-UP KITS

By Jeff Smith

Working on a Holley four-barrel is fairly easy. These carburetors have a simple modular design, making repairs easy even for the beginner. Often the hardest part is finding the right parts when it's time for a re-tune. Rebuild kits supply gaskets and other consumable parts, but not the jets, accelerator pump cams, squirters or other tuning components. You can obtain these parts separately through Holley, but ordering all that stuff is a pain.

Barry Grant Fuel Systems specializes in high-performance Holley carburetors and understands the frustration of searching for those crucial parts. The company has created a series of Performance Pack and Tuning Kit packages that include all the important pieces you need to either upgrade your Holley or put that final tweak to the tune-up.

There are far more options in the Barry Grant Fuel Systems catalog than we have room to detail here, so we chose just a couple of the more popular part numbers. One handy kit worth having is the Jet Change Pack that offers five bowl and metering-block gaskets along with 30 bowl-screw gaskets. The selection of parts in these kits is excellent, and these kits are a great starting point for assembling a fully stocked Holley tune-up box. After you've built your Holley stash, you'll wonder how you ever got along without it, and you'll instantly become extremely popular with all your parts-poor friends. **HR**

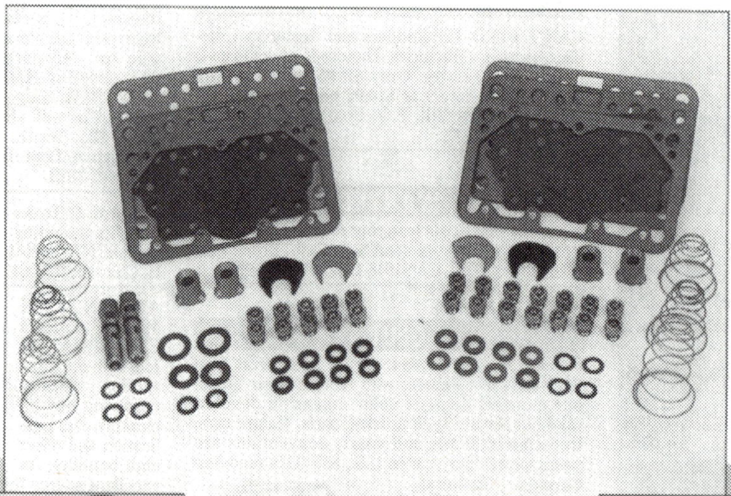

Barry Grant's Phase 1 kit is the most basic. It offers metering-block and bowl gaskets, two accelerator pump cams and squirters, a small assortment of jets and three lighter vacuum diaphragm springs. The 1850 kit also includes larger needles and seats for increased fuel flow. The Phase 1 kit for the 1850 600-cfm carb is on the left (BG part No. 190060), and the Phase 1 kit for the 3310 750-cfm Holley (BG part No. 190061) is on the right.

The Phase 2 kit for the 3310 Holley converts the 3310-2, 3310-3 and 3310-4 750s from the secondary metering plate to the metering block. Another great addition to your Holley tune-up box is a can of Barry Grant's Non-Stick Carburetor Gasket Spray (part No. 171027), which prevents torn gaskets when removing the float bowl or metering blocks.

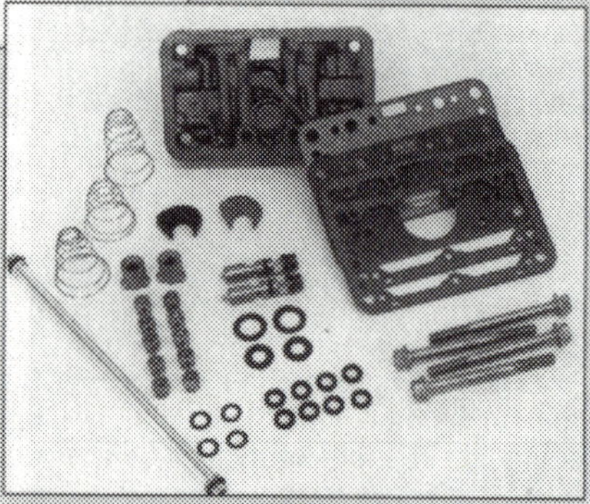

The Phase 2 Performance Pack is the next upgrade for either the 1850 or the 3310. The kit for the 1850 (part No. 190050) includes all the pieces in the Phase 1 kit plus a secondary metering block that allows jet changes so you don't have to fumble with those secondary metering plates. A longer secondary feed pipe and longer bowl screws complete the package.

SOURCE

Barry Grant Fuel Systems
Dept. HR09
Route 1, Box 1900
Dahlonega, GA 30533
706/864-8544

HOT ROD: Vol. 48 No. 9. Copyright 1995 by Petersen Publishing Company. All Rights Reserved. Published monthly by Petersen Publishing Company, 6420 Wilshire Blvd., Los Angeles, CA 90048-5515. Second Class postage paid at Los Angeles, CA 90052, and at additional mailing offices. Subscription rates for 1 year (12 months): U.S., APO, FPO and U.S. Possessions $19.94. Canada $28.83 (price includes surface mail postage to Canada and GST Reg. # R124683855). All other countries $28.94. For subscription inquiries within the U.S. please call 1 (800) 800-4681. For inquiries outside the U.S. call 1 (303) 678-0354. Canada Post International Publications Mail (Canadian Distribution) Sales Agreement No. 546372. *Trademark registered. **HOT ROD (ISSN 0018-6031). POSTMASTER:** Please send address changes to HOT ROD, P.O. Box 56249, Boulder, CO 80322-6249.

Private Pilot and Recreational Pilot

Ground School Course

Michael J. Culhane

Accelerated Aviation Training

Richmond B.C.

All chart excerpts in this publication are for examination purposes only and must not be used for navigation.

The air regulations in this publication have been reproduced in plain English, to facilitate learning. The use of this format does not relieve pilots of their responsibility to ensure they are knowledgeable and up to date on the actual legislation in its original format.

ISBN 1-895801-34-6

INTRODUCTION

Congratulations on your purchase of a CULHANE aviation training manual! Since 1988, our manuals have helped tens of thousands of pilots to succeed on their Transport Canada writtens. This course will prepare you for the Canadian written examination for the Private Pilot License (PPL) and for the Recreational Pilot Permit (RPP). The course has been divided into the following sections:

1.00 Air Law
2.00 Engines and Airframes
3.00 Theory of Flight
4.00 Flight Instruments
5.00 Aviation Weather
6.00 Navigation
7.00 Flight Operations and Human Factors
8.00 Transport Canada Sample Exam

In sections 1 to 7 of this course, all required background material has been provided. Study and learn these subjects, and complete all examples and exercises provided. When you feel you know the material, move on to the sample exam at section 8, and write the exam under simulated test conditions. After completing the exam, score yourself and look up the subjects needing further review by cross referencing and re-learning the subject matter. With the cross referencing in the answer key, you will be able to quickly identify any subjects requiring further study.

To maximize your scoring potential on the Transport Canada written, and as a part of completing this course, you should also obtain and complete **Private Pilot and Recreational Pilot Written Test Book, 1999 Edition**. The written test book contains a further 6 Transport Canada sample exams, all of which are cross referenced to this course.

An important note for RPP users: this manual has been produced to the PPL standard. I have done this because in terms of mandatory subject areas stipulated by Transport Canada, the required subjects for the RPP and PPL are virtually identical, and very few differences exist. Therefore, those preparing for the RPP should simply complete this course "as if" they were preparing for the PPL.

For your convenience, all Transport Canada study topics from the *Study and Reference Guides* have been included and cross referenced to the subjects in this manual: see pages 169 (PPL) and 175 (RPP).

With your flying school's permission, the study time spent completing this course can be applied towards the mandatory ground school times as supervised self study. If it is your intention to apply your study times towards your ground school requirements, obtain your school's approval and insure that you and your instructor keep track of and enter your home study times by subject, date and time in your Pilot Training Record (PTR) - a PTR excerpt is provided at page viii for this purpose.

To complete the necessary calculations in this course, and for the purposes of the Transport written, I recommend you obtain an electronic flight computer, such as the "Pathfinder" model. While the old fashioned E-6B type manual computers can certainly be used to complete the required calculations, they are more difficult to use, slower, less accurate, and more prone to user error. Users of this course should also obtain a basic pocket calculator having a square root key. Such a calculator is invaluable in quickly adding up sets of distances, weights etc. and the square root key is useful for several formulas.

While every effort is made to insure our information is up-to-date, from time to time amendments are necessary and I welcome any comments you may wish to provide that could enhance the content or accuracy of this course. Please contact me directly at once if you have any suggestions. However, while I welcome such feedback, I am not available for tutoring or coaching on the phone as you work through the exercises in this course: please contact a local instructor should such assistance be required.

Thank you for purchasing this course - good luck!

Michael J. Culhane
ATPL BSc BA LLB

December, 1998. Updated March, 1999.

Accelerated Aviation Training
5980 Forsythe Crescent
Richmond BC V7C 2C3 CANADA
Ph: 604-279-0179 FAX: 604-279-0161
E-mail: info@culhanebooks.com

www.culhanebooks.com

TABLE OF CONTENTS

TABLE OF CONTENTS

ABBREVIATIONS

AAIR	Annual Airworthiness Information Report
AAS	Airport Advisory Service
ACC	Area Control Centre
AD	Airworthiness Directive
ADCUS	Advise Customs Service
ADF	Automatic Direction Finder
ADIZ	Air Defence Identification Zone
AES	Atmospheric Environment Service
AFM	Aircraft Flight Manual
AGL	Above Ground Level
AIC	Aeronautical Information Circular
A/I	Attitude Indicator
AIP	Aeronautical Information Publication
AIRAC	Aeronautical Information Regulation and Control (ICAO)
AIRMET	Pilot's Meteorological Advisory
AIS	Aeronautical Information Services
ALT	Altimeter
AME	Aircraft Maintenance Engineer
AOE	Airport of Entry
AOM	Aircraft Operating Manual
ARCAL	Aircraft Radio Control of Aerodrome Lighting
ARU	Airspace Reservation Unit
ASDE	Airport Surface Detection Equipment
ASI	Airspeed Indicator
ASL	Above Sea Level
ASR	Airport Surveillance Radar
ATC	Air Traffic Control
ATF	Aerodrome Traffic Frequency
ATIS	Automatic Terminal Information Service
ATPL	Airline Transport Pilot Licence
ATS	Air Traffic Services
ATZ	Aerodrome Traffic Zone
AWBS	Aviation Weather Briefing Service
AWIS	Aviation Weather Information Service
AWOS	Automated Weather Observation System
BKN	Broken
CARs	Canadian Aviation Regulations
C of A	Certificate of Airworthiness
C of G	Center of Gravity
C of R	Certificate of Registration
CB	Cumulonimbus
CAME	Civil Aviation Medical Examiner
CARS	Community Aerodrome Radio Station
CARs	Canadian Aviation Regulations
CAS	Calibrated Airspeed
CASRP	Confidential Aviation Safety Reporting Program
CAVOK	Ceiling and Visibility OK
CAVU	Ceiling and Visibility Unlimited
CDA	Canadian Domestic Airspace
CD	Coefficient of Drag
CFI	Chief Flying Instructor
CFR	Airport Crash Fire Fighting and Rescue Service
CFS	Canada Flight Supplement
CH	Compass Heading
CL	Coefficient of Lift
CLR	Clear
COM	Company Operations Manual
CPL	Commercial Pilot License
CVFR	Controlled VFR
CZ	Control Zone
DAH	Designated Airspace Handbook
DC	Direct Current
DEV	Deviation
DF	Direction Finding
DFTE	Designated Flight Test Examiner
DH	Decision Height
DME	Distance Measuring Equipment
DOT	Department of Transport
DRCO	Dial-Up Remote Communications Outlet
DT	Daylight Saving Time
DVFR	Defence VFR
E	East
EET	Estimated Elapsed Time
EGT	Exhaust Gas Temperature
ELT	Emergency Locator Transmitter
ETA	Estimated Time of Arrival
ETE	Estimated Time Enroute
EWH	Eye to Wheel Height
FA	Area Forecast
FD	Upper Winds and Temperature Forecast
FL	Flight Level
FPM	Feet per Minute
FSS	Flight Service Station
FTU	Flight Training Unit
GNSS	Global Navigation Satellite System
GPH	Gallons Per Hour
GPS	Global Positioning System
GS	Groundspeed
HF	High Frequency
H/I	Heading Indicator
HND	Above Ground Level
HPa	Hectopascal
Hz	Hertz
IAS	Indicated Airspeed
IATRA	Type Rating - Aeroplane Written Examination
ICAO	International Civil Aviation Organization
IFR	Instrument Flight Rules
ILS	Instrument Landing System
ISA	International Standard Atmosphere
JBI	James Brake Index
KGS	Kilograms
KHZ	Kilohertz

ABBREVIATIONS

KTS	Knots
L/D	Lift Drag Ratio
LF	Low Frequency
LOP	Line of Position
LORAN	Long Range Navigation
MANOT	Missing Aircraft Notice
MB	Millibar
MEA	Minimum Enroute Altitude
METAR	Aviation Routine Weather Report
MC	Medical Certificate
MDA	Minimum Descent Altitude
MF	Mandatory Frequency
MF	Medium Frequency
MFC	Minimum Flight Crew
MH	Magnetic Heading
MHZ	Megahertz
MP	Manifold Pressure
MPH	Miles per Hour
MSL	Mean Sea Level
MTCA	Military Terminal Control Area
MVFR	Marginal VFR
N	North
NDA	Northern Domestic Airspace
NDB	Non Directional Beacon
NM	Nautical Mile
NORDO	No Radio
NOTAM	Notice to Pilots
OAT	Outside Air Temperature
OPP	Ontario Provincial Police
O/T	Other Times
OTT	Over the Top
OVC	Overcast
P	Air Density
PAPI	Precision Approach Path Indicator
PDM	Pilot Decision Making
PIC	Pilot in Command
PIREP	Pilot Report
PLH	Personnel Licensing Handbook
POH	Pilot Operating Handbook
PPAER	Private Pilot - Aeroplane written examination
PPC	Pilot Proficiency Check
PPL	Private Pilot License
PPR	Prior Permission Required
PROG	Prognosis
PSR	Primary Surveillance Radar
PSTAR	Student Pilot Permit and Air Regulations Exam
PTR	Pilot Training Record
QNH	Altimeter Setting
RAAS	Remote Aerodrome Advisory Service
RCMP	Royal Canadian Mounted Police
RCO	Remote Communications Outlet
RIC	Radiocommunication Information Circular
RONLY	Receiver Only
RPM	Revolutions Per Minute
RPP	Recreational Pilot Permit
RPPAE	Recreational Pilot Permit - Aeroplane Written Examination
RSC	Runway Surface Condition
S	Area of Airfoil
SAR	Search and Rescue
SCATANA	Security Control of Air Traffic
SCT	Scattered
SDA	Southern Domestic Airspace
SFOC	Special Flight Operations Certificate
SIGMET	Significant Meteorological Report
SIRO	Simultaneous Intersecting Runway Operations
SLP	Sea Level Pressure
SM	Statute Mile
SPP	Student Pilot Permit
SSR	Secondary Surveillance Radar
SVFR	Special VFR
TACAN	Tactical Air Navigation
TAF	International Aerodrome Forecast
TAS	True Airspeed
TC	Transport Canada
TC	Turn Coordinator
TCA	Terminal Control Area
TCU	Terminal Control Unit
TROWAL	Trough of Warm Air Aloft
TSB	Transportation Safety Board of Canada
TWB	Transcribed Weather Broadcast
TWR	Control Tower
UDF	UHF Direction Finding
UHF	Ultra High Frequency
UNICOM	Universal Communications
UTC	Universal Time Coordinated
V	Velocity
VAR	Variation
VASIS	Visual Approach Slope Indicator
VCS	Vehicle Control Service
VDF	VHF Direction Finding
VFR	Visual Flying Rules
VHF	Very High Frequency
VNC	Visual Navigation Chart
VOR	VHF Omni-directional Range
VORTAC	Combined VOR and TACAN
VOT	VOR Test
VSI	Vertical Speed Indicator
VTA	VFR Terminal Area Chart
VUDF	VHF/UHF Direction Finding
W	West
WAC	World Aeronautical Chart
WCA	Wind Correction Angle
Z	Co-ordinated Universal Time

GROUND SCHOOL TRAINING RECORD

Transport Canada requires 40 hours of ground school instruction for the Private license. The CARs define "ground school instruction" as "classroom type instruction generally given to one or more persons and covering an organized program of lectures, homework, self paced study programs, and examinations following an approved curriculum". This course is an organized self study program, consisting of text, study exercises and sample exam questions, following the Transort Canada study and reference guides. To have your home study using this course count towards the required 40 hours, you must first insure that you are taking this course under the overall supervision and approval of your school. Then, enter your home study time below as you complete this course. When you have completed this course, have your flying school transfer the entries below to your Pilot Training Record (PTR). NOTE: Transport Canada does not specify a specific ground school requirement for the RPP, but it would be wise for RPP applicants to keep track of study times in case of upgrading to PPL at a future time, as the home study times could be transferred at a later date for the PPL.

Subject Key (use for entering times in PTR extract below)

1. Air Law (licensing requirements, Canadian Aviation Regulations, air traffic services, air traffic procedures).
2. Airframes, Engines, and Aircraft Systems.
3. Aerodynamics and Theory of Flight.
4. Flight Instruments.
5. Meteorology.
6. Navigation, Radio Aids and Electronic Theory.
7. Flight Operations, Human Factors and Pilot Decision Making.
8. Sample Examination Practise.

RECORD OF GROUND SCHOOL – DOSSIER DU COURS AU SOL

YEAR – ANNÉE		SUBJECT – SUJET	TIME HEURE	YEAR – ANNÉE		SUBJECT – SUJET	TIME HEURE
MONTH MOIS	DAY JOUR	BROUGHT FORWARD – REPORTÉ →		MONTH MOIS	DAY JOUR	BROUGHT FORWARD – REPORTÉ →	
CERTIFIED CORRECT, CFI OR DELEGATE CERTIFIÉ JUSTE, CFI OU REPRÉSENTANT		TOTAL					
CERTIFIED CORRECT, STUDENT CERTIFIÉ JUSTE, ÉLÈVE						TOTAL →	

ACKNOWLEDGMENTS

I wish to thank those pilots I have trained over the years ... the classroom has been an excellent environment for learning, not only for students but also for the instructor. I am grateful to those users of my training manuals who have made recommendations towards improving the content, accuracy and effectiveness of my training manuals.

I wish to thank the staff at Transport Canada, who over the years have been extremely helpful in handling my inquires. Transport Canada's study and reference guides have been invaluable for me in isolating the required background subjects, and in maintaining a organizational structure.

My thanks to those dealers, wholesaler and retail, that have carried and promoted our product line.

I am appreciative of the flying schools, clubs and instructors who have chosen to endorse my manuals and have adopted them in ground training programs.

The Canadian Owner's and Pilot's Association (COPA), Aviation Canada magazine, and Airways magazine have in the past made favourable editorials and I am thankful for their endorsements.

My friends Bill Walker, Pierre Gavillet, and Stuart and Laurel Walker have been energetic in their support and have provided useful comments and recommendations on the format and style used in my publications.

Our printers at Budget Printing in Vancouver have always been speedy to complete our orders which in turn has been of great assistance in completing and shipping out our own orders promptly.

This publication has been created using Adobe Pagemaker, Adobe Illustrator, and Adobe Photoshop software. My thanks to Apple Computers Inc. and to Adobe Systems Inc. for their outstanding DTP products.

My family deserves mention for their support and assistance throughout: thanks Mom, Dad, Ruth, Steve, Dave, Martha, Tara, Laura, Reiner, and Noriko.

Above all and most importantly I wish to thank Patrizia Rossi who has been of invaluable assistance in the design, layout, production and completion of this publication.

1 Air Law and Procedures

1.01 Regulation Making and Compliance

Regulation Making and Administration

The Parliament of Canada has the constitutional authority for enacting laws and regulations affecting aeronautics. The AERONAUTICS ACT is federal legislation and is the overriding authority under which all other rules are set out. The Canadian Aviation Regulations (CARs) are made under the authority of the Aeronautics Act, and the Minister of Transport ("the Minister") has authority to make standards and to publish them in the CARs.

TRANSPORT CANADA (TC) is the governmental authority responsible for administrative control and enforcement of the Aeronautics Act and CARs.

The Minister shall not make a standard or an amendment to a standard unless the Minister has undertaken consultations with interested persons concerning the standard or the amendment in accordance with the procedures specified in CARAC Management Charter & Procedures. [CARAC refers to the Canadian Aviation Regulation Advisory Council - an industry/government advisory group involved in making recommendations for aviation regulations.] No standard or amendment to a standard shall come into effect less than 30 days after it is made, unless urgently required to insure safety.

Definitions

Know the following definitions:

AEROBATIC MANEUVER means a maneuver where a change in the attitude of the aircraft results in a bank angle greater than 60°, an abnormal attitude or an abnormal acceleration not incidental to normal flying.

AERODROME means any area of land, water (including frozen surfaces) or other supporting surface used or designed, prepared, equipped or set apart for use either in whole or in part for the arrival and departure, movement or servicing of aircraft and includes any buildings, installations and equipment in connection with it.

AEROPLANE means a power-driven heavier-than-air aircraft that derives its lift in flight from aerodynamic reactions on surfaces that remain fixed during flight.

AIRPORT means an aerodrome in respect of which an airport certificate has been issued by TC.

AIRWORTHINESS DIRECTIVE (AD) means an instruction issued by the Minister or by a civil aviation authority responsible for an aeronautical product type design that mandates a maintenance or operation action to insure that an aeronautical product conforms to its type design and is in a condition for safe operation.

AIRWORTHY means an aircraft is in a fit and safe state for flight and in conformity with its type design.

APRON means a part of an aerodrome, other than the maneuvering area, that is intended to be used for the loading and unloading of passengers and cargo, the refuelling, servicing, maintenance and parking of aircraft and the movement of aircraft, vehicles and persons engaged in services necessary for those purposes.

AIRCRAFT CATEGORY, in respect of pilot licencing, refers to the classification of an aircraft as an aeroplane, balloon, glider, gyroplane, helicopter or ultra-light aeroplane. In respect of aircraft certification, aircraft category is a grouping of aircraft based upon intended use or operating limitations such as normal, utility, aerobatic, commuter and transport.

AIRCRAFT CLASS, in relation to the classification of aeroplanes, means aeroplanes having similar operating characteristics to single-engined aeroplanes, multi-engined aeroplanes, centre-line thrust aeroplanes, land aeroplanes or sea aeroplanes.

AIR TIME means, with respect to keeping technical records, the time from the moment an aircraft leaves the surface until it comes into contact with the surface at the next point of landing.

CONTROLLED AERODROME means an aerodrome at which an air traffic control unit is in operation.

CONTROLLED AIRSPACE means an airspace of fixed dimensions that is so specified in the *Designated Airspace Handbook (DAH)* and within which air traffic control service is provided.

FLIGHT TIME means the time from the moment an aircraft first moves under its own power for the purpose of taking off until the moment it comes to rest at the end of the flight.

FLIGHT VISIBILITY means the visibility forward from the cockpit of an aircraft in flight.

GROUND VISIBILITY, in respect of an aerodrome, means the visibility at that aerodrome as contained in a weather observation reported by an air traffic control unit, FSS/CARS station, AWOS station or air operator radio station.

MANEUVERING AREA means that part of an aerodrome, other than an apron, that is intended to be used for the take-off and landing of aircraft and for the movement of aircraft associated with take-off and landing.

MOVEMENT AREA means a part of an aerodrome that is intended to be used for the surface movement of aircraft, and includes the maneuvering area and aprons.

OPERATOR, in respect of an aircraft, means the person that has possession of the aircraft as owner, lessee (renter) or otherwise.

OPERATOR, in respect of an airport, means the holder of an airport certificate issued under the CARs, or the person in charge of the airport, whether as employee, agent or representative of the holder of the certificate.

OWNER, in respect of an aircraft, means the person who has legal custody and control of the aircraft.

Aviation Documents and Inspections

The owner or operator of an aircraft shall, on reasonable notice given by the Minister, make the aircraft available for inspection in accordance with the notice.

Canadian aviation document holders shall produce their document for inspection in accordance with the terms of a demand made by a peace officer, an immigration officer or the Minister.

Where a Canadian aviation document has been suspended or cancelled, the person to whom it was issued shall return it to the Minister immediately after the effective date of the suspension or cancellation.

Recording systems, including computer records and microfiche, that do not comprise entries on paper may be used to comply with the record-keeping requirements of the CARs, if measures are taken to insure that the records contained in the recording systems are protected by electronic or other means, against inadvertent loss or destruction and against tampering and a copy can be printed on paper and given to the Minister on reasonable notice.

[Reference: CARs 101.01, 103.01, 103.02, 103.03, 103.04]

1.02 Aviation Medicals

No person shall exercise or attempt to exercise the privileges of a permit, licence or rating unless the person holds a valid MEDICAL CERTIFICATE (MC) of the appropriate category.

CIVIL AVIATION MEDICAL EXAMINERS (CAME) are physicians who are authorized to examine the medical fitness of candidates for issue or revalidation of pilot and other Canadian civil aviation personnel permits and licences and who report their findings to the Regional Aviation Medical Officer.

MCs are issued by Transport Canada following receipt of a successful medical examination report from a CAME.

The purpose an aviation medical is to determine if a candidate meets the medical standards as specified in the CARs.

An applicant for an aviation medical must complete a TC medical form that requires both the signature of the applicant and of the CAME. The applicant's signature authorizes release of medical information when considered necessary to determine the applicant's eligibility for a medical. The candidate must also sign a declaration of medical history.

It is an offence under the CARs for a candidate to make any false representation in respect of the candidate's medical condition for the purpose of obtaining a permit or licence.

If a candidate is renewing a medical, if successful, the examiner will write "fit" on the endorsement on the reverse side of the candidate's last issued MC. This will constitute a temporary revalidation of the permit or licence for a period of three months commencing on the date of the examination. Such revalidation may only be performed once on any MC. The ultimate authority for the validity of a medical certificate rests with TC and any medical can be suspended where considered appropriate.

For an applicant who has been unsuccessful in obtaining a medical, or where a medical has been suspended, there are various avenues of appeal for reassessment, or further assessment by medical specialists. As well, the Minister has discretionary powers to issue a restricted or limited medical, provided that it is in the public interest and not likely to affect safety. Before issuing a restricted medical, the Minister may require an applicant to undergo a practical test or assessment in respect of the functions of a flight crew member, or may require any necessary medical examination to determine whether the applicant meets the appropriate medical fitness requirements.

No person holding a medical shall exercise the privileges of his or her licence or rating if that person suffers from any illness, injury or disability, or is taking a drug, or is receiving medical treatment, that could impair the holder's ability to safely exercise that person's privileges.

MEDICAL CATEGORY refers to the medical category classification relating to the degree of medical fitness necessary for the various aviation permits, licences and ratings. There are 4 aviation medicals issued: Category 1, Category 2, Category 3 and Category 4. Each medical has detailed specifications, and the following summary is a highlight of some significant requirements of each.

Category 1 Medical

A Category 1 medical is the minimum standard required for an ATPL or CPL. Although this advanced medical is not required for PPL or RPP applicants, it is wise for those seeking a career in aviation insure that they can pass the Category 1 standard, and I would recommend that such persons, even at the PPL or RPP level complete this MC. The following standards apply to a Category 1 medical:

a) ELECTROCARDIOGRAM (ECG) is required at first medical, within the two years preceding the medical between ages 30 and 40, and within the 12 months preceding the exam after age 40.
b) AUDIOGRAM is required at initial medical and at the first medical after age 55.
c) VISUAL REQUIREMENT shall be at least 20/30 in each eye separately, with or without the use of correcting lenses.
d) CHEST X-RAY is required at initial examination only in cases where there is a history or clinical evidence of respiratory problems.
e) VALIDITY of a Category 1 medical is 12 months when under 40 years of age, 6 months when 40 years of age or over. The medical validity period is calculated from the first day of the month following the date of the medical examination.

Category 2 Medical

A Category 2 medical is the required standard for an Air Traffic Controller Licence and Flight Engineer Licence.

Category 3 Medical

A Category 3 medical is the required standard for Private Pilot Licence or Student Pilot Permit. The following standards apply:

a) ECG is required at the first examination after the age of 40 years and again within the 5 years preceding the examination.
b) AUDIOGRAM is only required if the applicant cannot hear from each ear separately the medical examiner whispering softly from 6 feet away.
c) VISUAL REQUIREMENT shall be at least 20/30 in each eye separately, with or without the use of correcting lenses.
d) CHEST X-RAY is required at initial examination only in cases where there is a history or clinical evidence of respiratory problems.
e) VALIDITY of a Category 3 medical for a PPL is 24 months when under 40 years, 12 months when 40 years or over. The medical valid period is calculated from the first day of the month following the date of medical examination.

Category 4 Medical

A Category 4 medical is the required standard for a Recreational Pilot Permit - Aeroplane. Basically, the applicant completes a medical declaration form which is then signed by any physician (not necessary to be a TC approved CAME). The following standards apply:

a) ECG is required for the first examination after the age of 50 years, and every 4 years thereafter.
b) HEARING REQUIREMENT is not allocated.
c) VISUAL REQUIREMENT shall be at least 20/30 in each eye separately, with or without the use of correcting lenses.
d) CHEST X-RAY is not required.
e) VALIDITY of a Category 4 medical for an RPP is 60 months if under 40 years of age, and 24 months if 40 years of age or more. The medical validity period is calculated from the first day of the month following the date of medical examination.

[Reference: CARs 404.01, 404.03, 404.04, 404.05, 404.06, 424.01, 424.04, 424.11, 424.12, 424.17]

1.03 Student Pilot Permit (SPP)

An applicant for a Student Pilot Permit (Aeroplane) shall have reached his or her 14th birthday and have a valid Category 1 or 3 Medical, or a completed medical declaration form in the case of a recreational pilot (Medical Category 4).

The applicant must obtain at least 90% on the PSTAR exam. The PSTAR covers CARs, air traffic control clearances and instructions, air traffic control procedures, Special VFR, AIP circulars and supplements. The PSTAR exam is made up by the Flight Training Unit (FTU) from the *Study and Reference Guide: Student Pilot Permit TP 11919E*. NOTE: To prepare for the PSTAR written, obtain our *CULHANE Student Pilot and Air Regs Written Test Book, 1999 Ed.*, which contains all PSTAR questions, re-organized into practise exams with cross references to this course.

The FTU is required to certify that the applicant has reached a satisfactory standard of knowledge with a statement of assurance that the applicant obtained at least 90%. The PSTAR exam must be corrected to 100% by the ground or flying instructor, and the FTU staff should insure that the applicant thoroughly reviews any weak areas.

An applicant shall present a pilot logbook certified by a flight instructor that the applicant has reached a satisfactory standard of experience and skill to complete solo flight. Once these requirements have been met and demonstrated to an authorized person, a Student Pilot Permit (SPP) in the appropriate category will be issued. Since only one category of aircraft may be entered on the SPP by an authorized person, a student completing training in two different categories of aircraft, such as aeroplane and helicopter, at the same time, shall obtain a second SPP and medical if required for the other category.

SPP holders are restricted to day VFR flying in Canada, and they may not carry passengers. All flights must be authorized by a flying instructor.

Once an SPP expires, it cannot be re-validated with a medical, rather, the entire procedure in terms of completing the PSTAR etc. must be re-done.

[Reference: CARs 401.19, 421.19]

1.04 Radiotelephone Operator's Restricted Certificate

Anyone operating radio transmitter equipment installed in a Canadian registered aircraft must hold a valid Radiotelephone Operator's Restricted Certificate (Aeronautical). The certificates are issued by authorized flight instructors or a radio regulations inspector. Applicants for a radio certificate are given an oral exam on proper radio procedure, declaration of an emergency, phonetic alphabet etc.

The authoritative study source for the certificate is the *Study Guide for the Radiotelephone Operator's Restricted Certificate (Aeronautical), Radiocommunication Information Circular RIC - 21*. As well, most of the required topics for the radio certificate are in the AIP COM section and in this manual.

No student pilot undergoing training should complete a solo flight until he or she has obtained a radio certificate, because the required aircraft radio station license states that only those having a radio operator's certificate are authorized to operate the aircraft radio.

[Reference: Radio Communications Act Regulations, s. 33]

1.05 Recreational Pilot Permit (RPP)

An applicant shall have reached his or her 16th birthday. The required medical standard is at least Category 4. The RPP is maintained by a valid Category 1, 3, or 4 MC.

The applicant shall obtain at least 60% on the TC Recreational Pilot Permit written - RPPAE. The RPPAE written is a sectionalized exam that covers air law, navigation, meteorology and general aeronautical knowledge. The specific ground study subjects are contained in the TC publication *Study and Reference Guide, Recreational Pilot Permit - Aeroplane TP 12467E.* All TC subjects from the RPP study and reference guide, along with cross references, are provided at p. 177.

The required flight training is as follows:

a) an applicant shall complete a course of not less than 25 hours RPP flight training, which shall be under the supervision and direction of a qualified flight instructor on an aeroplane operating with a certificate of airworthiness;
b) the flight training shall include not less than 15 hours dual flight time, including not less than 2 hours cross-country flight time with a cross country flight of two legs of not less than 30 minutes each;
c) at least 5 hours solo;
d) successful completion of a flight test with a Designated Flight Test Examiner (DFTE).

The standards to be reached on the RPP flight test are indicated in the TC publication *Flight Test Standards, Recreational Pilot Permit TP 12475E.*

RPP Privileges

The holder of an RPP may, under day VFR, act as PIC of an aeroplane of a class and type in respect of which the permit is endorsed with a rating, provided that the aeroplane is single engine, no more than 4 person capacity, and no more than one passenger is carried. As well, an RPP holder may, under day VFR, act as pilot in command of an ultra light aeroplane.

[Reference: CARs 401.22, 421.22]

1.06 Private Pilot Licence

PPL Licencing requirements

An applicant for a PPL must be at least 17 years old, and must have completed a Category 1 or Category 3 MC. The licence is maintained by a valid Category 1 or 3 MC.

A 40 hour ground school course must be completed consisting of instruction on the following subjects:

a) CARs;
b) aerodynamics and theory of flight;
c) meteorology;
d) airframes, engines and systems;
e) flight instruments;
f) radio and electronic theory;
g) navigation;
h) flight operations;
i) licensing requirements;
j) human factors, including pilot decision making.

The specific ground study subjects for the PPL are contained in the TC publication *Study and Reference Guide, Private and Commercial Pilot Licences TP 2476E.* All TC subjects from the PPL study and reference guide, along with cross references, are provided at page 169.

At least 60% must be obtained on the TC Private Pilot written examination for aeroplanes (PPAER). The PPAER exam is sectionalized into air law, navigation, meteorology, and aeronautics - general knowledge. Each section must be passed.

The following flight experience is required:

a) 45 hours total private pilot flight training under the direction and supervision of the holder of a flight instructor rating;
b) at least 17 hours dual instruction, including three hours cross-country and 5 hours instrument time. Note: not more than two hours instrument time may be flight simulator time, which shall be in addition to the 45 hours total time and 17 hours dual time required;
c) at least 12 hours solo flight time, including 5 hours cross-country time, with a triangular flight of not less than 150 nautical miles, which includes two full stop landings at points other than the point of departure.

The PPL flight test must be successfully passed with a score of at least 73. The standards to be reached in the flight test are indicated in the TC publication *Flight Test Guide Private and Commercial Pilot Licenses - Aeroplane Category TP 2655E.*

PPL Privileges

The holder of a PPL, may act as PIC or co-pilot of an aeroplane of a class and type in respect of which the licence is endorsed with ratings, and act as PIC of an ultra-light aeroplane.

[Reference: CARs 401.26, 421.26]

1.07 Night Rating

For PPL pilots, a night rating is required to operate an aircraft at night (the night rating cannot be added to an RPP). The required flight training is a minimum of 20 hours of pilot flight time which shall include a minimum of:

a) 10 hours total night;
b) 5 hours minimum night dual, including two hours night cross country;
c) 5 hours minimum night solo, including 10 take-offs, 10 circuits and 10 landings;
d) 10 hours dual instrument time. 5 hours simulator time may be applied, but any simulator time cannot be used to reduce the dual and solo aircraft time.

There is no written exam or flight test for a night rating.

[Reference: CARs s. 401.42, 421.42]

1.08 VFR Over-The-Top Rating

A VFR OTT rating may be endorsed to a PPL.

The training required is 15 hours of dual instrument time, of which a maximum of 5 hours may be simulator. The training must be provided by a licenced flying instructor. No flight test or written examination is required. Basically, the training covers the same instrument flying elements found in the CPL, including full panel, partial panel, recovery from unusual attitudes and radio navigation aids. The specific training requirements are found in the TC publication *Instructor Guide, VFR Over-the-Top Rating TP 12775E.*

The privilege of a VFR OTT rating is to fly VFR above cloud layers, during the cruise portion of flight, subject to specific weather restrictions (see s. 1.15). As well, additional aircraft instrumentation and navigation equipment is required beyond that required for basic VFR (see s. 1.15).

[Reference: CARs 401.44, 401.45, 421.44, 602.116]

1.09 Aeroplane Class Ratings

A PPL, RPP, CPL and ATPL can all be endorsed with an aeroplane class rating.

Seaplane Class Rating

Conversion training to add seaplane privileges are:

a) 7 hours of training including 5 hours dual;
b) at least 5 take-offs and landings as sole occupant of aeroplane (as PIC for two-pilot crew aircraft);
c) training on taxiing, sailing, docking, takeoffs and landings;
d) experience on glassy water, rough water and crosswinds (recommended);
e) conversion training can be provided by holder of CPL or ATPL with at least 50 hours seaplane PIC.

Landplane Class Rating

Conversion training for landplane privileges are:

a) 3 hours of training;
b) at least 2 hours dual instruction;
c) training to include taxiing, landings, including crosswind and takeoffs;
d) conversion training can be provided by holder of CPL or ATPL with at least 50 hours landplane PIC.

Multi-Engine Class Rating

No specific number of hours of training are needed to obtain a multi engine rating. For training licenced pilots, the instructor must hold a valid CPL or ATPL endorsed for the type of aeroplane and at least 50 hours of multi-engine PIC experience with at least 10 hours PIC on type. The training shall be conducted in accordance with the *Flight Instructor Guide, Multi-Engine Class Rating - TP 11575E.* Generally, about 7 - 10 hours of dual will be required to complete a multi-engine rating.

There is no written exam required for a multi-rating.

A flight test must be successfully completed with a TC inspector or DFTE. The required standards to be met in the flight test are provided in the TC publication *Flight Test Guide, Multi-Engine Class Rating TP 219E.*

[Reference: CARs 101.01, 401.38, 421.38, 425.21]

1.10 Aeroplane Type Ratings

Minimum Flight Crew Requirements

The CARs state the minimum flight crew (MFC) to operate various aircraft. Basically, simpler aircraft have a single pilot MFC, while more advanced aircraft have a two pilot MFC. Examples of aircraft with a MFC of one pilot are:

Cessna 150, 152, 170, 172, 180, 210, 310
Piper 140, 151, 180, PA-28, PA-30, Malibu
Grumman Cheetah, Cougar, Goose, Mallard
De Havilland Beaver, Single Otter
Cessna Citation I/SP (high performance),
Piaggio Avanti (high performance)

Examples of aircraft with a MFC of two pilots are:

Cessna Citation I, II, III, IV
All Lear Jet Models
Airbus A-310, A-320, A-340
B-707, B-727, B-737, B-747
L-1011
Canadair Challenger
De Havilland Dash 7, Dash 8

Blanket Aeroplane Type Ratings

A BLANKET AIRCRAFT TYPE RATING is a rating endorsed on a pilot license that entitles the pilot to fly any aircraft described in the blanket rating. A blanket aircraft type rating is obtained automatically by completing the requirements for a license (dual and solo, flight test, written exam), and is normally combined with an aeroplane class rating. For example, a private pilot having completed the PPL on a single engine land aeroplane would be issued the following blanket rating automatically:

All single pilot, non high performance, single engine land aeroplanes.

Individual Type Ratings

An INDIVIDUAL TYPE RATING is required for aircraft not included in a blanket type rating, such as aeroplanes with a two pilot MFC, high performance aeroplanes, medical restrictions or other categories where the blanket rating is not applicable. An individual type rating is indicated by an aircraft type designator endorsed on a pilot's licence.

Type Ratings: High Performance Aeroplanes

In some circumstances, although an aircraft is designated single pilot, it may be too advanced to operate without special training. This kind of aircraft is classified as a HIGH PERFORMANCE AEROPLANE, which is any aeroplane with a MFC of one pilot having a Vne (never exceed speed) of 250 knots or greater, or a Vso (stall speed, landing configuration) of 80 knots or greater.

To obtain an individual type rating for a high performance aeroplane requires:

a) at least 200 hours total flight time in airplanes;
b) a course of ground and flight training acceptable to the Minister on the airplane type;
c) within the 12 months preceding the application for the type rating, performing a qualifying flight under the supervision of a TC inspector or with the holder of a CPL or higher pilot license having at least 50 hours PIC flight time on the airplane class of which 10 hours are on the airplane type.

Type Ratings: Two Pilot MFC Aeroplanes

To obtain an individual aircraft type rating for a two-pilot MFC aeroplane requires:

a) at least 250 hours airplane flight time;
b) a course of ground school and flight training acceptable to the Minister;
c) within the previous 12 months, successfully passing the Type Rating- Aeroplane written exam (IATRA) with a score of at least 70%;
d) a PILOT PROFICIENCY CHECK (PPC) must be successfully passed on the airplane type within the previous 12 months. A PPC is a flight test to determine the pilot is fully competent at operating the aircraft in both normal and emergency situations. NOTE: for commercial operations with an air operator, often the pilot will require a PPC which must be renewed every year, even where no PPC would be required for non-commercial operation of the same aircraft. For example, a pilot with a PPL, IFR rating and Multi-rating could fly a Piper Twin Seneca on a blanket rating IFR with passengers non-commercially (no PPC required), but for the same operations commercially, a PPC would be mandatory (don't worry about these PPC details until/unless you complete a CPL)!

[Reference: CARs 401.40, 421.40]

1.11 Pilot Licence Currency and Recency

Requirement to Hold Licence

No person shall act as a flight crew member or exercise the privileges of a flight crew permit, licence or rating unless the proper permit, licence or rating has been obtained.

While exercising the associated privileges of a permit, licence or rating, the holder must have it her/her possession the appropriate permit, licence or rating and must be able to produce it along the an appropriate and valid MC.

Flight Crew Members of Aircraft Registered in Contracting States Other Than Canada

No person shall act as a flight crew member or exercise the privileges of a flight crew licence in Canada in an aircraft registered in a contracting state unless the person holds, and can produce while so acting or while exercising such privileges, a valid Canadian flight crew licence or a flight crew licence or a foreign licence validation certificate issued under the laws of the contracting state.

5 Year Recurrency

No pilot shall exercise the privileges of his or her licence/permit/rating unless the holder has acted as PIC or co-pilot of an aircraft within the 5 years preceding the flight. To regain currency, within the 12 months preceding the flight, the holder must complete the appropriate TC written and complete a FLIGHT REVIEW (same as flight test) in the appropriate aircraft with a licenced flight instructor and the instructor certifies in the holder's pilot logbook that the holder meets the required skills for the licence/permit/rating.

24 Month Recurrency

Within the 24 months preceding a flight, all licence/permit holders are required to complete a RECURRENT TRAINING PROGRAM, which can consist of:

a) attendance at a TC safety seminar;
b) participation in a TC approved recurrent training program covering human factors, meteorology, flight planning and navigation and regulations;
c) completion TC annual self-paced study program found in the current *Aviation Safety Newsletter*;
d) completion of a training program for a PPC OR completion of a pilot permit, licence or rating including a night rating, VFR OTT rating, IFR rating, multi-engine rating, flight instructor rating, landplane or seaplane rating;
e) completion of the written examination for a permit licence or rating.

Carrying Passengers

For any flight if passengers are carried, the pilot must have completed at least 5 take-offs and landings in the same category and class of aircraft within the previous 6 months, by day or night if the flight is by day, and by night if the flight is by night.

Pilot Logbooks

Anyone holding a pilot license must keep a logbook, to prove compliance with experience requirements and to prove dual/solo when applying for a license. Entries must be periodically certified by the flying school CFI or aircraft owner. A pilot logbook must have at least:

a) pilot's name;
b) date of flight;
c) type of aircraft and registration;
d) whether flight dual, PIC or co-pilot;
e) flight conditions: day, night, VFR, IFR;
f) point of departure, arrival, intermediate stops;
g) flight time.

[Reference: CARs 401.03, 401.04, 401.05, 401.08, 421.05]

1.12 Reimbursement of Flight Costs

General

A COMMERCIAL AIR SERVICE is defined as any use of aircraft for hire or reward.

HIRE OR REWARD is defined as any payment, consideration, gratuity or benefit, directly or indirectly charged, demanded, received or collected by any person for the use of an aircraft.

No person shall operate a commercial air service unless the person holds a valid air operator certificate issued by TC, certifying that the air operator is adequately equipped and able to conduct a safe operation and covering the various conditions under which the commercial air service may be provided.

RPP holders are not entitled to any flight cost reimbursement. PPL holders may obtain reimbursement for flight costs in limited circumstances.

Private Pilot Flight Cost Reimbursement

The holder of a PPL may receive reimbursement for costs incurred in respect of a flight where

a) the holder is the owner or operator of the aircraft;
b) the holder conducts the flight for purposes other than hire or reward;
c) the holder carries passengers only incidentally to the purposes of the flight;
d) the reimbursement is provided only by the passengers flown, and is for the purpose of sharing costs for fuel, oil and fees charged against the aircraft is respect of the flight, as applicable.

The holder of a PPL may receive reimbursement from the holder's employer for costs incurred in respect of a flight where

a) the holder is employed on a full-time basis by the employer for purposes other than flying;
b) the holder accepts remuneration for flying on the employer's business only when the flying is incidental to the execution of the holder's duties;
c) the remuneration, in the case of an aircraft owned by the holder, is paid at a rate based on distance travelled or hours flown that does not exceed the total of the holder's direct operating costs and the fees charged against the aircraft is respect of the flight;
d) the remuneration, in the case of a rental aircraft, is a reimbursement that does not exceed the total of the holder's rental costs, direct operating costs and the fees charged against the aircraft in respect of the flight.

The holder of a PPL may receive remuneration from a charitable, a not-for-profit or a public security organization in respect of a flight conducted by the holder as a volunteer for that organization where the remuneration

a) in the case of an aircraft owned by the holder, is paid at a rate based on distance travelled or hours flown that does not exceed the total of the holder's direct operating costs and the fees charged against the aircraft in respect of the flight;
b) in the case of a rental aircraft, is a reimbursement that does not exceed the total of the holder's rental costs, direct operating costs and the fees charged against the aircraft in respect of the flight. NOTE: Be extremely careful if sharing expenses with a passenger on a private flight: review the actual CARs provisions carefully and insure that the passenger is truly being brought INCIDENTALLY - not preplanned, as a chance event and subordinate to the purpose of the flight. Violation proceedings are often brought against pilots who, by accepting monies (or other reward) from passengers, in fact are providing an illegal commercial air service. [Reference: Aeronautics Act s. 101, CARs 401.28]

1.13 General Operating and Flight Rules

Reckless or Negligent Operation of Aircraft

No person shall operate an aircraft in such a reckless or negligent manner as to endanger or be likely to endanger any person's life or property. [602.01]

Fitness of Flight Crew Members

No operator of an aircraft shall require any person to act as a flight crew member and no person shall act as a flight crew member, if either the person or the operator has any reason to believe, having regard to the circumstances of the flight to be made, that the person is suffering or is likely to suffer from fatigue, or is otherwise unfit to perform properly the person's duties as a flight crew member. [602.02]

Alcohol or Drugs - Crew Members

No person shall act as a crew member of an aircraft within 8 hours after consuming an alcoholic beverage, while under the influence of alcohol or while using any drug that impairs the person's faculties to the extent that the safety of the aircraft or of persons on board the aircraft is endangered in any way. [602.03]

Alcohol or Drugs - Passengers

INTOXICATING LIQUOR means a beverage that contains more than 2.5% proof spirits. No person shall consume on board an aircraft an intoxicating liquor unless the intoxicating liquor has been served to that person by the operator of the aircraft or, where no flight attendant is on board, has been provided by the operator of the aircraft.

No operator of an aircraft shall provide or serve any intoxicating liquor to a person on board the aircraft, where there are reasonable grounds to believe that the person's faculties are impaired by alcohol or a drug to an extent that may present a hazard to the aircraft or to persons on board the aircraft. No operator of an aircraft shall allow a person to board the aircraft, where there are reasonable grounds to believe that the person's faculties are impaired by alcohol or a drug to an extent that may present a hazard to the aircraft or to persons on board the aircraft. The operator of an aircraft my allow a person whose faculties are impaired by a drug to board an aircraft, where the drug was administered in accordance with a medical authorization and the person is under the supervision of an attendant. [602.04]

Compliance with Instructions

Every passenger on board an aircraft shall comply with instructions given by any crew member respecting the safety of the aircraft and of persons on board the aircraft. Every crew member on board an aircraft shall, during flight time, comply with the instructions of the PIC or of any person whom the PIC has authorized to act on behalf of the PIC. [602.05]

Smoking

No person shall smoke on board an aircraft during take-off or landing or when directed not to smoke by the PIC. No person shall smoke in an aircraft lavatory. No person shall tamper with or disable a smoke detector installed in an aircraft lavatory without permission from a crew member or the operator of the aircraft. [602.06]

Aircraft Operating Limitations

No person shall operate an aircraft unless it is operated in accordance with the operating limitations set out in the AFM or other authorized document. [602.07]

Portable Electronic Devices

No operator of an aircraft shall permit the use of a portable electronic device on board an aircraft, where the device may impair the functioning of the aircraft's systems or equipment. No person shall use a portable electronic device on board an aircraft except with the permission of the operator of the aircraft. [602.08]

Fuelling with Engines Running

No persons operating an aircraft shall permit the fuelling of the aircraft while an engine used for the propulsion of the aircraft is running and passengers are on board the aircraft or are embarking or disembarking unless it is done so in accordance with detailed requirements specified in the appropriate air operator standards. [602.09]

Starting and Ground Running of Aircraft Engines

No person shall start an engine of an aircraft unless:

a) a pilot's seat is occupied by a person who is competent to control the aircraft;
b) precautions have been taken to prevent the aircraft from moving.

No person shall leave an engine of an aircraft running unless

a) a pilot's seat is occupied by a person who is competent to control the aircraft;
b) where no persons are on board the aircraft, precautions have been taken to prevent the aircraft from moving, and the aircraft is not left unattended. [602.10]

Aircraft Icing

CRITICAL SURFACES means the wings, control surfaces, propellers, horizontal stabilizers, vertical stabilizers or any other stabilizing surface of an aircraft and, in the case of an aircraft that has rear mounted engines, includes the upper surface of its fuselage.

No person shall conduct or attempt to conduct a take-off in an aircraft that has frost, ice or snow adhering to any of its critical surfaces. A person may conduct a take-off in an aircraft that has frost adhering to the underside of its wings that is caused by cold-soaked fuel, if the take-off is conducted in accordance with the aircraft manufacturer's instructions for take-off under those conditions.

Where conditions are such that frost, ice or snow may reasonably be expected to adhere to the aircraft, no person shall conduct or attempt to conduct a take-off unless, for non airline operators, the aircraft has been inspected immediately prior to take-off (by PIC, flight crew member or trained person designated by operator) to determine whether any frost, ice or snow is adhering to any of its critical surfaces or the operator has established an approved aircraft inspection program and the dispatch and take-off of the aircraft are in accordance with that program.

Where, before commencing take-off, a crew member of an aircraft observes that there is frost, ice or snow adhering to the wings of the aircraft, the crew member shall immediately report that observation to the PIC, and the PIC or a flight crew member designated by the PIC shall inspect the wings of the aircraft before take-off. Before an aircraft is de-iced or anti-iced, the PIC of the aircraft shall insure that crew members and passengers are informed of the decision to do so. [602.11]

Overflight of Built-Up Areas or Open Air Assemblies of Persons during Take-offs, Approaches and Landings

Except if conducting a take-off, approach or landing at an airport or military aerodrome, no person shall conduct a take-off, approach or landing in an aircraft during which the aircraft will overfly a built-up area or an open-air assembly of persons, unless the aircraft is operated at an altitude from which, in the event of an engine failure or any other emergency necessitating an immediate landing, it would be possible to land the aircraft without creating a hazard to persons or property on the surface. [602.12]

Take-Offs, Approaches and Landings within Built-up Areas of Cities and Towns

Unless otherwise permitted (ie aerial work with licenced air operator), no person shall conduct a take-off, approach or landing in an aircraft within a built-up area of a city or town, unless that take-off, approach or landing is conducted at an airport or a military aerodrome.

A person may conduct a take-off or landing in an aircraft within a built-up area of a city or town at a place that is not located at an airport or a military aerodrome where the place is not set apart for the operation of aircraft, the flight is conducted without creating a hazard to persons or property on the surface and the aircraft is operated for the purpose of a police operation that is conducted in the service of a POLICE AUTHORITY (RCMP, OPP, Sûreté du Québec, Canadian Coast Guard, or other municipal/regional police force established under provincial legislation) or for the purposes of saving human life. [602.13]

Minimum Altitudes and Distances

Except where conducting a take-off, approach or landing, under conditions where low altitude flight is permitted (see below) no person shall operate an aeroplane over a built-up area or over an open air assembly of persons unless the aircraft is operated at an altitude from which, in the event of an emergency necessitating an immediate landing, it would be possible to land the aircraft without creating a hazard to persons or property on the surface, and, in any case, at an altitude that is not lower than 1,000 feet above the highest obstacle located within a horizontal distance of 2,000 feet from the aeroplane. In other circumstances, at a distance less than 500 feet from any person, vessel, vehicle or structure. [602.14]

Permissible Low Altitude Flight

A person may operate an aircraft at altitudes and distances less than those minimums specified above where the aircraft is operated at altitudes and distances that are no less than necessary for the purposes of the operation in which the aircraft is engaged, the aircraft is operated without creating a hazard to persons or property on the surface and the aircraft is operated

a) for a police operation that is conducted in the service of a police authority;
b) for saving human life;
c) for fire-fighting or air ambulance operations;
d) for administration of the fisheries legislation;
e) for national or provincial park administration;
f) for flight inspection.

A person may operate an aeroplane, to the extent necessary for aerial work in an aerial work zone, less than the 1,000 feet minimum. As well, an aeroplane can be operated at less than the 500 feet minimum, where the aircraft is operated without creating a hazard to persons or property on the surface and the aircraft is operated for the purpose of aerial application or aerial inspection, aerial photography conducted by the holder of an air operator certificate, or flight training conducted by or under the supervision of a qualified flight instructor. [602.15]

Right of Way - General

The PIC of an aircraft that has the right of way shall, if there is any risk of collision, take such action as is necessary to avoid collision, regardless of who has the legal right of way.

Where the PIC of an aircraft is aware that another aircraft is in an emergency situation, the PIC shall give way to that other aircraft.

When two aircraft are converging at approximately the same altitude, the PIC of the aircraft that has the other on its right shall give way, except as follows:

a) a power-driven, heavier-than-air aircraft shall give way to airships, gliders and balloons;
b) an airship shall give way to gliders and balloons;
c) a glider shall give way to balloons;
d) a power-driven aircraft shall give way to aircraft that are seen to be towing gliders or other objects or carrying a slung load.

When two balloons operating at different altitudes are converging, the PIC of the balloon at the higher altitude shall give way to the balloon at the lower altitude.

Where an aircraft is required to give way to another aircraft, the PIC of the first-mentioned aircraft shall not pass over or under, or cross ahead of, the other aircraft unless passing or crossing at such a distance as will not create a risk of collision.

Where two aircraft are approaching head-on or approximately so and there is a risk of collision, the PIC of each aircraft shall alter its heading to the right.

An aircraft that is being overtaken has the right of way and the PIC of the overtaking aircraft, whether climbing, descending or in level flight, shall give way to the other aircraft by altering the heading of the overtaking aircraft to the right, and no subsequent change in the relative positions of the two aircraft shall absolve the PIC of the overtaking aircraft from this obligation until that aircraft has entirely passed and is clear of the other aircraft.

Where an aircraft is in flight or maneuvering on the surface, the PIC of the aircraft shall give way to an aircraft that is landing or about to land.

The PIC of an aircraft that is approaching an aerodrome for the purpose of landing shall give way to any aircraft at a lower altitude that is also approaching the aerodrome for the purpose of landing.

The PIC of an aircraft at a lower altitude for landing purposes shall not overtake or cut in front of an aircraft at a higher altitude that is in the final stages of an approach to land.

No person shall conduct or attempt to conduct a take-off or landing in an aircraft until there is no apparent risk of collision with any aircraft, person, vessel, vehicle or structure in the take-off or landing path. [602.19]

Right of Way - Aircraft Maneuvering on Water

Where an aircraft on the water has another aircraft or a vessel on its right, the PIC of the first-mentioned aircraft shall give way.

Where an aircraft on the water is approaching another aircraft or a vessel head-on, or approximately so, the PIC of the first mentioned aircraft shall alter its heading to the right.

The PIC of an aircraft that is overtaking another aircraft or a vessel on the water shall alter its heading to keep well clear of the other aircraft or the vessel. [602.20]

Avoidance of Collision

No person shall operate an aircraft in such proximity to another aircraft as to create a risk of collision. [602.21]

Towing

No person shall operate an aeroplane that is towing an object unless the aeroplane is equipped with a tow hook and release control system that meet the applicable standards of airworthiness. [602.22]

Dropping of Objects

No person shall create a hazard to persons or property on the surface by dropping an object from an aircraft in flight. [602.23]

Formation Flight

No person shall operate an aircraft in formation with other aircraft except by pre-arrangement between the PICs of the aircraft, or where the flight is conducted within a control zone, the PICs and the appropriate ATC unit. [602.24]

Entering or Leaving an Aircraft in Flight

No person shall enter or leave an aircraft in flight except with the permission of the PIC of the aircraft. No PIC of an aircraft shall permit a person to enter or leave the aircraft during flight unless the persons leaves for the purpose for making a parachute descent or the flight is authorized pursuant to a SFOC or in accordance with aerial work operator certificate requirements. [602.25]

Parachute Descents

Except in accordance with a SFOC issued specifically for a parachute descent, no PIC of an aircraft shall permit, and no person shall conduct a parachute descent from the aircraft in or into controlled airspace or an air route, or over or into a built-up area or an open air assembly of persons. [602.26]

Aerobatic Maneuvers - Prohibited Areas and Flight Conditions

No person operating an aircraft shall conduct aerobatic maneuvers

a) over a built up area or an open air assembly of persons;
b) in controlled airspace, except in accordance with a SFOC;
c) when flight visibility is less than three miles;
b) below 2,000 feet AGL, except in accordance with a SFOC. [602.27]

Aerobatic Maneuvers with Passengers

No person operating an aircraft with a passenger on board shall conduct an aerobatic maneuver unless the PIC of the aircraft has engaged in at least 10 hours dual flight instruction in the conducting of aerobatic maneuvers or 20 hours conducting aerobatic maneuvers and at least on hour of conducting aerobatic maneuvers in the preceding 6 months. [602.28]

Compliance with Air Traffic Control Instructions and Clearances

AIR TRAFFIC CONTROL CLEARANCE means an authorization issued by an ATC unit that authorizes an aircraft to proceed within controlled airspace in accordance with the conditions specified by that unit. An ATC clearance is initiated by the pilot in the sense that the pilot has requested something, and a clearance will always have some form of the word "CLEAR" within it. EXAMPLE: "Cessna 172 ALFA BRAVO CHARLIE, you are cleared to land on runway 26".

An AIR TRAFFIC CONTROL INSTRUCTION means a directive issued by an air traffic control unit for air traffic control purposes. EXAMPLE: "Piper 140 ALFA BRAVO CHARLIE, remain west of the control zone, expect landing clearance shortly".

The PIC of an aircraft shall comply with and acknowledge, to the appropriate ATC unit, all of the air traffic control instructions directed to and received by the PIC and comply with all of the air traffic control clearances received and accepted by the PIC. In the case of a VFR flight, the PIC shall read back to the appropriate ATC unit the text of any air traffic control clearance received, when so requested by the ATC unit.

The PIC of an aircraft may deviate from an ATC clearance or an ATC instruction to the extent necessary to carry out a collision avoidance maneuver where the maneuver is carried in accordance with an Airborne Collision Avoidance System (ACAS) or a Traffic Alert and Collision Avoidance System (TCAS), or in response to a Ground Proximity Warning system (GPWS) on board the aircraft. The PIC of an aircraft shall, as soon as possible after initiating such a collision avoidance maneuver, inform the appropriate ATC unit of the deviation and immediately after completing the collision avoidance maneuver, comply with the last ATC clearance received and accepted by, or the last air traffic control instruction received and acknowledged by the PIC.

No PIC need comply with any ATC clearance or instruction where to do so would jeopardize aviation safety. An unsafe clearance should be immediately rejected by the PIC, and an alternate clearance should be requested. [602.31, AIP RAC 1-16]

[Reference: CARs sections and AIP references as specified]

1.14 Aircraft Registration and Airworthiness

Aircraft Registration

No person shall fly an aircraft, except a hang glider, unless it is registered. All aircraft (except hang gliders) must have the nationality marks and registration marks affixed in a clean and visible manner. The nationality mark will be either a "C" or a "CF" followed by a hyphen and the registration marks. EXAMPLE: CF-ABC, C-GABC.

Canadian registered aircraft must have attached two fireproof plates. The AIRCRAFT IDENTIFICATION PLATE bears the manufacturer's name and model, aircraft serial number, type approval and date and place of manufacture, and is to be located to the hull or fuselage where it is visible from the ground, either adjacent to and aft of the rear-most entrance door on the fuselage surface near the tail surfaces. A MARK PLATE bears the aircraft nationality and registration marks and on airplanes, is affixed in a prominent position near the main entrance, but is no longer required.

Aircraft Ownership

Except for hang gliders, model aircraft and military aircraft, all aircraft in Canada must be registered. The CERTIFICATE OF REGISTRATION (C of R) is a required document and must be carried on board the aircraft. Where the owner changes their address, or where there is a change of ownership of an aircraft, TC must be notified within 7 days. Where there is a change in ownership, Part 4 of the C of R (pink copy) acts as an interim C of R for the new owner and is valid for 60 days. To apply for a C of R in his/her name, the new owner must submit a bill of sale or other evidence of ownership, Part 2 of the C of R (white copy) and the required fee.

Aircraft Airworthiness

FLIGHT AUTHORITY means a CERTIFICATE OF AIRWORTHINESS (C of A), special certificate of airworthiness, flight permit or validation of a foreign airworthiness document, all of which attest to an aircraft's fitness for flight. No person shall operate an aircraft in flight unless a flight authority is in effect in respect of the aircraft, the aircraft is operated in accordance with the conditions set out in the flight authority and the flight authority is carried on board the aircraft.

A STANDARD C OF A is issued for normal aircraft types (such as Cessna 172, Piper 140 etc.) and SPECIAL C OF A, for situations such as amateur built aircraft. FLIGHT PERMITS are issued for aircraft in aeronautical research (experimental flight permit) or when a C of A isn't valid but the aircraft is capable of safe flight (specific purpose flight permit, such as ferry flights or importation flights).

A C of A or Flight Permit remains valid continuously provided the proper on-going maintenance inspections, compliance with Airworthiness Directives (AD's) and rectification of defects are being carried out. Aircraft owners and operators are responsible for compliance with all AD's that are applicable to their aircraft: if an AD is not complied with, the C of A is out of force.

Within 5-6 weeks of the C of A or flight permit anniversary date, TC sends an ANNUAL AIRWORTHINESS INFORMATION REPORT (AAIR) to the owner, which the owner completes and returns. Submission of the AAIR is mandatory, regardless of whether or not the C of A or flight permit is being renewed. The pink copy of the AAIR shall be retained with the C of A or flight permit.

Availability of Aircraft Flight Manual (AFM)

No person shall conduct a take-off in an aircraft for which an AFM is required by airworthiness standards, unless the AFM (or AOM where authorized) is available to the flight crew members.

Markings and Placards

No person shall conduct a take-off in an aircraft in respect of which markings or placards are required by the applicable standards of airworthiness unless the markings or placards are affixed to the aircraft or attached to a component of the aircraft in accordance with those standards.

Unserviceable and Removed Equipment

No person shall conduct a take-off in an aircraft that has equipment that is not serviceable or from which equipment has been removed if, in the opinion of the PIC, aviation safety is affected. A person may conduct a take-off in an aircraft that has equipment that is not serviceable or from which equipment has been removed where the aircraft is operated in accordance with the conditions of a flight permit that has been issued specifically for that purpose.

Technical Records

Every owner of an aircraft shall keep the following technical records in respect of the aircraft:

a) a journey log;
b) a separate technical log for the airframe and each installed engine;

c) an approved empty weight and balance report.

Every person who makes an entry in a technical record shall make the entry accurately, legibly and in a permanent manner, and enter the person's name and signature or employee identifier and date the entry.

The owner of an aircraft shall insure that all necessary measures are taken to protect the technical records for the aircraft from damage and loss. Every person who brings into use a new volume of an existing technical record shall make the entries relating to the preceding volume that are necessary to insure that an unbroken chronological record is maintained.

Where a person alters an entry on a technical record for the purpose of correcting the entry, the person shall do so by striking out the incorrect entry in such a manner that the underlying information remains legible, and inserting the correct entry together with the date of the alteration, the reason for the alteration if necessary, the persons's name and signature or employer identifier or electronic user code.

Journey Log Requirements

The particulars required to be entered in a journey log are set out in the CARs (ie PIC, date, air time, flight time, defects etc.). No person shall make a single entry in a journey log in respect of a series of flights unless the aircraft is operated by the same PIC throughout the series, or an approved daily flight record is used in a flight training operation.

The owner of an aircraft shall retain every entry in a journey log for a period of not less than one year. A three year retention period applies where the aircraft is registered to a Canadian and the journey log is used for the purposes of recording particulars of aircraft flight time.

No person shall take-off an aircraft unless the journey log is on board the aircraft. A person may conduct a take-off in an aircraft without carrying the journey log on board where it is not planned that the aircraft will land and shut down at any location other than the point of departure.

Technical Records other than the Journey Log

Engine parts, propeller and airframe parts and components installed and maintenance work must be recorded in the appropriate technical record. The particulars of the maintenance work including the name and identification number of the person who made the original entry must be recorded.

Transfer of Records

Every owner of an aircraft who transfers title of an aircraft, airframe, engine, propeller or appliance to another person shall, at the time of transfer, also deliver to that person all of the technical records that relate to that aeronautical product.

[Reference: CARs Part II, 605.03 - 605.10, 605.92 - 97, AIP LRA]

1.15 Visual Flight Rules

Minimum Visual Meteorological Conditions for VFR Flight in Controlled Airspace

CONTROLLED AIRSPACE means an airspace of fixed dimensions that is so specified in the DAH and within which air traffic control service is provided.

No person shall operate an aircraft in VFR flight within controlled airspace unless

a) the aircraft is operated with visual reference to the surface;
b) flight visibility is not less than three miles;
c) the distance of the aircraft from cloud is not less than 500 feet vertically and one mile horizontally;
d) where the aircraft is operated within a control zone, when reported, ground visibility is hot less than three miles and except when taking off or landing, the distance of the aircraft from the surface is not less than 500 feet. [602.114]

Minimum Visual Meteorological Conditions for VFR Flight in Uncontrolled Airspace

No person shall operate an aircraft in VFR flight within uncontrolled airspace unless

a) the aircraft is operated with visual reference to the surface;
b) where the aircraft is operated at or above 1,000 feet AGL, during the day, flight visibility is not less than one mile, during the night, flight visibility is not less than three miles, and, in either case, the distance of the aircraft from cloud is not less than 500 feet vertically and 2,000 feet horizontally;
c) where the aircraft is an aeroplane and is operated at less than 1,000 feet AGL, during the

day, flight visibility is not less than two miles, except if otherwise authorized in an air operator certificate. During the night, flight visibility shall be not less than three miles, and during both day and night, the aircraft must be operated clear of cloud. [602.115]

VFR Over-The-Top (VFR OTT)

An aircraft may be operated in VFR OTT flight during the cruise portion of the flight during day if

a) the aircraft is operated at a vertical distance from cloud of at least 1,000 feet;
b) where the aircraft is operated between two cloud layers, the vertical distance between the layers is at least 5,000 feet;
c) flight visibility at the cruising altitude of the aircraft is at least 5 miles;
d) the weather at the aerodrome of destination is forecast to have a sky condition of scattered cloud or clear and a ground visibility of 5 miles or greater with no forecast of precipitation, fog, thunderstorms or blowing snow, and those conditions are forecast to exist for the period from one hour before to two hours after the ETA (if based on TAF) and for the period from one hour before to three hours after the ETA (if based on FA where no TAF is available).

Additional aircraft equipment is required for VFR OTT operations (see below). A CPL or ATPL licence holder may fly VFR OTT (VFR OTT operations are not permitted for RPP holders, and PPL pilots need a VFR OTT rating). [602.116]

Special VFR Flight

An aircraft may be operated in special VFR flight within a control zone if

a) weather conditions preclude compliance with normal VFR requirements;
b) flight visibility is not less than one mile where the aircraft is an aeroplane;
c) the aircraft is operated clear of cloud and with visual reference to the surface at all times;
d) authorization to do so has been requested and obtained from the appropriate ATC unit.

When aerodrome traffic permits, an ATC unit shall authorize a PIC to operate an aircraft in special VFR flight within a control zone if

a) the PIC requests authorization to operate the aircraft in special VFR flight;
b) when reported, the ground visibility within the control zone is not less than one mile, where the aircraft is an aeroplane;
c) the aircraft is equipped with radiocommunication equipment capable of maintaining communication with the appropriate ATC unit;
d) during night, the authorization is for the purpose of allowing the aircraft to land at the destination aerodrome. [602.117]

Day VFR Flying Equipment

DAYLIGHT is the period from 1/2 hour before sunrise to 1/2 hour after sunset, or when the centre of the Sun's disc is less than 6° below the horizon. The minimum equipment for day VFR are as follows:

a) airspeed indicator;
b) an altimeter;
c) magnetic compass;
d) a reliable and functioning timepiece;
e) engine tachometer for each engine and for each rotor that has limiting speeds established by the manufacturer;
f) oil pressure indicator for each engine employing an oil pressure system;
g) a coolant temperature indicator for each liquid cooled engine;
h) oil temperature gauge for each air cooled engine having a separate oil system;
i) a manifold pressure gauge for each reciprocating engine used to power a helicopter;
j) a means for the flight crew, when seated at the flight controls to determine the fuel quantity in each main fuel tank and if the aircraft employs retractable landing gear, the position of the landing gear;
k) two-way radio if operating in Class B, C or D airspace or within an MF area or ADIZ;
l) appropriate radio navigation equipment if the aircraft is operated within Class B airspace. [605.14]

For VFR OTT operations, the aircraft must also have a pitot de-ice control, gyroscopic direction indicator, attitude indicator, turn and slip indicator or turn coordinator, two-way radio and adequate radio navigation equipment (ie VOR, ADF). [605.14, 605.15]

Night VFR Flying Equipment

NIGHT is from 1/2 hour after sunset to 1/2 hour before sunrise or when the centre of the Sun's disc is more than 6 degrees below the horizon. In addition to the required day VFR equipment, for night VFR the following is also required:

a) turn and bank indicator, or turn coordinator;
b) gyroscopic heading indicator;
c) a means of illuminating flight instruments;
d) a means of determining the aircraft heading that is independent of magnetic north, when operating in the NDA;
e) navigation lights (projection requirements: 110° for left (red) and right (green) position lights, 140° for rear position light (white), 30° above and below horizontal plane for rotating beacon (flashing red or white/red strobes);
f) landing light when carrying passengers;
g) flashlight at each crew member's station;
h) spare fuses. [605.16]

Use of Position and Anti-Collision Lights

No person shall operate an aircraft in the air or on the ground at night, or on water between sunset and sunrise, unless the aircraft position lights and anti-collision lights are turned on. Anti-collision lights may be turned off where the PIC determines that, because of operating conditions, doing so would be in the interests of safety. [605.17]

Continuous Listening Watch

Where an aircraft is equipped with radiocommunication equipment, the PIC shall insure that

a) a listening watch is maintained on the appropriate frequency;
b) where communications are required, communication is established with an ATC unit, FSS or CARS station as applicable, on that appropriate frequency. [602.137]

VFR 2-Way Communications Failure

VFR Comm Failure within Class B/C/D Airspace
Where there is a two-way radio communication failure between the controlling ATC unit and a VFR aircraft while operating in Class B, C or D airspace, the PIC shall leave the airspace via the shortest route (land at the airport if already within a control zone). If transponder equipped, squawk 7600 and inform ATC as soon as possible of the actions taken. [602.138]

The pilot should then land at a suitable aerodrome following the NORDO arrival procedures:

a) approach the traffic circuit of aerodrome from upwind side, join crosswind at circuit height and join circuit on downwind leg. Keep vigilant for traffic and if at a controlled airport, look for light signals;
b) before turning final, look for traffic;
c) landing clearance should be given on final approach, if not, pull up and make another circuit;
d) taxi to ramp (look for light signal at controlled airport).

VFR Comm Failure Outside Class B/C/D Airspace
If a VFR 2-way radio failure is experienced, continue to fly in VFR weather conditions and land at the nearest suitable aerodrome, if transponder equipped, select code 7600 and complete a NORDO arrival as above.

If no suitable nearby aerodrome is available, the pilot may continue into and enter Class B, C or D airspace, squawk 7600, use the arrival procedures above and advise ATC afterwards. [602.138, AIP RAC 4-10]

[Reference: CARs and AIP references specified]

1.16 Canadian Airspace Structure

Canadian Domestic Airspace (CDA)

CDA includes all airspace over the Canadian land mass, the Canadian Arctic, Canadian Archipelago and high seas areas within airspace boundaries.

Fig. 1-1: Canadian Domestic Airspace.

CDA is geographically divided into the Southern Domestic Airspace and the Northern Domestic Airspace. In the SDA, magnetic track is used to determine cruising altitude for direction of flight, while true track is used in the NDA due to erratic indications from the magnetic compass.

The CDA is further divided vertically into the Low Level Airspace, which consists of all of the airspace below 18,000 feet ASL, and the High Level Airspace which consists of all airspace from 18,000 feet and above.

Low Level Controlled Airspace

A CONTROL ZONE is airspace designated around certain airports to control IFR and VFR aircraft. The normal radius is 7 NM, although some control zones may have a 5 NM or 3 NM radius. The normal height is 3,000 feet AAE (Above Aerodrome Elevation).

A CONTROL AREA EXTENSION is designated around busy airports to provide additional controlled airspace for IFR aircraft. They are normally based at 2,200 feet AGL and extend up to 18,000 feet ASL. The outer portions may be based at higher levels.

A TERMINAL CONTROL AREA (TCA) is provided to service IFR aircraft. Aircraft operating within a TCA are subject to certain rules. A TCA is normally circular centered on the primary aerodrome with a 45 NM radius from 9,500 feet to 35 NM at 2,200 feet to 12 NM at 1,200 feet. A military TCA (MTCA) is the same as a TCA, except that it will be found at a military aerodrome.

A TRANSITION AREA provides additional airspace for IFR operations, and is normally based at 700 feet AGL, extending upwards to the base of overlying controlled airspace. The area around an aerodrome will normally be 15 NM radius, but shall be of sufficient size to contain all of the aerodrome published instrument approach procedures.

Airspace Classifications

CDA is divided into classifications Class A through G. The rules for operating in a particular type of airspace depends upon the classification of that airspace.

Class A
Controlled airspace. Only IFR flight is permitted. Consists of controlled, high level airspace from 18,000 feet to FL 600.

Class B
Consists of controlled, low level airspace above 12,500 feet, or the MEA (Minimum Enroute Altitude), whichever is higher, to, but not including 18,000 feet. May also include a control zone and associated terminal control area. Flights may be IFR or VFR. VFR aircraft must have a clearance from ATC prior to entering Class B airspace. Position reports are required. Aircraft must have a 2-way radio and appropriate navigation equipment to maintain airway/air route tracking. Where weather falls below VFR, VFR aircraft must leave Class B airspace by the shortest route, by exiting horizontally or vertically, and inform ATC as soon as possible. If in a control zone, a request for Special VFR may be made.

Class C
Controlled airspace. IFR and VFR permitted. VFR aircraft require a clearance to enter. 2-way radio required. Class C control zones and terminal Class C airspace. Traffic information and conflict resolution may be provided. Class C becomes Class E if appropriate ATC unit not in operation.

Class D
Controlled airspace. IFR and VFR permitted. VFR aircraft must establish contact with ATC to enter. 2-way radio required. Traffic information and conflict resolution may be provided. Class D control zones and terminal Class D airspace. Class D becomes Class E when appropriate ATC unit not in operation.

Class E
Class E airspace is designated where there is a need for controlled airspace, but it does not meet the criteria for Class A, B, C or D. VFR/IFR aircraft may enter. Includes low level airways, control area extensions, transition areas or control zones established without an operating control tower.

Class F
Airspace where limitations must be put on aircraft operations. May be Class F ADVISORY (VFR aircraft should avoid unless participating in the activity) or Class F RESTRICTED (advance permission required to enter). Class F airspace is published on VFR and IFR charts. It includes alert areas, danger areas, rocket ranges, restricted areas, forest fire restrictions, military activity areas. May be controlled or uncontrolled, or a combination of both. Check NOTAMS for details on whether a particular Class F airspace is in operation. Identification code groups: CYA = Advisory (A = Acrobatics, T = training, F = aircraft test area, P = Parachute, S = Soaring, M = military operations, H = hang gliding). CYR = Restricted. CYD = Danger. B.C. = 100, Alberta = 200, Sask =

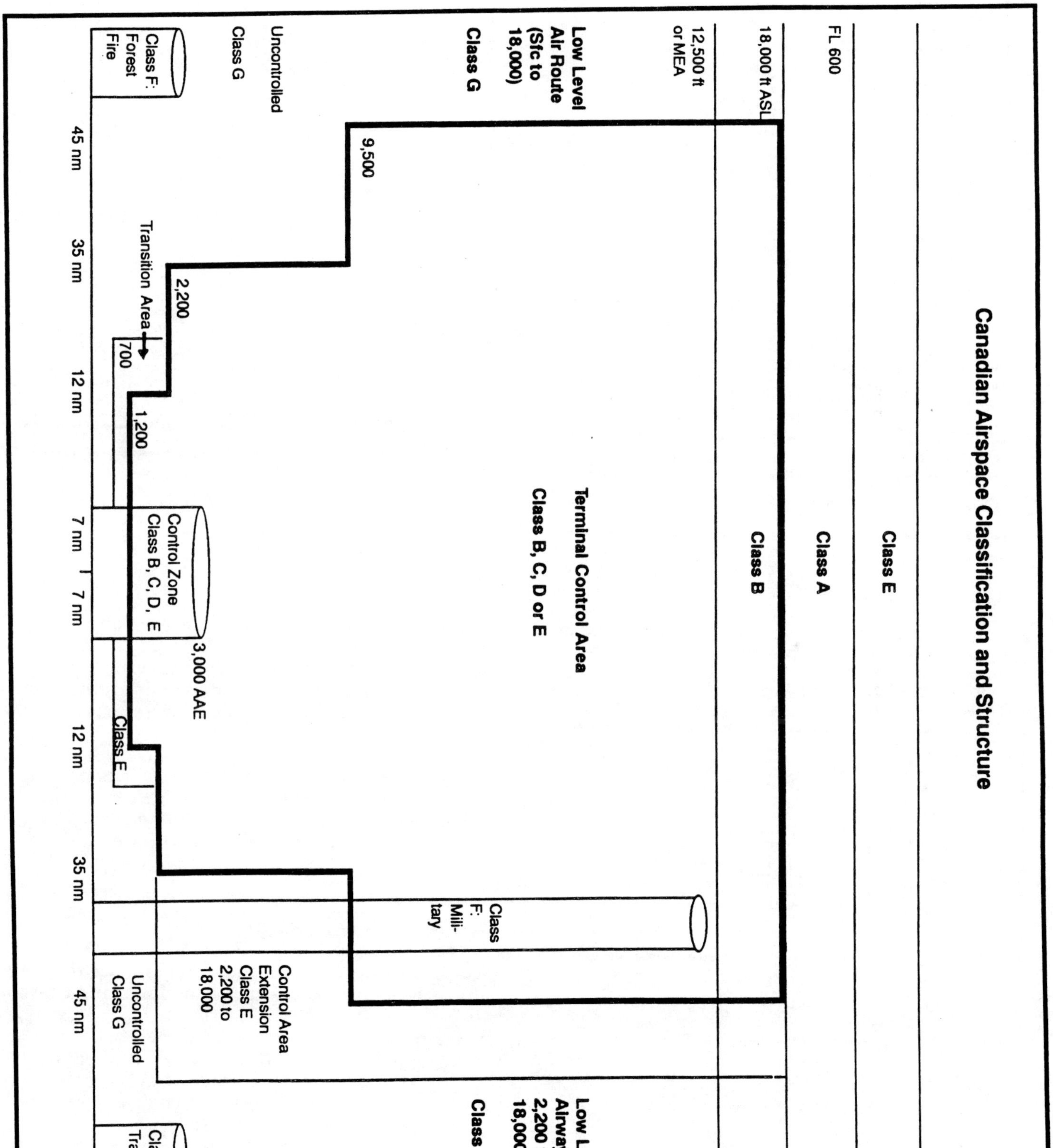

Fig. 1-2: Canadian Airspace Structure.

300, Man. = 400, Ont. = 500, Que. = 600, Atlantic = 700, Yukon = 800, NWT = 900. EXAMPLE: CYA 113 (A) means Class F Advisory airspace No. 113 in B.C. for acrobatics.

Class G
Airspace not designated A, B, C, D, E, or F. ATC has no authority. Flight information and alerting of SAR services provided by ATS. All uncontrolled domestic airspace. Includes all LOW LEVEL AIR ROUTES.

Airway and Air Route Structure

Low Level Airways are controlled airspace from 2,200 feet AGL to, but not including 18,000 feet ASL. VHF airways are 4 NM wide, either side of centerline, increasing where 4.5° lines intersect 4 NM width. LF/MF airways are 4.34 NM wide, either side of centerline, increasing where 5° lines intersect 4.34 NM width. Air routes have the same dimensions as airways, the differences being that air routes are in uncontrolled airspace and commence at the surface.

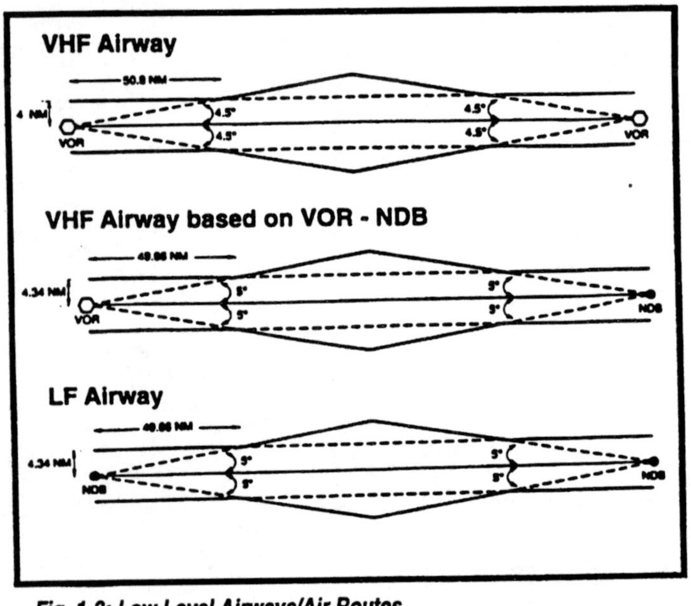

Fig. 1-3: Low Level Airways/Air Routes.

Flight Information Regions (FIR)

An FIR is airspace of defined dimensions extending upwards from the surface, in which flight information service (ie NOTAMS) and alerting services are provided. The FIRs are Vancouver, Edmonton, Winnipeg, Toronto, Montreal, Moncton and Gander.

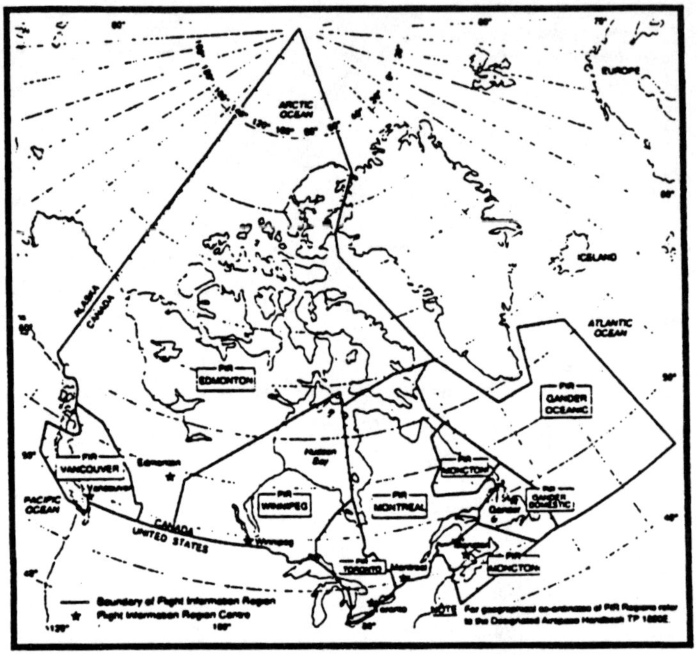

Fig. 1-4: Flight Information Regions.

1.17 Altimeter Setting Procedures

Altimeter Setting Region

When operating within the Altimeter Setting Region, the aircraft altimeter is to be set to the current altimeter setting of the nearest station. When approaching an aerodrome for landing, set the aircraft altimeter to that of the aerodrome if it is obtainable.

Unless otherwise authorized by ATC, when flying from the Altimeter Setting Region into the Standard Pressure Region, set the altimeter to 29.92" immediately following entry into the Standard Pressure Region.

Standard Pressure Region

Within the Standard Pressure Region, set the altimeter to 29.92", except in the following cases:

a) Departure: set to airport altimeter setting (airport elevation if unavailable). Immediately prior to reaching cruise flight level, set 29.92".
b) Arrival: immediately prior to commencing a descent for landing at an aerodrome located in the Standard Pressure Region, set the altimeter to the current altimeter setting of the aerodrome of intended landing, if it is available. When conducting a holding procedure prior to landing at an aerodrome within the Standard

Pressure Region, do not set the aircraft altimeter to the current altimeter setting of the aerodrome of intended landing until immediately prior to descending below the lowest flight level from which the holding procedure is conducted. CURRENT ALTIMETER SETTING is from observation within previous 90 minutes. When flying from the Standard Pressure Region into the Altimeter Setting Region, set the altimeter to the appropriate current altimeter setting immediately prior to entry into the altimeter setting region.

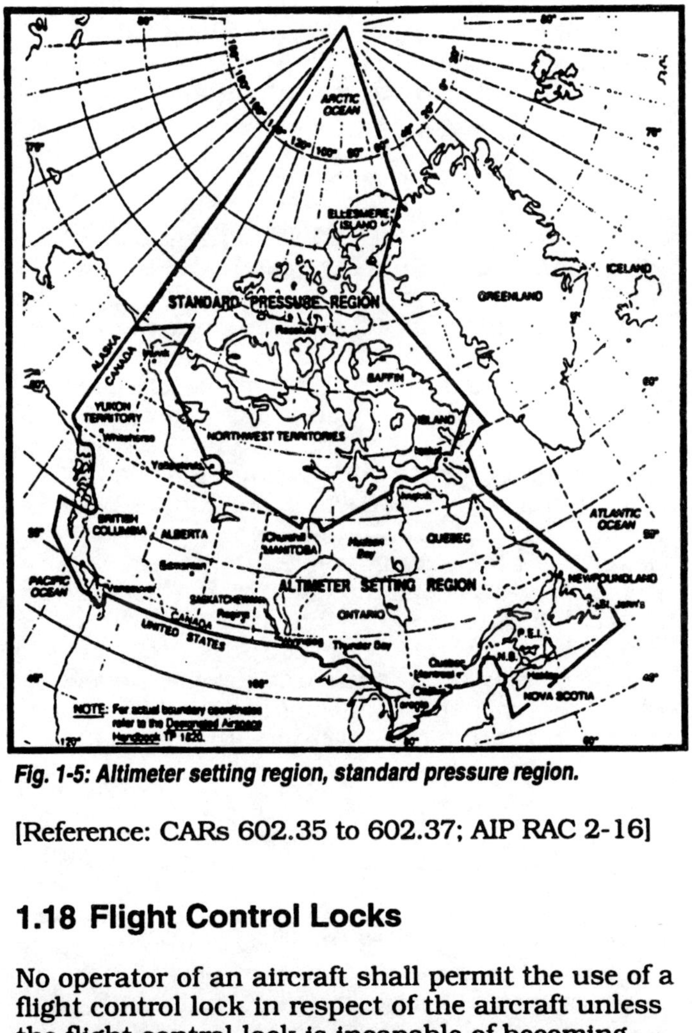

Fig. 1-5: Altimeter setting region, standard pressure region.

[Reference: CARs 602.35 to 602.37; AIP RAC 2-16]

1.18 Flight Control Locks

No operator of an aircraft shall permit the use of a flight control lock in respect of the aircraft unless the flight control lock is incapable of becoming engaged when the aircraft is being operated. As well, flight control locks must have an unmistakable warning so that the person operating the aircraft will know that the control lock is engaged.

[Reference: CARs 605.29]

1.19 Transponder Operation

Radar service by ATC improves aircraft spacing and enhances safety. A transponder is a device on an aircraft which enhances the aircraft's target on a controller's radar scope. Radar traffic information is provided to pilots on the basis of the "clock" system. For example, if a controller advises a pilot "traffic at 6 O' clock" the traffic would be directly behind, based on the aircraft TRACK (path over the ground) as observed on the controller's radar. Radar service also provides navigation assistance to VFR flights.

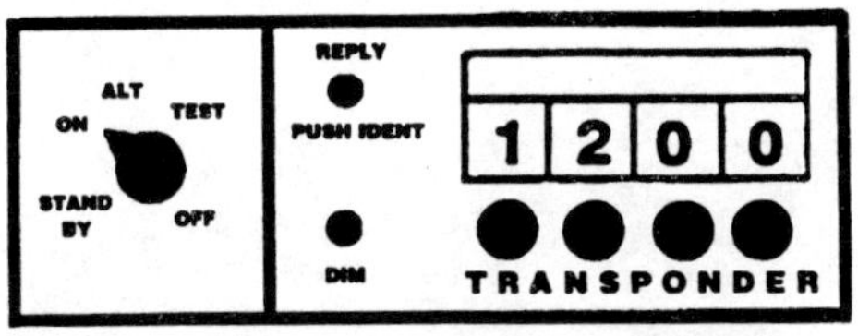

Fig. 1-6: Face of typical MODE A/MODE C transponder.

When pilots receive ATC instructions concerning transponder operation, they must operate their transponders as directed by ATC. ATC radar units have an alarm system that activates when the pilot selects code 7700 - EMERGENCY, code 7600 - COMMUNICATIONS FAILURE, and code 7500 - HIJACK. The transponder should be on "STANDBY" mode for taxi, and switched to "ON" just before take-off. The "IDENT" button should only be selected when specifically requested by ATC.

If no specific instructions are given by ATC concerning which code to select, when VFR, select code 1200 when at or below 12,500 feet ASL, and code 1400 when above 12,500 feet ASL.

Conventional transponders are referred to as MODE A. MODE C capability means the transponder has automatic pressure altitude reporting equipment (encoding altimeter). MODE A/C transponders are required for operations in all Class A, B and C airspace, and all Class E airspace between 10,000 feet and 12,500 feet ASL within RADAR coverage (see Fig. 1- 7 below). As well, a MODE A/C transponder is required when operating within the following Class D TCA's: Winnipeg Int'l, Ottawa/Macdonald-Cartier Int'l, Québec/Jean Lesage Int'l, St-Hubert and Halifax Int'l. A MODE A/C transponder is mandatory at the following Class E airspace aerodromes: Saskatoon, Regina, Thunder Bay, Moncton, Gander and St. John's.

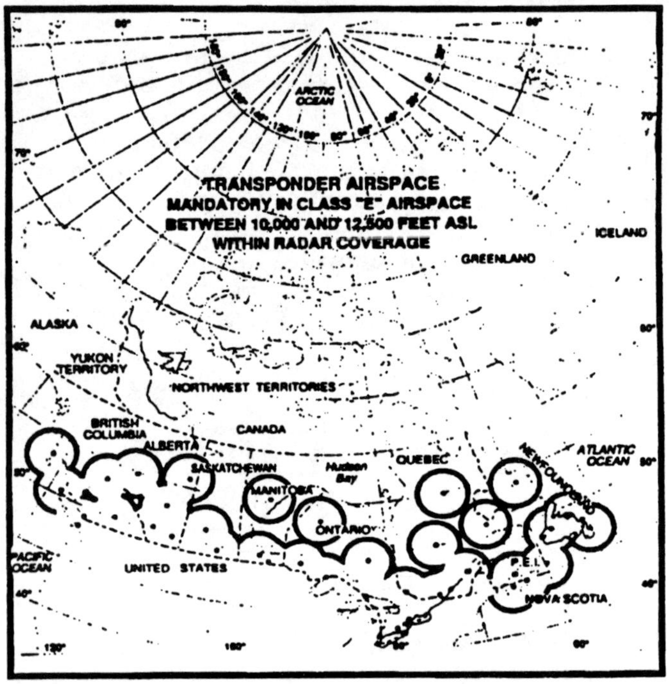

Fig. 1-7: Transponder Airspace.

When the transponder or the automatic pressure altitude reporting equipment fails during flight where its use is mandatory, an aircraft may be operated to the next airport of intended landing, and thereafter, to complete an itinerary or to a repair base, if authorized by ATC.

ATC may, on receipt of a specific request, authorize an aircraft not equipped with a functioning transponder to operate in airspace where its use is mandatory. Approval is subject to traffic conditions.

[Reference: CARs 605.35; AIP RAC 1-18 to 1-21]

1.20 Flight Preparation, Flight Plans, Flight Itineraries

General Rules

The PIC of an aircraft shall, before commencing a flight, be familiar with the available information that is appropriate to the intended flight. The PIC of an aircraft shall, before commencing a flight, be familiar with the available weather information that is appropriate to the intended flight.

No pilot shall operate an aircraft in VFR flight unless a VFR FLIGHT PLAN or a VFR FLIGHT ITINERARY has been filed, except where the flight is conducted within 25 NM of the departure aerodrome.

A flight plan shall be filed with an ATC unit, an FSS or CARS station. A flight itinerary shall be filed with a responsible person, an ATC unit, an FSS or CARS station. A flight plan or flight itinerary shall be filed by sending, delivering or otherwise communicating it or the information contained in it and receiving acknowledgment that the information has been received.

Responsible Person

A RESPONSIBLE PERSON means an individual who has agreed with the person who has filed a flight itinerary to insure that the following are notified if the aircraft is overdue: an ATC unit, FSS station, CARS or rescue coordination centre.

Overdue Aircraft Notification

An aircraft is considered OVERDUE on a flight plan immediately after any search and rescue notification time specified on the flight plan, and if no time has been specified, within one hour after the last reported ETA. An aircraft is considered overdue on a flight itinerary immediately after any search and rescue notification time specified in the flight itinerary, and if no time has been specified, within 24 hours after the last reported ETA. Any person who assumes responsibilities with respect to an aircraft and who has reason to believe that the aircraft is overdue, or any other person who has been directed by that person to do so, shall immediately, by the quickest means available, notify an ATC unit, FSS/CARS or a rescue coordination centre and provide, to the best of the person's knowledge, all the available information concerning the overdue aircraft that may be requested.

Changes

Where the PIC wishes to make any changes to a VFR flight plan or flight itinerary while enroute in respect of the route of flight, duration of flight or destination aerodrome, the pilot shall notify as soon as practicable an ATC unit, an FSS/CARS station, or the responsible person of the intended change.

Closing

To close a flight plan, an ARRIVAL REPORT must be submitted to an ATS unit no later than the search and rescue time specified in the flight plan, and if not specified, within one hour of the last reported ETA. If a flight plan is not closed, search and rescue will be initiated. An arrival report must include the registration marks, type of flight plan/ itinerary, departure aerodrome, date and time of arrival, and arrival aerodrome.

To close a flight itinerary, an arrival report must be filed as above, either with ATC/FSS/CARS or the responsible person, not later than the SAR time specified or where not specified, within 24 hours after the last reported ETA.

Travel Outside Canada

A VFR flight plan must be filed prior to conducting any flight between Canada and a foreign state. Where the flight is to the US from a Canadian airport very close to the border, the flight plan should be filed at least on hour in advance in order to facilitate adequate coordination and data transfer. Transborder flights should be made only to the point of customs entry, and thereafter, another flight plan should be filed. For flights to nations other than the US, use the ICAO format.

Intermediate Stops

Except for transborder flights, a single VFR flight plan or flight itinerary may be filed including one or more intermediate stops en route, provided the stop will be of short duration for purposes such as boarding passengers or refuelling. Each intermediate stop is indicated by repeating the name of the stopping point and its duration in the route section of the flight plan or itinerary. When an intermediate stop has been planned, the EET calculated should include all intermediate stops to final destination. EXAMPLE: For a 30 minute stop at Montreal, Dorval, put in Route: CYUL (0+30) CYUL.

VFR Flight Plan/Flight Itinerary Format

Review Fig. 1-8 which shows a sample VFR flight plan. The same form is used for a Canadian Flight Plan/Itinerary, IFR Flight Plan/Itinerary, VFR/IFR Composite Flight Plan/Itinerary and ICAO Flight Plan. The following is a summary for completion of a VFR Flight Plan or VFR Flight Itinerary.

Aircraft Identification

The aircraft identification (call sign) is inserted in Box 7. EXAMPLE: CF-ABC

Flight Rules and Type of Flight

In Box 8, first enter the flight rules. V = VFR. Then enter the type of flight in the first character: C = Controlled VFR, D = Defence VFR, E = Defence Flight Itinerary, F = Flight Itinerary. Second character: G = General Aviation. EXAMPLE: FG = flight itinerary, general aviation.

Number and Type of Aircraft and Wake Turbulence Category

The first part of Box 9 is for the number of aircraft (for VFR flights leave blank). Then enter the type designator. For wake turbulence category, use no entry for non-TCAS aircraft with a MCTOW of less than 300,000 lbs.

Equipment

In Box 10, enter the communications, navigation and transponder equipment in the order COM - NAV - SSR. For COM/NAV equipment, N = no navigation or communications equipment or the equipment is unserviceable, S = Standard, VHF communications radios, ADF, VOR, and ILS. Other codes: C = LORAN C, D = DME, F = ADF, L = ILS, O = VOR, V = VHF, G = GNSS. For SSR equipment, N = Nil, A = Mode A, C = Mode C. EXAMPLE: a Cessna 172 with a VHF radio, VOR, ADF, GPS and Mode C transponder would be abbreviated to C172 under "type" and V/FOG/C under "equipment" in Box 10.

Departure Aerodrome

In Box 13, enter the departure aerodrome using the 3 or 4 digit character from the CFS. EXAMPLE: London, ON = CYXU, Valcourt, QB = SQ3.

Departure Time

Enter anticipated time of departure from departure aerodrome in hours and minutes UTC.

Cruising Speed, Altitude, Level, Route

In Box 15, under "cruising speed", insert the flight planned TAS in knots. N = knots. EXAMPLE: N0090 = 90 Knots TAS. For "cruising level", insert the planned cruising level for the first or the whole portion of the route to be flown. A = altitude in hundreds of feet ASL. EXAMPLE: A060 = 6,000 feet ASL.

Transport Canada / Transports Canada

CANADIAN FLIGHT PLAN / ITINERARY / **PLAN / ITINÉRAIRE DE VOL CANADIEN**

ICAO FLIGHT PLAN / **PLAN DE VOL OACI**

PRIORITY / PRIORITÉ: <<≡ FF → ADDRESSEE(S) / DESTINATAIRE(S): <<≡

FILING TIME / HEURE DE DÉPÔT → ORIGINATOR / EXPÉDITEUR <<≡

SPECIFIC IDENTIFICATION OF ADDRESSEE(S) AND/OR ORIGINATOR / IDENTIFICATION PRÉCISE DU(DES) DESTINATAIRE(S) ET/OU DE L'EXPEDITEUR

3 MESSAGE TYPE / TYPE DE MESSAGE: <<≡ (FPL
7 AIRCRAFT IDENTIFICATION / IDENTIFICATION DE L'AÉRONEF: – C.F.-A.A.T.
8 FLIGHT RULES / RÈGLES DE VOL: – V
TYPE OF FLIGHT / TYPE DE VOL: G <<≡

9 NUMBER / NOMBRE: –
TYPE OF AIRCRAFT / TYPE D'AÉRONEF: C.1.8.5
WAKE TURBULENCE CAT / CAT. DE TURBULENCE DE SILLAGE: /
10 EQUIPMENT / EQUIPEMENT: – SG/C <<≡

13 DEPARTURE AERODROME / AÉRODROME DE DÉPART: – C.Y.H.U
TIME / HEURE: 1.6.0.0 <<≡

15 CRUISING SPEED / VITESSE DE CROISIÈRE: – N.0.1.2.0
ALTITUDE, LEVEL / NIVEAU: →
ROUTE / ROUTE: CYJN (+15) CYJN
<<≡

16 DESTINATION AERODROME / AÉRODROME DE DESTINATION: – S.S.3.
TOTAL EET / DURÉE TOTALE ESTIMÉE (DAYS/JOURS, HRS., MINS): 0.0.4.0
SAR (HRS, MINS):
ALTN AERODROME / AÉRODROME DE DÉGAGEMENT: →
2ND ALTN AERODROME / 2ÈME AÉRODROME DE DÉGAGEMENT: → <<≡

18 OTHER INFORMATION / RENSEIGNEMENTS DIVERS: – 0
)<<≡

19 ENDURANCE / AUTONOMIE (HR, MIN): – E/ 0.4.3.0
PERSONS ON BOARD / PERSONNES À BORD: → P/ 0.0.3
EMERGENCY RADIO / RADIO DE SECOURS: → R/ UHF [X] VHF [V] ELBA [E] ELT [A.F]

SURVIVAL EQUIPMENT / ÉQUIPEMENT DE SURVIE: → S / POLAR / POLAIRE [X] DESERT / DÉSERT [X] MARITIME / MARITIME [M] JUNGLE / JUNGLE [X]
JACKETS / GILETS DE SAUVETAGE: → J / LIGHT / LAMPES [L] FLUORES / FLUORES [X] UHF [X] VHF [X]

DINGHIES / CANOTS: → D / NUMBER / NOMBRE: 0.1 → CAPACITY / CAPACITÉ: 0.0.6 → COVER / COUVERTURE: [X] → COLOUR / COULEUR: YELLOW <<≡

AIRCRAFT COLOUR AND MARKINGS / COULEUR ET MARQUES DE L'AÉRONEF: A / WHITE WITH BLUE TRIM
WHEELS / ROUES [X] SEAPLANE / HYDRAVION [X] SKIS [ ] AMPHIBIAN / AMPHIBIE [X]

REMARKS / REMARQUES: → [X] <<≡

AN ARRIVAL REPORT WILL BE FILED WITH - UN COMPTE RENDU D'ARRIVÉE SERA NOTIFIÉ À: MONTREAL FSS

NAME AND PHONE NUMBER OR ADDRESS OF PERSON(S) OR COMPANY TO BE NOTIFIED IF SEARCH AND RESCUE ACTION INITIATED / NOM ET NUMÉRO DE TÉLÉPHONE OU ADRESSE DE LA (DES) PERSONNE(S) OU COMPAGNIE À AVISER SI DES RECHERCHES SONT ENTREPRISES: ACCELERATED AVIATION 6042790179

PILOT-IN-COMMAND / PILOTE COMMANDANT DE BORD: C / C. YAEGER
PILOT'S LICENCE No / Nº DE LICENCE DU PILOTE: 123456) <<≡

FILED BY / DÉPOSÉ PAR

SPACE RESERVED FOR ADDITIONAL REQUIREMENTS / ESPACE RÉSERVÉ À DES FINS SUPPLÉMENTAIRES

Fig. 1-8: Sample VFR Flight Plan. Aircraft is a Cessna 185 Amphibian with wheels and floats equipped with VHF radios and VOR, ADF, ILS, GPS and MODE C transponder. Departure will be at 1600 UTC from CYHU (St. Hubert, QB), speed 120 kts to CYJN (St. Jean, QB), where at 15 minute stopover is planned. Then, the flight will continue to destination SS3 (Les Cedars, QB). Estimated flight time, inclusive of stopover is 40 minutes, no other information. Fuel endurance is 4 hours and 30 minutes. There are 3 persons on board. The aircraft has a type AF fixed ELT, capable of transmission over VHF frequency 121.5 MHz. Aircraft is equipped with lighted life jackets and sea survival equipment, including a yellow life raft. Aircraft is white with blue trim, no remarks. Arrival report will be filed with Montreal FSS.

For "route", include changes of speed, altitude changes, and airway numbers or points along route. Flight is assumed to be direct unless DCT is used for points between other points named. EXAMPLE: For a routing of Calgary (International) - Red Deer - Edmonton Municipal (30 minute stopover) -Edson (destination) enter CYYC under Box 13, then YQF - YXD (+ 30) YXD under Box 15 - Route, then YET under Box 16.

Destination Aerodrome, EET, SAR, Alternate

In Box 16, enter the 3 or 4 letter code for the destination aerodrome as found in the CFS. Enter the Estimated Elapsed Time (EET), which may also include the number of days. The pilot can insert own SAR time, to a maximum of 24 hours beyond EET. No alternate is required on a VFR fight plan/itinerary.

Other Information

Insert "0" in Box 18 if no other information applies. ADCUS should be inserted for a transborder flight, as well as number of US citizens/Canadian citizens and pilot's name.

Supplemental Information

Enter fuel endurance in hours and minutes in Box 19. Enter total number of persons on board.

Under "emergency radio", put an "X" over the "U" if not equipped with UHF emergency frequency (243.0 MHz), put an "X" through the V if not equipped with VHF emergency frequency (121.5 MHz). If ELT is not available, cross out the "E" under ELBA. Under ELT, insert the ELT type. Type A or AD ELT = automatic ejectable or automatic deployable. Type F or AF = fixed or automatic fixed ELT. Type AP = automatic portable. Type W or S = water activated or survival (a floating, free ELT). Type P = Personnel (non fixed).

For Survival Equipment, Jackets and Dinghies, cross out types of survival equipment not carried.

If no remarks, cross out the "N" in the box.

Under arrival report, include with who or what facility the flight plan will be closed. Enter pilot name, licence number and person to be notified if SAR initiated.

[Reference: CARs 602.70 - 79, 602.145; AIP RAC 3-3]

1.21 Pre-Flight and Fuel Requirements

Carry-on Baggage, Equipment and Cargo

No person shall operate an aircraft with carry-on baggage, equipment or cargo on board unless the carry-on baggage, equipment and cargo are

a) stowed in a bin, compartment, rack or other location that is certified in accordance with the aircraft type certificate in respect of the stowage of carry-on baggage, equipment or cargo; or
b) restrained so as to prevent them from shifting during movement of the aircraft on the surface and during take-off, landing and in-flight turbulence.

No person shall operate an aircraft with carry-on baggage, equipment or cargo on board unless

a) the safety equipment, the normal and emergency exits that are accessible to passengers and the aisles between the flight deck and a passenger compartment are not wholly or partially blocked by carry-on baggage, equipment or cargo;
b) all of the equipment and cargo that are stowed in a passenger compartment are packaged or covered to avoid possible injury to persons on board;
c) where the aircraft is type-certificated to carry 10 or more passengers and passengers are carried on board, no passenger's view of any "seat belt" sign, "no smoking" sign or exit sign is obscured by carry-on baggage, equipment or cargo except if an auxiliary sign is visible to the passenger or another means of notification of the passenger is available, all passenger service cars and trolleys are securely restrained during movement of the aircraft on the surface, take-off and landing, and during in-flight turbulence where the PIC or in charge flight attendant has directed that the cabin be secured, and all of the video monitors that are suspended from the ceiling of the aircraft and extend into an aisle are stowed and securely restrained during take-off and landing;
d) all of the cargo that is stowed in a compartment to which crew members have access is stowed in such a manner as to allow a crew member to effectively reach all parts of the compartment with a hand-held fire extinguisher. [602.86]

Fuel Requirements

No PIC of an aircraft shall commence a flight or, during flight, change the destination aerodrome set out in the flight plan or flight itinerary, unless the aircraft carries the required fuel.

An aeroplane operated in VFR flight shall carry an amount of fuel that is sufficient to allow the aircraft,

a) when operated during the day, to fly to the destination aerodrome and then to fly for a period of 30 minutes at normal cruising speed;
b) when operated at night, to fly to the destination aerodrome and then to fly for a period of 45 minutes at normal cruising speed.

Every aircraft shall carry an amount of fuel that is sufficient to provide for

a) taxiing and foreseeable delays prior to take-off;
b) meteorological conditions;
c) foreseeable air traffic routings and traffic delays;
d) landing at a suitable aerodrome in the event of loss of cabin pressurization or, in the case of a multi-engined aircraft, failure of any engine, at the most critical point during the flight; and
e) any other foreseeable conditions that could delay the landing of the aircraft. [602.88]

Passenger Briefings

The PIC of an aircraft shall insure that all of the passengers on board the aircraft are briefed before take-off with respect to the following, where applicable:

a) the location and means of operation of emergency and normal exits;
b) the location and means of operation of safety belts, shoulder harnesses and restraint devices;
c) the positioning of seats and the securing of seat backs and chair tables;
d) the stowage of carry-baggage;
e) where the aircraft is unpressurized and it is possible that the flight will require the use of oxygen by the passengers, the location and means of operation of oxygen equipment;
f) any prohibition against smoking.

The PIC of an aircraft shall insure that all of the passengers on board the aircraft are briefed, in the case of an over-water flight where the carriage of life preservers, individual flotation devices or personal flotation devices is required, before commencement of the over-water portion of flight, with respect to the location and use of those items.

The PIC of an aircraft shall, before take-off, insure that all of the passengers on board the aircraft are provided with information respecting the location and use of

a) first aid kits and survival equipment;
b) any ELT that is required to be carried;
c) any life raft that is required to be carried. [602.89]

[Reference: CARs sections specified]

1.22 NOTAM

NOTAMs give information relating to the establishment or changes to facilities such as VOR, ADF, airport services, hazards and procedures, which pilots must receive in a timely fashion. NOTAMs can be obtained by telephone, or from an FSS.

The teletype format of a NOTAM is as follows:

(1) GG CYZZLF
(2) 141736 CYOWYF
(3) 990007 NOTAMN CYOW OTTAWA INTL
(4) CYOW ILS 07 U/S TIL APRX 9901191200.

Meaning of the above NOTAM:

(1) Teletype priority and group address.
(2) Origination time. Day, hour in UTC, and issuing location: 14th day of month, time 17:36 UTC, Ottawa International Airport.
(3) 990007 = NOTAM continuity number (the 7th NOTAM of that airport for 1999). NOTAMN = New NOTAM. NOTAMR is Revised, NOTAMC is Cancelled, NOTAMJ is for a JBI NOTAM, NOTAMQ is for a Query. If APRX is used, there must be a follow-up NOTAM to cancel the original NOTAM.
(4) 9901191200 = year, month, day and hour in UTC. The ILS at Ottawa Int'l for runway 07 will remain unserviceable until Jan 19, 1999 at 1200 UTC.

1.23 AIP Canada

The Canadian Aeronautical Information Publication (AIP) is a pre-flight reference document providing summarized rules and procedures for operation of aircraft in CDA. It also contains some important CARs in original format.

Where the term SHOULD is used either in the AIP

or any regulation, it means that pilots should conform to the applicable procedure, while the word SHALL means that the applicable procedure is mandatory and is supported by regulatory authority.

Remember that the final authority for any rule or procedure is not the AIP but the original regulation or rule in its legislative version in the *Aeronautics Act*, CARs or other enactment. If any doubt exists or for the final legal authority on a rule or regulation, refer to these regulatory sources and not the AIP.

The AIP incorporates regular revisions and like any other flight chart or publication, users should insure their copy is up-to-date.

While permanent changes are published as AIP amendments, temporary changes of operational significance, containing extensive text or graphics and of long duration (three months or longer) will be published as an AIP Supplement. An AIP Supplement should not be used to duplicate information better contained in the AIP itself. AIP Supplements should be kept in the AIP binder in the appropriate section for as long as all or some of the contents remain valid.

An AERONAUTICAL INFORMATION CIRCULAR (AIC) provides advance notification of major changes to legislation, regulations, procedures or purely administrative matters where the text is not part of the AIP.

An AVIATION NOTICE is used to disseminate information which is mainly of interest to specific TC regions or when the information is of a one-time nature and the notice will have served its purpose on first reading.

AIRAC Canada is a notice which is issued weekly by AIS to provide advance notification to chart makers and producers of aeronautical information concerning changes within the CDA. This notice insures that all users of the CDA have the same information on the same date.

[Reference: AIP MAP 6-1]

1.24 Aircraft Accidents

The objective of aviation safety investigation is to prevent recurrence of aviation accidents or incidents. The Transportation Safety Board of Canada (TSB) is responsible for accident investigation. A team of investigators is on 24 hours standby duty.

An AVIATION OCCURRENCE means any accident or incident associated with the operation of aircraft.

A REPORTABLE AVIATION ACCIDENT means an accident resulting directly from the operation of an aircraft, where

a) a person sustains a serious injury or is killed as a result of being on board the aircraft, coming into contact with any part of the aircraft or its contents, or being directly exposed to the jet blast or rotor downwash of the aircraft;
b) the aircraft sustains damage or structural failure adversely affecting the structural strength, performance or flight characteristics of the aircraft and normally requiring major repair or replacement of an affected component part, or
c) the aircraft is missing or inaccessible.

A REPORTABLE AVIATION INCIDENT means an incident resulting directly from the operation of an airplane having a maximum certificated takeoff weight greater than 5,700 kgs, or from the operation of a rotorcraft having a maximum certificated takeoff weight greater than 2,250 kgs where

a) an engine fails or is shut down as a precaution;
b) a transmission gearbox malfunction occurs;
c) smoke or fire occurs;
d) difficulties in controlling the aircraft are encountered owing to any aircraft system malfunction, weather phenomena, wake turbulence, uncontrolled vibrations or operations outside the flight envelope;
e) the aircraft fails to remain within the intended landing or takeoff area, lands with all or part of the landing gear retracted or drags a wing tip, an engine pod or any other part of the aircraft;
f) any crew member is unable to perform the crew member's duties as a result of a physical incapacitation that poses a threat to the safety of any person, property or the environment;
g) depressurization requiring emergency descent;
h) a fuel shortage requiring a diversion or priority;
i) the aircraft is fuelled with the wrong fuel;
j) a collision, a risk of collision or a loss of separation occurs;
k) a crew member declares an emergency or requires priority handling by ATC or the standing by of emergency response services;
l) a slung load is released unintentionally or as a precautionary or emergency measure;
m) release in/from aircraft of dangerous goods.

After declaring an emergency, the pilot may be requested to submit a written report, and if so requested, he/she must submit it within 48 hours.

Where an aviation occurrence has occurred, the TSB should be notified as soon as possible. Details on the information required and appropriate phone numbers can be found in the GEN section of the AIP. The accident site should be left undisturbed except to rescue people or minimize dangers such as fire.

The CONFIDENTIAL AVIATION SAFETY REPORTING PROGRAM (CASRP) allows anyone involved in aviation to confidentially report his/her experiences or aviation safety concerns to the TSB.

[Reference: AIP GEN 3-1 to 3-5]

1.25 Distress and Urgency Signals

A pilot may use any means at his or her disposal to attract attention to obtain help.

Distress Signals

If grave and imminent danger threatens and immediate assistance is required, signal SOS in Morse code (... --- ...), and/or use the spoken word MAYDAY, and/or fire rockets or red lights at short intervals and/or or use parachute flares. To cancel a MAYDAY, state MAYDAY (once), ALL STATIONS (3 times), this is ______ (aircraft ident), station called, time, name of station in distress, DISTRESS TRAFFIC ENDED, station called, OUT.

Urgency Signals

If a pilot wishes to advise he/she is compelled to land, but immediate assistance is not required, he/she should switch the landing light on and off repeatedly, as well as switch on and off the navigation lights repeatedly (in a way that would make it clear the lights were not flashing normally). If a pilot has an urgent message to send concerning the safety of a ship, aircraft or person, he/she should signal the code group XXX (-..- -..- -..-) and/or use the spoken word PAN.

[Reference: AIP SAR 4-1, Radio Info Circ. RIC - 21]

1.26 Aircraft Seats, Belts & Harnesses

Seat and Safety Belt Requirements

INFANT means a person under two years of age. [101.01]

No person shall operate an aircraft other than a balloon unless it is equipped with a seat and safety belt for each person on board the aircraft other than an infant.

The seat and safety belt requirement does not apply to a person operating an aircraft that was type certificated with a safety belt designed for two persons.

A safety belt when required shall include a latching device of the metal-to-metal type. [605.22]

Restraint System Requirements

An aircraft may be operated without being equipped with safety belts as above in respect of the following persons if a restraint system is secured to the primary structure of the aircraft is provided for each person who is

a) carried on a stretcher or in an incubator or other similar device;
b) carried for the purpose of parachuting from the aircraft;
c) required to work in the vicinity of an opening in the aircraft structure. [605.13]

Shoulder Harness Requirements

No person shall operate an aeroplane, other than a small aeroplane (a "small aeroplane" is one that is under 12,500 lbs gross weight at take-off) manufactured before July 18, 1978, unless each front seat or, if the aeroplane has a flight deck, each seat on the flight deck is equipped with a safety belt that includes a shoulder harness.

No person shall operate a small aeroplane manufactured after December 12, 1986, the initial type certificate of which provides for not more than 9 passenger seats, excluding any pilot seats, unless each forward or aft-facing seat is equipped with a safety belt that includes a shoulder harness.

A seat with a safety belt and shoulder harness is mandatory for each person on board an aircraft involved in aerobatic flight, aerial application or aerial inspection operations conducted at altitudes below 500 feet AGL. [605.24]

General Use of Safety Belts and Restraining Systems

The PIC of an aircraft shall direct all of the persons on board the aircraft to fasten safety belts

a) during movement of the aircraft on the surface;
b) during take-off and landing;
c) at any time during flight that the PIC considers it necessary that safety belts be fastened.

These PIC directions also apply to the use of a child restraint system, a restraining system used by a person engaged in parachute descents and a restraint system used by a person working in the vicinity of an opening in the aircraft structure.

Where an aircraft crew includes flight attendants and the PIC anticipates that the level of turbulence will exceed light turbulence, the PIC shall immediately direct each flight attendant to discontinue duties relating to service, secure the cabin, and occupy a seat and fasten the safety belt provided.

Where an aircraft is experiencing turbulence and the in-charge flight attendant considers it necessary, the in-charge flight attendant shall direct all of the passengers to fasten their safety belts, and direct all of the other flight attendants to discontinue duties relating to service, to secure the cabin and to occupy their seats and fasten the safety belts provided. Where such directions have been made by the in-charge flight attendant, he/she shall so inform the PIC. [605.25]

Use of Passenger Safety Belts and Restraining Systems

Where the PIC or the in-charge flight attendant directs that safety belts be fastened, every passenger who is not an infant shall

a) insure that the passenger's safety belt or restraint system is properly adjusted and securely fastened;
b) if responsible for an infant for which no child restraint system is provided, hold the infant securely in the passenger's arms;
c) if responsible for a person who is using a child restraint system, insure that the person is properly secured.

No person shall be responsible for more than one infant. [605.26]

Use of Crew Member Safety Belts

The crew members on an aircraft shall be seated at their stations with their safety belts fastened

a) during take-off and landing;
b) at any time that the PIC directs;
c) in the case of crew members who are flight attendants, at any time that the in-charge flight attendant so directs.

Where the PIC directs that safety belts be fastened by illuminating the safety belt sign, a crew member is not required to fasten his/her safety belt pursuant to a direction from the PIC (under other than for take-off or landing),

a) during movement of the aircraft on the surface or during flight, if the crew member is performing duties relating to the safety of the aircraft or of the passengers on board;
b) where the aircraft is experiencing light turbulence, if the crew member is a flight attendant and is performing duties relating to the passengers on board;
c) if the crew member is occupying a crew rest facility during cruise flight and the restraint system for that facility is properly adjusted and securely fastened.

The PIC shall insure that at least one pilot is seated at the flight controls with safety belt fastened during flight time. [605.27]

Child Restraining Systems

No operator of an aircraft shall permit the use of a child restraint system on board the aircraft unless

a) the person using the child restraint system is accompanied by a parent or guardian who will attend to the safety of the person during the flight;
b) the weight and height of the person using the child restraint system are within the range specified by the manufacturer;
c) the child restraint system bears a legible label indicating the applicable design standards and date of manufacture;
d) the child restraint system is properly secured by the safety belt of a forward-facing seat that is not located in an emergency exit row and does not block access to an aisle;
e) the tether strap is used according to the manufacturer's instructions or, depending on the seat design, secured so as not to pose a hazard to the person using the child restraint system or to any other person.

Where a seat incorporates design features to reduce occupant loads, such as the crushing or separation of certain components, and the seat is in compliance with the applicable design standards, no person shall use the tether strap on the child restraint system to secure the system.

Every passenger responsible for a person using a child restraint system shall be

a) seated in a seat adjacent to the seat to which the child restraining system is secured;
b) familiar with the manufacturer's installation instructions for the child restraining system;
c) familiar with the method of securing the person in the child restraint system and of releasing the person from it. [605.28]

[Reference: CARs sections specified]

1.27 Oxygen Equipment Requirements

Oxygen Supply: Unpressurized Aircraft

No person shall operate an unpressurized aircraft unless it is equipped with sufficient oxygen dispensing units and oxygen supply to comply with these requirements:

a) an oxygen supply must be available for all crew members and 10% of passengers (not less than one passenger), for the entire period of flight exceeding 30 minutes at cabin pressure altitudes above 10,000 feet ASL but not exceeding 13,000 feet ASL;
b) an oxygen supply must be available for all persons on board the aircraft for the entire period of flight at cabin pressure altitudes above 13,000 feet ASL. For aircraft operated in an air transport service, the period of flight oxygen availability must be at least one hour.

Use of Oxygen

Where an aircraft is operated at cabin pressure altitudes above 10,000 feet ASL but not exceeding 13,000 feet ASL, each crew member shall wear an oxygen mask and use supplemental oxygen for any part of the flight at those altitudes that is more than 30 minutes in duration. Where an aircraft is operated at cabin pressure altitudes above 13,000 feet ASL, each person on board the aircraft shall wear an oxygen mask and use supplemental oxygen for the duration of the flight at those altitudes.

[Reference: CARs 605.31, 605.32]

1.28 Emergency Radio Frequency and Visual Interception Signals

The emergency frequency is 121.5 MHz (243.0 MHz military). All aviation 2-way radios must be capable of communicating on 121.5.

When operating within sparsely settled areas, 121.5 must be continuously monitored or when operating a Canadian aircraft over water more than 50 NM from shore (unless carrying out communications on other frequencies or cockpit duties/equipment limitations do not permit monitoring two VHF frequencies). When an aircraft is intercepted by a military aircraft, the pilot should advise ATC of the interception, attempt to contact the intercepting aircraft on 121.5 MHz, squawk 7700 and follow visual instructions from the intercepting aircraft.

Series	Intercepting Aircraft Signal	Meaning	Intercepted Aircraft Response	Meaning
1.	DAY - Rocking wings from a position in front and, normally, to the left of intercepted aircraft and, after acknowledgement, a slow level turn, normally to the left, on to the desired heading. NIGHT - Same and, in addition flashing navigational lights at irregular intervals. DAY OR NIGHT - Meteorological conditions or terrain may require the intercepting aircraft to take up a position in front and to the right of the intercepted aircraft and to make the subsequent turn to the right. If the intercepted aircraft is not able to keep pace with the intercepting aircraft, the latter is expected to fly a series of race-track patterns and to rock its wings each time it passes the intercepted aircraft.	You have been intercepted. Follow me.	AEROPLANES: DAY - Rocking wings and following. NIGHT - Same and, in addition flashing navigational lights at irregular intervals. HELICOPTERS: DAY or NIGHT - Rocking aircraft, flashing navigational lights at irregular intervals and following.	Understood, will comply.
2.	DAY or NIGHT - An abrupt breakaway manoeuvre from the intercepted aircraft consisting of a climbing turn of 90 degrees or more without crossing the line of flight of the intercepted aircraft.	You may proceed.	AEROPLANES: DAY or NIGHT - Rocking wings. HELICOPTERS: DAY or NIGHT - Rocking aircraft.	Understood; will comply.
3.	DAY - Circling aerodrome, lowering landing gear and overflying runway in direction of landing or, if the intercepted aircraft is a helicopter, overflying helicopter landing area. NIGHT - Same and, in addition, showing steady landing lights.	Land at this aerodrome.	AEROPLANES: DAY - Lowering landing gear, following the intercepting aircraft and, if after over-flying the runway landing is considered safe, proceeding to land. NIGHT - Same and, in addition showing steady landing lights (if carried). HELICOPTERS: DAY or NIGHT - Following the intercepting aircraft and proceeding to land, showing a steady landing light (if carried).	Understood, will comply.

Fig. 1-9: Signals initiated by intercepting aircraft.

Series	Intercepted Aircraft Signal	Meaning	Intercepting Aircraft Response	Meaning
4.	AEROPLANES: DAY - Raising landing gear while passing over landing runway at a height exceeding 300 m (1,000 feet) but not exceeding 600 m (2,000 feet) above AAE, and continuing to circle the aerodrome.	Aerodrome you have designated is inadequate.	DAY or NIGHT - If it is desired that the intercepted aircraft follow the intercepting aircraft to an alternate aerodrome, the intercepting aircraft raises its landing gear and uses the Series 1 signals prescribed for intercepting aircraft.	Understood, follow me.
	NIGHT - Flashing landing lights while passing over landing runway at a height exceeding 300 m (1,000 feet) but not exceeding 600 m (2,000 feet) AAE, and continuing to circle the aerodrome. If unable to flash landing lights, flash any other lights available.		If it is decided to release the intercepted aircraft, the intercepting aircraft uses the Series 2 signals prescribed for intercepting aircraft.	Understood, you may proceed.
5.	AEROPLANES: DAY or NIGHT - Regular switching on and off of all available lights but in such a manner as to be distinct from flashing lights.	Cannot comply.	DAY or NIGHT - An abrupt breakaway manoeuvre from the intercepted aircraft consisting of a climbing turn of 90° or more without crossing the line of flight of the intercepted aircraft.	Understood.
6.	AEROPLANES: DAY or NIGHT - Irregular flashing of all available lights. HELICOPTERS: DAY or NIGHT - Irregular flashing of all available lights.	In distress.	DAY or NIGHT - An abrupt breakaway manoeuvre from the intercepted aircraft consisting of a climbing turn of 90° or more without crossing the line of flight of the intercepted aircraft.	Understood.

Fig. 1-10: Signals initiated by intercepted aircraft.

[Reference: CARs 602.143, 602.144; AIP COM 5-8, SAR 4-4]

1.29 Emergency Locator Transmitter (ELT)

General

No person shall operate an aircraft unless it is equipped with one or more ELT's as required by the aircraft type and area of operation. An ELT required to be carried shall be armed as specified in the AFM, AOM, POH or equivalent document provided by the manufacturer.

ELT's help search crews locate downed aircraft. They send out a signal on 121.5 MHz and/or 243.0 MHz. The signal can be detected by SARSAT and COSPAS satellites, or by any aircraft listening over the emergency frequencies.

ELT Categories

There are 5 types of ELT:

a) TYPE A or AD - Automatic ejectable or Automatic Deployable. Automatically ejects from the aircraft and activates during crash force.
b) TYPE F or AF - Fixed or Automatic Fixed. Activates based on inertia switch when aircraft subject to crash force. Most GA aircraft use this type.
c) TYPE AP - Automatic Portable. This type is similar to Type F or AF except that the antenna is integral to the unit for portable operation.
d) TYPE P - Personnel. This type has no fixed mounting and does not transmit automatically.
e) TYPE W or S - Water Activated or Survival. Transmits automatically when immersed in water. It is waterproof and floats and should be tethered to life rafts or survivors.

An aircraft may be operated without an ELT on board where the aircraft is

a) a glider, balloon, airship, ultra-light aeroplane or gyroplane;
b) registered under the laws of another state, equipped with a serviceable radio transmitter that is approved by that state for search and rescue purposes, having a distinctive siren like sound capable of communication on frequencies 121.5/243.0 MHz;
c) operated by the holder of a flight training unit operating certificate, engaged in flight training and operated within 25 NM of the aerodrome of departure;
d) engaged in a flight test;
e) a new aircraft engaged in flight operations incidental to manufacture, preparation or delivery of the aircraft.

ELT Activation

No person shall activate an ELT except in an emergency. A person may activate an ELT during the first 5 minutes of any hour for a duration of not more than 5 seconds for the purpose of testing it. Where an ELT has been inadvertently activated during flight, the PIC of the aircraft shall insure that the nearest ATC unit, FSS or CARS is so informed as soon as possible and the ELT is switched off.

[Reference: CARs 605.38, 605.39, 605.40]

1.30 Air Traffic Services (ATS)

Air Traffic Control Services

The following air traffic control and information services are provided through area control centres, terminal control units and control towers:

a) Airport Control Service. Provided by airport control towers to aircraft and vehicles on the maneuvering area of an airport and to aircraft operating in the vicinity of an airport.
b) Area Control Service. Provided by ACCs to IFR and CVFR flights operating within specified control areas.

c) Terminal Control Service. Provided by IFR units (ACCs) or TCUs to IFR and CVFR flights operating within specified control areas.
d) Terminal Radar Service. An additional service provided by IFR units to VFR aircraft operating within Class C airspace.
e) Alerting Service. To notify appropriate organizations for aircraft in need of search and rescue services, or to alert crash equipment, ambulances, doctors and any other safety services.
f) Airspace Reservation Service. Includes the services of the Airspace Reservation Unit (ARU) and ACCs in providing reserved airspace for specified air operations in controlled airspace, and in providing information concerning these reservations and military activity areas in controlled and uncontrolled airspace.
g) Aircraft Movement Information Service. Provided by ACCs for the collection, processing and dissemination of aircraft movement information for use by air defence units relative to flights operating into or within the ADIZ.
h) Customs Notification Service (ADCUS). Provided, on request, by ATC units for advance notification of customs officials for transborder flights at specified ports of entry.
i) Flight Information Service. Provided by ATC units to assist pilots of aircraft by supplying information concerning known hazardous flight conditions. This information will include data concerning unfavourable flight conditions and other known hazards; which may not have been available to the pilot prior to take off or which may have developed along the route of flight.

ATC service has been established primarily for the prevention of collisions and the expediting of traffic. ATC service takes precedence over flight information service, but every effort is made to provide flight information and assistance. Flight information is made available, whenever practicable, to any aircraft in communication with an ATC unit, prior to take-off or in flight, except where such service is provided by the aircraft operator. Many factors such as traffic volume, controller workload, communications frequency congestion and radar limitations may prevent a controller from providing this service.

Flight Service Stations

FSS stations provide the following services:

a) En Route Flight Information Service. Continuous monitoring of assigned frequencies permits pilots the communications access to obtain and pass flight information, or to report emergencies should the need arise. In addition, the FSS relays IFR position reports and ATC clearances in areas where aircraft are beyond the communications range of the ATC facility responsible.
b) Airport Advisory Service (AAS). The FSS provides advisory information consisting of wind, preferred or active runway, time (departures only), altimeter, aircraft traffic, ground traffic, and other information to assist pilots to execute safe and expeditious departures and arrivals at uncontrolled airports.
c) Remote Aerodrome Advisory Service (RAAS). At aerodromes where the advisory is provided through RCO's, the service is referred to as a RAAS. RAAS consists of weather reports including wind and altimeter settings (METAR), the active or preferred runway (if known), field condition reports, NOTAM, PIREPs, and known aircraft and vehicle traffic. Since RAAS is a remote service provided by an FSS in another location, the pilot is responsible to comply with mandatory aerodrome traffic frequency procedures.
d) Vehicle Control Service (VCS). The FSS specialist controls vehicles operating on the maneuvering area of airports with a co-located tower and an FSS during the hours when the tower is closed. This service is also available at airports without a tower during the operating hours of the FSS. VCS is not available at sites served by RAAS.
e) Flight Plan Service. The FSS provides a preflight service which includes the provision of weather, NOTAM, RSC/JBI and other information. It accepts and processes flight plans and flight itineraries. An aviation information display is also maintained and is easily accessible to pilots. It assists pilots in compiling all the information essential to planning a safe flight.
f) Surface Weather Observing Service. The observation, recording, and dissemination of surface weather data, including specials, is performed by the FSS for aviation purposes.
g) Aviation Weather Information Service (AWIS). The FSS provides pertinent aviation weather information tailored to accommodate pilots at the preflight and en route stages. The service permits specialists to assist pilots in making decisions and calculations based on weather determinants.
h) Aviation Weather Briefing Service (AWBS). This is a fully interpretive preflight and en route weather briefing service provided toll-free by selected FSSs in each region. These sites are equipped with a full suite of weather products, including satellite and radar imagery. Briefers are trained to adapt meteorological information to fit the needs of all aviation users and to provide consultations and advice on special weather problems. Flight documentation for

long range flights is also available, on request. This level of service is described as W1 in the CFS.

i) VFR Alerting Service. The FSS notifies SAR and conducts a communications search in the event that a VFR flight plan or flight itinerary is not closed within a specified time, or upon receiving an overdue report for an aircraft.
j) Aeronautical Broadcast Service. The FSS broadcasts weather and other information required by pilots to plan and complete flight safely.
k) Navigation Assistance Service. The FSS provides VDF assistance, at sites where the system is installed, to aircraft in emergency or potential emergency situations, or when requested by a pilot. Other navigation assistance may also be provided, depending on facilities available.
l) Navigation Aids Monitoring Service. The FSS monitors the status of navigation aids as assigned, and takes appropriate corrective and notification action should abnormal operation occur.
m) NOTAM Service. For designated locations and/or areas assigned to it, the FSS is the office responsible for co-ordination and dissemination of NOTAMs.
n) PIREPS. The FSS collects and distributes pilot reports of weather and other significant flight information.
o) Fixed Telecommunications Service. The FSS is connected to fixed telecommunications networks so that operational and administrative messages may be exchanged among FSS, other domestic and international aeronautical agencies, and aircraft in flight.
p) Paid Communications Service. On request from subscribing domestic and international air carriers, the FSS relays messages between an aircraft and its company for a cost recovery fee.

The majority of FSS stations provide flight information service 24 hours per day to the airports where they are located and to any number of assigned RCOs. Advisory services will also be provided from the FSS stations at controlled airports when the control towers are closed.

FSS stations give wind direction to the nearest 10° and wind speed/gusts to nearest 5 kts.

Remote Communications Outlets, Dial-Up RCOs

RCO's are remote VHF transmitters/receivers established as an extended communications capability to enable FSS specialists to provide flight information services in the vicinity of designated aerodromes. Flight Information Service En Route (FISE) is provided on the en route frequency and RAAS is provided on the MF.

An RCO can also be used to accept position reports and relay ATC clearances.

A Dial-Up Remote Communications Outlet (DRCO) is a standard RCO which has had a dial-up unit installed to connect the pilot with an ATS unit such as FSS via a normal telephone line. In this manner, the line is opened only after the communication has been initiated by the pilot or by ATS. The radio range of the RCO is unaffected by the conversion. To activate a DRCO, use the following procedure:

a) select the published RCO frequency on the aircraft radio;
b) key the radio microphone distinctly 4 times in a row, with not more than 1 second between keying. If successful, you will hear a dial tone, a series of touch tones, and a ringing signal;
c) if connected, there will be a recording that states "link established". Initiate the radio conversation as per standard radiotelephony practices, ie "Regina Radio, this is Grumman Cheetah Golf Alfa Bravo Charlie, over". The ATS specialist may take a few moments to reply.
d) a "call terminated" message indicates that the telephone line has been disconnected.

ATC Clearances, Instructions and Information

Whenever an ATC clearance is received and accepted by the pilot, compliance shall be made with the clearance. If a clearance is not acceptable, the pilot should immediately inform ATC of this fact, since acknowledgement of the clearance alone will be taken by a controller as indicating acceptance.

A pilot shall comply with an ATC instruction which is directed to and received by the pilot, provided aircraft safety is not jeopardized.

A clearance will be identified by the use of some form of the word CLEAR in its content. An instruction will always be worded in such a manner as to be readily identified, although the word "instruct" will seldom be included. Pilots shall comply with and acknowledge receipt of all ATC instructions directed to and received by them.

Always remember that control is predicated on known air traffic only, and when complying with clearances or instructions, pilots are not relieved of the responsibility for practising safe procedures and good pilot judgement.

A clearance or instruction is only valid WHILE IN CONTROLLED AIRSPACE. Pilots crossing between controlled and uncontrolled airspace should pay close attention to the terrain and obstacle clearance requirements.

ATS personnel routinely inform pilots of conditions, observed by others or by themselves, which may affect flight safety and are beyond their control, such as observed airframe icing and bird activity. These are meant solely as assistance or reminders to pilots and are not intended in any way to absolve the pilot of the responsibility for the safety of the flight.

RADAR Services

General
The use of radar increases airspace utilization by allowing ATC to reduce the separation interval between aircraft. In addition, radar permits an expansion of flight information services, such as traffic information, radar navigation assistance and information on chaff drops, bird activity and severe weather information. Due to limitations inherent in all radar systems, it may not always be possible to detect aircraft, weather disturbances, etc. Where radar information is derived from secondary surveillance radar (SSR) only (without associated primary radar coverage) it is not possible to provide traffic information on aircraft that are not transponder equipped or to provide some of the other flight information.

Before providing radar services, ATC will identify the aircraft concerned either through the use of position reports, identifying turns, or the use of transponders. Pilots will be notified whenever radar identification is established or lost.

Pilots are cautioned that radar identification of their flight does not relieve them of the responsibility for collision avoidance or terrain or obstacle clearance. ATC assumes responsibility for terrain or obstacle clearance when vectoring en route IFR and CVFR flights, and IFR aircraft being vectored for arrival until the aircraft is within the final approach area.

Radar Vectoring
Radar vectoring is used when necessary for separation purposes, when required by noise abatement procedures, when requested by the pilot, or whenever vectoring will offer operation advantages to the pilot or the controller. When vectoring is initiated, the pilot will be informed of the location to which the aircraft is being vectored. Pilots will be informed when radar vectoring is terminated, except when an arriving aircraft is vectored to the final approach course or to the traffic circuit.

When an aircraft is vectored to final approach or to the traffic circuit, the issuance of approach clearance indicates that normal navigation should be resumed.

Normally, radar service will be continued until an aircraft leaves the area of radar coverage, enters uncontrolled airspace, or is transferred to an ATC unit not equipped with radar. The pilot will be informed when radar service is terminated.

Wake Turbulence Separation
Wake turbulence has its greatest impact on departure and arrival procedures, however, pilots should not assume that it will only be encountered in the vicinity of aerodromes. Caution should be exercised if a flight is conducted anywhere behind and less than 1,000 feet below a large aircraft. Aircraft wake turbulence groups are as follows:

a) Group 1 (Heavy). All aircraft with a maximum take-off weight of 300,000 lbs or more.
b) Group 2 (Medium). Aircraft with a maximum take-off weight of between 12,500 and 300,000 lbs.
c) Group 3 (Light). Aircraft with a takeoff weight up to 12,500 lbs inclusive.

For radar departures, controllers generally use the following radar separation minimum between a preceding IFR/VFR aircraft and a vectored aircraft directly behind it less than 1,000 feet below:

Heavy behind a Heavy	4 miles
Medium behind a Heavy	5 miles
Light behind a Heavy	6 miles
Light behind a Medium	4 miles

For non-radar departures, controllers will apply a two minute separation interval to any aircraft that takes off into the wake of a known heavy aircraft if the aircraft concerned commences the take-off from the threshold of the same runway, or any following aircraft departs from the threshold of a parallel runway that is located less than 2,500 feet away from the runway used by the preceding heavy aircraft. ATC does not apply this two minute spacing interval between a light following a medium aircraft in the above circumstances, but will issue wake turbulence advisories to light aircraft.

Controllers will apply a three minute separation interval to any aircraft that takes off into the wake of a known heavy aircraft, or a light aircraft that takes off into the wake of a known medium aircraft if the following aircraft starts it take-off roll from an intersection or from a point further along the runway than the preceding aircraft, or if the controller has reason to believe that the following aircraft will require more runway length for take-off than the preceding aircraft. ATC uses separation intervals of up to three minutes if the projected flight paths of any following aircraft will cross that of a preceding heavy aircraft. Regardless, ATC will not guarantee that wake turbulence will not be encountered.

If a pilot is advised to wait before being given clearance for take-off due to wake turbulence, the pilot may request a PILOT WAIVER of wake turbulence separation. Although controllers are not permitted to initiate waivers to wake turbulence separation minimums, they will issue take-off clearance to pilots who have waived wake turbulence requirements on their own initiative. In some situations a pilot waiver may not be granted such as with intersection take-offs of a light or medium behind a heavy, or a light or medium take-off following a heavy missed approach.

[Reference: AIP RAC 1-1 to 1-6, 1-11, 1-12, 1-16, 1-17, 4-1, 4-2, 6-5, 6-6]

1.31 Operations at Controlled Airports

Departure

Normally, a two-way radio is required. Some controlled airports have ATIS (Automatic Terminal Information Service). An ATIS broadcast is an hourly recorded message which continuously broadcasts routine information, such as ceiling, visibility, winds, temperature, altimeter setting, active runway, NOTAMS and other pertinent information. Where an airport has an ATIS frequency, the normal procedure for departure is to first obtain the ATIS on the appropriate frequency. Next, contact ground control for taxi clearance to the active runway, informing ground control that you have the latest ATIS (giving the ATIS phonetic code). To check your radio signal strength, ask ATC for a RADIO CHECK - ATC will give you a number from 1 to 5: 1 = unreadable, 2 = readable now and then, 3 = readable with difficulty, 4 = readable, 5 = perfectly readable.

Once you receive taxi clearance, remember that if you are cleared to taxi to the runway in use, you may cross non-active runways or taxiways to get there without further clearance. A HOLD SHORT clearance is used for operations at intersecting active runways. Pilots must read back any hold short clearances so the controller is assured an aircraft will not enter the wrong runway. When ready for departure, obtain take-off clearance on the tower frequency. If you are cleared for an IMMEDIATE TAKE-OFF, taxi onto the runway and take-off in one continuous movement. After departure, contact the tower to inform ATC once you have cleared the control zone.

Special procedures and extreme vigiligance for traffic are required for operations at busier airports having Simultaneous Intersecting Runway Operations (SIRO), where midway through one runway will be another runway crossing it. For example, at a SIRO airport, an aircraft may be cleared to take-off in advance of the intersecting runway, with an immediate turn for traffic avoidance.

Arrival

To land at a controlled airport, first obtain the ATIS broadcast. Prior to entering the control zone, contact ATC over the tower frequency giving your aircraft identification, type, position, altitude, ATIS received, transponder code (if appropriate) and intentions. The controller will then clear you to a position within the control zone, either to a leg of the circuit pattern, or to a geographic position from which you will receive further clearance.

In a standard circuit pattern, arrivals are from the upwind side, or at 45° to the downwind leg. Aircraft should enter the circuit pattern at the published circuit altitude in the CFS. Controllers may alter these procedures, such as by clearing an arriving aircraft to enter the circuit at left base, for example. The reverse procedure applies for a right hand circuit pattern. The CFS provides special procedures, type of facilities available, obstructions, airport frequencies and other relevant information. At SIRO airports, special clearances may be given such as the requirement to complete landing prior to the intersecting runway.

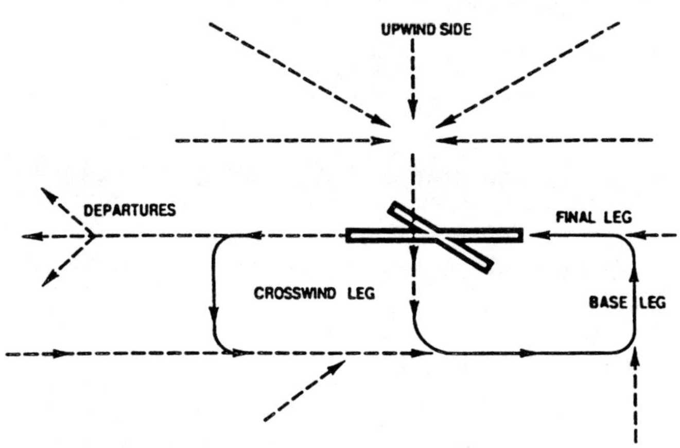

Fig. 1-11: Standard circuit pattern, controlled airport.

Lights and Visual Signals

An ATC unit may use directional light signals to communicate with aircraft such as NORDO aircraft, or to assist aircraft having radio difficulties. A pilot shall, where practicable, acknowledge all clearances and instructions received by visual signals, by rocking the wings in flight or by making a single flash of the landing light.

Directional Signals: Ground

Signal	Meaning
Green flashes	Cleared to taxi
Steady green	Cleared to take-off
Red flashes	Taxi clear of landing area
Steady red	Stop
Flashing white	Return to starting point
Blinking Runway lights	Ground personnel to clear area

Directional Signals: Air

Signal	Meaning
Green flashes	Return for landing
Steady green	Cleared to land
Red flashes	Airport unsafe, do not land
Steady red	Give way to other aircraft and continue circling
Red pyrotechnic	Do not land for time being
Bursting red and green stars	You are in vicinity of a prohibited, danger or restricted area: alter course.

[Reference: AIP RAC 4-2, 4-6, 4-7, 4-11, 4-20]

1.32 Operations at Uncontrolled Airports

An uncontrolled airport is an airport that does not have a control tower or an airport where the tower is not in operation. To enhance safety, all aircraft operating at an uncontrolled airport must communicate on a common frequency and follow a common reporting procedure, when operating within a specified area of an airport. Aircraft may communicate with an FSS, UNICOM, CARS or broadcast intentions to other aircraft in the area.

Mandatory Frequency (MF) and Aerodrome Traffic Frequency (ATF)

An MF is assigned for some uncontrolled airports. When operating within the specified area for an MF, pilots must remain on the MF and make all transmissions/broadcasts on that frequency.

An ATF is designated for uncontrolled airports that do not meet the criteria for an MF. The specific MF or ATF is published in the CFS. Where no UNICOM exists, the ATF will be 123.2. Example CFS entry:

ATF - tfc 118.2 06Z - 14Z 5 NM 6900 ASL

Meaning of example: ATF is 118.2 for traffic operating from 0600 UTC to 1400 UTC, when within 5 NM and below 6,900 feet ASL.

To communicate using an ATF or MF, either direct your call to a station, or broadcast your call to local air traffic (if the station does not answer or there is no station for the airport). Remember the following procedures when using an MF/ ATF:

a) Departure - report departure intentions before moving onto the runway. After departure, monitor the ATF/MF until well clear of the specified area (normally 5 - 10 NM).
b) Arrival - report position, altitude, arrival procedure intentions and estimated time of landing well before entering the specified area. Maintain a listening watch, report joining the circuit pattern, report when on final approach, and report when clear of the active runway.
c) Maintain a listening watch at all times when within the specified area.

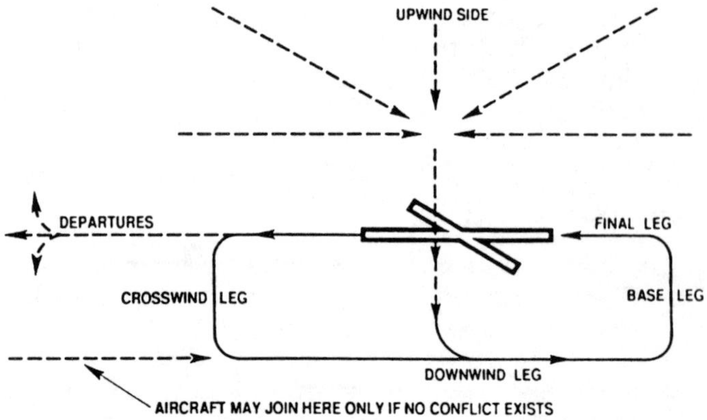

Fig. 1-12: Standard circuit pattern at uncontrolled aerodrome, arrival and departure procedures.

Arrivals

To enter the circuit for landing at an uncontrolled aerodrome, join from the upwind side in the marked area, turn crosswind as indicated, and then join the downwind leg at 1,000 feet AGL, or, join straight in on the downwind leg at circuit altitude. If it is necessary to cross the airport prior to joining the circuit, such as to study the windsock, the crossover should be carried out well above circuit height, and the letdown should be completed on the upwind side.

Departures

To depart from an uncontrolled aerodrome, depart as indicated, keeping a lookout for aircraft entering on the downwind leg.

NOTE: The above standard procedures for operations at uncontrolled aerodromes may be altered where appropriate. Consult the CFS for the aerodrome used.

Noise Operating Criteria

Many aerodromes have operating restrictions to reduce noise including preferential runways, hours of operation, approach and departure procedures etc. Always consult the CFS to check for such restrictions for the aerodrome involved.

[Reference: AIP RAC 4-15 to 4-20]

1.33 VFR Enroute Procedures

Pilots are encouraged to file position reports to the nearest FSS, or broadcast position reports over 126.7 MHz when in uncontrolled airspace. A position report should provide the following:

a) identification.
b) position.
c) time over.
d) altitude.
e) VFR flight plan/itinerary.
f) destination.

Unless taking off or landing, no aircraft is to be flown except at an altitude that will permit an emergency landing. Aircraft are not to be flown over built-up areas at less than 1,000 feet above the highest obstruction within a horizontal radius of 2,000 feet. Lower flight is permitted when over unpopulated areas when there will not be a hazard to persons or property. The dangers of low flying cannot be over-emphasized. There are numerous accidents every year caused by aircraft striking man made structures such as towers and transmission lines.

No person shall fly an aircraft at a height of less than 2,000 feet over an aerodrome except for taking off or landing, or as directed by ATC. Aircraft operating in Class B airspace must be on a CVFR flight plan. Such aircraft will then basically operate in accordance with IFR procedures while in Class B airspace (with the exception that CVFR aircraft must remain in VFR weather conditions at all times).

Special procedures apply to aircraft operating in Class C terminal control areas. Generally, such areas surround the busier airports in Canada having a mix of IFR and VFR traffic. For the specific procedures involved at a given terminal, consult the appropriate VTA chart and CFS. To establish initial contact with a Class C terminal control unit, provide the following:

a) aircraft type and identification.
b) position in relation to published call-up point.
c) altitude.
d) intended destination and route.
e) transponder code and ATIS received.

FUR FARMS may be marked with chrome yellow and black stripes painted on pylons or roofs. Avoid overflying these farms less than 2,000 feet. As well, herds of reindeer, caribou, moose or muskoxen should be overflown at altitudes of at least 2,000 feet.

[Reference: AIP RAC 1-27, 5-1 to 5-4]

1.34 VFR Holding Procedures

In busy airspace, ATC may ask VFR aircraft to ORBIT visually over a geographic location until they can be cleared to the airport. Terrain and collision avoidance are the pilot's responsibility.

[Reference: AIP RAC 4-9]

1.35 Operational & Emergency Equipment

General

No person shall operate a Canadian registered aircraft unless the operational and emergency equipment required is carried on board.

Requirements for Power Driven Aircraft

No person shall conduct a take-off in a power-driven aircraft, other than an ultra-light aeroplane, unless the following operational and emergency equipment is carried on board:

a) a checklist or placards that enable the aircraft to be operated in accordance with the limitations in the AFM, AOM, POH or any equivalent document provided by the manufacturer. The checklist or placards shall enable the aircraft to be operated in normal, abnormal and emergency conditions and shall include a pre-start check, pre-take-off check, post take-off check, pre-landing check, and emergency procedures. The emergency procedures shall include emergency operation of fuel, hydraulic, electrical and mechanical systems (where applicable), emergency operation of instruments and controls (where applicable), engine inoperative procedures and any other procedure that is necessary for aviation safety;
b) where the aircraft is operated in VFR OTT, night VFR flight or IFR flight, all of the necessary current aeronautical charts and publications covering the route of the proposed flight and any probable diversionary route;
c) a hand held fire extinguisher in the cockpit that is suitable for extinguishing the fires that are likely to occur, designed to minimize the hazard of toxic gas concentrations and readily available in flight to each flight crew member;
d) a timepiece that is readily available to each flight crew member;
e) for night operations, a flashlight that is readily available to each crew member;
f) a first aid kit.

The above checks and emergency procedures shall be performed and followed where applicable.

Survival Equipment - Flights over Land

No person shall operate an aircraft over land unless there is carried on board survival equipment, sufficient for the survival on the ground of each person on board, given the geographical area, the season of the year and anticipated seasonal climatic variations, that provides the means for

a) starting a fire;
b) providing shelter;
c) providing or purifying water;
d) visually signalling distress.

This equipment is not required in these cases:

a) multi-engined aircraft operated south of 66° 30' north latitude in IFR flight within controlled airspace or along designated air routes;
b) aircraft operated by an air operator, where the aircraft is equipped with equipment specified in the air operator's COM;
c) aircraft that are in a geographical area where and at a time of year when the survival of the persons on board is not jeopardized. [602.61]

Life Preservers and Flotation Devices

No person shall conduct a take-off or a landing on water in an aircraft or operate an aircraft over water beyond a point where the aircraft could reach shore in the event of an engine failure, unless a life preserver, individual flotation device or personal flotation device is carried for each person on board.

Life preservers, and individual and personal flotation devices shall be stowed in a position that is easily accessible to the person for whose use it is provided, when that person is seated. [602.62]

Life Rafts, Survival Equipment - Flights over Water

No person shall operate over water a single-engined aeroplane, or a multi-engined aeroplane that is unable to maintain flight with any engine failed, at more than 100 NM, or the distance that can be covered in 30 minutes of flight at the cruising speed filed in the flight plan or flight itinerary, whichever distance is the lesser, from a suitable emergency landing site unless LIFE RAFTS are carried on board and are sufficient in total rated capacity to accommodate all of the persons on board.

Life rafts where required shall be stowed so that they are easily accessible for use in the event of a ditching, installed in conspicuously marked locations near an exit and equipped with an attached survival kit, sufficient for the survival on water of each person on board the aircraft, given the geographical area, the season of the year and anticipated seasonal climatic variations, that provides a means for providing shelter, providing or purifying water, and visually signalling distress. [602.63]

[Reference: CARs sections specified]

1.36 Security Control of Air Traffic

To permit identification and control of aircraft in the interest of national security, the Air Defence Identification Zone (ADIZ) has been established.

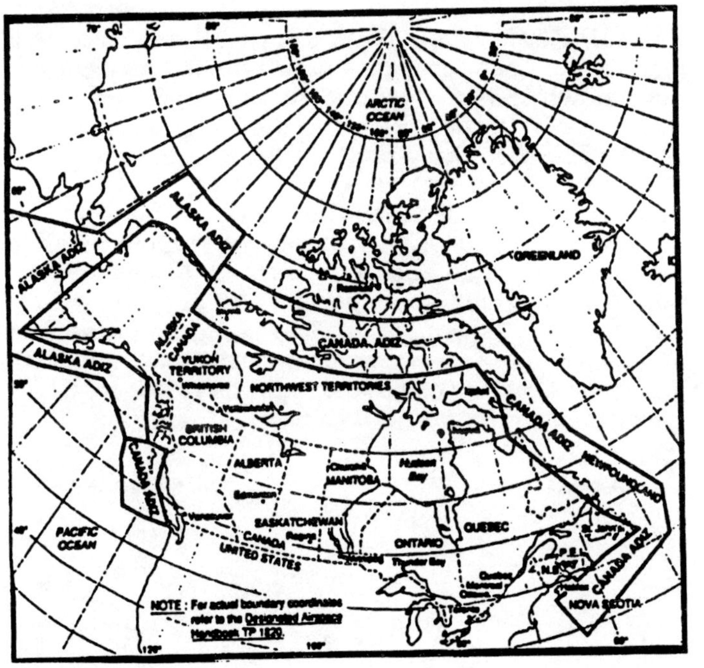

Fig. 1-13: ADIZ.

In times of war, further rules will come into effect, which are termed "SCATANA" rules. SCATANA rules establish priorities among aircraft, for example, aircraft that are engaged in active continental defence missions would be termed "priority 1". Occasionally, a SCATANA TEST will be broadcast to aircraft from communications facilities to be used in times of war. A pilot receiving a SCATANA test should acknowledge having received the broadcast and resume normal operations.

All aircraft operating in the ADIZ must have a 2-way radio and maintain a listening watch on the appropriate frequency. Pilots must file a Defence Flight Plan or Defence Flight Itinerary. The estimated time of penetration of the ADIZ boundary (or arrival at a reporting point) must be provided and revised if more than +/- 5 minutes or +/- 20 NM of estimate.

[Reference: CARs s. 602.145, 602.146; AIP RAC 1-18, RAC 2-20, 3-6]

1.37 Cruising Altitudes

The appropriate cruising altitude is determined on the basis of the height of obstructions, aircraft performance and weather conditions. Cruising altitudes are based on the aircraft TRACK (path flow over the ground). The MAGNETIC TRACK is used in the SDA, and the TRUE TRACK is used in the NDA. The cruising altitude requirement applies to aircraft operating above 3,000 feet AGL.

VFR Cruising Altitudes (below 18,000 ft ASL):

Aircraft Track	
000° - 179°	180° - 359°
3,500	4,500
5,500	6,500
7,500	8,500
9.500	10,500
11,500	12,500
etc.	etc.

IFR/CVFR Cruising Altitudes (below 18,000 ft ASL):

Aircraft Track	
000° - 179°	180° - 359°
3,000	4,000
5,000	6,000
7,000	8,000
9,000	10,000
11,000	12,000
etc.	etc.

Some airways or air routes use cruising altitudes that are an exception to the above requirements: refer to the CFS or IFR charts for specific requirements. The exceptions do not apply to VFR flights.

As well, the cruising altitudes are not required for aerial survey operations, but such flights must be authorized by ATC in advance.

[Reference: CARs 602.34; AIP RAC 2-2]

1.38 Aerodromes and Airports

General

When an aerodrome is closed permanently, the operator of the aerodrome shall remove all of the markers and markings installed at the aerodrome.

The operator of an aerodrome, other than a water aerodrome, shall install red flags or red cones along the boundary of an unserviceable movement area.

Where a runway or part of a runway is closed, the operator of the aerodrome shall place closed markings consisting of a large letter "X" on the runway as follows:

a) for runways longer than 4,000 feet, a closed marking shall be located at each end of the closed runway or part thereof and additional closed markings shall be located on the closed runway at intervals not exceeding 1,000 feet (the size of the "X" is approximately 36 metres X 14.5 metres);
b) for runways between 1,500 feet and 4,000 feet, a closed marking shall be located at each end of the closed runway or part thereof and an additional closed marking of the same dimensions shall be located on the closed runway or part thereof at a point equidistant from the two markings (these closed markings are to be 1/2 the size of the markings in a) above);
c) where the runway is 1,500 feet or less in length, a closed marking shall be located at each end of the closed runway or part thereof (1/2 size of markings in a) above).

Where a taxiway or part of a taxiway is closed, the operator of the aerodrome shall place on each end of the closed taxiway, or part thereof, a closed marking (the size of the "X" is approximately 9 metres X 9 metres).

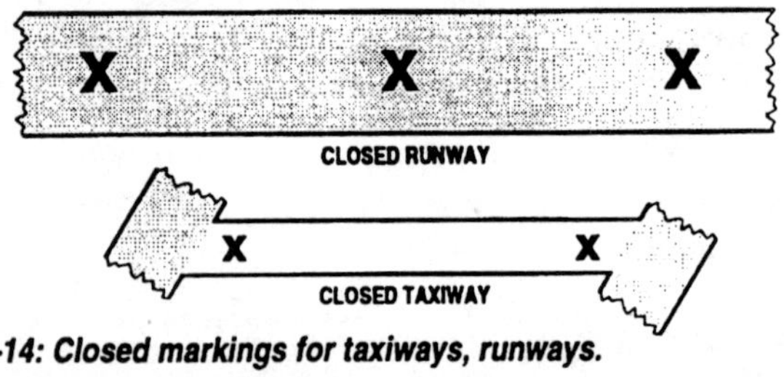

Fig. 1-14: Closed markings for taxiways, runways.

Where a maneuvering area of part thereof is closed permanently, the operator of the aerodrome shall obliterate all of the markings that indicate that the maneuvering area or part thereof is open, and paint the required closed markings. Where the surface of a maneuvering area or part thereof is snow covered or otherwise unsuitable for painting or where the closure is not permanent, closed markings may be applied by means of a conspicuously coloured dye or may be constructed from a suitable conspicuously coloured material or product. [301.04]

At airports, a DRY TRANSPORT CANADA WIND SOCK reacts to wind as follows: horizontal = 15+ kts, 5° below horizontal = 10 kts, 30° below horizontal = 6 kts. [AIP AGA 5-11]

Aerodrome Lighting

Except for police operations or to save human life, no person shall conduct a landing or take-off at night with unless the aerodrome is lighted in accordance with CARs requirements. [602.40]

The MINIMUM AERODROME LIGHTING for an aerodrome used for night operations is two parallel rows of white lights (or retro-reflective markers capable of reflecting aircraft lights) visible for at least two NM for an aircraft in the air. If retro reflective markers are used, since they are intended only for use when the aircraft is on final approach, a fixed white or strobe light shall be positioned at each end of the runway to assist pilots in locating the aerodrome and aligning the aircraft with the runway. At least 8 lights or markers are required on each side of the runway, spaced not more than 200 feet apart and having a total distance of at least 1,400 feet.

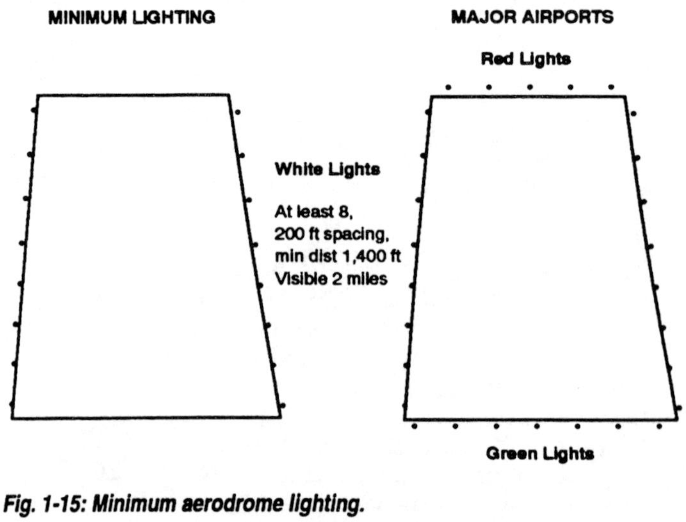

Fig. 1-15: Minimum aerodrome lighting.

UNSERVICEABLE AREAS within the movement area of an aerodrome being used at night are marked by steady burning red lights or red retro-reflective markers or floodlighting outlining the perimeter of the unserviceable area(s).

Where a TAXIWAY is used at night, the operator of the aerodrome shall indicate each side of the taxiway with aline of fixed blue lights or blue retro-reflective markers placed so that the two lines of lights or markers are parallel and the distance between adjacent lights or markers in each line is not more than 200 feet.

AIRCRAFT PARKING AREAS shall be marked with blue lights, blue retro-reflective markers or flood-lighting at intervals not exceeding 200 feet.

Many aerodromes are equipped with a flashing white beacon light to assist pilots in locating the aerodrome at night.

The operator of an aerodrome may display flare pots to provide temporary lighting for the landing or take-off of aircraft.

Aircraft Radio Control of Aerodrome Lighting (ARCAL) is a pilot activated runway lighting system based on clicking the microphone a designated number of times over a designated VHF frequency.

Refer to the CFS for specific aerodrome lighting used at a given location.

[Reference: CARs 301.07; AIP AGA 7-1, 7-8, 7-12]

Prohibitions

No person shall

a) walk, stand, drive a vehicle, park a vehicle or aircraft or cause an obstruction on the movement area of an aerodrome/airport, except in accordance with permission given by the operator of the aerodrome/airport, and, where applicable, by the appropriate ATC unit or FSS;
b) tow an aircraft on an active movement area at night unless the aircraft displays operating wingtip, tail and anti-collision lights or is illuminated by lights mounted on the towing vehicle;
c) park or otherwise leave an aircraft on an active maneuvering area at night unless the aircraft displays operating, tail and anti-collisions lights or is illuminated by lanterns suspended from the wingtips, tail and nose of the aircraft;
d) operate any vessel, or cause any obstruction, on the surface of any part of a water area of an aerodrome/airport that is to be kept clear of obstructions in the interest of aviation safety, and when ordered, by signal or otherwise by ATC/FSS or the operator, shall leave the area and/or not approach the area;
e) knowingly remove, deface, extinguish or interfere with a marker, marking, light or signal that is used at an aerodrome/airport for the purpose of air navigation, except with permission from the operator or ATC/FSS;
f) at a place other than an aerodrome/airport, knowingly display a marker, marking, light or signal that is likely to cause a person to believe that the place is an aerodrome;
g) knowingly display at or in the vicinity of an aerodrome/airport a marker, marking, sign, light or signal that is likely to be hazardous to aviation safety by causing glare or by causing confusion with or preventing clear visual perception of a marking, sign, light or signal that is required by the CARs;
h) allow a bird or other animal that is owned by the person or that is in the person's custody or control to be unrestrained within the boundaries of an aerodrome/airport except for the purpose of controlling other birds or animals at the aerodrome as permitted by the operator;
i) discharge a firearm within or into an aerodrome/airport without the permission of the aerodrome/airport operator. [301.08, 302.10]

Fire Prevention

No person shall, while at an aerodrome or airport, smoke or display an open flame

a) on an apron;
b) on an aircraft loading bridge or on a gallery or balcony that connects or overhangs an apron;
c) in an area where smoking or the presence of an open flame is likely to create a fire hazard.

The operator of an aerodrome/airport may, in writing, authorize maintenance or servicing operations on an apron that involve the use, production or potential development of an open flame or a spark where the operations are conducted in a manner that is not likely to create a fire hazard.

The operator of an aerodrome/airport may permit smoking in an enclosed building or shelter located on an apron where such smoking is not likely to create a fire hazard that could endanger persons or property. [301.09, 302.11]

1.39 Forest Fire Operations

Forest first areas are given a NOTAM. Such areas will be closed to normal aircraft operations. Aircraft are not to be operated within 5 NM and below 3,000 feet AGL of the limits of a forest fire area.

Reference: CARs 601.15 - 17, AIP RAC 2-16]

1.40 Required Documents on Aircraft

For every flight, the following documents must be carried on board the aircraft:

a) Flight authority (ie C of A with AAIR, flight permit etc.).
b) Certificate of registration.
c) Aircraft journey log.
d) Weight and balance data.
e) Aircraft radio license.
f) Interception procedures.
g) Aircraft manual.
h) Pilot license.
i) Pilot medical.
j) Pilot radiotelephone operator's certificate.
k) Proof of liability insurance.

The technical logs must never be carried on board the aircraft.

1.41 Phonetic Alphabet

Upon initial contact with ATC, or upon initial broadcast over an ATF or MF, the pilot shall state the aircraft type, followed by the last four letters of the registration, spoken in phonetics. Subsequent communications may be abbreviated to the last three letters of the registration, if this abbreviation is initiated by ATC. The phonetic alphabet should be memorized.

A - Alfa. B - Bravo. C - Charlie. D - Delta. E - Echo. F - Foxtrot. G - Golf. H - Hotel. I - India. J - Juliett. K - Kilo. L - Lima. M - Mike. N - November. O - Oscar. P - Papa. Q - Quebec. R - Romeo. S - Sierria. T - Tango. U - Uniform. V - Victor. W - Whiskey. X - X-ray. Y - Yankee. Z - Zulu.

[Reference: AIP COM 5-2]

1.42 Aircraft Maintenance Requirements

Aircraft are required to be maintained in accordance with CARs maintenance requirements. All airworthiness limitations and airworthiness directives in respect of a particular aircraft must be complied with. Depending on the operations involved, such as private, commercial, air operator etc., a MAINTENANCE SCHEDULE is required to be established and complied with for every aircraft. Following maintenance, an AME is required to sign a MAINTENANCE RELEASE, which means the aircraft is considered airworthy, and this releases the aircraft for return to service.

A licensed pilot, who is an owner/operator but not an AME, may perform ELEMENTARY WORK on the aircraft. This is limited to piston powered, non-commercial, unpressurized aircraft with a loaded gross weight of less than 5,700 Kgs (12,566 lbs). The details of the elementary work carried out shall be entered in the relevant maintenance records and signed by the owner or operator who performed the work. Examples are:

a) tire replacement;
b) safety belt replacement;
c) checking cylinder compression;
d) changing fuel, oil or filters;
e) replacing fuses, bulbs and reflectors;
f) cleaning, replacement of fluids or lubrication not involving disassembly.
g) cleaning and replacing spark plugs.

These are only general examples. More complex work must be carried out by a licensed AME. No maintenance release is required following elementary work.

Following an ABNORMAL OCCURRENCE, no aircraft shall be flown until it has been inspected for damage. Where no disassembly is involved, the inspection may be done by the PIC. Should any damage be present, the aircraft must be inspected and/or repaired by a qualified person (ie AME).

[Reference: CARs 605.84, 605.85, 605.86, 605.88]

1.43 Munitions of War

No person shall carry weapons, ammunition or other equipment designed for war unless the aircraft is a Canadian aircraft or the Minister has authorized the carriage of such equipment.

[Reference: CARs 606.01]

1.44 Aircraft Liability Insurance

Aircraft owners are required to carry liability insurance. The coverage amounts vary depending on the aircraft weight and operations. For example, owners of private registered aircraft under 2,300 lbs require public liability insurance of at least $100,000.00. All aircraft owners/operators must carry proof of insurance on board the aircraft.

[Reference: CARs 606.02]

NOTES

NOTES

2 Engines, Airframes Systems

2.01 Airframes

An airframe consists of a fuselage, wings and tail section. Airframes are constructed of alloy steels, high carbon steels, honeycomb sandwich type construction, composites, wood or fabric. Materials used in airframe construction must withstand compression, tension, torsion, shearing and bending.

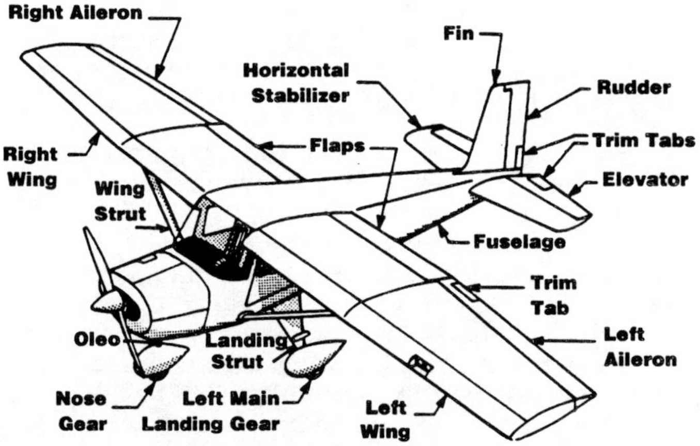

Fig. 2-1: Basic parts of an airframe.

The FUSELAGE is one of the main structural units of the airplane. It houses the crew, passengers, cargo, instruments, and other essential equipment. The fuselage in modern airplanes is made up of a combination of TRUSS and MONOCOQUE design. In the truss type construction, strength and rigidity result from joining tubing (steel or aluminum) to produce a series of triangular shapes, called trusses. In monocoque construction, rigs, formers, and bulkheads of varying sizes give shape and strength to the stressed-skin fuselage.

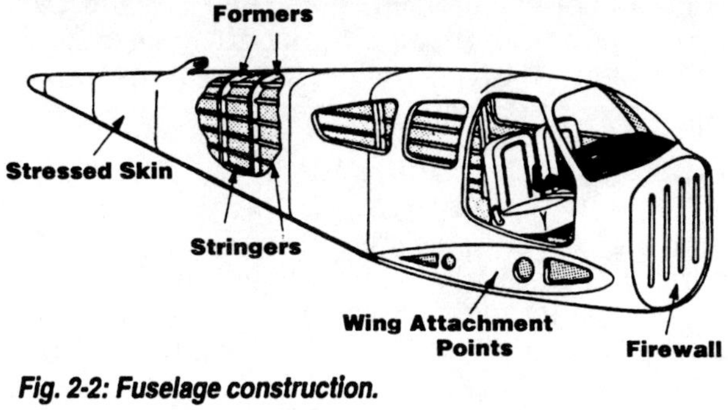

Fig. 2-2: Fuselage construction.

With single engine airplanes, the engine is normally attached to the front of the fuselage. The FIREWALL is a partition that is fireproof and divides the rear of the engine and cockpit or cabin to protect the pilot and passengers from accidental fires. The firewall is usually made of fire resistant, stainless steel.

The wings are airfoils that are attached to each side of the fuselage and are the main lifting surfaces that support the airplane in flight. The main types of wings are CANTILEVER and SEMICANTILEVER.

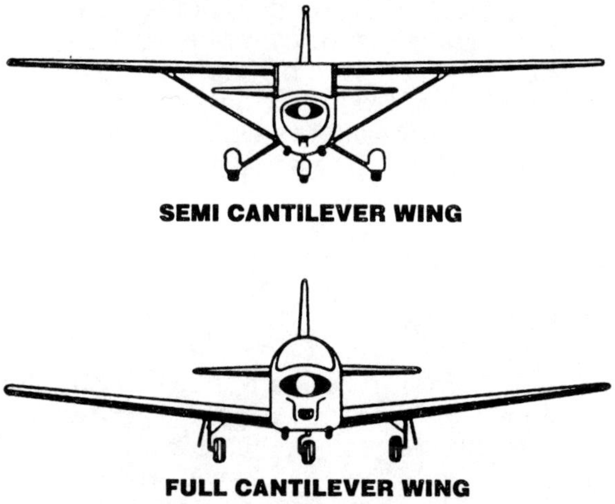

Fig. 2-3: Semicantilever and full cantilever wings.

The cantilever wing requires no external bracing and the stress is carried by internal wing spars, ribs, and stringers. In this type of wing, the skin or metal wing covering is constructed to carry most of the wing stresses. Treated aluminum alloy is most commonly used as the wing covering. Cantilever construction is typical of Piper training aircraft.

The semicantilever wing is braced both externally by mean of wing struts attached to the fuselage, and internally by spars and ribs. This construction is typical of Cessna training aircraft.

The principal structural parts of the wing are SPARS, RIBS, and STRINGERS. These are reinforced by trusses, I-beams, tubing, or other appropriate devices. The wing ribs determine the shape and thickness of the wing. In most modern airplanes, the fuel tanks are either an integral part of the wing's structure, or consist of flexible containers mounted inside of the wing structure.

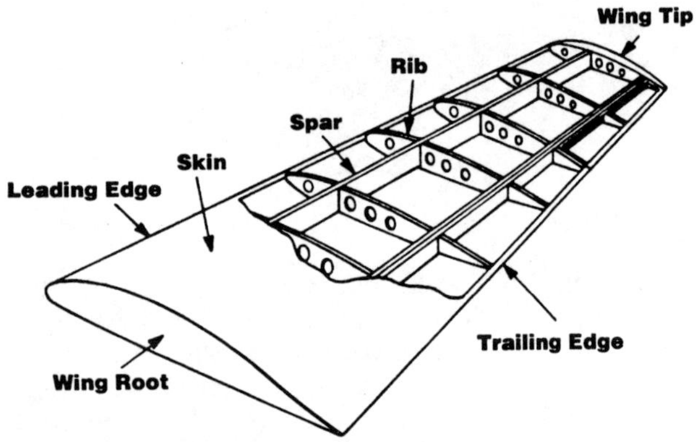

Fig. 2-4: Wing construction.

The EMPENNAGE is commonly known as the tail section and includes the vertical fin or stabilizer and the horizontal stabilizer, the movable surfaces including the rudder and rudder trim tabs, as well as the elevator and elevator trim tabs. In some aircraft, such as Piper training aircraft, the entire horizontal surface of the empennage can be adjusted from the cockpit as a complete unit for the purpose of controlling the pitch attitude or trim of the airplane. Such designs are referred to as STABILATORS.

2.02 Landing Gear

The function of landing gear is to take the shock of landing and allow maneuvering on the ground. Most light training aircraft are fixed gear, either nose-wheeled or tail-wheeled. As well, most training aircraft have a nosewheel and are referred to as TRICYCLE GEAR AIRPLANES. A nosewheel gear arrangement has the following advantages:

a) Allows more forceful application of the brakes during landings at high speeds without resulting in the airplane nosing over.
b) It permits better forward visibility for the pilot during takeoff, landing, and taxiing.
c) It tends to prevent ground looping (swerving) by providing more directional stability during ground operation, since the airplane's Center of Gravity (CG) is forward of the main wheels. The forward CG, therefore, tends to keep the airplane moving forward in a straight line rather than ground looping.

The main landing gear assembly consists of two main wheels and struts. Each main strut is attached to the primary structure of the fuselage or the wing. With the tailwheel type of airplane, the two main struts are attached to the airplane slightly ahead of the airplane's CG. In the nosewheel type, the two main struts are attached to the airplane slightly to the rear of the airplane's CG.

The landing gear shock-struts may be either self-contained hydraulic units or flexible spring-like structures that support the airplane on the ground and protect the airplane's structure by absorbing and dissipating the shock loads of landing and taxiing over rough surfaces. Many airplanes are equipped with what is known as OLEO STRUTS, the basic parts of which are a piston and a cylinder. The lower part of the cylinder is filled with hydraulic fluid and the piston operates in this fluid; the upper part is filled with air. Several holes in the piston permit fluid to pass from one side of the piston to the other as the strut compresses and expands and forces the piston up and down.

More advanced aircraft have retractable gear systems, which are operated by mechanical, hydraulic, electrical or pneumatic means. The main advantage of a retractable landing gear system is that it reduces airframe drag and thereby improves performance.

The BRAKES are used for slowing, stopping, holding, or steering the airplane. Brake systems are normally independent, having their own hydraulic fluid reservoir and are entirely independent of the airplane's main hydraulic system. Brake systems are powered by master cylinders similar to those used in cars. The system consists of a reservoir, 2 master cylinders, mechanical linkage which connects each master cylinder with its corresponding brake pedal, connecting fluid lines, and a brake assembly in each main landing gear wheel. Each master cylinder is actuated by toe pressure on the respective brake pedal. The master cylinder builds up pressure by the movement of a piston inside a sealed, fluid-filled cylinder. The resulting hydraulic pressure is transmitted through the fluid line which is connected to the brake assembly in the wheel.

Prior to flight, brake discs should be checked for even wear and no signs of cracking or pitting.

2.03 Engines

The engine is often referred to as the POWERPLANT, because in addition giving the aircraft power for forward motion, it also provides energy for electrical, hydraulic and pneumatic systems.

Light training aircraft utilize RECIPROCATING ENGINES, where pressures from burning and expanding gases cause a piston to move up and down in an enclosed cylinder. This reciprocating motion of the piston is transferred through a connecting rod into rotary motion by a crankshaft, geared to a propeller.

The basic parts of a reciprocating engine are the crankcase, cylinders, pistons, connecting rods, valves, spark plugs, and crankshaft. In the head or top of each cylinder are 2 valves and 2 spark plugs. One of these valves opens and closes a passage leading from the carburetor (or induction manifold) and is called the INTAKE VALVE. The other opens and closes a passage leading to the outside atmosphere (or exhaust manifold) and is called the EXHAUST VALVE. Inside each cylinder is a movable PISTON which is attached to a CRANKSHAFT by means of a CONNECTING ROD. When the rapidly expanding gases (resulting from the heat of combustion of the fuel-air mixture ignited by spark plugs) push the piston down within the cylinder, it causes the crankshaft to rotate.

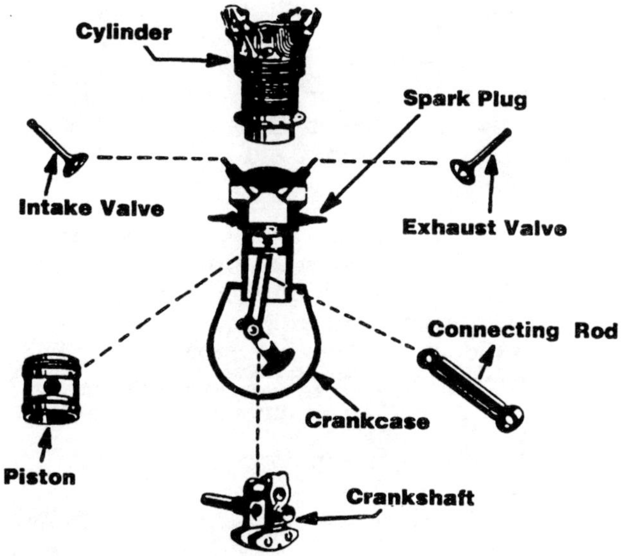

Fig. 2-5: Basic components of the Engine.

The main types of light aircraft engines are defined on the basis of the arrangement of the cylinders. The main classifications are HORIZONTALLY OPPOSED, IN-LINE and RADIAL.

COMBUSTION is achieved by igniting fuel-air within the cylinder. Most aircraft piston engines operate on a 4-stroke system. The valves are timed so that they open slightly early and close slightly late, which is termed VALVE LEAD or VALVE LAG. This timing improves engine efficiency.

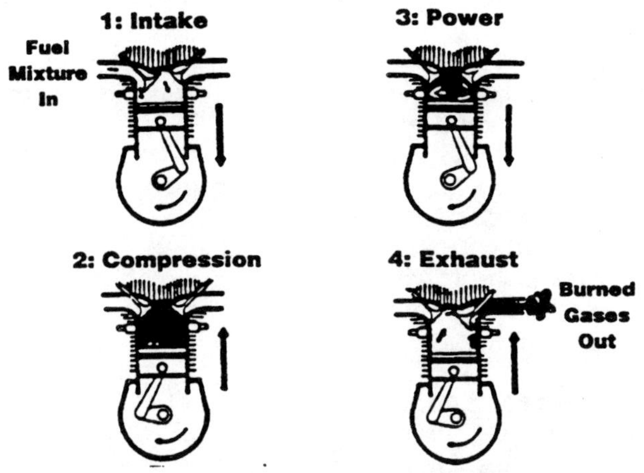

Fig. 2-6: Strokes of a 4-stroke internal combustion engine.

TOP DEAD CENTER refers to the highest position of travel of the piston. BOTTOM DEAD CENTER refers to the lowest position of travel of the piston. COOLING is generally achieved by air coming in contact with baffles or fins added to the engine cylinders. Cooling is enhanced on some aircraft with the addition of COWL FLAPS, which increase the amount of incoming air passing over the engine. When taxiing downwind, the engine may overheat due to loss of airflow over engine. The aircraft should be turned into wind to enhance cooling (do not shut down engine).

The following engine instruments should be known:

a) TACHOMETER. Indicates speed of engine crankshaft rotation, calibrated in Revolutions Per Minute (RPM).

Fig. 2-7: Tachometer.

b) OIL PRESSURE GAUGE. Indicates oil pressure being supplied to moving engine parts by lubricating system. It warns of impending engine failure which may result from an exhausted oil supply, failure of oil pump, broken oil lines etc.
c) OIL TEMPERATURE GAUGE. Indicates temperature of oil entering the engine. With a very low oil temperature, the oil may be so thick that it is not able to lubricate the internal moving engine parts. With an excessively high temperature the lubricating oil tends to thin out

and decreases the lubricating qualities of the oil.

d) FUEL QUANTITY GAUGES. Allow the pilot to determine the approximate amount of fuel in tanks. Pilots should ALWAYS complete a visual check of fuel prior to flight and never rely on fuel quantity gauges.

e) FUEL PRESSURE GAUGE. Indicates the difference in pressure between the fuel and the air being supplied to the engine, and therefore whether fuel is being supplied and the actual pressure at which it is being forced into the engine. Fuel pressure gauges are very important for low wing aircraft: without fuel pressure from engine driven or electrical fuel pump, fuel will not reach engine.

f) MANIFOLD PRESSURE GAUGE. Indirectly indicates the power output of the engine by measuring the pressure of the air in the fuel/air induction manifold. The higher the manifold pressure, the greater the power being developed by the engine. By means of the throttle, the pilot can control the pressure in the manifold, and thus the power. The manifold pressure gauge is calibrated in inches of mercury, and is found in airplanes equipped with a supercharger or a constant speed propeller.

Fig. 2-8: Manifold Pressure Gauge.

g) CYLINDER HEAD TEMPERATURE GAUGE, Indicates the temperature of the cylinder head of the hottest cylinder (usually one of the rear ones in a horizontally-opposed engine) and gives the pilot a means of determining whether the engine is operating at normal or excessive engine temperatures. It is an important instrument for high compression/high power engines. Operating the engine at a temperature higher than it was designed for will cause loss of power, excessive oil consumption, and damage to the cylinder walls, pistons and valves.

h) EXHAUST GAS TEMPERATURE (EGT) GAUGE. Indicates temperature at exhaust manifold. An important instrument to determine whether proper combustion has taken place. The leaner the mixture, the hotter the EGT. The richer the mixture, the cooler the EGT. Can also be useful in leaning out mixture at altitude, where lean mixture is established just to rich side of peak EGT.

2.04 Aviation Fuels

OCTANE RATING refers to the ability of a fuel to burn slowly and expand evenly, rather than explode suddenly. Fuel grades are coloured according to their octane rating:

Fuel	Colour
80/87	Red
100 LL (low lead)	Blue
100/130 (high lead)	Green

The aircraft operating manual specifies the correct grade of fuel for the aircraft. Insure that the proper fuel grade is utilized. If the proper grade of fuel is not available, use the next higher grade, but NEVER use a lower grade. Using an octane rating fuel that is too low for the aircraft will result in DETONATION, which is when the fuel combusts too rapidly and will cause serious damage to the engine. Detonation can also be caused by overly hot engine operating temperatures. Detonation is hard to detect, but generally there will be a noticeable loss of power, engine overheating, and if allowed to continue, engine damage.

PRE-IGNITION is caused by hot spots or carbon particles inside the piston igniting the fuel/air mixture before the spark plug fires. One important difference between detonation and pre-ignition is that if conditions for detonation exist in one cylinder, they may exist in all cylinders; but pre-ignition may exist only in one or two cylinders.

VAPOUR LOCK occurs in conditions of high temperature when fuel is vaporized in the fuel line, resulting in fuel blockage. FUEL DENSITY changes with temperature: the colder the temperature, the heavier the fuel. Generally, consider conventional grades to weigh 6.0 lbs per US gallon or 1.6 lbs per litre. [NOTE: 1 Imperial Gallon of fuel weighs 7.2 lbs. The old imperial gallon is not generally used as a measurement because a) all standard training aircraft specify capacity and performance in U.S. gallons and b) in Canada, fuel is purchased in litres].

FUEL ADDITIVES improve performance. TETRA-ETHYL LEAD is added to aviation fuels to get better anti-detonation qualities. ETHYLENE DIBROMIDE is added to fuel to act as a cleaning agent to prevent spark plug electrode oxidation.

2.05 Fueling Precautions

Fuel can become contaminated by water and other foreign substances. It is therefore recommended that pilots top up the aircraft fuel tanks after flight. Always drain a fuel sample and inspect it for proper colour, water, sediment and brilliance.

The aircraft and fueling equipment through which fuel passes require BONDING. Bonding prevents sparks by equalizing electrical potentials. The hose nozzle must be bonded to the aircraft before the tank cap is removed in over-wing fueling. Grounding of the fuel service vehicle and bonding of the service vehicle and hose nozzle to the aircraft before fueling will dissipate static electricity.

If it is necessary to refuel from a drum, be sure to use a proper filter/water separator and do not drain the fuel from the bottom of the drum, because this is where sediments and water accumulate. Pilots should only use a chamois as a filter in an emergency because the fibres from the chamois will clog the fuel system and may cause an engine failure.

2.06 Fuel Systems

Simple fuel systems are normally GRAVITY FEED, which means that the force of gravity alone pushes the fuel to the engine, and the fuel cells are located at a higher position on the aircraft than the engine. In aircraft having low wing tanks, a FUEL PUMP is needed. The main fuel pump is normally engine driven, and an ELECTRIC FUEL PUMP is employed as a backup. A FUEL PRIMER is a small hand pump which is used to pump fuel into the engine prior to starting in cold weather. Care should be taken not to over-prime, because the engine can become flooded, resulting is a risk of backfire or engine fire.

FUEL VENTS prevent a vacuum forming inside the fuel tanks. Venting is accomplished with either a fuel vent designed in the fuel cap, or an over-flow vent placed under the fuel tank. Venting insures proper air pressure within the tanks for a steady fuel flow.

FUEL SUMPS are located at the bottom of fuel tanks and lines from which water and other sediment can be drained. FUEL STRAINERS are located at the low point in the fuel lines between the fuel selector and the carburetor, and trap water and sediment in a container which can be drained.

2.07 Ignition Systems

MAGNETOS supply electrical current to the spark plugs. The magnetos are driven by the engine crankshaft, and are independent of the aircraft battery or electrical system. To start the engine, the aircraft battery sends power to the starter, which via a series of gears, rotates the crankshaft and the magnetos are engaged. Then, the armature of the magneto produces current and the sparks for ignition of the fuel-air in each cylinder. The starter system disengages after start, and the battery no longer contributes to the actual operation of the engine. There are two magnetos to supply DUAL IGNITION (two spark plugs per cylinder). The pilot can select the LEFT MAG, RIGHT MAG or BOTH (normally BOTH should be selected). Each magneto supplies one set of spark plugs, resulting in more efficient combustion, better engine performance and safety (should one magneto fail, the other will continue to deliver current, with only a slight loss of RPM). SHIELDED IGNITIONS SYSTEMS utilize wire braids around the spark plug leads and a steel shell around the spark plug itself to contain radiated energy and thus prevents radio disruption. If the magneto grounding wire is broken, the mags can continue to supply electrical current when the mags are in the OFF position - so a live mag check at shutdown is a good procedure.

2.08 Oil Systems

The functions of oil in a light aircraft engine are:

a) Cooling.
b) Sealing.
c) Lubrication.
d) Cleaning.

Oils are graded according to viscosity. For engines of moderate horsepower, use the following grades:

Grade	Season
120	Summer
100	Fall or Spring
80	Winter
65	Arctic

Hot engine oil is pumped through an OIL COOLER having baffles which reduces oil temperature and thereby aids in engine cooling. An OIL FILTER is used to trap sediments and impurities. The PRESSURE RELIEF VALVE provides constant oil pressure. There are 2 methods of lubrication: FORCE FEED (dry sump) and SPLASH (wet sump). In force feed lubrication, oil is sprayed under pressure. In splash lubrication, oil is contained in a sump or reservoir and splashes over the engine as the crankshaft rotates.

MINERAL OIL has no additives and is used to break-in new engines. DETERGENT OIL has various additives for cleaning, dispersion and suspension of particles. Never add detergent oil to mineral oil (drain the mineral oil first). Never use automotive oil in an aircraft engine.

When flying, be sure to monitor the oil pressure gauge and oil temperature gauge. If the oil pressure is too low, oil is not being delivered to all parts of the engine, and the flight should be terminated as soon as possible. If the oil temperature is too high, the oil may not be viscous enough to lubricate all parts of the engine. Refer to the owner's manual for the correct range of oil temperature and oil pressure.

2.09 The Carburetor and Fuel Injection

The function of the carburetor is to mix the correct ratio of fuel to air, to vaporize it and to deliver the mixture to the cylinders. The ratio of fuel to air is governed by weight (not volume). The ratio for best power is 14 pounds air to 1 pound fuel.

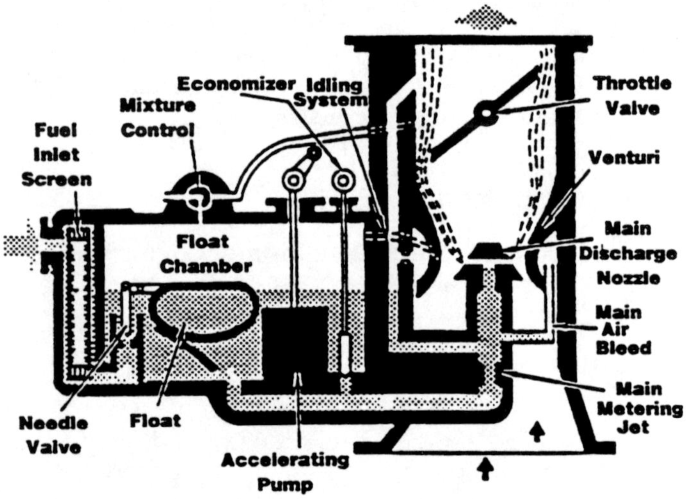

Fig. 2-9: Cutaway view of aircraft carburetor.

As the incoming air passes through the venturi, its speed increases and a low pressure area is created. As the nozzle sprays fuel into this low pressure area, the fuel is vaporized. The throttle controls the butterfly valve, which regulates the amount of fuel/air mixture entering the engine and thereby controls the power output.

The MIXTURE control governs the amount of fuel that is sent through the discharge nozzle. A high proportion of fuel to air is a RICH mixture, and a low proportion of fuel to air is a LEAN mixture. Adjust the mixture control as specified in the operating handbook. When the mixture is too lean, the engine will run hot (in cruise flight, if the mixture is too lean, the engine may backfire). When the mixture is too rich, the engine burns too much fuel.

CARBURETOR HEAT is used to prevent or melt ice which may form in the carburetor. Carburetor ice forms when the outside air is humid and the temperature drop within the venturi is enough to form ice. Carburetor ice is possible in temperatures from -6° C to +32° C.

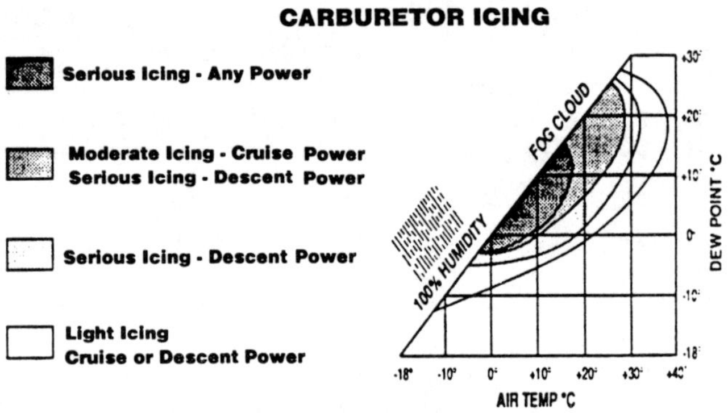

Fig. 2-10: Atmospheric conditions for carb ice.

An indication of carburetor ice is a drop in power. [On an aircraft having a fixed pitch propeller, a decrease in power is indicated by a drop in RPM, while aircraft having a constant speed propeller show a decrease in MANIFOLD PRESSURE.] The carburetor heat control is used to melt the ice. Using carburetor heat results in a reduction of power and a more rich mixture, because the warm, heated air being sent to the carburetor is less dense. Carburetor heat should only be used when ice is suspected or is likely to form (such as at reduced power settings), because it results in hot, unfiltered air going into the engine, reducing efficiency and supplying dirt (avoid taxiing with carb heat on).

In a FUEL INJECTED engine, a fuel injection nozzle supplies a fine spray of fuel which is discharged into the air stream entering the intake manifold. The result is that a carburetor is unnecessary, add all of the problems associated with carburetor ice do not exist. Negative aspects of fuel injection systems are vapor lock during ground operations on hot days and difficulty starting a hot engine.

ALTERNATE INDUCTION AIR is used in fuel injected engines to provide an alternate source of intake air should the primary intake experience a blockage (for example, caused by buildup of airframe ice during flight into freezing rain).

2.10 Electrical Systems

Light aircraft electrical systems are either 12 or 24 volts, direct current (DC). Either a GENERATOR or ALTERNATOR is incorporated to supply electrical current to the system and re-charge the battery. The advantage of an alternator is that it supplies electrical current at low RPM, while the older generator requires high RPM to function adequately. Power is supplied through one or more BUS BARS, and the MASTER SWITCH controls this power to all circuits. All electrical circuits are protected by fuses or circuit breakers. An AMMETER indicates the flow of current, in amperes, from the alternator to the battery or from the battery to the electrical system. When the engine is operating and the master switch is on, the ammeter shows the charging rate applied to the battery. If the alternator is not functioning or the electrical load exceeds the output of the alternator, the ammeter shows the battery discharge rate. The ammeter should always show on the "+" side of the "0" mark. If the needle is in the "-" side, electrical energy is being drawn from the battery instead of the alternator.

Fig. 2-11: Ammeter.

An OVER-VOLTAGE SENSOR is a warning light which, when illuminated, indicates that the alternator has been shut down due to an over-voltage condition, and the battery is supplying all electrical power. The over-voltage sensor may be reset by turning the master switch off and then back on again. If the warning light does not illuminate, normal alternator charging has resumed; however, if the light again illuminates, a malfunction has occurred and the flight should be terminated as soon as possible. The over-voltage warning light may be tested by momentarily turning off the ALT side of the master switch and leaving the BAT side on.

A VOLTMETER indicates the voltage in the electrical system. The VOLTAGE REGULATOR prevents the battery from being over-charged or the generator or alternator from overloading the system.

Prior to start, all electricals (flap, radios, instruments, lighting etc) are powered by the battery. After start, all electricals are powered by the generator or alternator, as the case may be. When starting, the starter switch activates a SOLENOID, which in turn activates a flow of current to the STARTER MOTOR, which causes the engine to turn over.

2.11 Supercharging and Turbocharging

A SUPERCHARGER is an internally driven compressor, powered directly by the engine. The purpose of supercharging is to maintain seal level rated horsepower as altitude increases. The supercharger is installed downstream from the carburetor, and it compresses the fuel-air mixture after it leaves the carburetor. Feeding the higher pressure fuel-air mixture into the intake manifold is FORCED INDUCTION, which results in a supercharged engine.

Forced induction can be used to increase normal sea level horsepower at low altitudes, or to make up for the loss of power at high altitude to obtain sea level horsepower. For a supercharger to operate effectively, the temperature of the fuel-air mixture must be maintained within a specified range.

In a TURBOCHARGER, an impeller is powered by re-direction of exhaust gases to increase the pressure of the intake air. The advantage of a turbocharger over a supercharger is that engine horsepower is not lost to drive the impeller. The turbocharger compresses intake air alone, not fuel-air as in the supercharger. The redirection of exhaust gases is effected by the WASTEGATE, which is a valve controlled by a governor. When the wastegate is open, the engine is normally aspirated. When the wastegate is closed, the engine is turbocharged.

2.12 Exhaust Systems

Aircraft heating systems often obtain heat energy from waste heat taken from the exhaust manifold. Therefore, pilots should carefully inspect the exhaust tubes for any indication of leakage so as to prevent the possibility of carbon monoxide poisoning.

2.13 Oxygen Systems

Supplemental oxygen allows for flight at higher altitudes in non-pressurized aircraft, and is required emergency equipment for high altitude aircraft. Some advanced, non-pressurized, high performing single engine aircraft may be equipped with oxygen bottles, tubes and masks for short flights above 10,000 feet. Users of such equipment should be thoroughly familiar with the mechanics and safety features of their equipment.

NOTES

3 Theory of Flight

3.01 Wing Design

CHORD is the distance from the leading edge to the trailing edge. WING SPAN is the distance from wing tip to wing tip. ASPECT RATIO is defined as the wing span divided by wing chord. A high aspect ratio wing has minimal induced drag and is found in sailplane design.

CAMBER refers to the relative curvature of a wing. High camber airfoils are for aircraft having slow speed and high lift. UPPER CAMBER refers to the curvature of the upper surface of the airfoil. LOWER CAMBER refers to the curvature of the lower surface. MEAN CAMBER refers to the mean line which is equidistant at all points between the upper and lower surfaces.

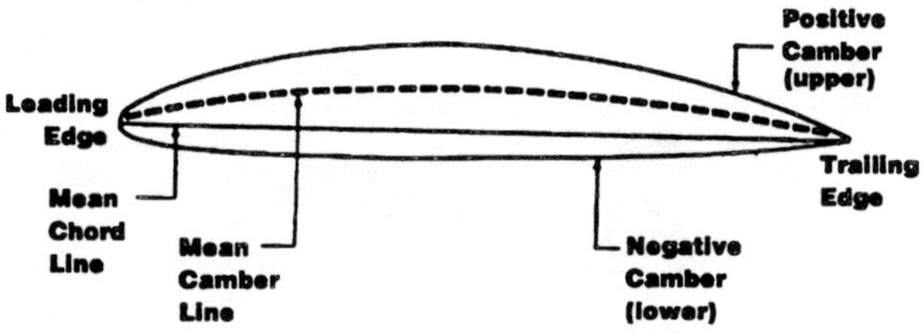

Fig. 3-1: Terminology of airfoil section.

WING PLANFORM refers to the shape of a wing when viewed from above (some wings are rectangular, while others are delta shaped, for example).

WING SLATS are miniature airfoils that open forward of the main airfoil to smooth out the airflow over the wing. At high angles of attack, slats automatically move forward ahead of the wing. WING SLOTS are openings in the leading edge of airfoils which allow air to pass through the forward portion of the wing at high angles of attack.

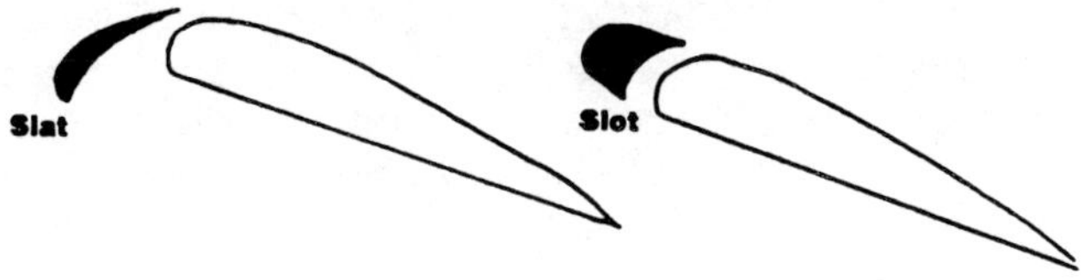

Fig. 3-2: Wing slats, wing slots.

SPOILERS disrupt lift by interfering with the smooth airflow on the upper surface of the wing. Spoilers are used either to assist in aileron control or are linked to brakes to disrupt lift on landing and enhance braking action. WING FENCES are small vertical fins on the upper airfoil that smoothen the airflow and improve handling.

DIHEDRAL is a design feature where the wing tips are higher than at the center of the wing. Dihedral enhances lateral stability because when a disturbance causes a wing to lower, such as in turbulence, the aircraft sideslips and the lowered wing will strike the air at a higher angle, resulting in greater lift and returning the lower wing to its original position.

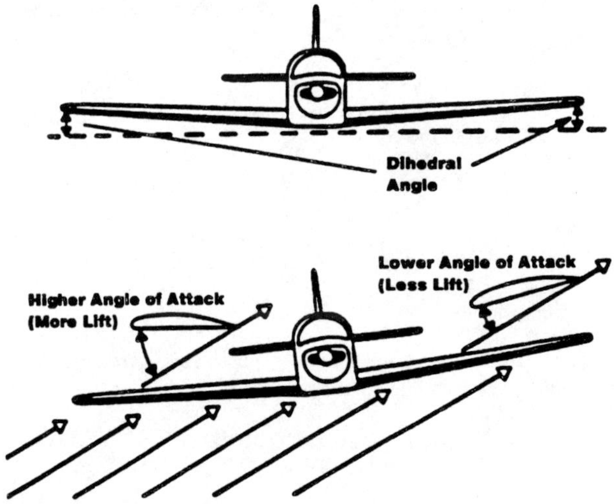

Fig. 3-3: Dihedral.

ANGLE OF INCIDENCE is the angle between the airfoil and the horizontal. It is the fixed angle at which the wing is mounted to the aircraft, as viewed from the longitudinal axis of the aircraft. Accordingly, the angle of incidence is constant and cannot be changed by the pilot.

FLAPS are high lift devices that produce the same amount of lift at lower speed, by increasing the camber of the airfoil. Flaps also increase drag. Flaps are useful for steepening the angle of descent and reducing stopping distance on landing. Flaps also reduce the aircraft stalling speed and improve forward visibility.

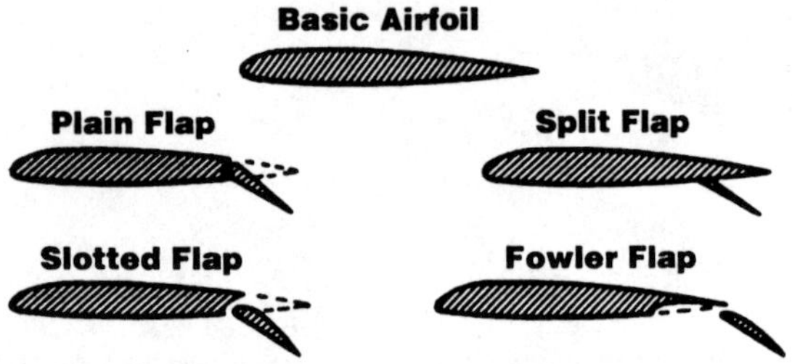

Fig. 3-4: Types of wing flaps.

WASHOUT is when the airfoil is twisted, so that the angle of incidence at the wingtip is less than at the fuselage. Washout improves stall characteristics, because at the stall, the inner portion of the airfoil stalls first, while the outer portion remains unstalled. The ailerons, at the outer portion of the wing, remain effective even though the inner wing is stalled.

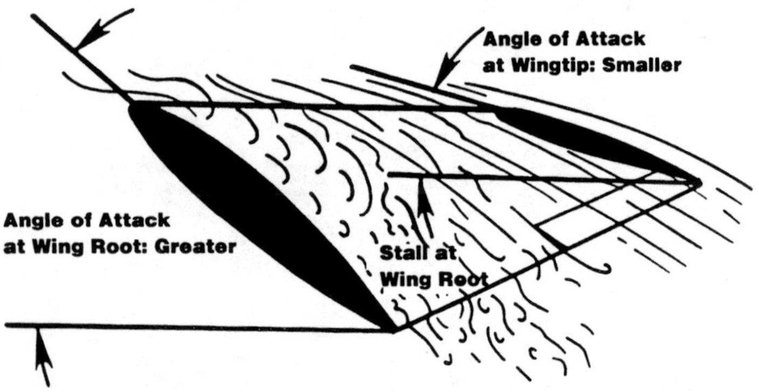

Fig. 3-5: Washout. Wing is twisted so that ailerons are still effective even though wing root is in stalled condition.

The effect of washout can be obtained with STALL STRIPS placed on the inner section of the leading edge.

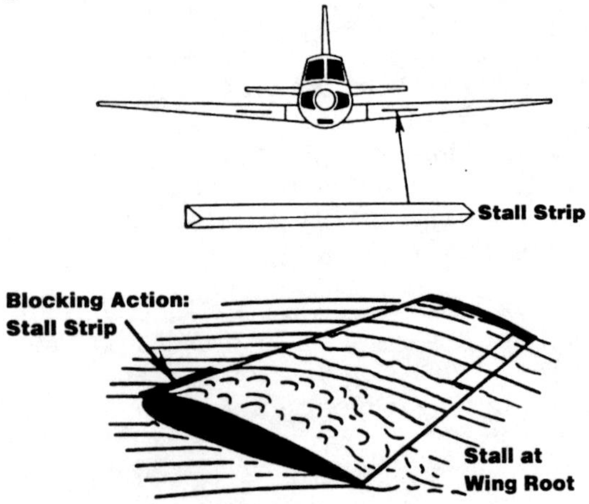

Fig. 3-6: Stall strip will insure that wing root will stall first.

STREAMLINING is when an object is re-shaped to minimize drag. An example of streamlining is with aircraft wheels: WHEEL PANTS are installed to reduce drag on fixed gear, light training aircraft (in C-172, wheel pants increase cruise airspeed by about 5 kts).

LAMINAR FLOW is the tendency of air to adhere to the wing as the wing moves through the air. The TRANSITION POINT of an airfoil is the point of the airfoil where the laminar layer becomes turbulent, resulting in high drag due to skin friction. A LAMINAR FLOW AIRFOIL (also referred as symmetric airfoil) has improved boundary layer characteristics: the separation point is further aft on the airfoil.

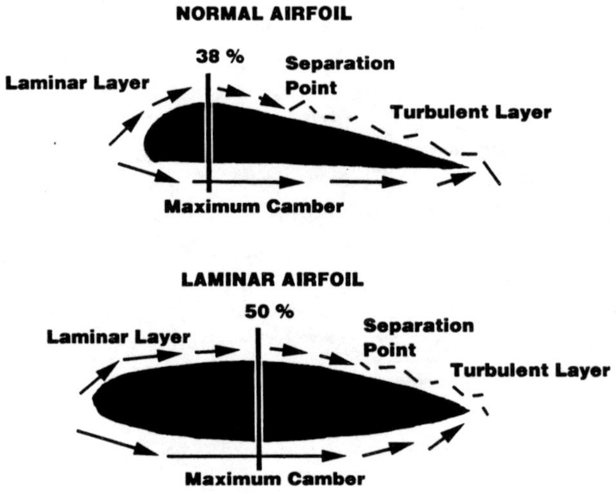

Fig. 3-7: Normal vs laminar airfoil.

3.02 Stability

STABILITY is the ability of an aircraft to remain in its position in flight, and to return to that position following a disturbance. Therefore, a stable airplane will tend to return to the original condition of flight if disturbed by a force such as turbulent air.

Stability is POSITIVE, NEUTRAL, or NEGATIVE. With positive stability, like a marble in a bowl, an object will return to its original position following a displacement. With neutral stability, like a marble on a flat surface, if the object is displaced, it will remain in its new position. With negative stability, like a marble placed on an upside-down bowl, even the slightest displacement of the object will cause the object to continue to move in the direction of the displacement.

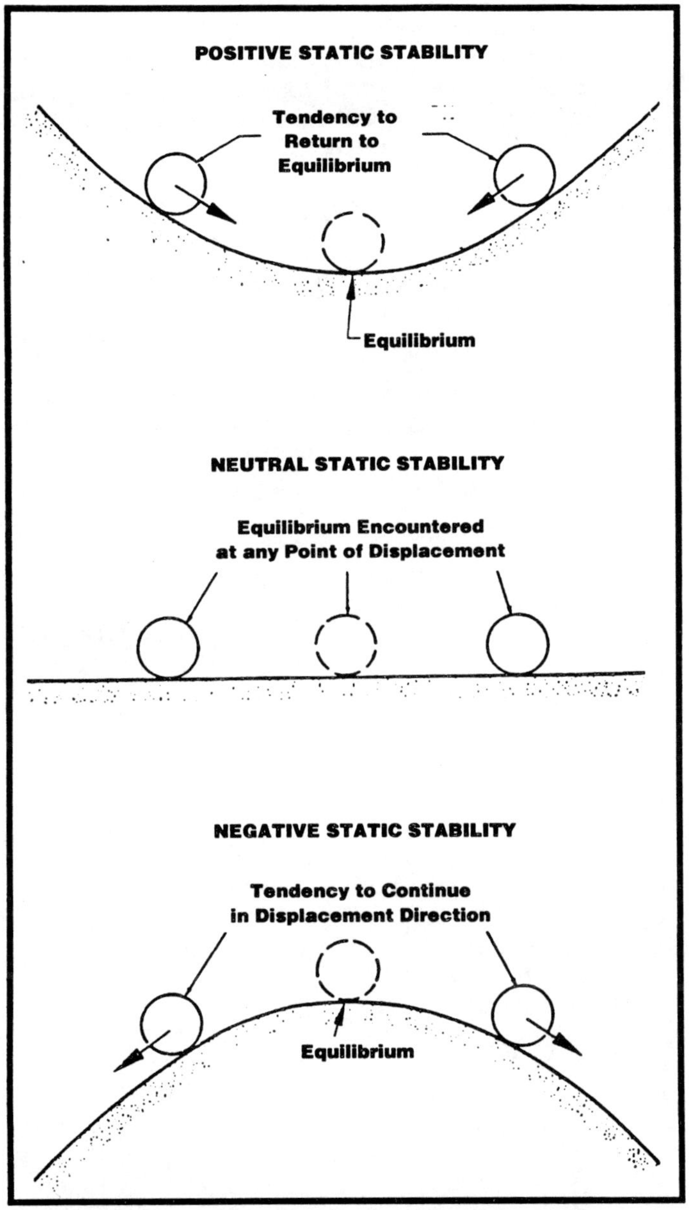

Fig. 3-8: Model for positive, neutral and negative static stability.

STATIC STABILITY is the initial tendency of an aircraft to return to its original position. An aircraft that is self-correcting shows POSITIVE STATIC STABILITY (also known as INHERENT STABILITY). With static stability, if an airplane's equilibrium is disturbed, forces will be activated which will initially tend to return the airplane to its original position. However, if the restoring forces are so great that they force the airplane beyond the original position and continue in that direction, the object displays NEGATIVE STATIC STABILITY.

DYNAMIC STABILITY is the overall tendency of the aircraft to return to its original position, following a series of dampening oscillations.

An aircraft should possess positive stability which is both static and dynamic.

Stability may be LONGITUDINAL (pitch), LATERAL (roll) or DIRECTIONAL (yaw). PITCH STABILITY is designed into light aircraft by having the center of gravity forward of the center of pressure, so the aircraft will have a natural nose down tendency. The horizontal stabilizer is placed at a negative angle of incidence so it receives a download to counteract the natural nose-down tendency.

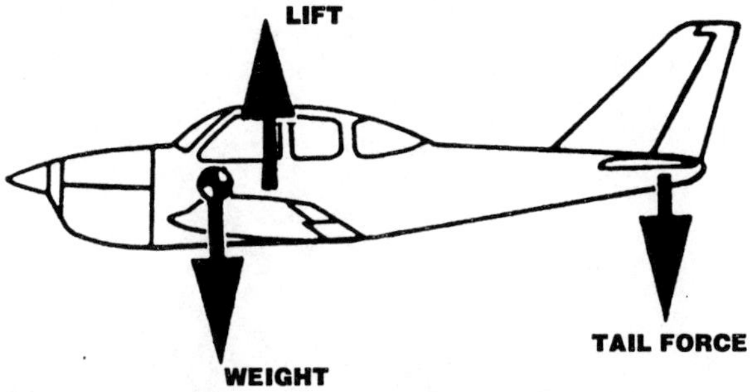

Fig. 3-9: Pitch stability.

YAW STABILITY is enhanced by the fin, due to the increased keel surface and resulting weather-cocking tendency. The surface of the fin acts similar to a weathervane and causes the airplane to weathercock into the relative wind.

LATERAL STABILITY is enhanced by dihedral.

3.03 Forces Acting on an Airfoil

The 4 forces acting on an airfoil in flight are LIFT, THRUST, DRAG and WEIGHT. Thrust comes from the engine and propeller. Weight is the downward force due to gravity. While in unaccelerated, steady flight, the attitude, direction and speed of the airplane will remain constant until one or more of the basic forces changes in magnitude. In unaccelerated, steady flight, the opposing forces are in EQUILIBRIUM.

When pressure is applied to the airplane controls, one or more of the basic forces change in magnitude and become greater than the opposing force, causing the airplane to accelerate or move in the direction of the applied force. For example, if power is applied while in level flight, and a climb attitude is established, the force of lift would increase during the time back elevator pressure is applied, but after a steady state climb is established, the force of lift would be approximately equal to the force of weight. The airplane does not climb because lift is greater than in level flight, but because thrust is greater than drag, and because a component of thrust is developed which acts upward, perpendicular to the flight path.

Lift results from the pressure differential between the top surface of the wing (high velocity, low pressure) and the lower surface of the wing (low velocity, high pressure). BERNOUILLI's THEOREM states that the total energy in a system remains constant, therefore, where there is an increase in velocity, there will be a decrease in pressure and vice versa. NEWTON'S LAWS OF MOTION state that a body in motion will continue in motion until acted upon by by some force, and for every action there is an equal and opposite reaction. As air passes over the airfoil, the air flows downward, which is termed DOWNWASH. Lift is produced when the air is deflected downward.

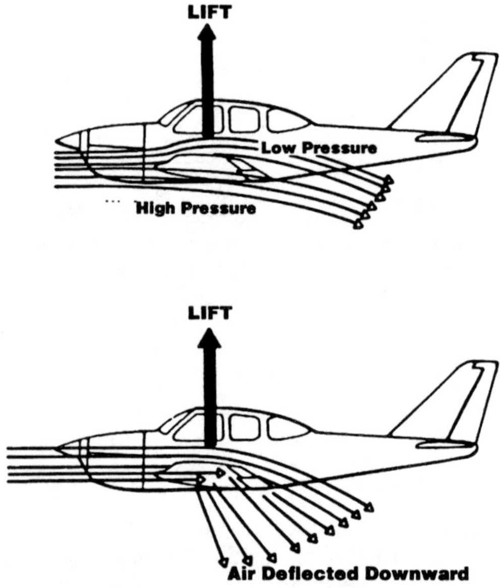

Fig. 3-10: Lift results from differential between high pressure under wing and low pressure over wing, and from deflection of air downward.

The LIFT FORMULA is used to calculate lift:

$$\text{Lift} = CL1/2pV^2S$$

CL = Coefficient of Lift (depends on design of airfoil)
P = Air Density
V = Velocity
S = Plan Area of Airfoil

EXAMPLE: According to the lift formula, what is the effect of doubling the velocity?

ANSWER: A fourfold increase in lift.

Lift acts upward and perpendicular to the relative wind. Although lift is generated over the entire wing, an imaginary point is established which represents the resultant of all lift forces, referred to as the CENTRE OF PRESSURE (CP).

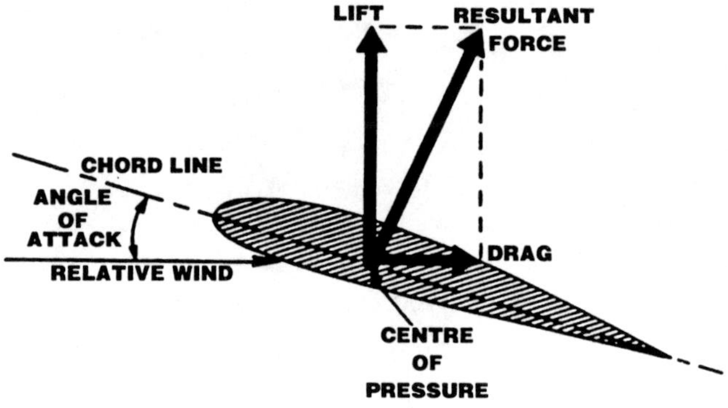

Fig. 3-11: Lift and drag on airfoil.

Drag is the resistance to forward motion. Drag acts parallel to and in the same direction as the relative wind. Note the following sources of drag:

a) Form drag. Results from shape of object passing through air.
b) Skin friction. Smooth surfaces cause less drag than rough surfaces. Skin friction is greater if the airplane surfaces are rough and dirty.
c) Profile drag. Profile drag = form drag + skin friction.
d) Induced drag. Drag caused by creation of lift, wing tip vortices.
e) Parasite drag. All drag except induced drag.

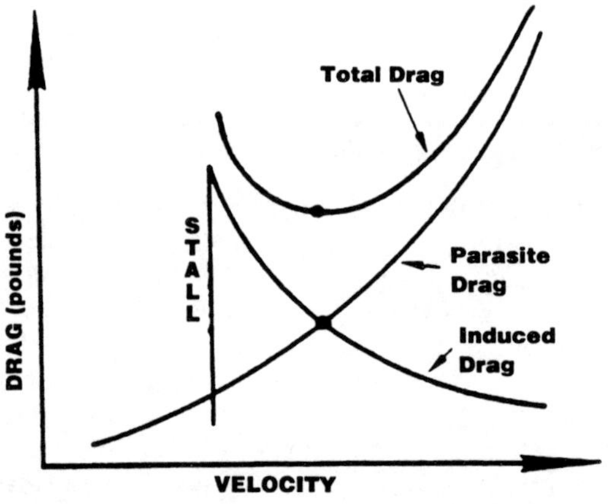

Fig. 3-12: Drag curves. Note that parasite drag increases with higher speeds, while induced drag decreases with high speeds.

The overall drag of an airfoil can be determined from the following formula:

$$\text{Drag} = CD1/2pV^2S$$

CD = Coefficient of Drag (depends on design of airfoil)
P = Air Density
V = Velocity
S = Plan Area of Airfoil

EXAMPLE: What is the relationship between air density and drag?

ANSWER: Drag is directly proportional to air density (increased air density = increased drag).

WING TIP VORTICES are produced when the airflow over the top surface of the airfoil (decreased pressure) comes in contact with the airflow on the undersurface of the wing (increased pressure), resulting in rotating air extending back from the wing tips. Wing tip vortices cause induced drag.

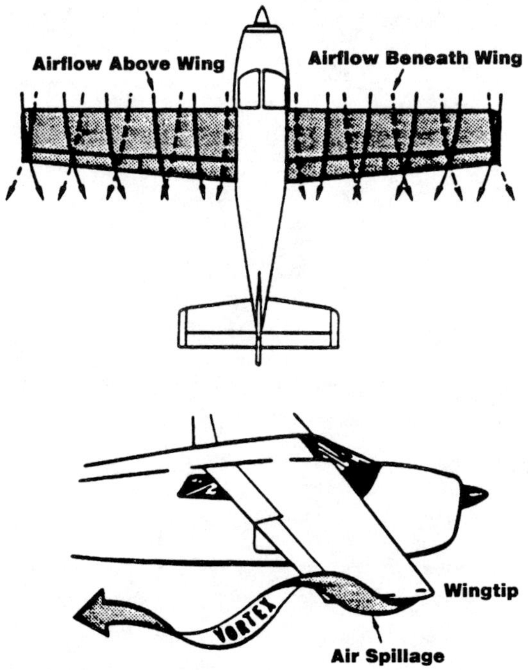

Fig. 3-13: Wing tip vortices.

3.04 Stalls

An airfoil stalls when the ANGLE OF ATTACK becomes excessive, such that the airfoil no longer produces lift, and drag becomes too great. Angle of attack is defined as the angle between the relative airflow and the chord line.

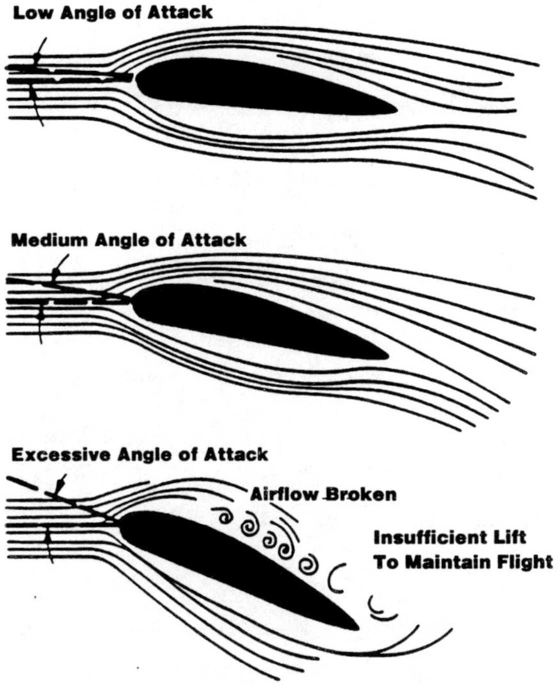

Fig. 3-14 Stalling angle and angle of attack.

For light aircraft, the stalling angle is about 16° angle of attack. An airfoil can stall at any airspeed, attitude or power setting. All that is required for a stall is that the angle of attack exceed the stalling angle. To unstall a wing, reduce the angle of attack with forward elevator pressure, and keep straight with rudder. Power should be applied to minimize the height loss.

When an aircraft is properly loaded, it stalls at an indicated airspeed (IAS) as specified in the owner's manual, when the stall is entered from slow, unaccelerated flight. The IAS stalling speed remains the same with an increase in altitude, while the true airspeed (TAS) stalling speed increases with an increase in altitude (see s. 4.02 for relationship between IAS and TAS).

The PRESSURE DISTRIBUTION of an airfoil refers to the pattern of pressure surrounding it. All lift forces act through the center of pressure. As the angle of attack is increased, the pressure distribution changes, so that the center of pressure moves forward. At the stalling angle, the center of pressure moves rearward.

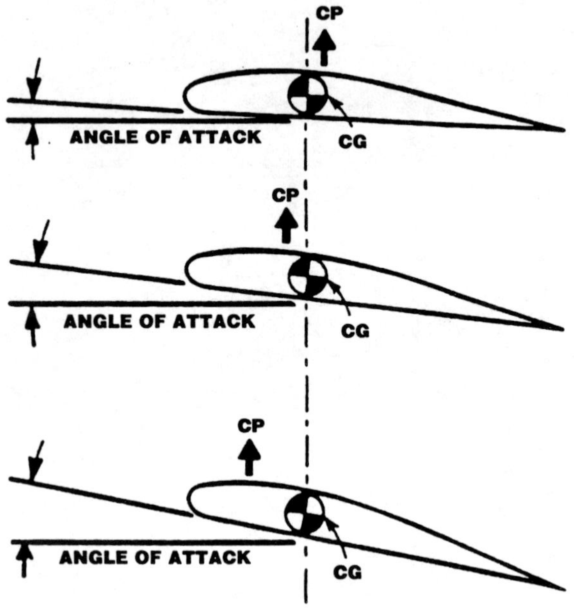

Fig. 3-15: Movement of Center of Pressure with increasing angle of attack.

3.05 Load Factor

Every airplane is designed and certified for a certain maximum weight during flight. It is important that the aircraft be loaded within the specified weight limits, because certain flight maneuvers will impose extra loads on the airplane structure which may impose stresses which will exceed the design capabilities of the airplane. Overstressing the airplane can also occur if the pilot engages in maneuvers creating high loads, regardless of how the airplane is loaded. These maneuvers not only increase the load that the airplane structure must support, but also increase the airplane's stalling speed.

According the the laws of physics, a mass will continue to move in a straight line unless some force intervenes, causing the mass to assume a curved path. The same applies to an airplane. Anytime an airplane moves in a curved path with a positive load, the load the wings must support will be greater than the aircraft weight. A positive load occurs when back pressure is applied to the elevator, and a negative load occurs when forward pressure is applied to the elevator. These loads imposed on a wing are referred to as LOAD FACTOR.

Load factor is a measure of the amount of weight an airfoil carries or "G" force. Load factor is increased by turning, abrupt deflection of controls and operations in turbulence. As the angle of bank increases, the amount of lift required to maintain level flight increases because of the increasing load factor.

When an aircraft is turning, the total lift is inclined to the inside of the turn, and this sets up new forces. The force to the inside of the turn is known as CENTRIPETAL FORCE, and the force acting to the outside of the turn is CENTRIFUGAL FORCE. Due to the acceleration created by the centripetal force, a turning aircraft is not in equilibrium.

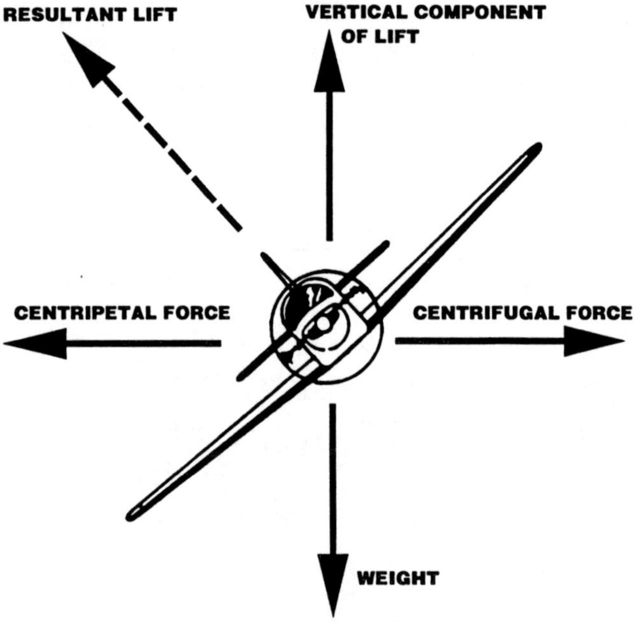

Fig. 3-16: Forces acting on an aircraft while turning.

As the angle of bank increases, the load factor is increased.

Angle of Bank	Load Factor
0°	1.00
15°	1.04
45°	1.41
60°	2.00
75°	4.00

For a 60° bank turn, the load factor is 2.0 and the airfoil must provide 2 times the amount of lift required for level flight:

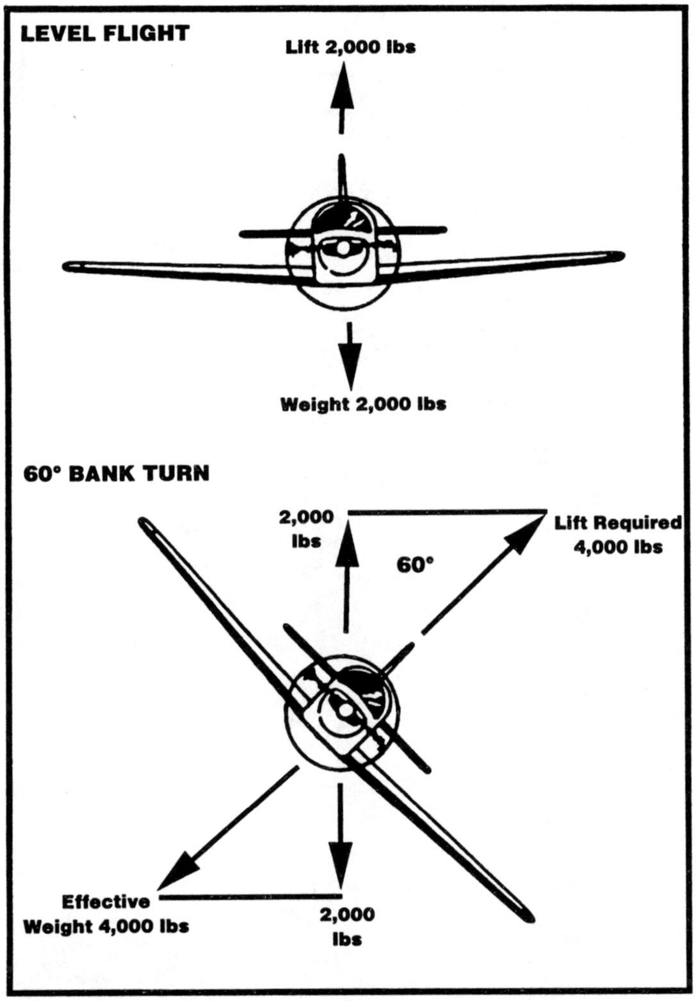

Fig. 3-17: Forces acting on an aircraft while turning, load factor for a 60° bank is 2.00.

The aircraft stall speed increases by the square root of the load factor. To calculate the stalling speed for a given angle of bank, use the following formula:

Vso in turn = Vso X √LOAD FACTOR

EXAMPLE: An aircraft stalls at 60 kts IAS in level flight. What will it stall at in a 60° bank turn?

ANSWER: 60 kts X √2.0 = 84.9 kts

NORMAL CATEGORY aircraft are generally certified for 3.8 limit load factor. Since the load factor squares as the speed doubles, tremendous loads may be imposed on structures by stalling an airplane at relatively high airspeeds. An airplane which has a normal unaccelerated stalling speed of 50 kts will be subjected to a load factor of 4 G's when forced into an accelerated stall at 100 kts. From this example, it can bee seen that normal design limitations can easily be exceeded.

The maximum speed at which an airplane can be safely stalled is the DESIGN MANEUVERING SPEED (Va). Va is an important reference point, because when operating below this speed, a damaging positive flight load should not be produced because the airplane should stall before the load becomes excessive. Any combination of flight control usage, including full deflection of the controls, or gust loads created by turbulence should not create an excessive air load if the airplane is operated below Va. Accordingly, the Va speed should be maintained when operating in rough or turbulent air. Va can be found in the Pilot's Operating Handbook or on a placard within cockpit. For a given aircraft, the Va will be slightly greater for increased weight. A rule of thumb that can be used to determine Va is to multiply the normal stalling speed by 1.7.

3.06 Propellers

The function of the propeller is to convert the turning movement of the crankshaft into thrust. As a propeller rotates, it moves the aircraft forward by moving a large mass of air backward. A propeller is similar to an airfoil, because it has an angle of attack and creates lift. Lift, in the case of a propeller, is thrust.

A propeller blade is twisted. The blade angle changes from the hub to the tip with the greatest angle of incidence at the hub and the smallest at the tip. The reason for this twist is to produce uniform lift from the hub to the tip. As the blade rotates there is a difference in the actual speed of the various portions of the blade. The tip of the blade travels faster than the part near the hub because the tip travels a greater distance than the hub in the same length of time.

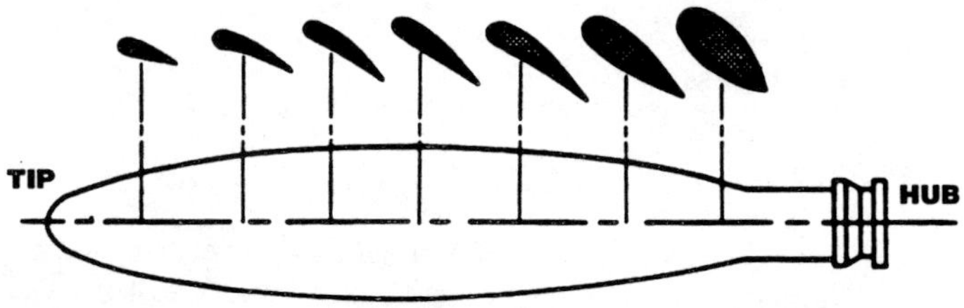

Fig. 3-18: Changes in propeller blade angle from hub to tip.

When the propeller rotates, it creates various forces resulting in yaw. TORQUE is caused by the propeller forcing the aircraft to rotate in the opposite direction, and results in a left turning tendency. ASYMMETRIC THRUST results in the downgoing portion of the propeller producing more pull than the upgoing portion, at high angles of attack. This is also known as "P-Factor". PRECESSION refers to the tendency of a propeller to act like a gyroscope: if the aircraft tail is raised, the nose yaws to the left (this effect is most apparent with tail-wheeled aircraft during the take-off roll). SLIPSTREAM is rotating air pushed back from the propeller, which strikes the fin from the left and causes a yaw to the left.

Generally, the adverse yaw caused by the propeller can be corrected for by design, such as off-setting the fin to reduce propeller slipstream. In flight, apply rudder to correct adverse yaw.

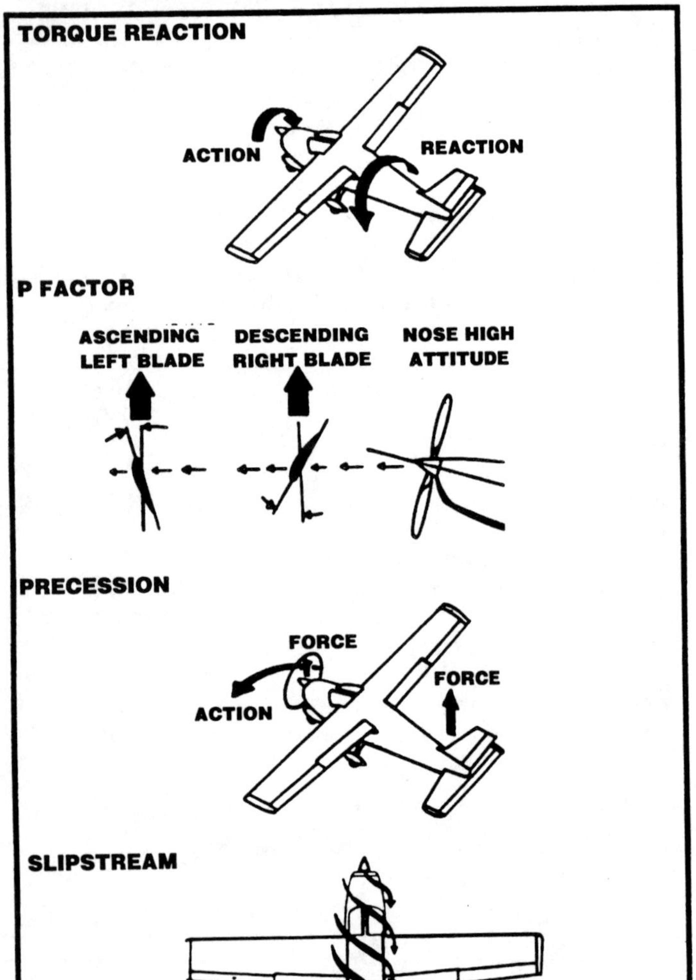

Fig. 3-19: Forces from a propeller that cause a left turning tendency.

BLADE ANGLE is the angle of the propeller blade. Most single engine training aircraft have FIXED PITCH propellers (constant blade angle). Other aircraft have VARIABLE PITCH propellers, which utilize a system in the propeller hub using oil pressure to vary the propeller blade angle. With CONSTANT SPEED propellers, the pilot selects a propeller RPM, and an oil regulated governor adjusts blade angle to maintain the selected RPM. FINE PITCH means small blade angles, and is used for take-off and climb. With COARSE PITCH, the propeller blade travels a greater forward distance per revolution. Coarse pitch is used for cruise.

3.07 Flight Controls

An aircraft has three axes of movement: the vertical axis (yaw control, rudder), the longitudinal axis (roll control, ailerons) and the lateral axis (pitch control, elevator).

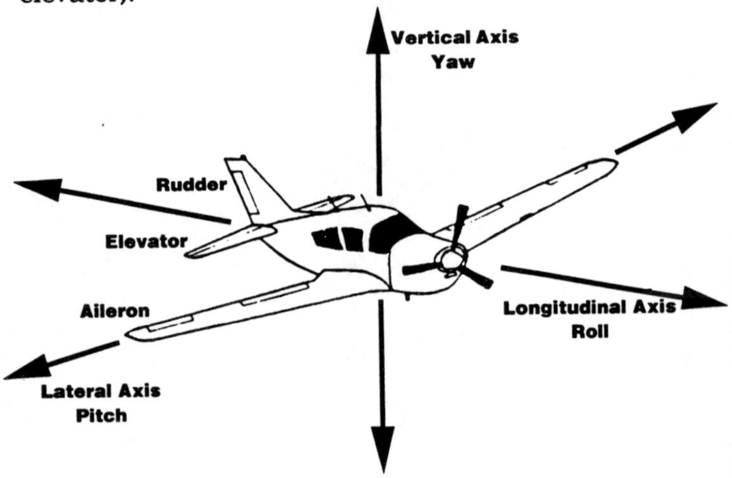

Fig. 3-20: Axes of movement.

The AILERONS control the airplane's movement about its longitudinal axis. Each aileron is hinged to the wing's rear spar and is linked to the other aileron by cables or rods so that when one aileron is deflected down, the opposite aileron moves up. When the pilot applies pressure to the left on the control column, the right aileron surface deflects downward and the left aileron deflects upward. The force exerted by the airflow on the deflected surfaces raises the right wing and lowers the left wing. This happens because the downward deflection of the right aileron changes the wing camber and increases the angle of attack and lift on that wing. Simultaneously, the left aileron moves upward and changes the effective camber, resulting in a decreased angle of attack, and less lift. Thus, decreased lift on the left wing and increased lift on the right wing causes the airplane to turn and bank to the left.

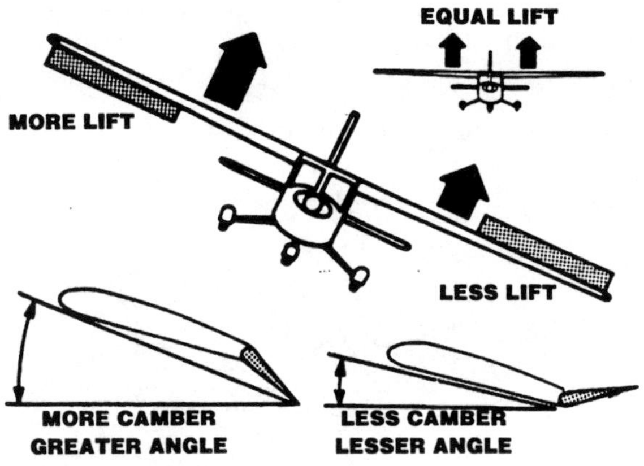

Fig. 3-21: How ailerons cause turn

AILERON DRAG is adverse yaw which results from greater pressure acting on the downgoing aileron than the upgoing aileron in a turn. The result of aileron drag is that the aircraft has a tendency to yaw in the direction of the downgoing aileron. Coordination of aileron and rudder minimizes the effect.

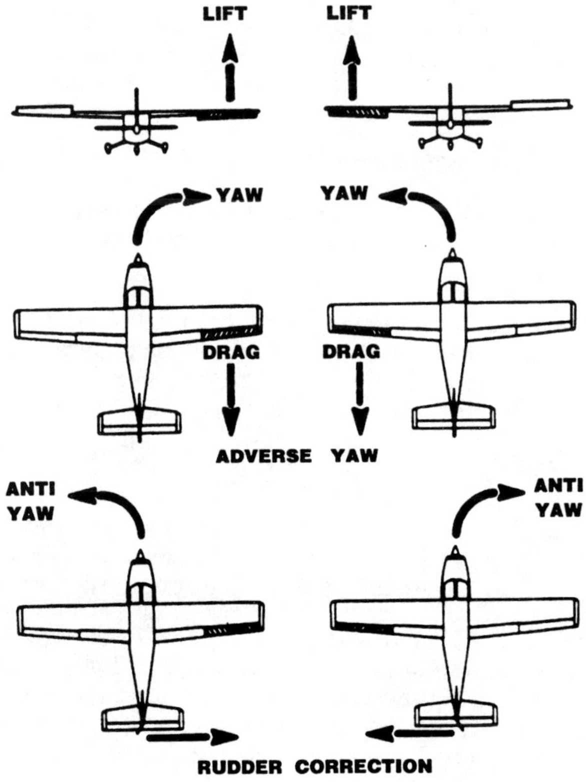

Fig. 3-22: Aileron drag causes adverse yaw.

DIFFERENTIAL AILERONS help to minimize aileron drag, because one aileron will be raised a greater distance than the other aileron is lowered for a given control movement. Since the raised aileron has more surface area exposed to the airflow (increasing drag) than the lowered aileron, the adverse yaw is reduced.

FRISE-TYPE AILERONS are specially designed to minimize aileron drag. Because the hinge of Frise aileron is offset, the leading edge of the downgoing aileron projects into the oncoming airflow, creating drag. This helps equalize the drag created by the lowered aileron on the opposite wing and thus reduces adverse yaw. The Frise-type aileron also forms a slot so that the air flows smoothly over the lowered aileron. This helps to make the aileron more effective at high angles of attack.

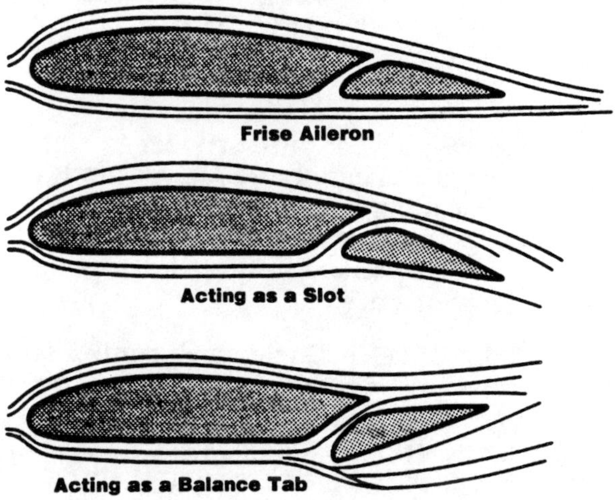

Fig. 3-23: Frise-Type Ailerons.

The ELEVATORS control the movements of the airplane about the aircraft lateral axis. Applying forward pressure on the control causes the elevator surfaces to move downward. The flow of air striking the deflected elevator surfaces exert an upward force, pushing the airplane's tail upward and the nose downward. In effect, the elevators are the angle-of-attack control. When back pressure is applied on the control column, the tail lowers and the nose rises, increasing the wing's angle of attack. On airplanes having a STABILATOR, the complete stabilator is moved in response to elevator control pressure, lowering or raising the stabilator leading edge, changing its angle of attack and lift. A CANARD is where the elevator assembly is mounted forward.

The RUDDER controls the movement of the airplane about its vertical or normal axis. This movement is called YAW. The rudder is a movable surface hinged to a fixed surface, the vertical stabilizer or fin. Movement of the rudder is controlled by the rudder pedals. When the rudder is deflected to one side, it protrudes into the airflow, causing a horizontal force to be exerted in the opposite direction. This pushes the tail of the airplane in that direction and yaws the nose in the desired direction.

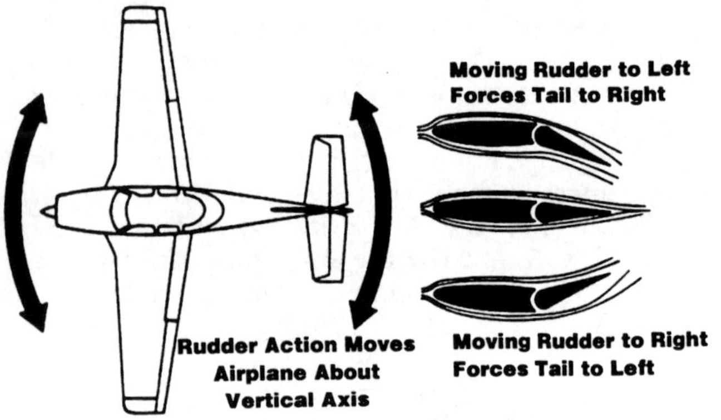

Fig. 3-24: Rudder control.

The primary purpose of the rudder in flight is to counteract the effect of adverse yaw and to help provide directional control. In flight the rudder does not turn the airplane; instead, the force of the horizontal component of wing lift turns the airplane when the wings are banked.

A TRIMMING DEVICE is a small control surface used to assist the pilot in maintaining the proper position of a primary control surface. Trim tabs are really small airfoils attached to or recessed into the trailing edge of the primary control surfaces. Trim tabs are used to maintain balance in straight and level flight and during other prolonged flight conditions without the pilot having to hold pressure on the controls. This is accomplished by deflecting the tab in the direction opposite to that in which the primary control surface must be held. The force of the airflow striking the tab causes the main control surface to be deflected to a position that will correct the unbalanced condition of the airplane.

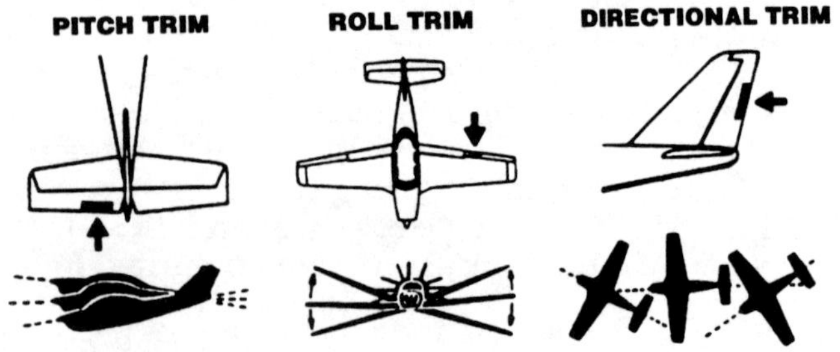

Fig. 3-25: Effect of trim tabs.

EXAMPLE: What would happen to the nose of an aircraft in flight, if the elevator trim tab were deflected down?

ANSWER: The nose of the aircraft would be raised. When trim tab is deflected down, elevator is raised, thereby lowering tail and raising nose.

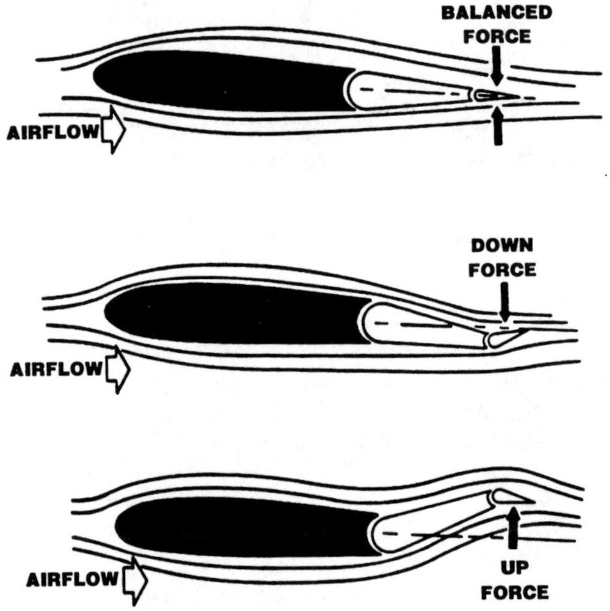

Fig. 3-26: Elevator trim tab function.

Trim tabs may be adjusted from the cockpit or in other older aircraft, may be adjusted only on the ground.

A BALANCE TAB looks like a trim tab, and is hinged at about the same point as would be a trim tab. The difference is that the balancing tab is coupled to the control surface by a rod, so that when the primary control surface is moved in any direction the tab automatically is moved in the opposite direction. In this manner the airflow striking the tab counterbalances some of the air pressure against the primary control surface and enables the pilot to more readily move and hold it in position.

A SERVO TAB is used primarily on large airplanes and functions similar to a trim tab. However, only the servo tab moves in response to movement of the pilot's flight control, and the force of the airflow on the servo tab then moves the primary control surface.

BALANCED CONTROLS assist the pilot in moving the controls. A STATIC BALANCE or MASS BALANCE is used to counteract FLUTTER and is placed ahead of the hinge line to balance the overhanging mass of the control. A balance can be INTERNAL (lead strip on inside of aileron, Cessna 152) or EXTERNAL (such as on DHC Beaver).

An AERODYNAMIC BALANCE utilizes the air striking the portion of the control surface that extends opposite to the direction of deflection of the primary surface, to assist the pilot in moving the control. An example is the HORN, on the upper portion of the rudder, which is behind the hinge line and creates a counter force to the rudder, assisting in maintaining the rudder position.

NOTES

NOTES

4 Flight Instruments

4.01 Pitot Static System

The pitot static system consists of the pitot tube, the static source and the various tubes connecting the airspeed indicator, the altimeter, and the vertical speed indicator. The openings of both the static source and the pitot tube should be checked during the preflight inspection for obstructions.

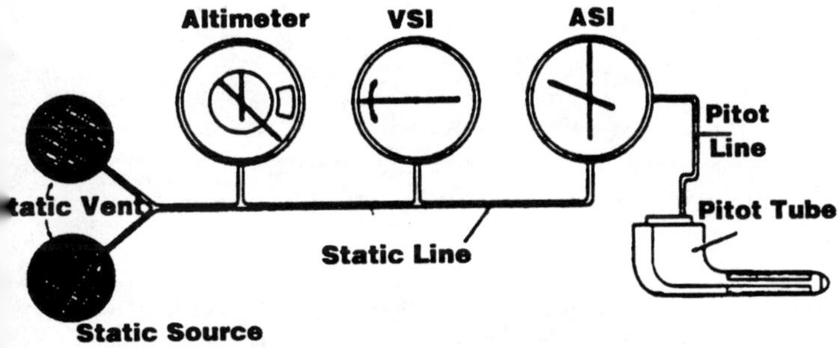

Fig. 4-1: Pitot static system - representation.

STATIC PRESSURE is atmospheric pressure acting upon an aircraft. It decreases with increased altitude. Static pressure is sensed though holes mounted on the side of the aircraft at a right angle to the relative airflow. Static or ambient air flows into the static source. DYNAMIC PRESSURE is pressure due to the motion of an aircraft. Dynamic or on-rushing air flows into the pitot tube.

Some aircraft are equipped with an ALTERNATE STATIC SOURCE, which is used if the normal static source becomes blocked. The pitot tube has a heat switch, which can be turned on if ice is present in the pitot tube.

4.02 Airspeed Indicator

The airspeed indicator (ASI) requires both the pitot and static source. It measures the differential between pitot pressure and static pressure to calibrate airspeed.

Speed is a rate of change of position or distance travelled per unit of time. Airspeed is normally expressed in knots (nautical miles per hour) or MPH. Note the following definitions:

a) IAS: Indicated Airspeed. Read directly off ASI.
b) CAS: Calibrated Airspeed. IAS corrected for instrument and position error. Refer to POH for table indicating corrections. This error is greatest at low airspeeds, in the cruising and higher airspeed ranges, IAS and CAS are about the same.
c) TAS: True Airspeed. CAS corrected for altitude and temperature changes. Use flight computer.

To find TAS, first read IAS, use airspeed calibration table to find CAS, then enter pressure altitude and temperature in flight computer to find TAS.

EXAMPLE: Determine TAS from the following.

Pressure Altitude	8,000 ft
Free Air Temp.	-10° C
Indicated Airspeed	99 Kts
Airspeed Calibration Error	-1 Kt

ANSWER: 109 Kts.

An ASI measures dynamic pressure and displays the forward speed of an aircraft through the air. It assumes that the surrounding air mass conforms to ICAO Standard Atmosphere (ISA) mean sea level conditions. With forward motion, the inner capsule expands, and the degree of expansion is translated to pointer movement.

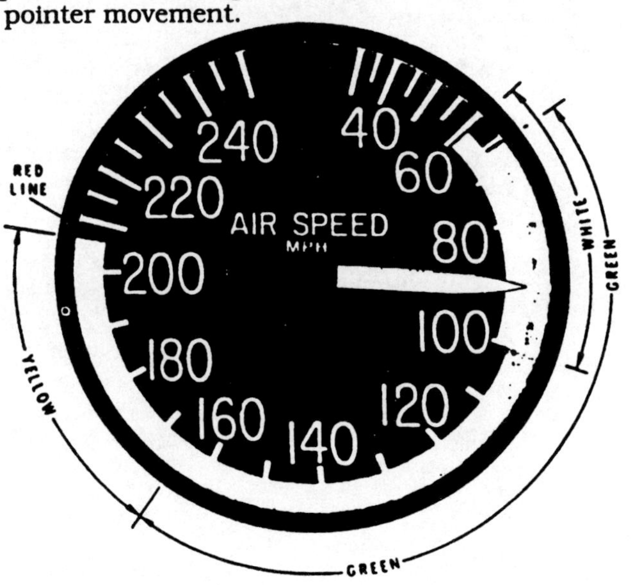

Fig. 4-2: Face of airspeed indicator.

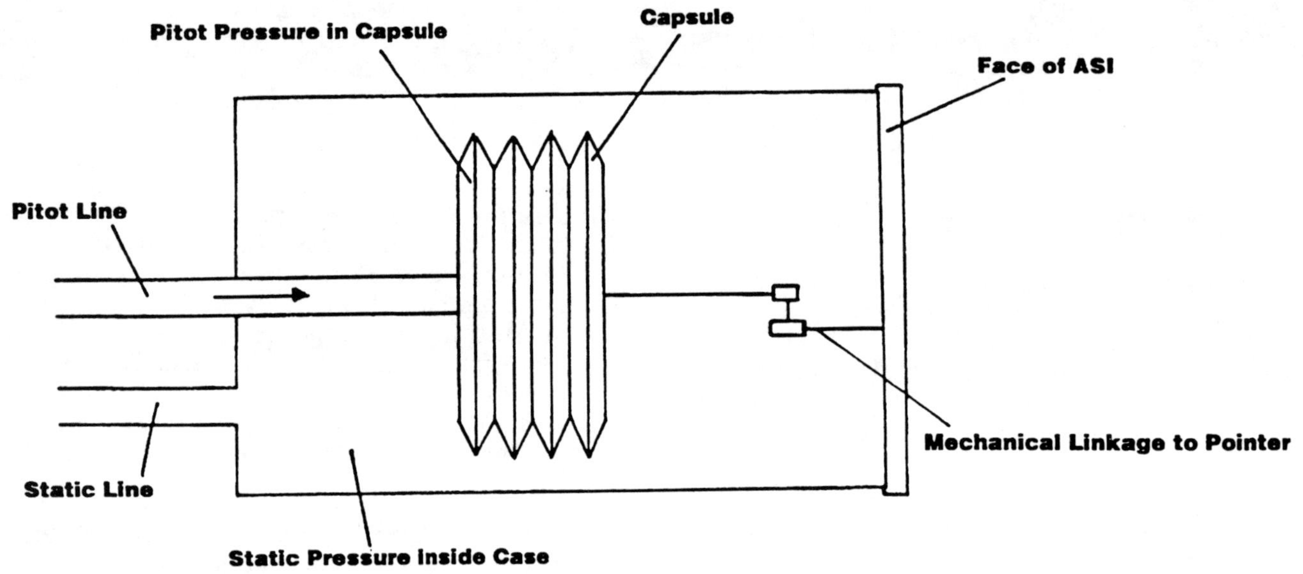

Fig. 4-3: Cutaway view of airspeed indicator. Dynamic air from pitot tube enters capsule, degree of expansion against surrounding static air is transmitted to pointer on face of instrument.

The airspeed indicator has various reference markings. In helicopters, a standard airplane ASI is normally used. It does not indicate airspeed in rearward flight. Various airplane colour codes appear. The green arc is the range of speeds for normal operations, with the top end being Vno (the maximum speed for normal operations). The yellow arc represents the caution range, and the red line the never exceed speed Vne.

Errors affecting the ASI are POSITION ERROR (caused by the position of the static ports sensing different readings) and DENSITY ERROR (caused by changes in altitude and temperature). COMPRESSIBILITY ERROR is a factor for high speed aircraft (at speeds greater than about 300 kts). If the pitot tube becomes blocked, the ASI needle will show a gradual increase in indicated airspeed with height, and it will act "like an altimeter". If the static source becomes blocked, the ASI will under-read in a climb and over-read in a descent.

A useful rule of thumb is that TAS increases by 2% the IAS per 1,000 ft gain in altitude.

EXAMPLE: IAS at 6,000 ft is 100 kts. What is the corresponding TAS?

ANSWER: TAS will be 112 kts. [2% X 6 = 12%. 12 kts + 100 kts = 112 kts].

The effect of increased TAS with altitude is very prominent when landing at a high elevation airport, because even though the IAS will remain the same, the TAS will be much greater and so will the groundspeed.

4.03 Vertical Speed Indicator

The VSI operates solely on the static pressure line. It measures the rate of change of atmospheric pressure when climbing or descending. The VSI is both a TREND instrument (indicates whether aircraft is climbing or descending) and a RATE instrument (gives actual rate of climb or descent in hundreds of feet per minute). A VSI has an aneroid capsule mounted inside a sealed case. The static pressure line is connected directly to the capsule, and the inside of the case is vented to static pressure through a metering device (calibrated leak). A linkage connects the capsule to the pointer.

Fig. 4-4: Face of vertical speed indicator.

Any change of aircraft altitude causes a pressure differential between the static air trapped inside the capsule and the static air that surrounds the capsule inside the case. The resulting expansion or contraction of the capsule produces pointer displacement.

A VSI is subject to LAG ERROR (reliable indications are not presented until pressure differential has stabilized, which may require as long as 6-8 seconds). The VSI is also subject to REVERSAL ERROR where abrupt changes in attitude may cause the VSI to momentarily show a change in the wrong direction.

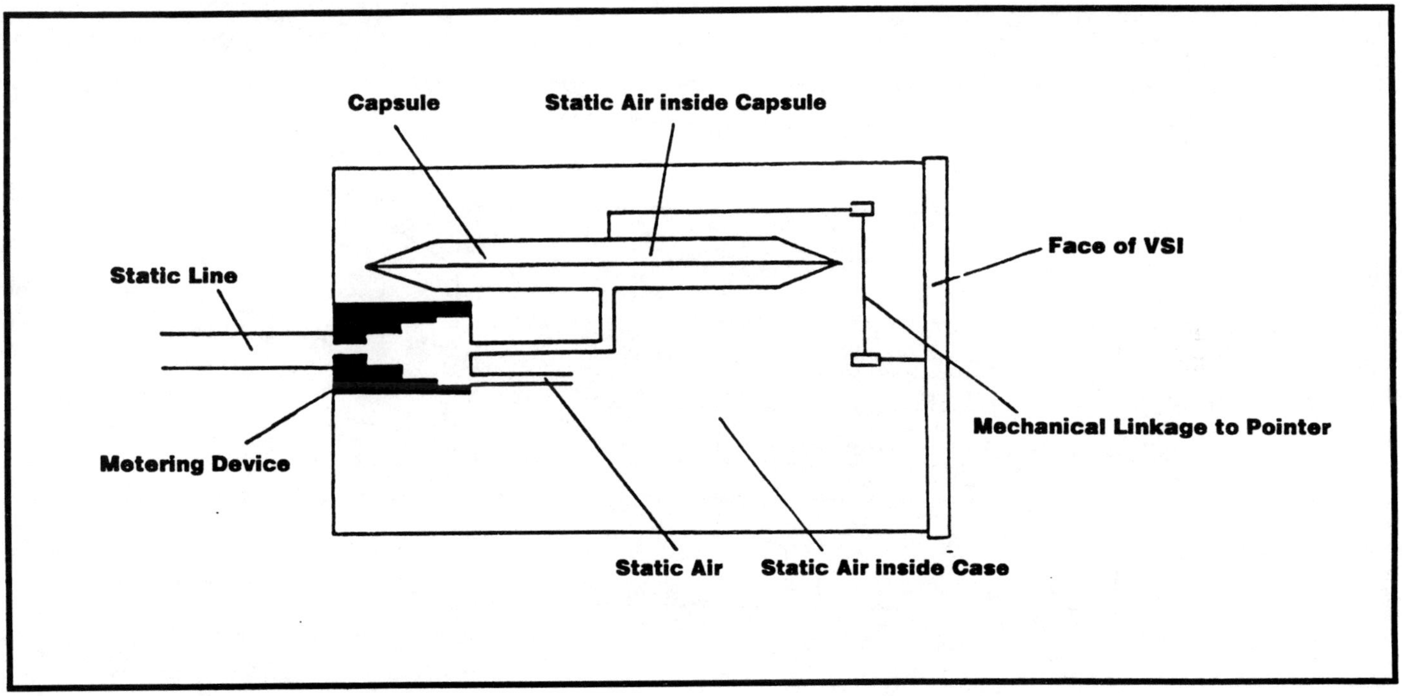

Fig. 4-5: Cutaway view of vertical speed indicator.

4.04 Altimeter

HEIGHT is vertical distance measured from a specified datum. ALTITUDE is vertical distance of an object in the air measured from mean sea level. ELEVATION is vertical distance of an object on the surface of the earth measured from mean sea level.

The altimeter is an aneroid barometer which measures changes in atmospheric pressure. Through connections these changes are recorded on the face of the instrument as changes in altitude. The altimeter setting is used to adjust the altimeter to read correctly in non-standard atmospheric conditions. The altimeter face has a long thin pointer representing hundreds of feet and a short, wide pointer indicating thousands of feet. A small triangle indicates 10 thousands of feet. The altimeter setting scale is graduated from 28.0 to 31.0 inches of mercury. A knob is on the left side of the instrument to adjust the altimeter setting.

Fig. 4-6: Face of altimeter.

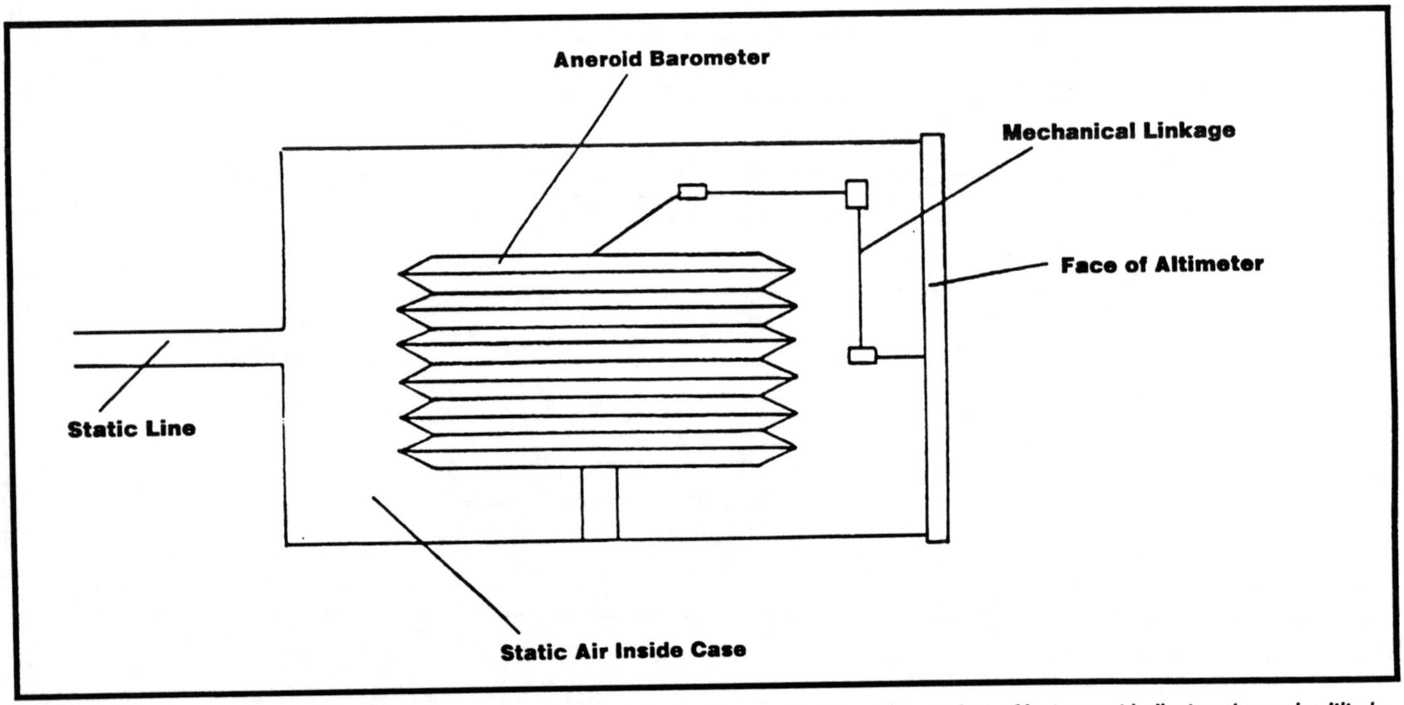

Fig. 4-7: Cutaway of altimeter. Expansion or contraction of aeneroid capsule transmitted to pointer on face of instrument indicates change in altitude.

On the inside of the altimeter is a partially evacuated capsule which responds to changes in air pressure like a barometer. The altimeter utilizes the static source of the pitot static system. As the aircraft climbs, the capsule expands, causing an indication of increased altitude on the face of the instrument. The altimeter setting mechanism allows selection of standard pressure 29.92 or the local airport altimeter setting.

INDICATED ALTITUDE is the altitude shown by an altimeter with the current altimeter setting applied. CALIBRATED ALTITUDE is indicated altitude corrected for instrument and position error. PRESSURE ALTITUDE is the altitude shown by an altimeter set to standard pressure (29.92 inches of mercury). DENSITY ALTITUDE is pressure altitude corrected for non-standard temperature and is calculated on the flight computer.

TRUE ALTITUDE is the actual altitude of an aircraft above mean sea level. ABSOLUTE ALTITUDE is the actual height of an aircraft above the surface.

To calculate true altitude using an E6B flight computer, follow these steps (standard electronic computers cannot be used for this calculation):

a) Calculate pressure altitude. If unknown, take difference between altimeter setting and 29.92, convert to feet. 1 inch Hg = 1,000 ft. Subtract from airport elevation if altimeter is HIGHER than 29.92, ADD to elevation if altimeter is LOWER than 29.92.
b) Set pressure altitude opposite OAT°C in true altitude window of flight computer.
c) Read true altitude on outer scale above indicated altitude on inner scale. If ground level is known, subtract elevation of ground station from true altitude to obtain absolute altitude.

EXAMPLE: Find true altitude.

Pressure altitude	10,000 ft
OAT °C	-13°C
Indicated altitude	12,000 ft
Ground Elev.	5,400 ft

ANSWER: 11,600. Set 10,000 ft pressure altitude opposite -13°C in true altitude window. True altitude is 11,600 opposite 12,000 on inner scale. Absolute altitude is 11,600 - 5,400 = 6,200 ft.

The ALTIMETER SETTING (QNH) is the observed barometric pressure at a selected datum, corrected for temperature, and reduced to mean sea level, assuming ISA conditions. Once applied to the altimeter subscale, the altimeter displays altitude above mean sea level. 1 inch on the altimeter subscale represents 1,000 feet change in altitude.

The altimeter has various errors. POSITION ERROR is caused by different indications resulting from changes in the static source angle to the airflow. In some aircraft, position error can be considerable. INSTRUMENT ERROR is caused by the aneroid capsule not assuming the precise size required for a particular pressure difference. MECHANICAL ERROR is caused by misalignment or slippage in the gears and mechanical linkage connecting the altimeter pointer. TEMPERATURE ERROR occurs when extremely cold temperatures prevail, causing the altimeter to over-read. ELASTIC ERROR, also called HYSTERESIS, is a lag error in altitude indications caused by the elastic properties of the materials used in the aneroid capsule. Elastic error occurs following large or rapid altitude changes. REVERSAL ERROR occurs during abrupt or rapid attitude changes and is only momentary in duration.

When the local altimeter setting has been set, the altimeter should indicate airport elevation, with a maximum error of +/- 50 feet.

A RADAR ALTIMETER or RADIO ALTIMETER provides a continuous indication of height above ground. The system accurately measures the distance between the ground and the aircraft in absolute altitude. The time interval between a transmitted and received radio signal is automatically converted to an altitude readout. The readout can be a logarithmic scale or digital. The main functions of a radar altimeter are to serve as a ground proximity warning, to provide a cross check for the barometric altimeter, and to indicate progress on final approach descent to MDA (Minimum Descent Altitude) or DH (Decision Height) on an IFR approach. Radar altimeters usually have a warning marker which can be set to a pre-determined altitude AGL to generate a visual or audio warning when the aircraft reaches the preset value. Errors may occur during flight over a surface into which radio waves can penetrate (ice, deep snow).

An ENCODING ALTIMETER is a specialized altimeter that sends out a pressure altitude readout to ATC and is part of a MODE C transponder system.

4.05 Magnetism & Magnetic Compass

The Earth is a magnet, having a north and south pole. Magnetic north is in a different location than geographic north. The difference between magnetic north and true north is VARIATION. The Earth's magnetic lines of force dip towards the poles, which is MAGNETIC DIP.

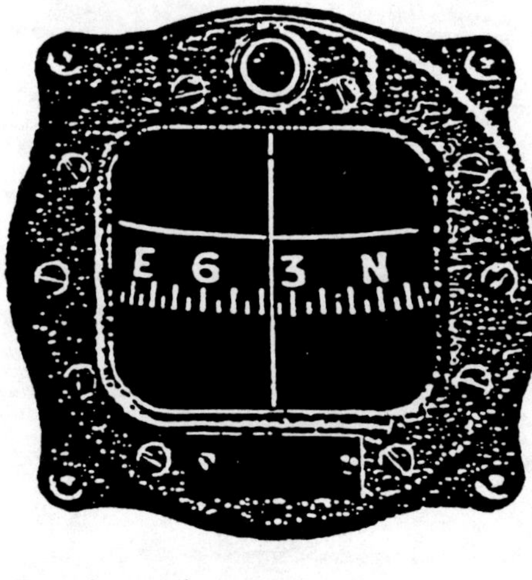

Fig. 4-8: Face of magnetic compass.

The aircraft magnetic compass is based upon two parallel magnetized needles, which are mounted on a float having a compass card. The float assembly is mounted on a pedestal, and is sealed in a case filled with kerosene.

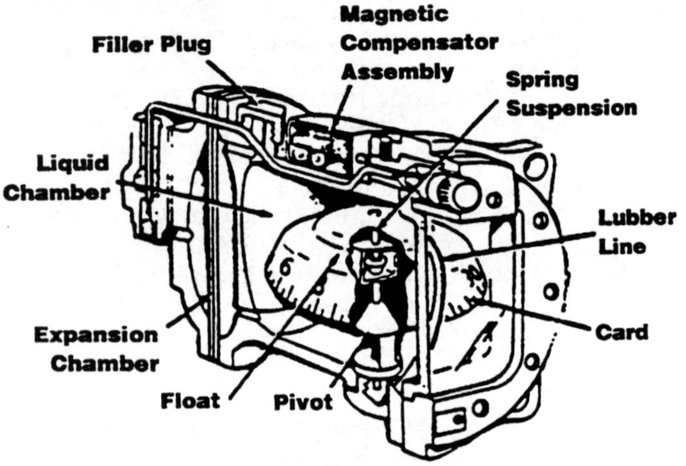

Fig 4-9: Components of magnetic compass.

Because the center of gravity of the compass is below the pivot point, it shows erroneous indications depending on whether the aircraft is accelerating or turning. These errors are caused by magnetic dip. NORTHERLY TURNING ERROR occurs as an aircraft turns to north or south. On turns to or from north, the compass lags behind the actual heading. On turns to or from south, the compass leads ahead of the actual heading.

ACCELERATION ERROR occurs as the aircraft accelerates or decelerates on east-west headings. As an aircraft accelerates such as during a transition to a climb as power is increased, the compass indicates slightly to the north. As an aircraft decelerates, such as following a power reduction prior to entering a descent, the compass indicates slightly to the south. Acceleration errors can be remembered with the word "ANDS": Accelerate = North indications, Decelerate = South indications. To avoid turning error and acceleration error, insure the aircraft is in level, unaccelerated flight prior to relying on indications from the magnetic compass.

EXAMPLE: What compass error results where an aircraft is turning from 330 to 360?

ANSWER: Turning error: compass lags behind actual heading.

EXAMPLE: Which compass error will be greater: a turn from 270 to 300, or from 300 to 330?

ANSWER: The turn from 300 to 330 will indicate greater turning error, because turning error is zero on E - W headings, and increases the closer the aircraft is to N or S.

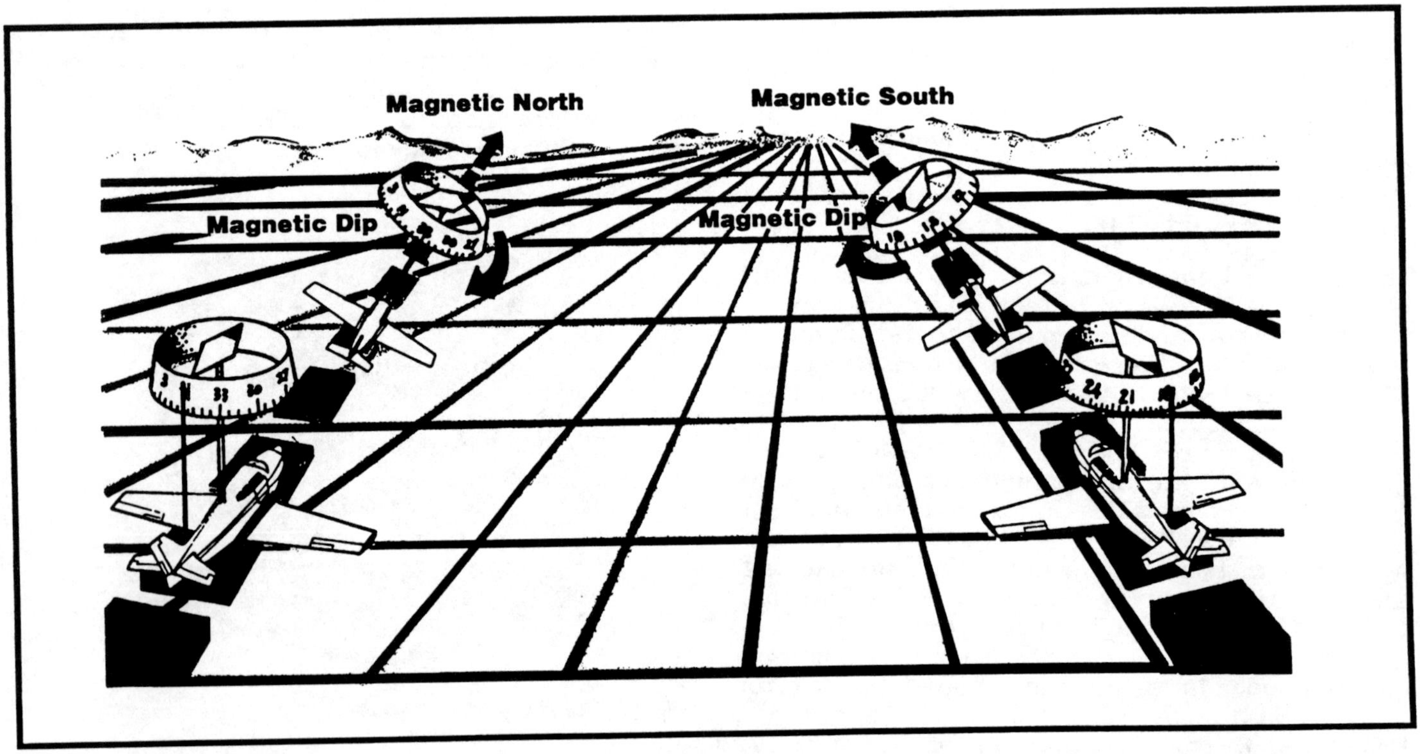

Fig. 4-10: Northerly turning error.

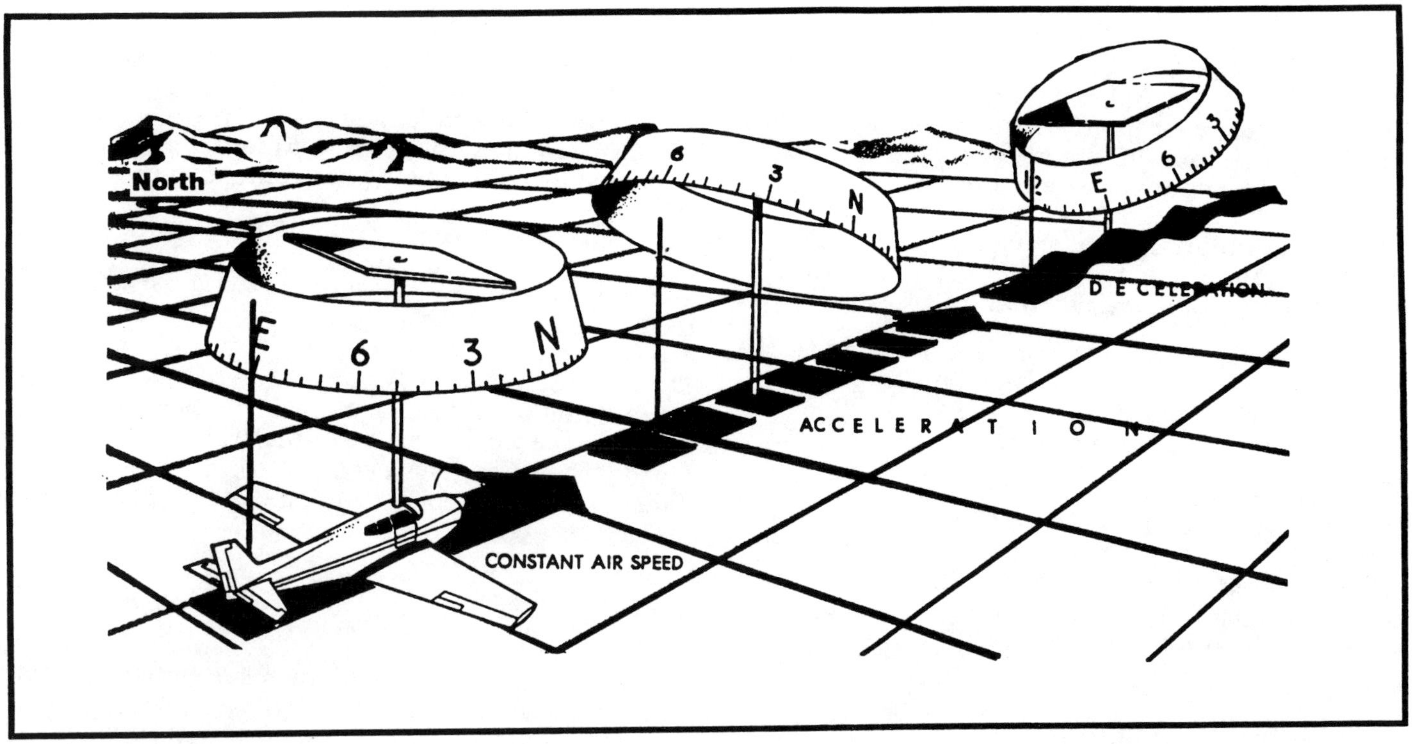

Fig. 4-11: Acceleration error.

The compass needle is affected by the magnetic fields generated by the electrical equipment and by metal components in the aircraft. These disturbances deflect the compass needles from alignment, and are termed DEVIATION. To reduce deviation, the compass is SWUNG, which means adjusted such that the compass indicates as correctly as possible. A compass must be swung once per year. The remaining deviation between magnetic heading and compass heading is accounted for on a COMPASS CORRECTION CARD. To fly compass headings, refer to the compass correction card for corrected headings to steer.

EXAMPLE: If the desired magnetic heading is 090°, and the compass deviation is -1°, what compass heading should be steered?

ANSWER: 089°.

COMPENSATION refers to adjusting the magnetic compass to reduce deviation error. Proper compensation of the compass is best performed by a properly qualified technician. Since the magnetic forces within the airplane change, because of landing shocks, vibration, mechanical work, or changes in equipment, the deviation should be checked as required.

To swing a compass, the helicopter is placed over a magnetic compass rose on the ground, the engine started, and electricals normally used (such as radios) are turned on. The helicopter is then aligned with magnetic north indicated on the compass rose and the reading shown on the compass is recorded on a deviation card. The helicopter is then aligned at 30° intervals and each reading is recorded. If the aircraft is to be flown at night, the lights are turned on and any significant changes in the readings are noted. If so, additional entries are made for use at night.

4.06 Gyroscopic Instruments

The attitude indicator, heading indicator, and turn coordinator/turn and slip indicator are controlled by gyroscopes. A gyroscope is a spinning mass on an axis. Gyroscopes are classified according to their relationship to the Earth's surface, either horizontal or vertical.

All gyroscopes exhibit RIGIDITY IN SPACE and GYROSCOPIC PRECESSION. Rigidity in space means that the axis always points to the same position in space, regardless of the position of the earth. Precession means that a force applied to the gyroscope wheel will be felt 90 degrees from the point of impact, in the direction of rotation.

Gyroscopic instruments are typically powered either by suction (heading indicator, attitude indicator) or electricity (turn coordinator/turn and bank indicator). The vacuum system spins the gyro by drawing a stream of air against the rotor vanes to spin the rotor. Either a venturi tube or an engine driven vacuum pump can be used to provide the suction required to spin the rotors. The amount of vacuum required is usually from 3.5" to 4.5" Hg.

The older venturi tube systems were based on a tube mounted on the exterior fuselage. The vacuum would only be effective at speeds of about 90 kts or more, and could not be used for IFR flight.

A typical modern vacuum system consists of an engine-driven vacuum pump, and air/oil separator, a vacuum regulator, a relief valve, an air filter, and tubing and necessary to complete the connections. A suction gauge on the instrument panel indicates the amount of vacuum in the system.

4.07 Attitude Indicator

The attitude indicator duplicates the natural horizon, and indicates pitch attitude (nose up/nose down) and bank attitude (left, right).

The attitude indicator utilizes the gyroscopic principles of rigidity in space (to establish a reference plane) and precession (to keep the gyro axis perpendicular to the Earth's surface). The gyroscope of an air-driven attitude indicator is vertically mounted so that its housing is free to pivot about the vertical axis. The gyro rotor is driven by an air jet which strikes buckets on the rotor.

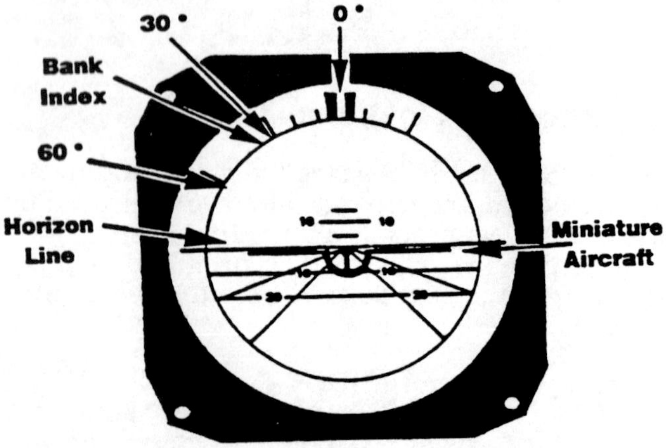

Fig. 4-12: Face of attitude indicator.

The horizon bar is linked to the gyroscope by a lever, which is attached to a pivot on the rear of the gyroscope frame and connected to the housing by a pin. The result is that as the aircraft climbs, dives, or banks, the instrument case rotates on the gyroscope gimbals while the bank index and horizon bar remain rigid, reflecting any movement of the aircraft about the pitch and bank axes. The attitude indicator has an correcting mechanism, to maintain the gyroscope's position.

The attitude indicator is subject to errors caused by bearing friction in the instrument. Other errors result from acceleration and deceleration, such that during take-off roll, for example, the attitude indicator falsely displays a climb. Deceleration has the opposite effect.

4.08 Heading Indicator

The operational principle of the heading indicator is similar to that of the attitude indicator, with the exception that the gyroscope axis of the heading indicator is horizontally mounted.

Fig. 4-13: Face of heading indicator

The gyro rotor is air driven. During turns, the casing turns about the gyro, which is geared to rotate an azimuth compass bearing card, indicating a change in direction.

Bearing friction and the rotation of the Earth cause the heading indicator to drift, so the heading indicator should be reset about every 15 minutes. Due to the errors associated with the magnetic compass, the heading indicator should only be set to the compass when in level flight.

4.09 Turn Coordinator

A turn coordinator is a combination of two instruments. A miniature aircraft indicates direction and rate of turn, roll rate and turn coordination. The turn coordinator is also a useful secondary bank reference. The ball is a separate instrument, called an INCLINOMETER, which indicates gravity, and whether the aircraft is in coordinated flight, is slipping or is skidding.

The TURN AND BANK INDICATOR is an older instrument which provides basically the same information as the turn coordinator (a turn needle is used instead of a miniature aircraft). The turn and bank indicator will display rate of turn and turn coordination, but not roll rate.

A RATE ONE turn is a turn which provides 360° change in direction in 2 minutes (3° per second).

Fig. 4-14: Face of turn coordinator.

Fig. 4-15: Face of turn and bank indicator.

The turn coordinator's construction is based on a laterally mounted, electric driven gyroscope. The gyro is free to tilt about the aircraft's longitudinal axis and is mechanically joined, through a linkage, to the turn pointer. A restraining spring links the gyro assembly to the instrument case. Mechanical stops prevent the gyro from tilting more than 45° either side of center.

The inclinometer consists of a black or white ball inside a sealed, curved glass tube filled with liquid to provide damping action and insure smooth operation of the ball. The ball is influenced by gravity and centrifugal force. During coordinated flight these two forces are in balance, and the ball rests in the central lower portion of the tube. When the forces become unbalanced, the ball moves away from the centre and indicates uncoordinated flight as a slip or skid.

In a SLIP, the rate of turn is too slow for the angle of bank, and the lack of centrifugal force causes the ball to be displaced to the inside of the turn. To correct, decrease the angle of bank, or increase the rate of turn, or both.

In a SKID, the rate of turn is too fast for the angle of bank, and excessive centrifugal force causes the ball to be displaced to the outside of the turn. To correct, increase the bank angle, or decrease the rate of turn, or both.

To visualize slipping and skidding, think of a car approaching a corner: if the car goes too fast, it skids. If the highway is banked, the car can approach the corner at a higher speed before skidding. A slip occurs when a car moves too slow on an ice covered, steeply banked corner. An aircraft follows the same principles.

EXAMPLE: A turn and bank indicator shows needle right, ball right. What control movements would correct the following indications to produce a properly coordinated turn?

ANSWER: The aircraft is slipping. To recover, reduce right bank, apply right rudder.

4.10 Instrument Pre-Flight Checks

To insure all flight instruments are functioning properly, confirm the following instrument indications while taxiing.

a) Airspeed indicator - zero.
b) Altimeter - airport elevation, correct setting.
c) Attitude indicator - level, zero bank.
d) Turn coordinator - for left turn, needle left, ball right; for right turn, needle right, ball left.
e) Heading indicator - set. For left turn, decreasing headings. For right turn, increasing headings.
f) Vertical speed indicator - zero.
g) Magnetic compass - shows known headings, free and swinging in turns.

4.11 Instrument Flying

NOTE: The following is only for written exam purposes. To become proficient at flight in IMC, a complete IFR training program is required.

Basic Principles

"Attitude Plus Power = Performance" means that the combination of attitude and power control, with cyclic and throttle, achieves a desired flight condition.

Instrument Grouping

Flight instruments are grouped into PITCH, BANK and POWER instruments. Instrument grouping is important because it allows a logical and efficient utilization of flight instruments. Those instruments which provide the most pertinent and essential information are called PRIMARY INSTRUMENTS. A primary instrument is usually one that is held at a constant value or indication. SUPPORTING or SECONDARY INSTRUMENTS back-up and supplement the information shown on the primary instrument.

A DIRECT INDICATING INSTRUMENT gives instantaneous attitude information: normally the attitude indicator. An INDIRECT INDICATING INSTRUMENT gives an indirect indication of pitch attitude. The altimeter, airspeed indicator and vertical speed indicator give indirect indications pitch attitude, and the heading indicator and turn coordinator give indirect indications of bank attitude.

Pitch Instruments

ATTITUDE INDICATOR. Gives a direct indication of nose up, level, or nose down pitch attitude. The most important pitch instrument, giving accurate, instantaneous pitch information.

ALTIMETER. Indirectly indicates pitch attitude, because if altitude changes, there must be a pitch change. Primary instrument for pitch (change in altitude from level flight is caused by pitch change).

AIRSPEED INDICATOR. Secondary pitch instrument (nose up attitude = decreased airspeed).

VERTICAL SPEED INDICATOR. Secondary pitch instrument (ie nose down attitude = descent). VSI is a trend (up or down) and a rate (feet per minute) instrument.

Bank Instruments

ATTITUDE INDICATOR. Gives a direct indication of banked attitude, giving direction and size of bank. The most important bank instrument, giving accurate, instantaneous bank information.

HEADING INDICATOR. Indirect instrument, because for a change in heading, there must be a bank. Primary instrument for bank (large banks will show faster change in heading).

TURN COORDINATOR. Provides rate of turn. Rate 1 turn = 360° in 2 minutes, or 3° per second. Indirect indication of bank. To calculate rate 1 turn, take 10% of airspeed and add 5 (mph) or 7 (kts). Result is bank angle required for rate 1 turn. Position of ball in inclinometer indicates whether in coordinated flight, slipping or skidding.

Power Instruments

RPM/MANIFOLD PRESSURE. Gives direct indication of power.

AIRSPEED INDICATOR.Secondary power instrument. In level flight, [power change results in airspeed change].

EXAMPLE: In straight and level flight, which instrument is primary for pitch, and which instrument is primary for bank?

ANSWER: In straight and level flight, the altimeter is primary for pitch, and the heading indicator is primary for bank.

EXAMPLE: Which instruments are considered primary and supporting for bank when establishing a level standard rate turn?

ANSWER: The attitude indicator in primary for bank as the bank is commenced, while the turn coordinator is secondary. Once established in the rate turn, the turn coordinator is primary for bank.

EXAMPLE: Which instruments are primary for pitch, bank and power when transitioning into a constant airspeed climb from straight and level flight?

ANSWER: As a transition is made into a climb at constant airspeed, the attitude indicator is primary for pitch, heading indicator is primary for bank and tachometer or manifold pressure is primary for power.

Fundamental Skills

Instrument Crosscheck
Scanning all instruments is essential for efficient control. The RADIAL SCAN method is to divide the scan into pitch, bank, and power, based on the A/I at centre.

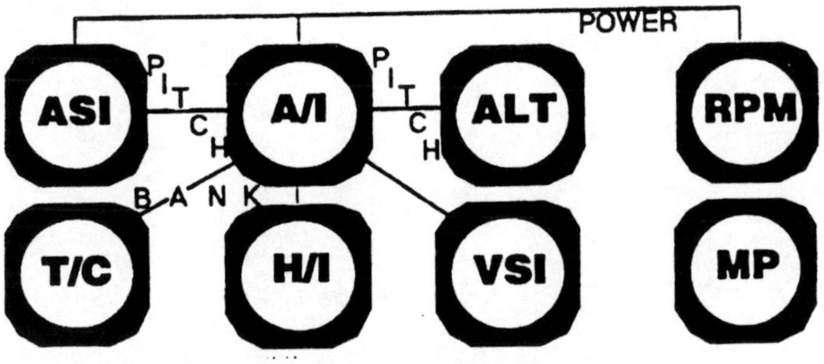

Fig. 4-16: Instrument cross check, standard T panel involves dividing scan into pitch, bank and power instruments.

Instrument Interpretation
All flight instruments must be observed carefully in order to properly interpret the correct instrument indication.

Aircraft Control
The correct control input must be applied for a desired flight condition.

Trim
Trimming after every attitude change is crucial to instrument flight. Trimming allows much easier control since the desired attitude will be maintained with much less pilot workload.

Basic Maneuvers

Climb
To enter a climb from cruise, first raise the wings of the miniature aircraft on the attitude indicator to the approximate climb attitude. Crosscheck the other instruments, then, when the airspeed needle approaches the selected climb speed, add full power and trim. To level off, anticipate the level off by 10% the rate of climb, then select the cruise attitude, reducing power and retrimming.

Descent
To enter a descent from cruise, reduce power to descent power, then maintain the level attitude on the attitude indicator. When the airspeed needle approaches descent speed, re-adjust the miniature aircraft to a slightly lower attitude, maintain descent speed and trim. Continue cross-checking. To level off, anticipate level off by 10% the rate of descent, then select cruise attitude, increase power and trim.

Turns
To enter a turn from a constant heading, first use the attitude indicator, and coordinate roll with rudder. Centralize the ailerons at the desired bank, and continue crosschecking. Trim if necessary. Commence the roll out when within 1/2 the angle of bank of the desired heading.

Unusual Attitudes

An unusual attitude (pitch or bank greater than 30°) can be entered due to turbulence, cockpit disorganization, vertigo etc. In an extreme unusual attitude, the gyroscopic instruments will not be reliable. To recover from an unusual attitude, first identify the attitude, correct the pitch and bank attitude, adjust power as required, then trim. All components are changed almost simultaneously with little lead of one over the other. The bank attitude is corrected with reference to the turn coordinator, then the ball is centered, the airspeed checked and corrected, then the power if required. Once the aircraft is near to straight and level flight, a thorough scan of all the instruments is then completed, and adjustments are made as necessary.

Partial Panel

Partial panel refers to controlling the aircraft without the attitude indicator or heading indicator. Basically, the other pitch and bank instruments must be relied on to control the aircraft. The basic skills are the same in terms of attitude and power control, but the instrument scan will have to be modified to extract the required information. For bank control, the turn coordinator is crucial, because the magnetic compass can only be relied on in level flight. Turning at a fixed rate one turn for a known number of seconds (timed turns) is more accurate. The airspeed indicator and altimeter will be crucial for pitch control.

NOTES

5 Aviation Weather

5.01 The Earth's Atmosphere

The atmosphere consists of 78% nitrogen, 21% oxygen, and 1% argon. Important physical properties are pressure, temperature, humidity, and radiation. The lowest layer of the earth's atmosphere is the TROPOSPHERE, from the ground to 60,000 feet at the equator, and to 35,000 feet at the poles. Above the troposphere lies the STRATOSPHERE. The TROPOPAUSE separates the troposphere from the stratosphere.

AIR DENSITY is the weight of air per cubic foot. ATMOSPHERIC PRESSURE is the weight of an imaginary column of air. Sea level pressure is 29.92 inches of mercury, and decreases 1.00 inch of mercury per 1,000 ft gain in height. The STANDARD ATMOSPHERE has an MSL pressure of 29.92 inches of mercury, MSL temperature of +15°C, and normal lapse rate of 1.98°C per 1,000 ft. STATION PRESSURE is the atmospheric pressure at a station.

Pressure varies horizontally. ISOBARS are lines of equal MSL pressure at 4 millibar intervals.

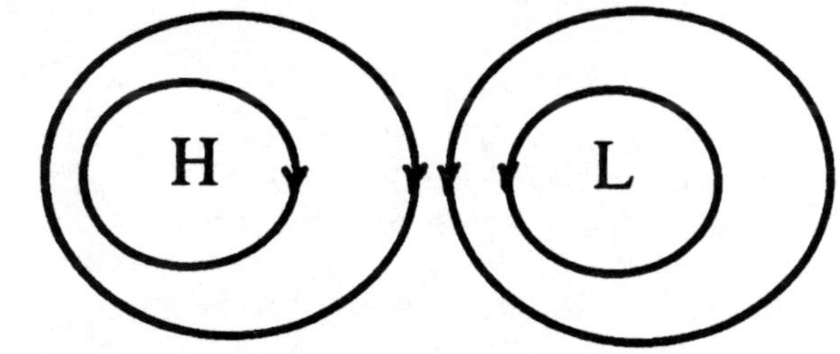

Fig. 5-1: High pressure and low pressure systems.

A HIGH or ANTI-CYCLONE is when atmospheric pressure increases towards the center. A LOW or CYCLONE is when atmospheric pressure decreases towards the center. Highs circulate clockwise, lows counterclockwise. A TROUGH is an elongated area of low pressure, similar to a valley. A RIDGE is an elongated area of high pressure, similar to a ridge. A COL is an neutral area between highs and lows.

5.02 Meteorological Aspects of Altimeters

PRESSURE ALTITUDE is the equivalent altitude in the standard atmosphere for a given pressure. To calculate pressure altitude, take the local altimeter setting and calculate the difference from 29.92 inches of mercury, then convert to feet, and add to airport elevation where the altimeter setting is lower than 29.92. [NOTE: pressure altitude can also be determined by setting an aircraft altimeter to 29.92 and reading off the resulting indication.]

EXAMPLE: Calculate pressure altitude.

1. Altimeter setting 29.62, airport elevation 3,000 ft.
2. Altimeter setting 30.12, airport elevation 1,500 ft.

ANSWERS:
1. 3,300 ft pressure altitude. [29.92 - 29.62 = +.30 = +300 ft. 300 + 3000 = 3,300]
2. 1,300 ft pressure altitude. [29.92 - 30.12 = -.20 = -200 ft. 1,500 - 200 = 1,300]

DENSITY ALTITUDE is pressure altitude corrected for non-standard temperature. To calculate density altitude, use a flight computer or koch chart (see s. 7.17 in this manual). High density altitude increases take-off distance and reduces the rate of climb. Conditions causing drastic reductions in aircraft performance are high airport elevation, high ambient temperature and high atmospheric humidity.

The altimeter must be set to the current local altimeter setting so that the it indicates the correct altitude. When flying from high pressure areas to low pressure areas, without re-setting the altimeter, the altimeter indicates high. A useful rule to remember is from HIGH TO LOW, LOOK OUT BELOW.

5.03 Temperature

The earth's atmosphere is heated from BELOW: the sun's radiation heats the earth, which in turn heats the lower levels of the atmosphere. CONVECTION CURRENTS occur where there is differential heating of the earth's surface. The normal rate of temperature decrease with altitude is 1.98°C per 1,000 ft (standard lapse rate). The temperature does not always drop with height: an ISOTHERMAL LAYER is when the temperature remains constant with increasing height. An INVERSION is when temperature increases with height. The CELSIUS SCALE is based on water freezing at 0°C and boiling at 100°C. The FAHRENHEIT SCALE is based on water freezing at 32°F and boiling at 212°F. To convert from °F to °C, or from °F to °C, use your flight computer: at the bottom of an E6B type computer will be a conversion scale. With an electronic computer, simply punch the appropriate buttons labeled °F or °C and CONV.

EXAMPLE: What is 10°C in °F?

ANSWER: 50°F.

5.04 Moisture

CONDENSATION is the change of state from water vapor (gas) to liquid water. RELATIVE HUMIDITY is the percentage of saturation of air for a given temperature. DEWPOINT is the temperature to which air must be cooled to reach 100% saturation. When the dewpoint temperature is equal to the ambient temperature, cloud forms. Most cloud forms when moist air rises, cools and condenses. SUBLIMATION is when moisture is added to the air by the change of state of ice to water vapour.

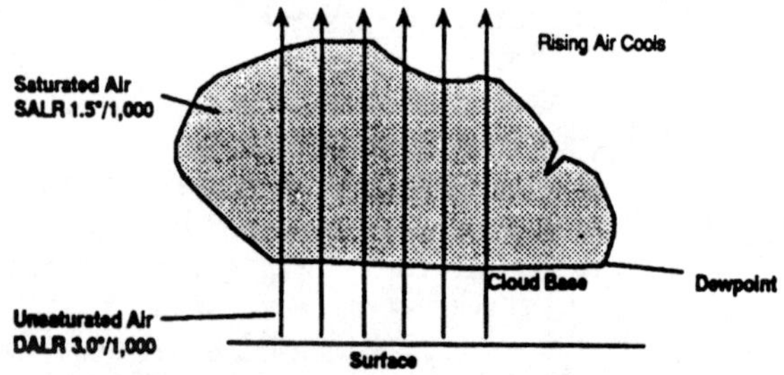

Fig. 5-2: Formation of convective cloud.

The DRY ADIABATIC LAPSE RATE (DALR) is 3.0°C/1000 ft. Use the DALR to find the cloud base. The SATURATED ADIABATIC LAPSE RATE (SALR) is 1.5°C/1000 ft. Use the SALR above the cloud base. The SALR is less than the DALR due to the heat release from condensation.

EXAMPLE: Surface temp. is 15°C, dewpoint is 9°C. Find cloud base and freezing level.

ANSWER: Cloud base = 2,000 ft (15° - 9° = 6°, use DALR). Freezing level = 8,000 ft (use SALR from cloud base for temperature drop of 9° - 0° = 9°).

5.05 Stability and Instability

Stable air resists vertical motion. Stable air has a shallow lapse rate. Unstable air favors vertical motion and has a steep lapse rate. The atmosphere is heated by RADIATION (reflection of Sun's rays on earth's surface), CONDUCTION (warm air contacting cold air), ADVECTION (horizontal movement of air) and CONVECTION (unequal surface heating). The atmosphere is cooled by RADIATION and ADVECTION. The lifting processes are CONVECTION, CONVERGENCE (excess air rises when pressure systems meet), MECHANICAL TURBULENCE (friction between air and ground), OROGRAPHIC LIFT (air moves up mountains) or FRONTAL LIFT (advancing cold front). SUBSIDENCE occurs when air sinks, and results in heating and compression, so that the relative humidity decreases and water evaporates.

5.06 Clouds

Clouds are classified as HIGH (above 20,000 feet), MIDDLE (6,500 feet to 20,000 feet), or LOW (having bases below 6,500 feet). The prefix CIRRO means high, and ALTO means middle. There are two basic forms of cloud: LAYER (also called STRATOFORM) or HEAP (also called CUMULOFORM). Layer clouds are horizontally developed, and heap clouds are vertically developed. Note the following cloud types:

High

a) Cirrus (Ci). "Horse tails".
b) Cirrostratus (Cs). Thin whitish veil; milky sky.
c) Cirrocumulus (Cc). Small rounded masses.

Middle

a) Altostratus (As). Uniform veil, grayish or bluish.
b) Altocumulus (Ac). Layer or series of flattened rounded masses, long rolls, undulations or waves.

Low

a) Stratus (St). Low layer of horizontal cloud.
b) Nimbostratus (Ns). Thick uniform dark gray or black layer from which precipitation falls.
c) Stratocumulus (Sc). Layer or series of patches of cloud with rounded masses of rolling appearance.

Vertically Developed

a) Cumulus (Cu). "Cotton ball" appearance.
b) Towering Cumulus (TCU). Cumulus to higher altitudes.
c) Cumulonimbus (Cb). Thunderstorm cloud.
d) Altocumulus Castellanus (ACC). Middle level cumulus cloud layer with protruding turrets.

Clouds named with NIMBO produce precipitation. Cumulonimbus cloud produces heavy, intermittent precipitation. Nimbostratus cloud produces heavy, continuous precipitation. Stratoform type cloud produces continuous precipitation, cumuloform cloud produces intermittent precipitation.

5.07 Turbulence

Turbulence is the result of friction between opposing air currents. Types of turbulence:

a) Convective Turbulence. Unequal surface heating.
b) Orographic Turbulence. Friction in air currents in mountainous terrain.
c) Wind Shear. Sudden change in wind speed or wind direction. Found near thunderstorms, causing downbursts termed MICROBURSTS. Also found in low level inversions.
d) Clear Air Turbulence. Found at high altitudes.
e) Mechanical Turbulence. Friction between air and ground as high winds blow over rough, broken terrain.

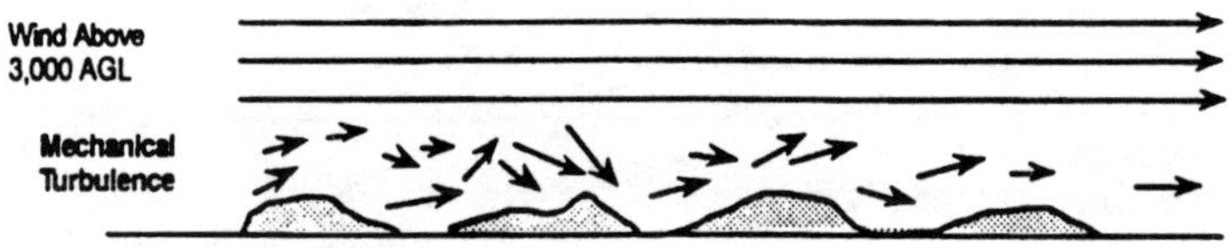

Fig. 5-4: Mechanical turbulence.

Turbulence is reported as LIGHT (slight changes in attitude or altitude), MODERATE (greater intensity than light turbulence, aircraft remains in positive control), SEVERE (large, abrupt changes in attitude or altitude, aircraft may be momentarily out of control), or EXTREME (aircraft violently tossed about, impossible to control). The degree of turbulence to expect can be estimated by the cloud type. Heap clouds (cumulus) indicate moderate to severe turbulence. Cumulo-nimbus and stacks of lens shaped clouds indicate severe turbulence. Under stratocumulus, light to moderate turbulence may be expected. Stratus clouds indicate smooth flying. Cumulus fractus is associated with strong, turbulent winds.

5.08 Wind

Wind is caused by pressure differences in the horizontal. Air flows from high pressure to low pressure. PRESSURE GRADIENT refers to the rate of change of atmospheric pressure in the horizontal. Where there is a steep pressure gradient, the isobars will be spaced closely together on a weather chart, and there will be high winds.

Because the earth rotates, air is deflected to the right in the northern hemisphere. This deflective force is termed CORIOLIS FORCE. It results in clockwise circulation at high pressure systems, and counterclockwise circulation in low pressure systems. At about 3,000 ft above ground level, wind tends to blow parallel to isobars. Below 3,000 feet, friction with the earth results in a change in direction and decrease in wind speed. In a climb, wind direction will VEER (clockwise change in direction) and wind speed will increase. In a descent, wind speed will decrease and wind direction will BACK (counter-clockwise change in direction).

According to GUY BALLOT'S RULE, in the northern hemisphere, if you stand with your back to the wind, the region of lower pressure will lie to your left.

A GUST is a rapid change of wind speed or wind direction, of brief duration. A SQUALL is a rapid change of wind speed or wind direction, of prolonged duration. A SEA BREEZE occurs during daytime when wind blows from sea to land. A LAND BREEZE occurs a night when wind blows from land to sea. A KATABATIC WIND blows down valleys at night or down ice-covered slopes by day. An ANABATIC WIND blows up a slope or valley by day.

MOUNTAIN WAVES occur when air crosses mountain ranges in North America, where the prevailing wind is from west to east. A typical location for mountain waves in Canada is over the foothills of the eastern Rockies. If the air has sufficient moisture, clouds will develop which distinguish the presence of mountain waves. A CAP CLOUD lies over the tops of mountain ranges and extends partially down the leeward slopes and indicates an extremely strong downdraft. A ROTOR CLOUD is formed downwind from each wave crest. The rotor cloud is simultaneously being formed and dissipated due to the rotation of the air, and may reach heights of up to 30,000 feet. Roll clouds resemble a long line of stratocumulus cloud lying parallel to the range. LENTICULAR CLOUD forms in the wave crests aloft, and is hundreds of miles long.

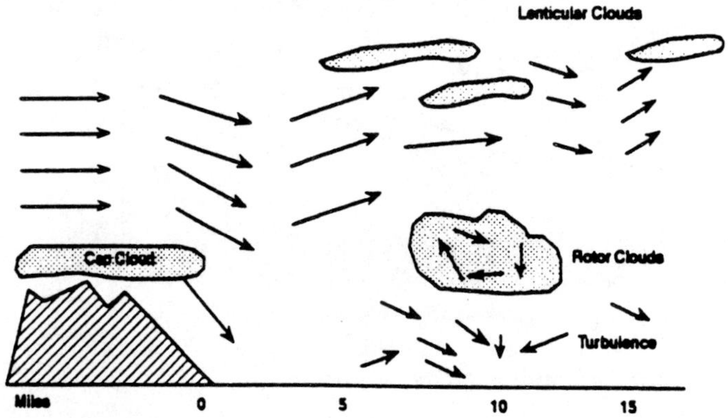

Fig. 5-5: Mountain waves.

The effects of mountain waves are strong downdrafts of up to 2000 FPM or more. Updraft areas can be used to gain altitude. Turbulence is most significant between the ground and the roll clouds. The altimeter tends to read high at the crests due to the low pressure created by the high winds.

WIND SHEAR occurs when the wind through which an aircraft is flying changes so rapidly that it causes a change in airspeed. Wind shears are either INCREASED PERFORMANCE SHEAR (airspeed increases due to entering increased headwind or decreased tailwind: lift increases) or DECREASED PERFORMANCE SHEAR (airspeed decreases due to entering decreased headwind or increased tailwind: lift decreases). Wind shear is either horizontal (due to low level inversion, changes in headwind or tailwind) or vertical (due to downbursts from thunderstorms, strong surface winds, passing fronts or mountain waves).

5.09 Air Masses

An air mass is a large body of air having uniform properties of moisture and temperature in the horizontal. There are four air masses that affect Canada's weather:

a) Continental Arctic (cA). Dry and cold. Forms over polar cap.
b) Maritime Arctic (mA). Moist and cold. Forms over polar oceans.
c) Maritime Polar (mP). Moist and temperate. Forms over temperate oceans.
d) Maritime Tropical (mT). Moist and warm. Forms over warm, semi-tropical oceans.

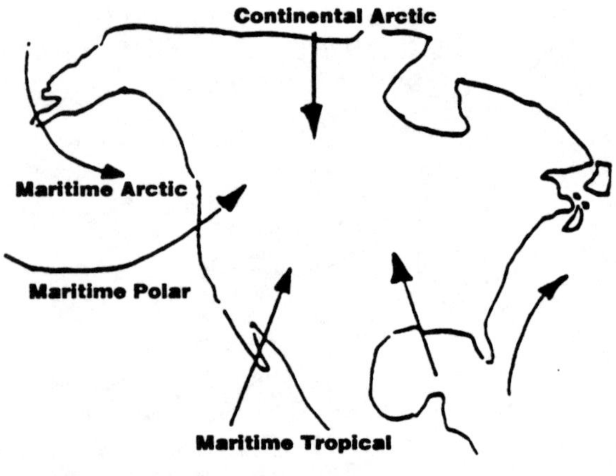

Fig. 5-6: Air masses affecting Canada.

Air masses become MODIFIED by taking on the characteristics of their source. For example, as the mT moves over the north Atlantic in summer, it is cooled and becomes mP.

5.10 Fronts

A FRONT is a transition zone between 2 air masses. A COLD FRONT is an advancing wedge of cold air, that displaces the warm air aloft. A WARM FRONT is an extensive slope of warm air, that over-rides the cold air. Frontal weather is determined by the stability and moisture content of the warm air mass.

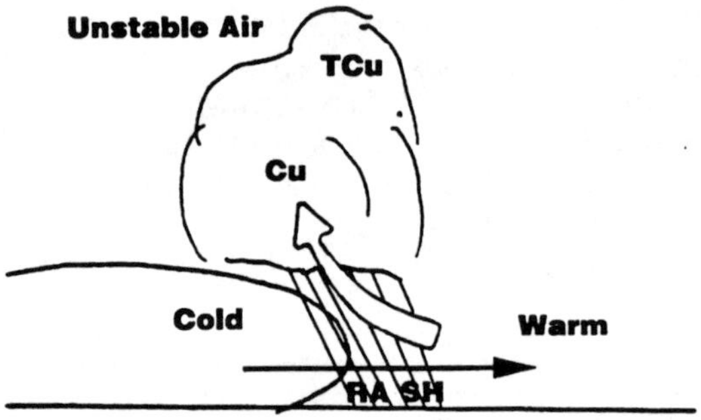

Fig. 5-7: Cross section of Cold Front.

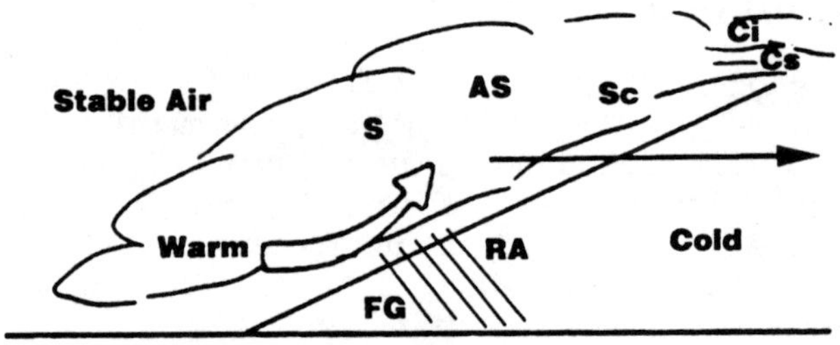

Fig. 5-8: Cross section of Warm Front.

When approaching a warm front from the cold air side, expect high cirrus, cirrostratus, with altostratus or altocumulus at mid level, followed by stratus or nimbostratus at lower levels. Depending upon the moisture and stability of the warm air mass, warm fronts can have Cb cells embedded in the main cloud deck.

In winter, a situation can develop where the freezing level is to the surface in the cold side of a warm front, while in the warm air sector the freezing level is much higher up. When precipitation falls through the front, the effect is in-flight freezing rain.

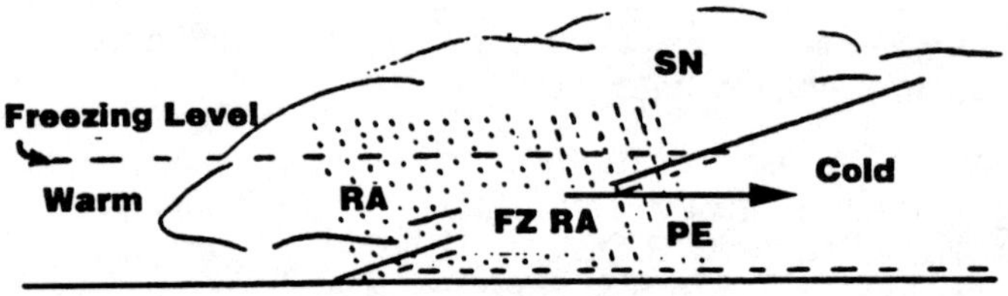

Fig. 5-9: Warm Front in winter. Note freezing level at surface in cold air, much higher in warm air.

A TROWAL is a "Trough of Warm Air Aloft", and is formed when a cold front advances on a warm front, undercutting and forcing the warm air up. A warm or cold OCCLUSION involves the same factors.

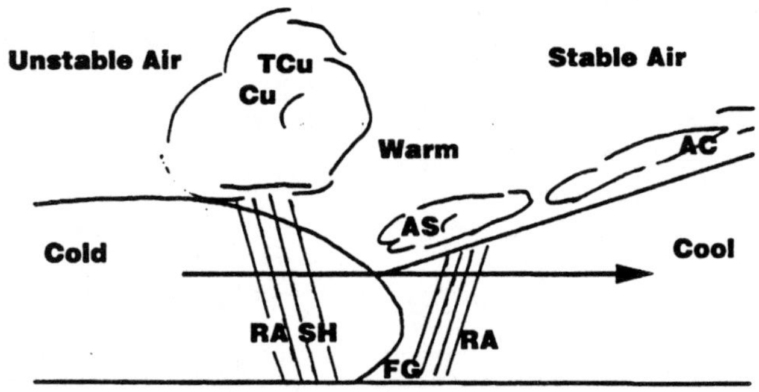

Fig. 5-10: Cold Front occlusion.

FRONTOGENESIS is when a front forms, FRONTOLYSIS is when a front dissipates. At a STATIONARY FRONT, the cold air mass does not move. An UPPER FRONT is produced when very cold air is trapped on the surface, and the frontal weather is aloft.

5.11 Airframe Icing

Airframe ice forms when flying in cloud with temperatures from +2°C to -40°C. CLEAR ICE is smooth, heavy ice which forms on the leading edge and spreads back. RIME ICE forms from small, super cooled water droplets and is opaque and granular. Clear ice is more likely at just below 0°C. Rime ice occurs at temperatures well below 0°C. HOAR FROST is a light, feathery, crystalline deposit that forms on parked aircraft at night due to sublimation.

5.12 Thunderstorms

The requirements for a thunderstorm are an adequate supply of moisture, a steep lapse rate and a lifting agent. The 3 stages are the developing, mature and dissipating stages.

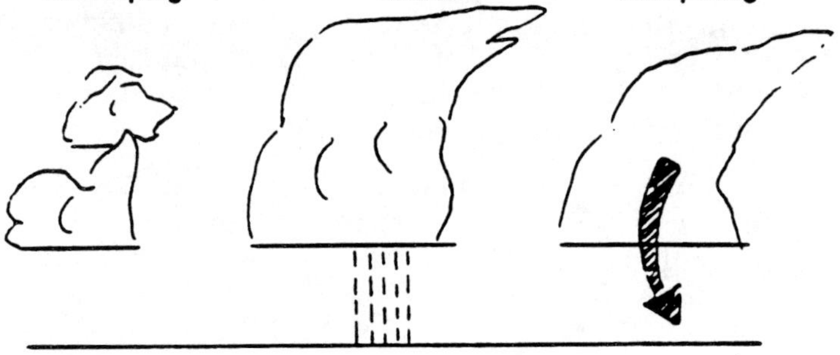

Fig. 5-11: Stages of a thunderstorm.

In the DEVELOPING STAGE, a cumulus type cloud grows and merges with other cumulus clouds. The base is about 4 miles wide and the height is up to about 20,000 feet. In this stage, the cloud consists mainly of updrafts. In the MATURE STAGE, rain or other precipitation such as hail begins to fall. Updrafts have reached their maximum speed. The top of the cloud reaches the tropopause and a characteristic anvil top forms due to the strong winds at the tropopause. As the rain falls, there are severe downdrafts. This air, when it hits the ground, spreads horizontally causing gusts, also known as a GUST FRONT. Ahead of the gusts are areas of downdrafts termed DOWNBURSTS. Occasionally, there are extreme downdrafts within the downbursts called MICROBURTS. Lighting occurs in the mature stage due to the movement of water, hail and snow within the cell, creating areas of differential electrical charge.

In the DISSIPATING STAGE of a thunderstorm, the dominant feature is a downdraft and most of the cloud begins to dissipate. The types of thunderstorms are:

a) Orographic. Forms in mountainous regions as air climbs over steep slopes.
b) Convective. Caused by solar heating or radiational cooling. Forms over land or water.
c) Frontal. Caused by frontal lift, for example, with fast moving cold front.

Thunderstorms have severe turbulence, heavy showers, hail and icing. Pilots of light aircraft should not fly anywhere near thunderstorm activity.

5.13 Squall Lines

A long line of thunderstorms is termed a squall line. Thunder and lightning frequently occur. During the passage of a line squall, there is a sudden wind change from southerly or south-westerly to north or north-westerly, together with a sudden drop in temperature and a rise in barometric pressure.

5.14 Hurricanes and Tornadoes

A hurricane is a destructive cyclonic storm that originates in tropical waters. Wind speed is great, and the movement of a hurricane is difficult to predict.

A TORNADO is a rotating tunnel shaped cloud linking the ground with the base of a violent thunderstorm. A WATERSPOUT is like to a tornado except it contacts water instead of land.

5.15 Surface Based Layers

A SURFACE BASED LAYER is a layer which contacts the ground, such as fog, haze, and smoke. The requirements for fog are high relative humidity, a cooling process and condensa-tion nuclei. Fog forms when air is cooled below the dewpoint, or moisture is added, raising the dewpoint. Fog types are:

a) Radiation Fog. Formed by radiation cooling on clear nights where relative humidity is high.
b) Advection Fog. Horizontal movement of warm, moist air over a cool surface, as in coastal areas.
c) Upslope Fog. Moist air moves up rising terrain.
d) Frontal Fog. Precipitation from warm front or cold front falls into colder air below.

5.16 Meteorological Services for Pilots

Weather information for pilots is available by telephone, computer terminal, weather briefing or radio broadcast. The most common weather service originates from the network of flight service stations. At FSS stations, pilots can obtain all weather information needed for flight. The Aviation Weather Information Service (AWIS) is the basic weather briefing service provided for pilots at FSS stations and is accesible at the station via telephone (collect calls accepted). AWIS is called W2 in the CFS.

The Aviation Weather Briefing Service (AWBS) is a more advanced weather briefing service that presently comes from AES offices and some FSS's. In addition to domestic flights, AWBS is used for long range flights and is called W1 in the CFS.

At several airports, Automatic Terminal Information Service (ATIS) information is available. ATIS is a recorded broadcast providing basic weather information for departure and arrival at an airport. Transcribed Weather Broadcasts (TWB) are recorded weather reports, forecasts and NOTAMS that are broadcast over various NDB stations. PATWAS refers to "Pilots Automatic Telephone Weather Answering Service" and is a recording service that pilots can select various reports and forecasts by pressing different keys on a touchtone telephone. Locations providing this service can be found in the CFS.

Aviation weather is also available from private funded DUATS computer terminals provided at some schools and FBO's. DUATS availability can be found in the CFS. DUATS can also be obtained via computer and modem, and there is a wide array of aviation weather now available on the INTERNET.

5.17 Area Forecast (FA)

An FA gives a prediction of weather expected throughout a given area for a 12 hour time interval. FA's are issued 4 times per day about 30 minutes before their valid period. The standard valid periods for an FA commence at 0000Z, 0600Z, 1200Z and 1800Z. Each new forecast replaces the last 6 hours of the preceding one. Z refers to universal time coordinated, or "UTC". An FA begins with the PROGNOSIS, which is a prediction for expected pressure systems throughout the entire area. The FA then focuses on various regions, providing expected CLOUDS AND WEATHER, ICING and TURBU-LENCE.

In an FA:

a) Heights are above sea level (ASL), unless noted with AGL abbreviations such as "ABV GND" or "CIGS . . . HND".
b) Heights are given in hundreds of feet.
c) Layers above 20,000 feet are indicated with an oblique "/", for example, /BKN = bkn above 20,000 ft ASL.
d) Visibility is given only if it will be 6 SM or less. Visibility over 6 SM is indicated as "6+".
e) Wind is given only if expected to be above 20 kts.
f) Cloud heights are given for bases and tops. For example, "40 SCT 60" means a cloud layer is based at 4,000 ft ASL, scattered, with its top at 6,000 ft ASL.

The CATEGORICAL OUTLOOK is found at the end of the FA and summarizes expected weather for the 12 hour period following the FA. The three outlook categories used to describe the expected ceiling and visibility are:

Category	**Ceiling**	**Visibility**
IFR	less than 1,000 ft	less than 3 miles
MVFR	1,000 ft - 3,000 ft	3 - 5 miles
VFR	more than 3,000 ft	more than 5 miles

For MVFR or IFR conditions, the cause will be indicated ie "MVFR FG FU" would mean the low visibilities/ceilings will be caused by fog and smoke. As well, the term "WND" in the categorical outlook indicates winds are expected to be greater than 20 kts, or gusts of 30 kts or more.

Fig. 5-12 shows the FA regions in Canada: all FA's will involve providing detail on these areas and regions. Fig. 5-13 provides a basic list of common FA abbreviations. In addition to these abbreviations, weather forecasters use the standard codes found in the TAFs and METARS.

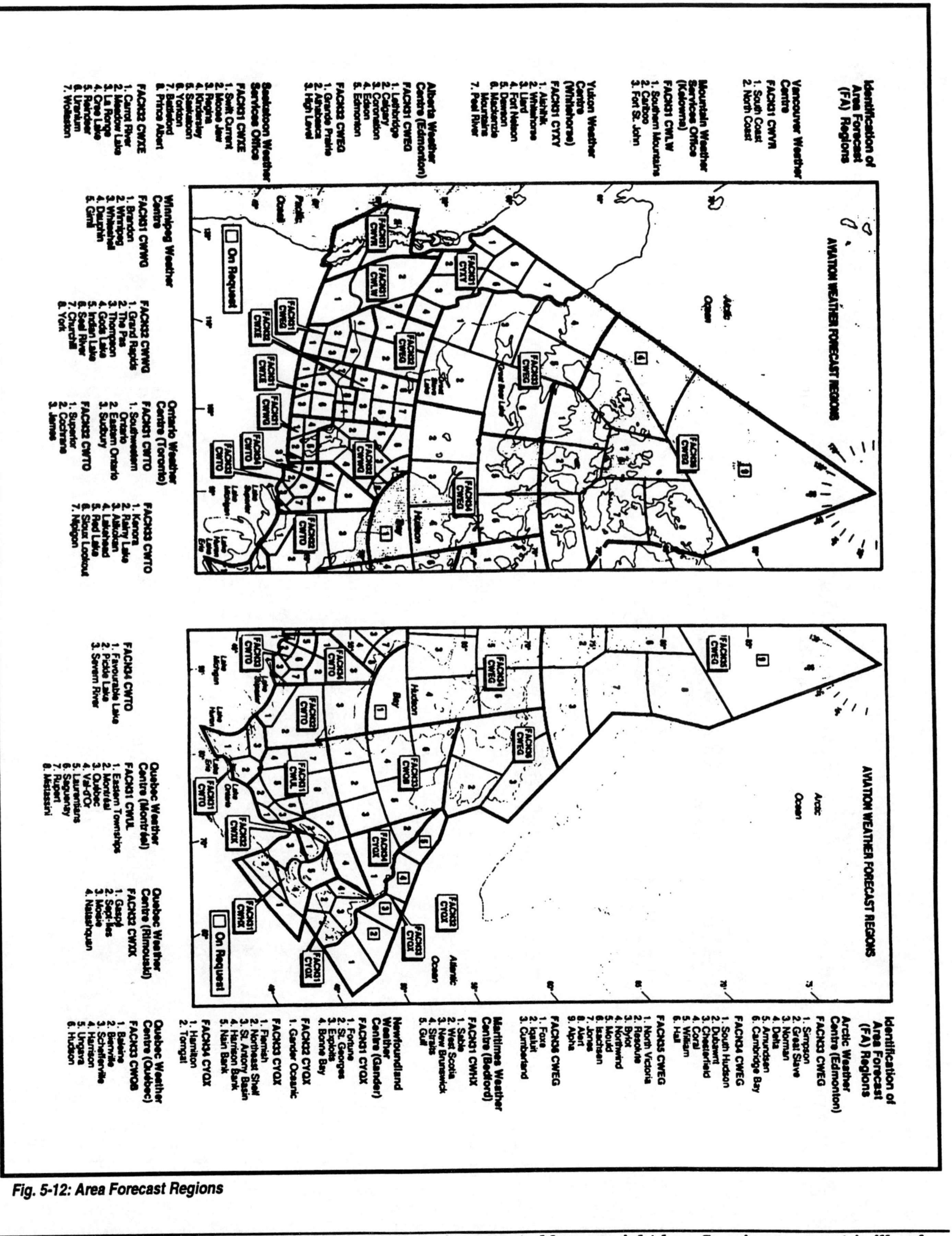

Fig. 5-12: Area Forecast Regions

ABT	About
ABV	Above
AC	Altocumulus
ACC	Altocumulus Castellanus
ACRS	Across
ACSL	Standing Lenticular Altocumulus
ACTV	Active
ADVY	Advisory
AFT	After
AFTN	Afternoon
AHD	Ahead
ALF	Aloft
ALG	Along
ALQDS	All quadrants
ALT	Altitude
ALTA	Alberta
AMS	Air Mass
APRCH	Approach
APRCHG	Approaching
ASL	Above Sea Level
ASSOCD	Associated
ATLC	Atlantic
ATTN	Attention
AURBO	Aurora Borealis
AVG	Average
BAFDY	Bay of Fundy
BECMG	Becoming
BFR	Before
BGN	Begin
BGNG	Beginning
BHND	Behind
BINOVC	Breaks in Over-cast
BKN	Broken
BL	Blowing
BLDG	Building
BLO	Below
BLZD	Blizard
BNDRY	Boundary
BR	Mist
BRF	Brief
BRFLY	Briefly
BRKS	Breaks
BTN	Between
CASCDS	Cascades
CAT	Clear Air Turb-ulence
CAVOK	Ceiling and Visibility OK
CB	Cumulonimbus
CC	Cirrocumulus
CF	Cumlus Fractus
CIG	Ceiling
CLD	Cloud
CLR	Clear
CLRG	Clearing
CMA	Continental-Maritime Arctic
CMA	Comma
CMPLX	Complex
CNTR	Centre
CNTRD	Centred
CONDS	Conditions
CONTRA	Condensation Trails
CONTG	Continuing
CONTUS	Continuous
CONTVD	Continental Divide
CONTG	Continuing
COR	Correction
CST	Coast
CU	Cumulus
CVCTV	convective
CVR	Cover
DCRG	Decreasing
DEG	Degree
DFUS	Diffuse
DISTNT	Distant
DNS	Dense
DNSLP	Downslope
DP	Deep
DP	Dewpoint
DPNG	Deepening
DR	Drifting
DRFTG	Drifting
DRPG	Dropping
DSIPTG	Dissipating
DTRT	Deteriorate
DURG	During
DVLPG	Developing
DZ	Drizzle
E	East
ELSW	Elsewhere
ELY	Easterly
EMBD	Embedded
ENDG	Ending
ENTR	Entire
EVE	Evening
FCST	Forecast
FG	Fog
FILG	Filling
FL	Flight level
FLO	Flow
FLRY	Flurry
FLWD	Followed
FLWG	Following
FM	From
FNT	Front
FNTGNS	Frontogenesis
FQT	Frequent
FRLVL	Freezing Level
FRMG	Forming
FROPA	Frontal Passage
FROSFC	Frontal Surface
FRQ	Frequent
FRZG	Freezing
FT	Feet
FTHLS	Foothills
FU	Smoke
FZ	Freezing
GASTR	Strait of Georgia
GGNBA	Georgian Bay
GLFALSK	Gulf of Alaska
GLFST-LAWR	Gulf of St. Law-rence
GND	Ground
GNLY	Generally
GRAD	Gradient
GRDLY	Gradually
GRBNKS	Grand Banks
GRTLKS	Great Lakes
GVG	Giving
HGT	Height
HI	High
HIER	Higher
HLSTO	Hailstones
HLTP	Hilltop
HND	Hundred
HR	Hour
HNSBA	Hudson Bay
HVY	Heavy
ICG	Icing
ICGIC	Icing in Cloud
ICGIP	Icing in Precipitation
IMDTLY	Immediately
INCOMP	Incomplete
INCRG	Increasing
INDEF	Indefininte
INOV	In the Vicinity of
INSTBY	Instabliity
INTMT	Intermittent
INTR	Interior
INTS	Intense
INTSFY	Intensify
INVDG	Invading
IP	Ice Pellets
IS	Island
ISLD	Island
ISOL	isolated
JDFSTR	Juan de Fuca Strait
JMSBA	James Bay
KT	Knot
LABRDR	Labrador
LAT	Latitude
LCL	Local
LFTG	Lifting
LGT	Light
LIFR	Low IFR
LK	Lake
LLWS	Low Level Wind Shear
LMT	Limit
LN	Line
LND	Land
LO	Low
LONG	Longitude
LTL	Little
LTGIC	Lightning in Cloud
LTNG	Lightning
LTR	Later
LVL	Level
LWR	Lower
LWRG	Lowering
LYR	Layer
MAN	Manitoba
MB	Millibars
MCKNZ	Mackenzie
MDFYD	Modified
MDT	Moderate
MI	Mile
MID	Middle
MOVG	Moving
MPH	Miles per hour
MRNG	Morning
MRTM	Maritime
MRTMS	Maritimes
MST	Moist
MSTR	Moisture
MTNS	Mountains
MVFR	Marginal VFR
MXD	Mixed
MXG	Mixing
N	North
NE	Northeast
NELY	Northeasterly
NFLD	Newfoundland
NGT	Night
NLY	Northerly
NML	Normal
NMRS	Numerous
NR	Near
NRLY	Nearly
NW	Northwest
NWLY	Northwesterly
OBSC	Obscured
OCLD	Occlude
OCLDG	Occluding
OCLN	Occlusion
OCNL	Occasional
OCNLY	Occasionally
OFSHR	Offshore
ONSHR	Onshore
ORGRHC	Orographic
OT	On Top of Clouds
OTLK	Outlook
OTRW	Otherwise
OVC	Overcast
OVR	Over
OVRNG	Overrunning
PAC	Pacific
PCPN	Precipitation
PD	Period
PRECDD	Preceded
PRECDS	Precedes
PRES	Pressure
PROG	Prognostic
PROG	Prognosis
PRSTG	Persisting
PSG	Passage
PSG	Passing
PTCHY	Patchy
PTLY	Partly
QS	Quasi-Stationary
QSTNRY	Quasi-Stationary
QUAD	Quadrant
QUE	Quebec
RA	Rain
RCH	Reach
RCKY	Rockies
RDG	Ridge
RFRMG	Reforming
RGN	Region
RMK	Remark
RMNG	Remaining
RPDLY	Rapidly
RPT	Repeat
RSG	Rising
RTRN	Return
RUF	Rough
RVR	River
S	South
SASK	Saskatchewan
SC	Stratocumulus
SCT	Scattered
SCTR	Sector
SE	Southeast
SELY	Southeasterly
SFC	Surface
SH	Shower
SHFT	Shift
SHFTG	Shifting
SHLW	Shallow
SHWR	Shower
SKC	Sky Clear
SLKRS	Selkirks
SLO	Slow
SLOLY	Slowly
SLPG	Sloping
SLY	Southerly
SMK	Smoke
SML	Small
SN	Snow
SNRS	Sunrise
SNST	Sunset
SNW	Snow
SPECI	Special
SPRDG	Spreading
SQL	Squall
ST	Stratus
STBL	Stable
STG	Strong
STGTN	Strengthen
STNRY	Stationary
STP	Stop
SVR	Severe
SVRL	Several
SW	Southwest
SWLY	Southwesterly
SXN	Section
SYS	System
TCU	Towering Cumulus
TEMP	Temperature
TEMPO	Temporary
THK	Thick
THKNG	Thickening
THN	Thin
THNC	Thence
THNG	Thinning
THRU	Through
THRUT	Throughout
THSD	Thousand
TIL	Until
TRML	Terminal
TROF	Trough
TROWAL	Trough of Warm Air Aloft
TRRN	Terrain
TSTM	Thunderstorm
TURBC	Turbulence
TWD	Toward
UNSTBL	Unstable
UNSTDY	Unsteady
UPR	Upper
UPSLP	Upslope
UTC	Co-ordinated Universal Time
V	Variable
VC	Vicinity
VCNTY	Vicinity
VLNT	Violent
VLY	Valley
VR	Veer
VRBL	Variable
VRISL	Vancouver Island
VS	Visual
VSBY	Visibility
W	West
WDLY	Widely
WK	weak
WLY	Westerly
WND	Wind
WNDSHFT	Windshift
WRM	Warm
WRN	Western
WS	Wind Shear
WV	Wave
WX	Weather
XCP	Except
XT	Extend
XTGD	Extending
XTRM	Extreme
XTSV	Extensive
YDA	Yesterday
YKN	Yukon
ZULU (Z)	Co-ordinated universal time

Fig. 5-13: List of common abbreviations used in area forecasts.

EXAMPLE:

```
FACN1 CWVR 221730
18 - 06
OTLK 06 - 18
HTS ASL UNLESS NOTED

PROG
FNTL ZONE QSTNRY ALG 51N LAT THRU PRD. ASSOCD
MRTM TROWAL BLUE RIVER-ALTA LAKE-JUST SE PORT
HARDY ROTATING SEWDS TO LIE 50 N OF LN CRANBROOK-
PENTICTON-COMOX BY 06Z. CMA WRM FNT 60 NE DEASE
LAKE-FORT ST JOHN MOVG NEWDS AT 5 - 10 KTS. MDT-STG
NWLY FLO ALF. MOIST FAIRLY STBL AMS OVR RGNS.

S CST RGN
CLDS AND WX. 5-15 SCT OCNL BKN 20 30 BKN V OVC 80
100-120 BKN 160 OCNL -RA. VSBYS LCLY 4-6 -RA FG. LWR
LYR BKN V OVC XTNSV VSBYS 1-4 -DZ FG ALG W AND N
CSTS VRISL. PASSES CLOSED.

ICG. LGT-MDT RIME ICGIC. FRLVL 90-100 SLPG TO NR 70
NWRN SXNS.

OTLK. MVFR CIGS XCP IFR CIGS FG ALG W CST VRISL.

SRN MTNS RGN
CLDS AND WX. 50-70 SCT V BKN SC CU 90
110 BKN 150 BKN LYRS 180 / BKN OCNL -SHRA. BCMG NW
OF TROWAL ... 50-70 SCT V BKN SC CU 90 110 BKN LYRS 160
/ SCT OCNL -SHRA. ADDNL 25-30 SCT 40 FRMG IN PCPN
FQTLY BKN ALG WLY SLPS ROCKY MTNS AHD OF
TROWAL. PASSES MARGINAL WRN SXNS...CLOSED THRU
ROCKIES.

ICG. LGT-MDT RIME ICGIC. LGT MXD IN CU TOPS. FRLVL
70-80 LWRG TO NR 50 NW OF TROWAL.

TURBC. LGT VCNTY CU.

OTLK. MVFR THRU PASSES ROCKYS AND VCNTY TROWAL.
```

QUESTIONS BASED ON ABOVE FA:

1. What is the moisture and stability of the air mass?
2. Where will the maritime TROWAL lie by 06Z?
3. What is the direction and speed of the continental arctic - maritime arctic front?
4. What is the lowest height of cloud expected for the South Coast region?
5. Where is the lowest expected visibility in the South Coast region? What is forecast wind?
6. What is the thickness of the second layer of cloud in the South Coast region?
7. What is lowest height of the freezing level in the Southern Mountains region?
8. What kind of icing is forecast for the Southern Mountains region?
9. What is the validity period of the FA and outlook?
10. What is the range of ceiling heights and visibility forecast after 06Z along the west coast of Vancouver Island?

ANSWERS:

1. Air mass is moist and fairly stable.
2. 50 NM north of a line from Cranbrook - Penticton - Comox.
3. North-eastwards at 5-10 knots.
4. 500 feet ASL.
5. West coast of Vancouver Island visibility will be 1 SM. Winds are less than 20 knots.
6. 5,000 feet. Bases of the second layer are at 3,000 feet, broken variable overcast, tops at 8,000 feet. Thickness is 8,000 feet - 3,000 feet = 5,000 feet.
7. 5,000 feet. Freezing level is at 7,000 - 8,000 feet, lowers to near 5,000 feet when NW of TROWAL.
8. Light mixed icing in cloud. Light mixed icing in Cumulus tops.
9. The FA is valid 1800 - 0600Z. The OUTLOOK is valid 0600 - 1800Z.
10. Ceilings below 1,000 feet, and visibilities below 3 SM in fog.

5.18 International Aerodrome Forecasts (TAF)

The TAF is an international weather format for an aerodrome forecast, describing the most probable weather conditions expected for an aerodrome. TAFs give weather within 5 NM of the centre of the runway complex. TAFs are scheduled 4 times daily for 12 or 24 hour periods at 05Z, 11Z, 17Z and 23Z. The information provided is as follows:

a) Message type. TAF or TAF AMD indicates regular or amended forecast. An amended TAF is issued when the current TAF no longer describes the on-going weather.

b) Station identifier. A 4 letter ICAO identifier is used to indicate the airport name.

c) Date and time of issue. This is a 6 digit code: the first two digits are the date, followed by the time of issue. TAFs are normally issued about 30 minutes before their valid period. Time is UTC.

d) Valid period. A 4 digit code represents the valid period. Normally 12 hours, although sometimes 24 hours and for an amended or corrected TAF a different period may be used.

e) Winds. The wind direction and speed is given to the nearest 10 degrees true and knots. Calm is 00000KT. VRB is for variable direction, less than 3 kts. Gusts, if applicable, is given in 2 digits. Where there is significant wind shear, the code WS is used, followed by the height of the shear and forecast wind speed and direction at that height. For example, WS015/27030KT means wind shear at 1500 ft AGL, 270°T at 30 kts.

f) Prevailing visibility in statute miles when 6 SM or less. P6SM means visibility greater than 6 SM.

g) Weather. "-" means light intensity, no sign is moderate, "+" is heavy. There are 7 descriptors which apply to forecast precipitation and/or obstructions to visibility: SH = showers, DR = drifting, FZ = freezing, MI = shallow, BL = blowing, BC = batches. Weather codes:

FC = Tornado/Water-spout/Funnel cloud	RA = Rain
	SG = Snow Grains
TS = Thunderstorm	PE = Ice Pellets
DZ = Drizzle	IC = Ice crystals
FG = Fog (< 5/8 SM)	SA = Sand
BR = Mist (> 5/8 SM)	SN = Snow
GS = Small Hail	HZ = Haze
FU = Smoke	GR = Hail
SS = Sandstorm	DU = Dust
VA = Volcanic Ash	SQ = Squall
PO = Dust/Sand whirls	DS = Duststorm

The code "VC" refers to elements withing 5 - 10 NM from the centre of the runway complex.

h) Sky condition. Sky cover amounts are cumulative, so layer amounts include the sum of any layers below. Sky condition is given in eighths or "oktas". SKC = sky clear, no cloud present. FEW = >0 - 2 oktas. SCT = scattered, 3 - 4 octas. BKN = broken, 5 - <8 octas. OVC = sky overcast, 8 octas. A CLOUD CEILING exists at the height of the first layer for which a coverage symbol of BKN or OVC is present. VV = vertical visibility, in hundreds of feet. The presence of a vertical visibility constitutes an OBSCURED CEILING. Cloud height is forecast in three digits in hundreds of feet. CB is added where needed.

i) Change groups. Used when change expected. The following can be used: FM0100Z indicates a change from 0100Z. BECMG 0608 indicates a change between 0600Z and 0800Z. TEMPO 1801 indicates fluctuations between 1800Z and 0100Z. PROB30 2022 indicates a 30% probability of an event between 2000Z and 2200Z.

EXAMPLE:

```
CYYZ TAF 200438Z 200505 05010KT 5SM BR SCT008 BKN100
    TEMPO 0510 2SM BR OVC005 FM1000Z 06010G15KT
    3/4SM -RA BR VV003 TEMPO 1420 1/2SM RA FG VV001
    FM2200Z 01015G25KT 2SM -RA BR OVC008
```

QUESTIONS BASED ON ABOVE TAF:
1. What date and time was the TAF issued?
2. What is the valid period of the TAF?
3. The lowest visibility is forecast for what time?
4. What is the lowest forecast ceiling for 0700Z?
5. What is the forecast surface wind at 2300Z?
6. What is the forecast visibility at 1000Z?
7. What is the forecast weather at 1000Z?
8. What is the forecast vertical visibility at 1000Z?
9. What time is the forecast lowest ceiling?
10. After what time are the strongest winds forecast?

ANSWERS:
1. 20th day of the month, at 0438Z.
2. From 0500Z on the 20th to 0500Z on the 21st.
3. From 1400Z to 2000Z, the visibility is forecast to fluctuate as low as 1/2 SM in rain and fog.
4. 500 feet AGL overcast.
5. 010° True at 15 kts, gusting to 25 kts.
6. 3/4 SM.
7. Light rain and mist.
8. 300 feet AGL.
9. 100 feet AGL from 1400Z to 2000Z.
10. After 2200Z.

Weather Station Codes

Although the naming of Canadian airports and weather stations can seem confusing, here is a brief explanation. Originally, in the 1930's, Canada used two letters for identification of a weather reporting station. Additionally, preceding the 2-letter code, was placed a Y (meaning "yes") where the reporting station was co-located with an airport, a W (meaning "without") where the reporting station was not co-located with an airport, and a U where the reporting station was co-located with an NDB. An X was used if the last 2 letters of the code had already been taken by another Canadian ident, and a Z was used if the locator could be confused with a US three letter ident. Vancouver International Airport, for example, was named YVR since there is a weather reporting station at the airport. The ICAO names are in a 4 letter format starting with a C for Canadian airports and a K for US airports. For example: the ICAO identifier for Vancouver International Airport is CYVR, and for Las Vegas, USA the ICAO ident is KLAV. See the CFS for the codes of other airports not having a weather station.

Fig. 5-14 provides a summary of Canadian airports having weather reporting stations.

XBE	Bearskin Lake, ON
XGR	Kangiqsualujjuaq, QB
XKS	Kasabonika, ON
XLB	Lac Brochet, MB
XPK	Pukatawagan, MB
XQU	Qualicum, BC
XSI	South Indian Lake, MB
XTL	Tadoule Lake, MB
YAA	Anahim Lake, BC
YAB	Arctic Bay, NWT
YAC	Cat Lake, ON
YAG	Fort Frances, ON
YAH	La Grande-4, QB
YAJ	Lyall Harbour, BC
YAM	Sault Ste Marie, ON
YAQ	Kasabonika, ON
YAR	Attawapiskat, ON
YAV	Winnipeg, St. Andrews MB
YAW	Halifax - Shearwater, NS
YAX	Lac du Bonnet, MB
YAY	St. Anthony, NF
YAZ	Tofino, BC
YBA	Banff, AB
YBB	Pelly Bay, NWT
YBC	Baie Comeau, QB
YBD	Bella Coola, BC
YBE	Uranium City, SA
YBG	Bagotville, QB
YBI	Black Tickle, NF
YBK	Baker Lake, NWT
YBL	Campbell River, BC
YBN	Borden, ON
YBP	Brooks, AB
YBQ	Tadoule Lake, MB
YBR	Brandon, MB
YBT	Brochet, MB
YBU	Nipawin, SA
YBV	Berens River, MB
YBW	Calgary - Springbank, AB
YBX	Blanc Sablon, QB
YCA	Cartwright, NF
YCB	Cambridge Bay, NWT
YCC	Cornwal, ON
YCD	Nanaimo, BC
YCE	Centralia-Huron, ON
YCG	Castlegar, BC
YCH	Chatham, NB
YCK	Colville Lake, NWT
YCL	Charlo, NB
YCN	Cochrane, ON
YCO	Coppermine, NWT
YCP	Blue River, BC
YCQ	Chetwind, BC
YCR	Cross Lake, MB
YCS	Chesterfield Inlet, NWT
YCT	Coronation, AB
YCW	Chilliwack, BC
YCX	Gagetown, NB
YCY	Clyde River, NWT
YCZ	Fairmont Hot Springs, BC
YDA	Dawson City, YT
YDB	Burwash, YT
YDC	Princeton, BC
YDE	Paradise River, NF
YDF	Deer Lake, NF
YDG	Chute des Passes, QB
YDI	Davis Inlet, NF
YDL	Dease Lake, BC
YDM	Ross River, YT
YDN	Dauphin, MB
YDO	Dolbeau-St-Methode, QB
YDP	Nain, NF
YDQ	Dawson Creek, BC
YDS	Desolation Sound, BC
YEA	Empress, AB
YED	Edmonton Nameo, AB
YEE	Midland-Huronia, ON
YEG	Edmonton Int'l, AB
YEK	Arviat, NWT
YEL	Elliot Lake, ON
YEM	Manitowaning, ON
YEN	Estevan, SA
YET	Edson, AB
YER	Fort Severn, ON
YEU	Eureka, NWT
YEY	Amos, QB
YEV	Inuvik, NWT
YFA	Fort Albany, ON
YFB	Iqaluit, NWT
YFC	Fredericton, NB
YFD	Brantford, ON
YFE	Forrestville, QB
YFH	Fort Hope, ON
YFJ	La Macaza, QB
YFO	Flin Flon, MB
YFR	Ft. Resolution, NWT
YFS	Ft. Simpson, NWT
YFT	Makkovik, NF
YFX	Fox Harbour, NF
YGB	Gilles Bay, BC
YGD	Goderich, ON
YGE	Golden, BC
YGG	Ganges Harbor, BC
YGH	Ft. Good Hope, NWT
YGK	Kingston, ON
YGL	La Grande, QB
YGM	Gimli, MB
YGO	Gods Narrows, MB
YGP	Gaspe, QC
YGQ	Geraldton, ON
YGR	Iles de la Madeleine, QB
YGS	Germansen Landing, BC
YGT	Igloolik, NWT
YGV	Havre St. Pierre, QB
YGW	Kuujjuarapik, QC
YGX	Gillam, MB
YGZ	Grise Fiord, NWT
YHA	Quaqtaq, QB
YHB	Hudson Bay, SA
YHD	Dryden, ON
YHE	Hope, BC
YHF	Hearst, ON
YHG	Charlottetown, PEI
YHH	Nemiscau, QB
YHI	Holman Island, NWT
YHK	Gioa Haven, NWT
YHM	Hamilton, ON
YHN	Hornepayne, ON
YHO	Hopedale, NF
YHP	Poplar Hill, ON
YHR	Chevery, QB
YHS	Sechelt, BC
YHU	Montreal - St. Hubert, QB
YHY	Hay River, NWT
YHZ	Halifax, NS
YIB	Atikokan, ON
YID	Digby, NS
YIF	St-Augustin, QB
YIK	Ivujivik, QC
YIO	Pond Inlet, NWT
YIV	Island Lake MB
YJA	Jasper, AB
YJF	Fort Liard, NWT
YJM	Fort St. James, BC
YJP	Fort Providence, NWT
YJQ	Bella Bella, BC
YJT	Stephenville, NF
YKA	Kamloops, BC
YKC	Collins Bay, SA
YKD	Aklavik, NWT
YKF	Kitchener - Waterloo-Guelph, ON
YKG	Kangiqsujuaq, QC
YKJ	Key Lake, SA
YKK	Kitkatla, BC
YKL	Schefferville, QB
YKO	Akulivik, QB
YKQ	Waskaganish, QB
YKT	Klemtu, BC
YKU	Chisasibi, QB
YKX	Kirkland Lake, ON
YKY	Kindersley, SA
YKZ	Toronto - Buttonville, ON
YLA	Aupaluk, QB
YLB	Lac La Biche, AB
YLC	Lake Harbour, NWT
YLD	Chapleau, ON
YLE	Lac la Martre, NWT
YLH	Lansdowne House, ON
YLJ	Meadow Lake, SA
YLL	Lloydminster, AB
YLO	Shilo, MB
YLP	Mingan, QB
YLQ	La Tuque, QB
YLR	Leaf Rapids, MB
YLT	Alert, NWT
YLU	Kangiqsualujjuaq, QB
YLW	Kelowna, BC
YMA	Mayo, YT
YMD	Mould Bay, NWT
YME	Metane, QB
YMG	Manitouwadge, ON
YMH	Mary's Harbour, NF
YMJ	Moose Jaw, SA
YML	Charlevoix, QB
YMM	Ft. McMurray, AB
YMN	Makkovik, NF
YMO	Moosonee, ON
YMP	Port McNeil, BC
YMQ	Montreal, QB
YMT	Chibougamau, QB
YMU	Umiujaq, QB
YMW	Maniwaki, QB
YMX	Montreal - Mirabel, QB
YMY	Ear Falls, ON
YNA	Natashquan, QB
YNC	Wemindji, QB
YND	Gatineau - Hull, QB
YNE	Norway House, MB
YNH	Hudson's Hope, BC
YNJ	Langley, BC
YNL	Points North Landing, SA
YNM	Matagami, QB
YNO	North Spirit Lake, ON
YNS	Nemiscau, QB
YOC	Old Crow, YT
YOD	Cold Lake, AB
YOG	Ogoki, ON
YOH	Oxford House, MB
YOJ	High Level, AB
YOO	Oshawa, ON
YOP	Rainbow Lake, AB
YOS	Owen Sound, ON
YOW	Ottawa, ON
YOY	Valcartier - Lecomte, QB
YPA	Prince Albert, SA
YPB	Port Alberni, BC
YPC	Paulatuk, NWT
YPD	Port Hawkesbury, NS
YPE	Peace River, AB
YPG	Portage La Prairie - Southport, MB
YPH	Inukjuak, QB
YPI	Port Simpson, BC
YPJ	Aupaluk, QB
YPK	Pitt Meadows, BC
YPL	Pickle Lake, ON
YPM	Pikangikum, ON
YPN	Port Menier, QB
YPO	Peawanuck, ON
YPP	Parent, QB
YPQ	Peterborough, ON
YPR	Prince Rupert, BC
YPS	Pemberton, BC
YPT	Pelee Island, ON
YPU	Puntzi Mountain, BC
YPW	Powell River, BC
YPX	Povungnituk, QB
YPY	Ft. Chipewyan, AB
YPZ	Burns Lake, BC
YQA	Muskoka, ON
YQB	Quebec, QB
YQC	Quaqtaq, QB
YQD	The Pass, MB
YQF	Red Deer, AB
YQG	Windsor, ON
YQH	Watson Lake, YT
YQI	Yarmouth, NS
YQK	Kenora, ON
YQL	Lethbridge, AB
YQM	Moncton, NB
YQN	Nakina, ON
YQQ	Comox, BC
YQR	Regina, SA
YQS	St. Thomas, ON
YQT	Thunder Bay, ON
YQU	Grande Prairie, AB
YQV	Yorkton, SA
YQW	North Battleford,SA
YQX	Gander, NF
YQY	Sydney, NS
YQZ	Quesnel, BC
YRA	Rae Lakes, NWT
YRB	Resolute, NWT
YRC	Chicoutimi, QB
YRD	Dean River, BC
YRF	Cartwright, NF
YRG	Rigolet, NF
YRI	Riviere-du-Loup, QB
YRJ	Roberval, QC
YRL	Red Lake, ON
YRM	Rocky Mountain House, AB
YRN	Rivers Inlet, BC
YRO	Ottawa - Rockcliffe, ON
YRP	Ottawa - Carp, ON
YRQ	Trois-Rivieres, AB
YRR	Stuart Island, BC
YRS	Red Sucker Lake, MB
YRT	Rankin Inlet, NWT
YRV	Revelstoke, BC
YSA	Sable Island, NS
YSB	Sudbury, ON
YSC	Sherbrooke, QB
YSD	Suffield, AB
YSE	Squamish, BC
YSF	Stony Rapids, SA
YSG	Snowdrift, NWT
YSH	Smiths Falls, ON
YSJ	Saint John, NB
YSK	Sanikiluaq, NWT
YSL	St. Leonard, NB
YSM	Ft. Smith, NWT
YSN	St. Catharines, ON
YSO	Postville, NF
YSP	Marathon, ON
YSQ	Atlin, BC
YSR	Nanisivik, NWT
YST	St. Therese Point, MB
YSU	Summerside, PEI
YSW	Sparwood, BC
YSY	Sachs Harbour, NWT
YSZ	Ste-Anne-des-Monts, QB
YTA	Pembroke, ON
YTB	Koala, NWT
YTE	Cape Dorset, NWT
YTF	Alma, QB
YTG	Sullivan Bay, BC
YTH	Thompson, MB
YTJ	Terrace Bay, ON
YTL	Big Trout Lake, ON
YTQ	Tasiujuaq, QB
YTR	Trenton, ON
YTS	Timmins, ON
YTZ	Toronto - Toronto Island, ON
YUB	Tuktoyaktuk, NWT
YUD	Umiujaq, QB
YUF	Pelly Bay, NWT
YUL	Montreal - Dorval, QC
YUS	Shepard Bay, NWT
YUT	Repulse Bay, NWT
YUW	Dewar Lakes, NWT
YUX	Hall Beach, NWT
YUY	Rouyn - Noranda, QB
YVB	Bonaventure, QB
YVC	La Ronge, SA
YVG	Vermilion, AB
YVK	Vernon, BC
YVM	Broughton Island, NWT
YVN	Cape Dyer, NWT
YVO	Val D'Or, QB
YVP	Kuujjuaq, QB
YVQ	Norman Wells, NWT
YVR	Vancouver, BC
YVT	Buffalo Narrows, SA
YVV	Wiarton, ON
YVZ	Deer Lake, ON
YWA	Petawawa, ON
YWB	Kangiqsujuaq, QB
YWG	Winnipeg, MB
YWH	Victoria Harbour, BC
YWJ	Deline, NWT
YWK	Wabush, NF
YWL	Williams Lake, BC
YWM	Athabasca, AB
YWO	Lupin, NWT
YWP	Webequie, ON
YWV	Wainwright, AB
YWY	Wrigley, NWT
YXC	Cranbrook, BC
YXD	Edmonton - Municipal, AB
YXE	Saskatoon, SA
YXH	Medicine Hat, AB
YXJ	Ft. St. John, BC
YXK	Rimouski, QB
YXL	Sioux Lookout, ON
YXN	Whale Cove, NWT
YXP	Pangnirtung, NWT
YXQ	Beaver Creek, YT
YXR	Earlton, ON
YXS	Prince George, BC
YXT	Terrace, BC
YXU	London, ON
YXX	Abbotsford, BC
YXY	Whitehorse, YT
YXZ	Wawa, ON
YYB	North Bay, ON
YYC	Calgary - Int'l, AB
YYD	Smithers, BC
YYE	Ft. Nelson, BC
YYF	Penticton, BC
YYG	Charlottetown, PEI
YYH	Taloyoak, NWT
YYJ	Victoria, BC
YYL	Lynn Lake, MB
YYM	Cowley, AB
YYN	Swift Current, SA
YYO	Wynyard, SA
YYQ	Churchill, MB
YYR	Goose Bay, NF
YYT	St. Johns, NF
YYU	Kapuskasing, QB
YYW	Armstrong, ON
YYY	Mont Joli, QB
YYZ	Toronto - Pearson Int'l, ON
YZD	Toronto - Downsview, ON
YZE	Gore Bay, ON
YZF	Yellowknife, NWT
YZG	Salluit, QB
YZH	Slave Lake, AB
YZP	Sandspit, BC
YZR	Sarnia, ON
YZS	Coral Harbour, NWT
YZT	Port Hardy, BC
YZU	Whitecourt, AB
YZV	Sept-Iles, QB
YZW	Teslin, YT
YZX	Greenwood, NS
YZY	Mackenzie, BC
ZAC	York Landing, MB
ZAM	Salmon Arm, BC
ZBA	Burlington Airpark, ON
ZBB	Boundary Bay, BC
ZBD	Ilford, MB
ZBF	Bathurst, NB
ZEE	Kelsey, MB
ZEG	Edmonton Centre, AB
ZEL	Bella Bella, BC
ZEM	Eastmain River, QB
ZFA	Faro, YT
ZFD	Fond du Lac, SA
ZFG	Pukatawagan, MB
ZFM	Fort McPherson, NWT
ZFN	Tulita, NWT
ZGF	Grand Forks, BC
ZGI	Gods River, MB
ZGR	Little Grand Rapids, MB
ZGS	Gethsemani, QB
ZHP	High Prairie, AB
ZJG	Jenpeg, MB
ZJN	Swan River, MB
ZKE	Kaschechewan, ON
ZKG	Kegaska, QB
ZLQ	Thicket Portage, MB
ZLT	La Tabatiere, QB
ZMD	Muskrat Dam, ON
ZML	108 Mile Airport, BC
ZMN	Pikwitonei, MB
ZMT	Masset, BC
ZNA	Nanaimo - Harbour, BC
ZNG	Poplar River, MB
ZNL	Nelson, BC
ZPB	Sachigo Lake, ON
ZPC	Pincher Creek, AB
ZPO	Pinehouse Lake, SA
ZQM	Moncton Centre, NB
ZQS	Queen Charlotte Is, BC
ZQX	Gandar Centre, NF
ZRJ	Round Lake, ON
ZSJ	Sandly Lake, ON
ZSN	South Indian Lake, MB
ZST	Stewart, BC
ZSW	Prince Rupert - Cove, BC
ZTA	Bloodvein River, MB
ZTB	Tete a la Baleine, QB
ZTM	Shamattawa, MB
ZUC	Ignace, ON
ZUE	Cape Parry, NWT
ZUL	Montreal Centre, QB
ZUM	Churchill Falls, NF
ZVL	Edmonton - Villeneuve, AB
ZVR	Vancouver Centre, BC
ZWG	Winnipeg Centre, MB
ZWH	Lac Brochet, MB
ZWL	Wollaston Lake, SA
ZYZ	Toronto Centre, ON
FVP	St. Pierre, France
FVM	Miquelon, France

Fig. 5-14: Weather Station Codes.

5.19 Upper Level Winds and Temperatures Forecasts (FD)

The FD forecasts wind direction and speed for various altitudes above a selected airport. FD's are prepared twice daily. Remember the following:

a) A minus sign ("-") is for temperatures below 0°C.
b) Wind direction and speed in degrees true/kts.
c) 9900 means "light and variable" winds.
d) "Low" winds are 18,000 and below, "high" winds are above 18,000 feet.

EXAMPLE:

FDCN1 CWAO 221520
FCST BASED ON 221200 DATA VALID 221800 FOR USE 15-21.

	3000	6000	9000	12000
YVR	2710	2813+03	2921+00	3028-03
YYF	2713	2715+02	2921+00	2927-05
YYE	9900	2708-02	2817-07	2826-12

QUESTIONS BASED ON ABOVE FD:
1. What is the date and time of issue?
2. What period of time is the FD intended for use?
3. What will the winds be above YVR at 3,000 feet?
4. What will temperature be at YYF at 6,000 feet?
5. What will wind direction be at YVR at 4,500 feet?
6. What will wind speed be at 4,500 feet above YYF?
7. What will temperature be at YYF at 4,500 feet?
8. What will the winds be above YYE at 3,000 feet?
9. In a climb from 6,000 feet to 9000 feet above YVR, will the wind direction veer or back?
10. In descent from 12,000 feet to 9,000 feet above YVR, will the wind direction veer or back? Will the wind speed increase or decrease?

ANSWERS:
1. 22nd day of month at 15:20Z.
2. 1500Z to 2100Z (6 hours).
3. 270°T at 10 knots.
4. 2°C.
5. 275°T. 3,000 ft wind is 270°, 6,000 ft wind is 280°. Interpolate 275° for 4,500 ft.
6. 14 knots. Interpolate.
7. +5°C. Interpolate.
8. Light and variable.
9. Veer. 280° T at 6,000 feet to 290° T at 9,000 feet = clockwise change.
10. Back (from 300° T to 290° T). Wind speed will decrease, from 28 knots to 21 knots.

5.20 Aviation Routine Weather Report (METAR)

The international METAR format is now used worldwide. Countries may make minor changes to the METAR format, for example, in North America, heights are in feet, not metres, and in the US, temperatures are in °F, not °C. The order and information presented in the METAR is as follows:

a) Message type. METAR is a routine observation, SPECI is for special METAR where there is a significant change off the hour.
b) Station identifier. Indicated by 4 letter ICAO identifier and time of observation in UTC (Z).
c) Wind. The first three digits are wind direction to the nearest 10° true, next two digits is speed. G indicates gusts. VRB is for variable direction less than 3 knots. Calm is 00000KT.
d) Prevailing visibility in SM and fractions.
e) Runway Visual Range (RVR). 10 minute average RVR given when visibility is 1 SM or less, and/or the RVR is less than 6000 ft. D means downward trend, U means upward, N is no change.
f) Present weather. Precipitation or obscuration with intensity and/or descriptor. Light intensity: "-", heavy "+", no sign = moderate. Descriptors: SH = showers, DR = drifting, FZ = freezing, MI = shallow, BL = blowing, BC = batches.
g) Weather codes:

FC =	Tornado/Water-spout/Funnel cloud	RA =	Rain
		SG =	Snow Grains
TS =	Thunderstorm	PE =	Ice Pellets
DZ =	Drizzle	IC =	Ice crystals
FG =	Fog (< 5/8 SM)	SA =	Sand
BR =	Mist (> 5/8 SM)	SN =	Snow
GS =	Small Hail	HZ =	Haze
FU =	Smoke	GR =	Hail
SS =	Sandstorm	DU =	Dust
VA =	Volcanic Ash	SQ =	Squall
PO =	Dust/Sand whirls	DS =	Duststorm

h) Sky condition. Layer amounts are cumulative and include the sum of any layers below. Cover is in eighths or "oktas". SKC = sky clear, no cloud present. FEW = >0 - 2 oktas. SCT = scattered, 3 - 4 oktas. BKN = broken, 5 - <8 oktas. OVC = overcast, 8 oktas. A CLOUD CEILING exists at the height of the first layer fwhich is BKN or OVC. VV = vertical visibility in hundreds of feet where there is an OBSCURED CEILING. Cloud height is in 3 digits in hundreds of feet.
i) Temperature and dewpoint. M means below zero.
j) Altimeter setting. "A" indicates inches of mercury, Q indicates hectopascals.
k) Recent weather (RE) since the last routine report but not at the time of observation.
l) Wind shear. Included where reported within

the designated runway.

m) Remarks, RMK. Cloud layer types and opacity in eighths of sky concealed, and sea level pressure (SLP) in 3 digit code: to convert to hectopascals, place 9 or 10 before code and decimal before last digit, whichever is closest to 1,000 hPa.

n) METAR AUTO or SPECI AUTO indicate weather observations taken from an automated station. AUTO reports should only be used by pilots if specifically certified for aviation use. Pilots should also use caution when dealing with ATUO reports because key information can be missing (such as altimeter settings). METAR AUTO reports are made on the hour, and SPECI AUTO observations are made where there have been significant changes. Some AUTO stations have an automated voice generator to give reported weather information. If a particular part of an AUTO sensor is malfunctioning, that part of the data will be missing from the report. If there is a discrepency between a human observation by ATC/FSS or weather station, the human information prevails. As well, human observers may add remarks to AWOS reports.

EXAMPLE:

CYVR METAR 040500Z 01002KT 30SM CT250 03/01 A3010 RMK SLP200
CYWG METAR 040500Z 36009KT 15SM -SN SCT015 OVC044 M11/M12 A3024 RMK SC2 SC2 SLP240
CYYZ METAR 040500Z AUTO CCA 09005KT 5SM BR OVC006 04/03 A2966 RMK SC5 SLP150

QUESTIONS BASED ON ABOVE METARS:

1. What time and date applies to the CYVR METAR?
2. What are the reported surface winds at CYVR?
3. What is the 0500Z reported visibility at CYWG?
4. What weather is reported at CYWG?
5. What is the reported ceiling at CYWG?
6. What is the 0500Z CYYZ temperature/dewpoint?
7. What is the CYWG temperature and dewpoint?
8. What is the reported weather at CYYZ?
9. Which report is taken from an auto station?
10. What is the reported altimeter setting at CYVR?

ANSWERS:

1. 0500Z on the 4th day of the month.
2. 010° T at 2 kts.
3. 15 SM.
4. Light snow.
5. Ceiling overcast, 4,400 feet AGL.
6. Temperature +4° C, dewpoint +3° C.
7. Temperature -11° C, dewpoint -12° C.
8. Mist.
9. CYYZ.
10. 30.10 inches hg.

5.21 SIGMETS

A SIGMET is a serious weather condition reported as a warning to aircraft in flight. Examples are active thunderstorms, squall lines, heavy hail, serious turbulence or icing, mountain waves, hurricanes, widespread volcanic ash, wind shear etc.

EXAMPLE:

WSCN CYYZ 171800
171800 - 172200
SIGMET A5. LN OF TSTMS OBSERVED ON WA RADAR NORTH BAY - MUSKOKA - LONDON TOPS TO 30 THSD MOVG E 20.

MEANING OF EXAMPLE:
SIGMET ALFA 5 from Toronto/Pearson airport, issued 17th day of month at 1800Z for period 1800Z to 2200Z, line of thunderstorms observed on weather radar from North Bay to Muskoka to London with maximum tops to 30,000 feet moving easterly at 20 kts.

5.22 PIREPS

PIREP means "pilot report". Pilots are encouraged to contact FSS and report significant observations made in flight. The codes used are as follows:

UA	PIREP
UUA	Urgent PIREP
OV	Over (3 letter navaid with radial and distance)
TM	Time
FL	Altitude or flight level in hundreds of feet
TP	Aircraft type
SK	Sky condition, cloud layers
WX	Weather and visibility
TA	Temperature, °C
TB	Turbulence
IC	Icing
RM	Remarks

EXAMPLE:

UA/OV VQC 045043 1342 FL090/TP BE80/SK OVC 075/TA - 18/IC LGT RIME BLO - 075 RM WIND COMP HEAD 010 MH 065 TAS 210

MEANING OF EXAMPLE:
Location - Stirling VORTAC 045° Radial, 43 NM at 1342Z at 9,000 ft, type of aircraft - Beech Queen Air; Sky Condition - overcast, tops 7,500 ft; air temperature -18°C; light rime icing below 7,500 ft; Remarks - headwind component 10 kts, magnetic heading 065, true airspeed 210 kts.

5.23 Surface Weather Charts

Surface weather charts or maps are made from reports of surface weather from various reporting stations. With various symbols, lower level weather conditions such as pressure distribution, air masses, fronts, precipitation and obstructions to vision can be seen. Some upper features such as TROWALS and upper fronts are shown, thereby giving an excellent representation for flight planning purposes. Surface charts based on actual data from existing reports are referred are SURFACE ANALYSIS CHARTS. SURFACE PROGNOSTIC CHARTS show future expected pressure patterns and frontal positions for a specific time in the future.

Prognostic Surface Charts

Prognostic surface charts are issued 48 and 36 hours before valid times 00Z and 12Z. Remember that the weather may be considerably different as flight time will usually not coincide with chart valid time. Review the various symbols used at Fig. 5-15.

EXERCISE: Convert the following surface weather chart symbols to plain language.

1.  (green) ____________
2.  (green) ____________
3.  (yellow) ____________
4. 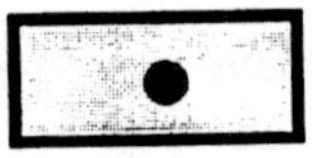 (green) ____________
5.  (green) ____________

ANSWERS:
1. Freezing drizzle.
2. Mixed intermittent rain and snow.
3. Fog.
4. Continuous rain.
5. Intermittent hail showers.

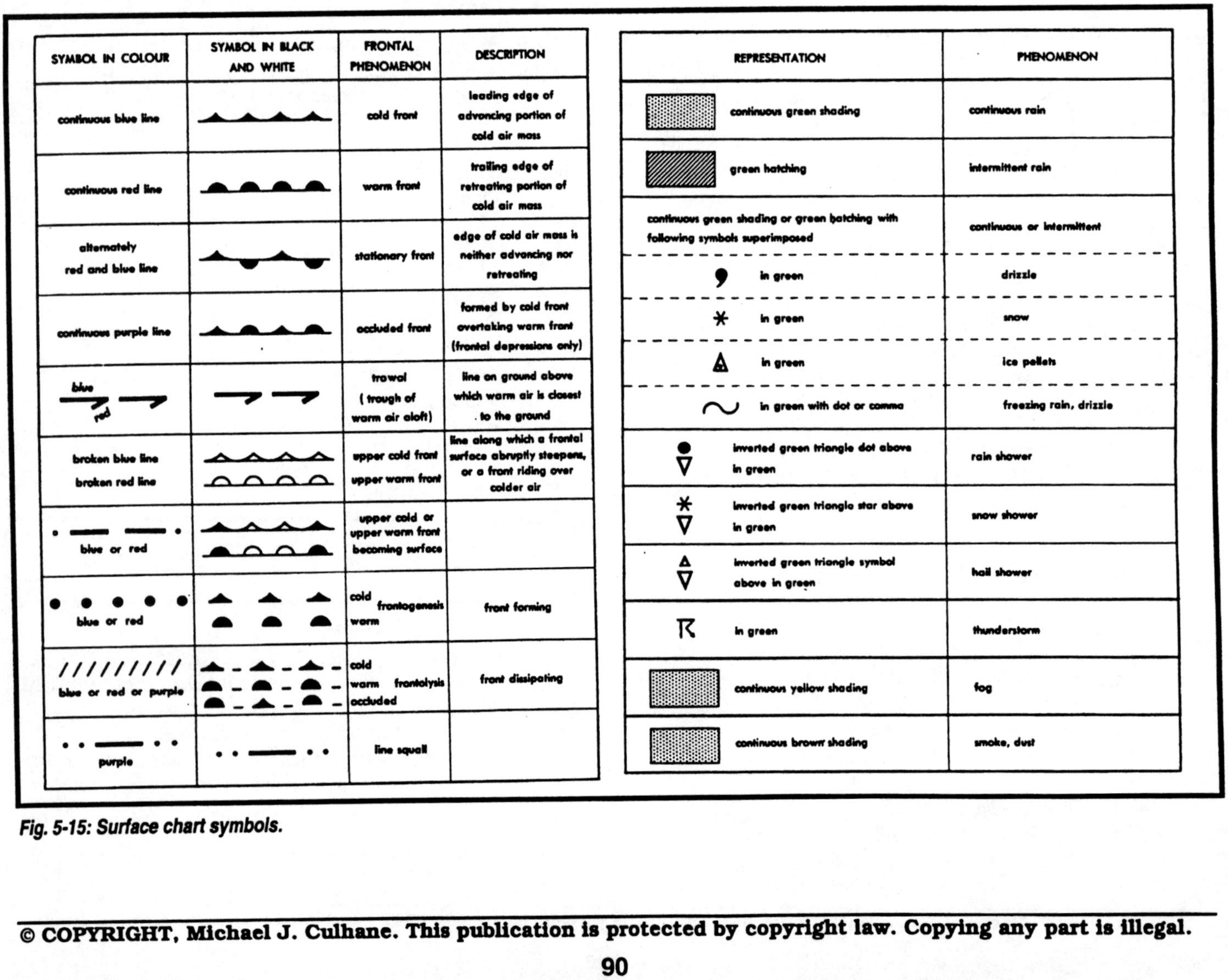

Fig. 5-15: Surface chart symbols.

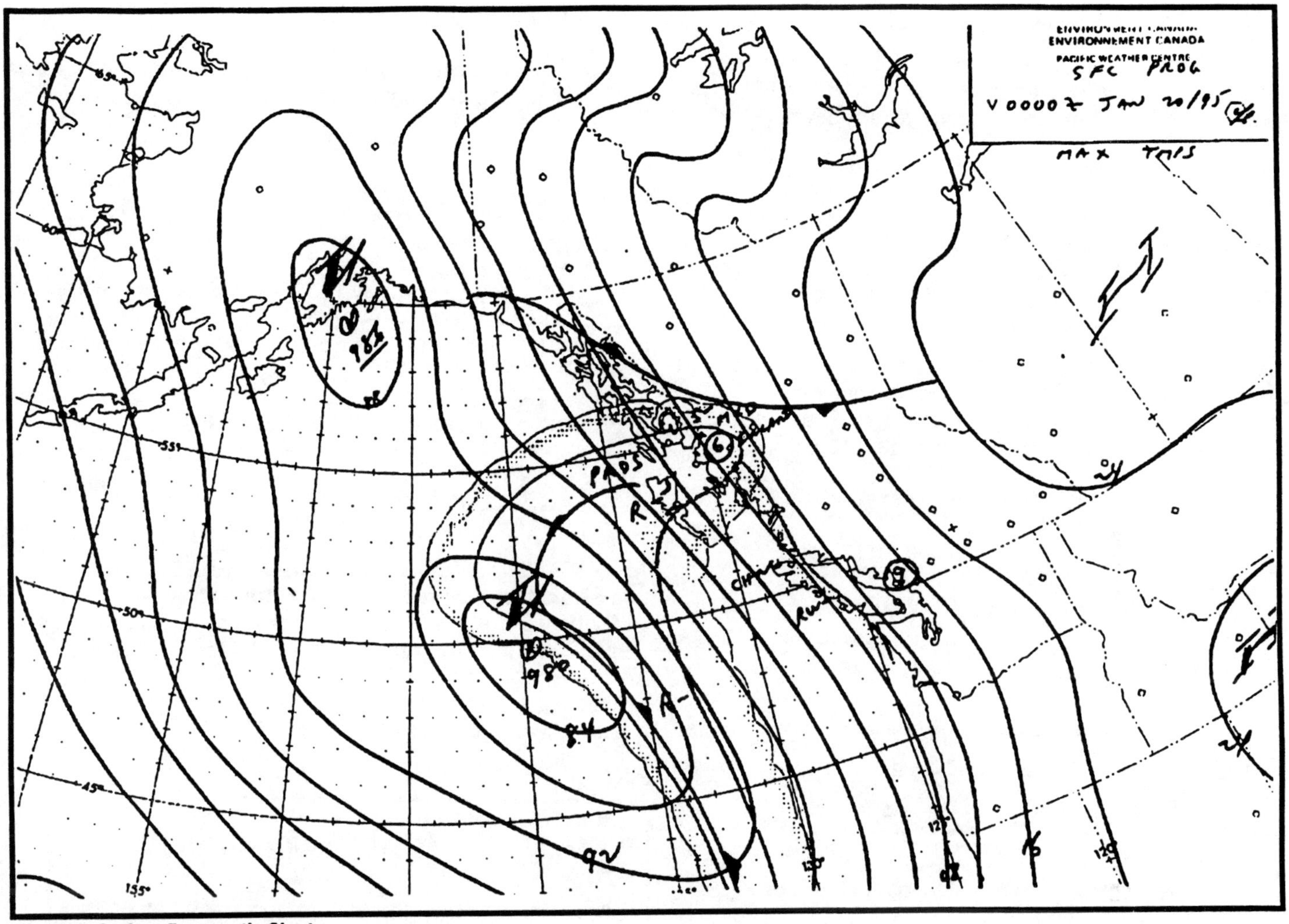

Fig. 5-16: Surface Prognostic Chart.

EXERCISE: Answer the following questions based on the surface prognostic chart above.

1. Is the chart above giving actual weather or predicted weather?
2. What time is the chart effective?
3. Assume the scalloped area off the coast is shaded green. What type of weather is anticipated in the shaded area?
4. What is the barometric pressure at the centre of the low off the BC coast?
5. What kind of frontal system is forecast to lie over mid to northern BC at 00Z?
6. Would you expect greater winds over the BC coast or over Saskatchewan, and why?
7. What is the general 3,000 ft wind direction forecast over southern Alberta at the US border and why?
8. What kind of frontal system lies over the Pacific ocean?

ANSWERS:

1. Predicted weather, because it is named "prognostic".
2. Chart is an estimate of conditions anticipated at 00Z, January 20, 1995.
3. Periods of continuous light rain and light mixed snow inland. Green shading areas indicate continuous precipitation.
4. 998.0 mb. Add 9 or 10, whichever is closer to 1000 mb.
5. Stationary front.
6. Faster winds over BC coast due to closer spacing of isobars.
7. Easterly. Winds circulate parallel to isobars, winds circulate clockwise about High located in middle of Sask.
8. Stationary.

Surface Analysis Charts

Surface analysis charts are considered to represent weather from the surface up to 3,000 feet, and weather visible from the surface at any level.

Since surface analysis charts are made from data at a certain time in the past, the information will already be a few hours old when it first comes out. They are based on observation times of 00Z, 06Z, 12Z and 18Z, and each chart is issued 2 - 3 hours after observations, therefore, the weather may be very different at the time of departure.

The STATION MODEL is used to enter data from the hourly reports onto the surface weather chart. The station model contains information such as winds, temperatures, cloud cover, humidity etc.

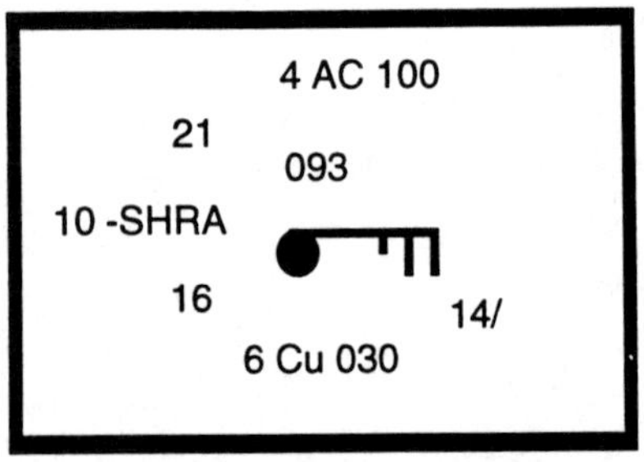

Fig. 5-17: Station model used in analysis charts.

The above station model shows the following information:

1. Total sky cover: overcast (by darkened circle).
2. Temperature: 21°C
3. Dewpoint: 16°C.
4. Wind: from 090° T at 25 knots. Shaft of pennant is direction toward station, each barb is 10 kts, each half barb is 5 knots. A second circle around the station means calm.
5. Present visibility and weather is 10 SM in light rain showers.
6. Middle cloud is 4/8 Altocumulus at 10,000 ft.
7. Sea level pressure 1009.3 mb. Use same rules to convert as in METAR.
8. Pressure change in past 3 hours: increased steadily 1.4 mb.
9. Low cloud: 6/8 Cumulus at 3,000 ft.

NOTE: Station information detail in surface analysis charts uses a complex system of numerical codes for cloud, winds and weather. For pilots, virtually all of the relevant station information detail would normally be obtained from a TAF and/or METAR. Station model detail is mainly for forecasters.

Refer to the surface analysis chart at Fig. 5-18 and answer the questions that follow.

5.24 Meaning of CAVOK

CAVOK means "ceiling and visibility OK" and refers to the following simultaneous conditions:

a) No cloud below 5,000 feet.
b) No cumulonimbus clouds.
c) Visibility at least 6 SM or more.
d) No precipitation, thunderstorms, shallow fog or low drifting snow.

CAVOK is used frequently in ATIS broadcasts or when indicating generally good VFR weather prevails. CAVOK provides no indication of prevailing winds.

CAVOK is not used in TAFs or METARs.

5.25 AIRMET

An AIRMET is a "Pilot's Meteorological Advisory" and is a short term weather advisory intended for aircraft in flight, to notify pilots of potentially hazardous weather conditions not described in the current FA and not requiring a SIGMET. Examples are freezing precipitation, moderate icing, moderate turbulence, surface winds over 20 knots etc.

EXAMPLE:

WACN1 CWHX 200700
AIRMET A1 ISSUED AT 0700Z AT CWHZ-AMEND FACN1 CWHX 200600 ISSUE

DC9 RPRTD MDT RIME ICG IN FZDZ ALG THE S CST OF NB AT 07Z. FZDZ EXPCD TO CONT UNCHGED TO 14Z.

MEANING OF EXAMPLE:

AIRMET from Maritime Weather Centre, time 0700 on 20th day of month. AIRMET A1 issued at 0700 Z at Maritime Weather Centre which amends the Area Forecast FACN1 issued at 0600Z.

A DC 9 aircraft reported moderate rime icing in freezing drizzle along the south coast of New Brunswick at 0700Z. The freezing drizzle is expected to continue unchanged until 1400Z.

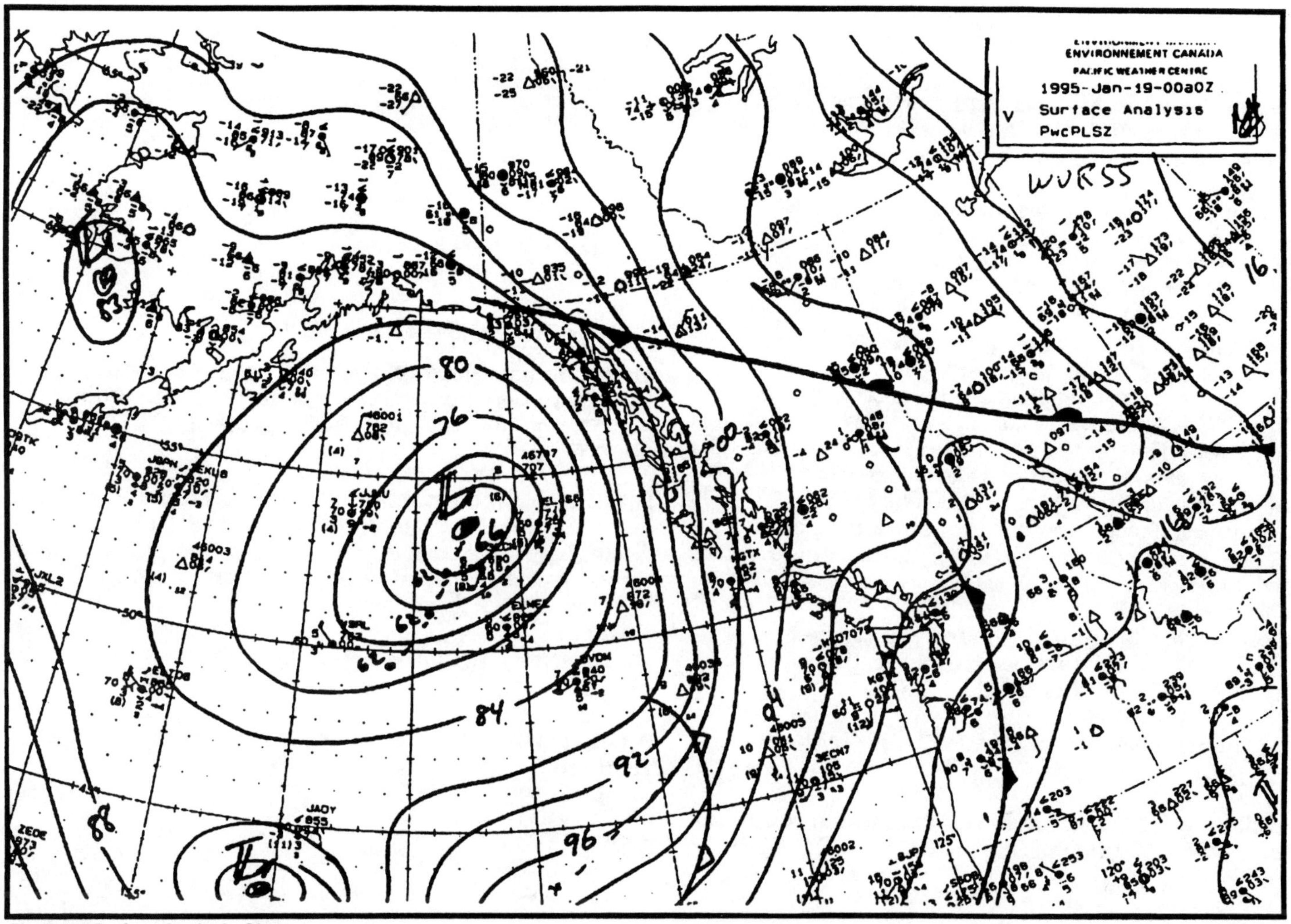

Fig. 5-18: Surface Analysis Chart.

EXERCISE: Based on the above surface analysis chart, answer the following questions.

1. Is the above chart based on forecast weather or actual weather?
2. What is the pressure at the centre of Low in the Pacific at about latitude 54°?
3. With reference to the low in question 2, in the last 24 hours, has it been increasing or decreasing in pressure, and what has its movement been?
4. Wind will be more or less parallel to the isobars shown on this chart at what altitude?
5. What time were the observations made upon which this chart is based?
6. What kind of a frontal systems lie over BC at 00Z?
7. Where will winds be faster, over west coast of BC or northern Alberta and why?
8. What are the approximate winds at 3,000 ft over the north coast of BC?

ANSWERS:

1. Actual weather - because it is called an analysis chart.
2. 966 mb (add 9 or 10 to 66, whichever is closer to 1,000 mb).
3. Increasing. Note the history of movement of the system is indicated by the 62 - 62 - 62 dots and trail to the lower left of the low, showing the pressure and movement. This cell is increasing in pressure and moving northeast. Each position represents the last 6 hours.
4. 3,000 ft.
5. 00Z on Jan 19, 1995.
6. Stationary front in northern BC, TROWAL in north south line and cold front at southern part at US border.
7. Winds will be faster over north coast of BC because isobars are spaced closer together.
8. Southerly - see isobars, lows circulate counterclockwise.

5.26 Upper Air Charts

Upper weather charts are either PROGNOSTIC (describing future, forecast weather) or ANALYSIS (describing existing conditions at a given time).

Upper charts are normally referenced to a key pressure level (in millibars or hectopascals) or flight level. Remember that in the same way that surface level charts are based on the location of MSL pressure or 1013.2 mb, the upper level charts will be based on the same principle, except that the pressure level is no longer sea level, but higher up. On upper charts, isobars are named CONTOUR LEVELS. Consider the contour lines on upper charts to provide the same information as isobars: where they are closely spaced, winds will be high, circulation about a high will be clockwise, counterclockwise about a low. Since temperatures get colder as you go further north, the contour levels will be lower toward the north.

Contour levels on upper charts are drawn in DECAMETRES, which are 10's of metres. For example, 300 decametres is 3,000 metres, and to convert to feet, simply multiply by 3.28, so 300 decametres will approximate 10,000 feet (9,840 feet).

Broken red lines are ISOTHERMS, which connect points of equal temperatures.

Upper Air Analysis Charts

Upper air analysis charts are produced from information coming from weather stations scattered around the continent. This data is used to produce the upper air analysis charts at 00Z and 12Z each day. There are three charts to 18,000 feet:

a) 850 MB ANALYSIS CHART (see Fig. 5-19). At about 5,000 feet. Includes contours at 60 meter intervals (solid lines) starting from the 1500 metre level, with isotherms at 5° C spacing (dashed lines) and low and high heights. Usually the fronts are shown in surface position.

b) 700 MB ANALYSIS CHART (see Fig. 5-20). At about 10,000 feet. Same information as 850 mb chart, but fronts not indicated. Contours spaced at 60 meter intervals.

c) 500 MB ANALYSIS CHART is used for flight planning near 18,000 feet (not required knowledge for PPL/RPP).

Wind speed is also indicated on upper charts by a system of pennants and feathers. The pennant is aligned with the feathers in the direction from which the wind is blowing. Each feather represents 10 kts, a half feather represents 5 kts, and a small darkened triangle represent 50 kts.

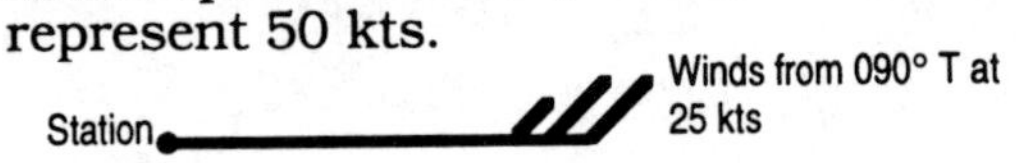

Complete the exercises following the 850 mb and 700 mb charts at Figs 5-19 and 5-20.

Upper Level Prognostic charts

The Canadian Meteorological Centre (CMC) provides significant weather prognostic charts for FL 100 to FL 250 and are based on the pressure levels 700 mb - 400 mb.

Upper level prog charts are useful to determine areas of forecast moderate to severe icing, moderate to severe turbulence, significant cloud layers, mountain waves, freezing level at 10,000 feet, and surface position of highs and lows.

Generally, these charts are used for higher, faster aircraft under IFR.

A detailed knowledge of upper level prog charts is not required for the PPL/RPP.

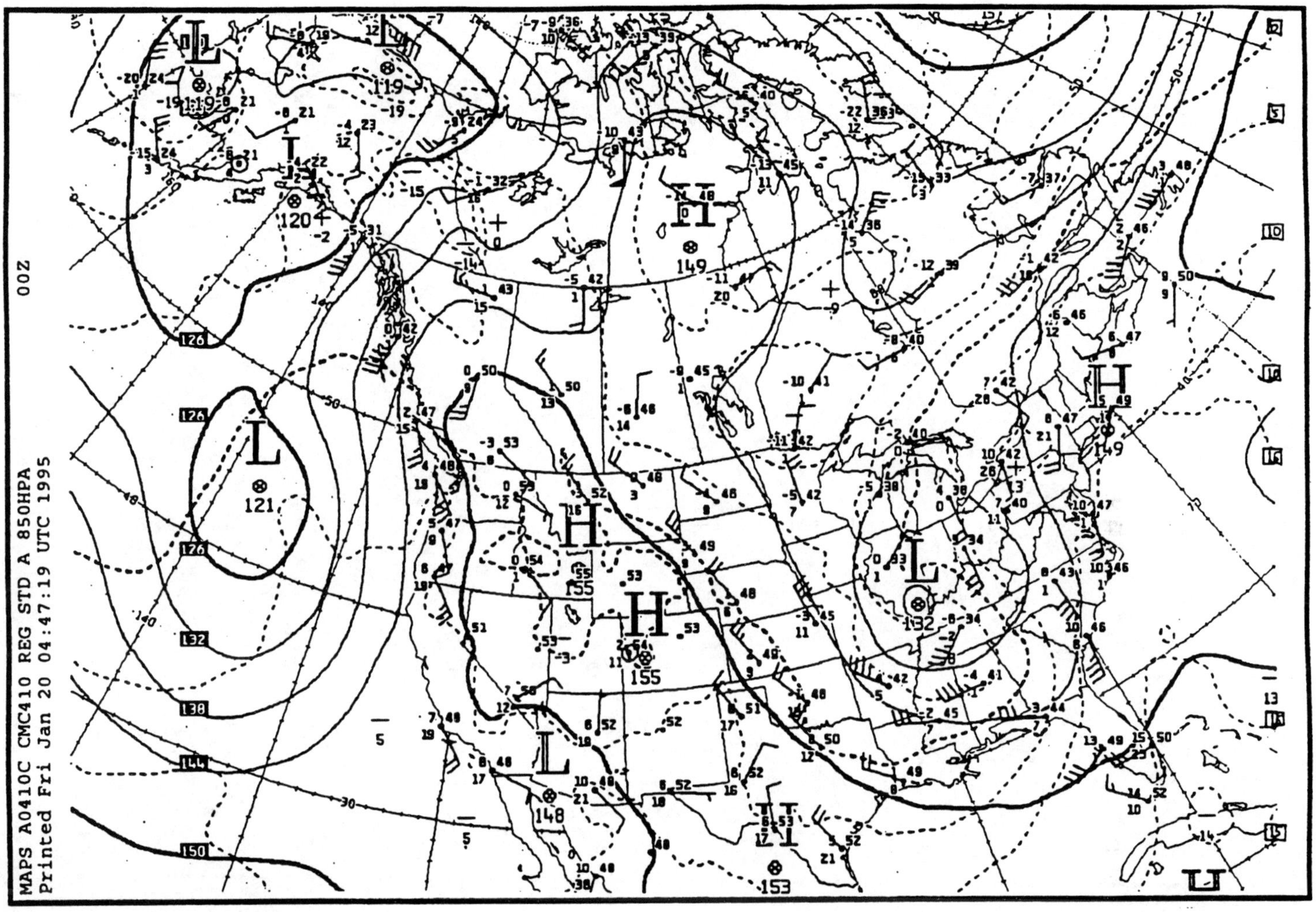

Fig. 5-19: 850 mb Analysis Chart.

EXERCISE BASED ON 850 MB ANALYSIS ABOVE:

1. What kind of pressure systems is centered to the south of the great lakes?
2. What does the number 121 indicate beneath the low pressure cell in the Pacific Ocean?
3. Why is the contour line from Florida through Alberta and back to Southern California darker than the other contour lines?
4. What is the height of the 850 mb contour level over Lake Winnipeg?
5. What is the temperature of the 850 mb pressure level over Lake Winnipeg?
6. What is the wind direction and wind speed at the 850 mb pressure level in central Saskatchewan?
7. What is the temperature and dewpoint of the 850 mb pressure level over central Saskatchewan?
8. What is the height of the 850 mb pressure level over central Saskatchewan?

ANSWERS:

1. Low pressure.
2. Height of 850 mb pressure level at centre of low = 1210 metres or 121 decametres.
3. The darkened contour line represents the 1500 metre level of the 850 mb contour and is where the contour lines start on this chart. See the "150" in the box on the western edge of this contour.
4. 1440 metres.
5. -10°C. See isotherms - dashed lines. 0°C is darker than the rest, then other isotherms are marked at 5 degree intervals.
6. Winds are from the north at 10 knots - see vertical pennant with one barb.
7. Temperature -6°C, dewpoint -20°C. See the figures -6 and 14: -6 is temperature in °C, 14 is the dewpoint depression, or how many degrees colder the dewpoint is than the temperature.
8. 1,460 metres. See the 46 figure.

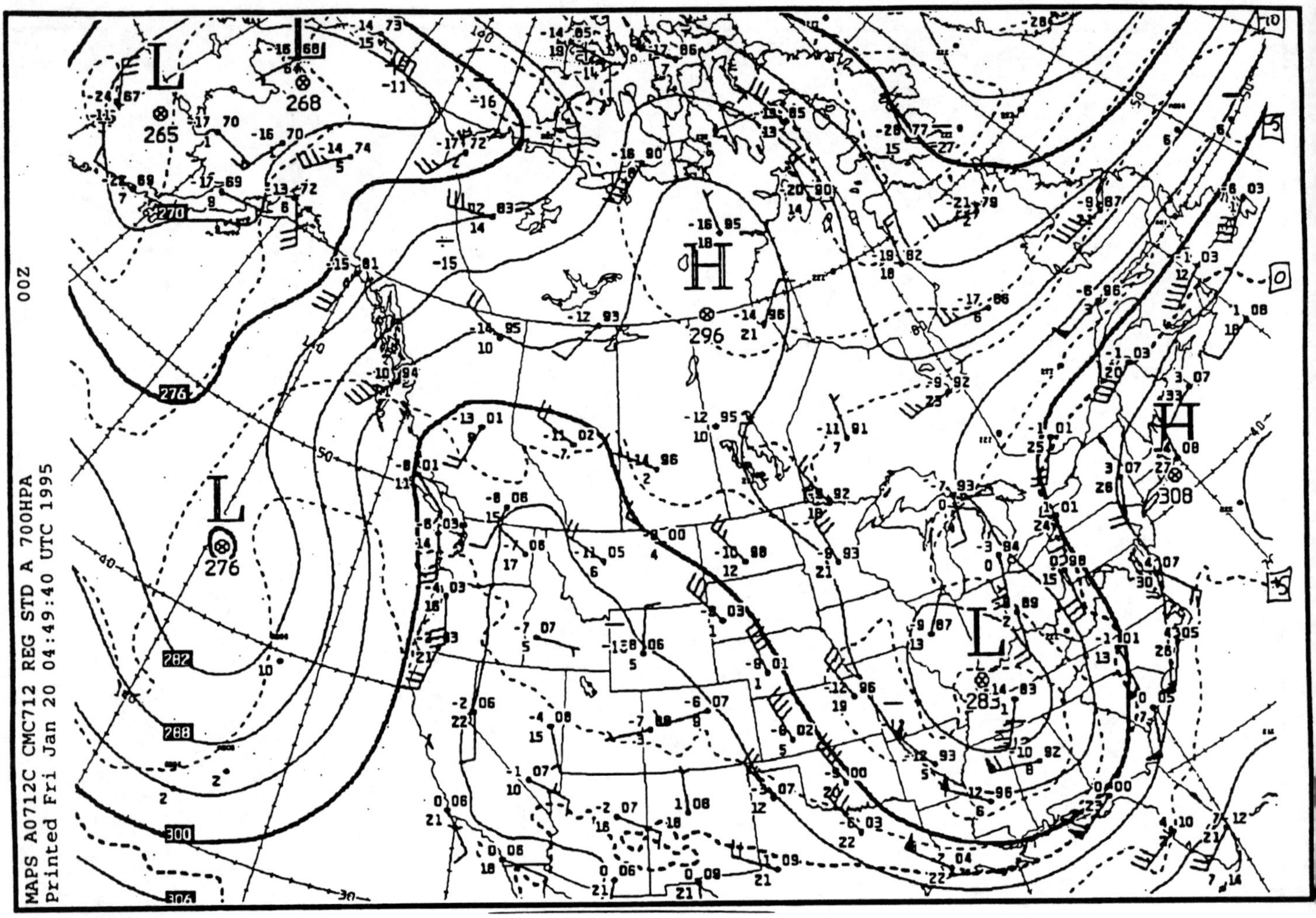

Fig. 5-20: 700 mb Analysis Chart.

EXERCISE BASED ON 700 MB ANALYSIS ABOVE:

1. When is the information presented on this chart based on, and is it based on actual or forecast weather?
2. What is the height of the 700 mb pressure level in feet for central Newfoundland?
3. What is the wind direction and speed at the 700 mb level in eastern Newfoundland?
4. What is the temperature and dewpoint at the 700 mb level for eastern Newfoundland?
5. What is the height of the 700 mb pressure level at eastern Newfoundland?
6. The 700 MB analysis chart shows the winds for what level in the atmosphere in terms of feet ASL?
7. What are the approximate 700 mb pressure level winds over the Gaspe?

ANSWERS:

1. 00Z (see left edge of chart). It is based on actual reported weather.
2. 9,840 ft. 3,000 metres is indicated by 300 symbol, multiply by 3.28 ft per metre.
3. Westerly winds of 40 knots. See pennant aligned from west in degrees true and 4 feathers for speed.
4. Temperature -6°C, dewpoint -7°C. See number -6 (temperature) and 1 (dewpoint depression).
5. 3030 Metres. See 03 just off 300 contour.
6. Approximately 10,000 feet ASL.
7. About 270°T at 45 kts, based on 50 kts from station to north and 40 knots from station to south and interpolating between the two.

NOTES

NOTES

6 Navigation

6.01 Definitions and Basic Concepts

A MERIDIAN is a line joining the north pole to the south pole. The PRIME MERIDIAN runs through Greenwich, England. LONGITUDE is the angular measurement of the number of meridians east or west of the prime meridian. The earth is divided into 360 meridians: 180° west and 180° east of the prime meridian. 180° opposite the prime meridian is the INTERNATIONAL DATE LINE. LATITUDE is measured north and south of the equator, 0 through 90°. The north pole, for example, is 90°N latitude.

Fig. 6-1: Latitude, Longitude.

A GREAT CIRCLE is a circle whose plane passes through the center of the earth. The equator is a great circle. A great circle is the shortest distance between two points on the globe. A RHUMB LINE i a curved line that crosses all meridians at a constant angle. The advantage to flying a rhumb line is that a constant heading can be maintained.

VARIATION is the angular displacement between true north and magnetic north. Variation is either east or west. An ISOGONAL LINE joins all points having equal variation. An AGONIC LINE joins all points having zero magnetic variation. To convert from degrees true to degrees magnetic, add variation west or subtract variation east. A useful phrase to remember is "FROM TRUE TO MAGNETIC, WEST IS BEST AND EAST IS LEAST".

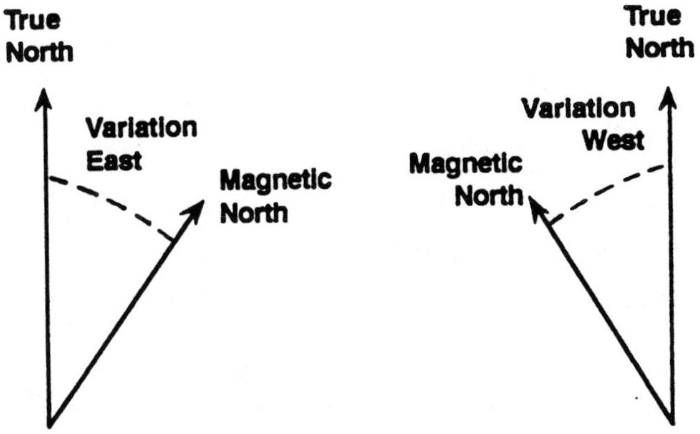

Fig. 6-2: Magnetic Variation is either East or West (unless at Agonic Line).

HEADING is the angular measurement of the longitudinal axis of the aircraft while in flight. TRUE HEADING is measured from true north, and MAGNETIC HEADING is measured from magnetic north. TRACK is the angular measurement of a line or path over the ground (the United States' term for track is COURSE). TRUE TRACK is measured from true north, and MAGNETIC TRACK is measured from magnetic north.

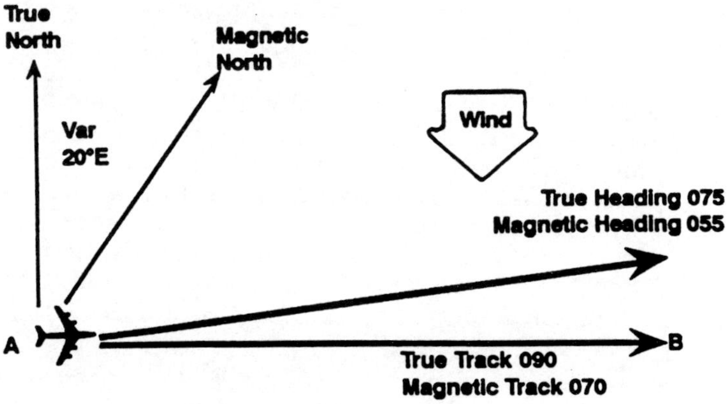

Fig. 6-3: True Heading, True Track, Magnetic Heading, Magnetic Track.

EXERCISE: Complete the following table.

	True	Variation	Magnetic
1.	090	2°W	___
2.	180	10°E	___
3.	___	22°W	055
4.	___	12°E	356
5.	185	___	180
6.	290	___	298

ANSWERS:
1. 092
2. 170
3. 033
4. 008
5. 5°E
6. 8°W

AIRSPEED is the speed of the aircraft moving through the air, either indicated (IAS) or true (TAS). GROUNDSPEED is the speed of the aircraft over the ground. A TRIANGLE OF VELOCITIES consists of the WIND vector (wind direction, wind speed), the AIR vector (true airspeed, true heading), and the GROUND vector (groundspeed and true track). DRIFT is the angular displacement of the aircraft from its heading to its track (downwind) caused by wind. WIND CORRECTION ANGLE is the correction in aircraft heading needed to counteract drift.

TRACK MADE GOOD is the actual path flown over the ground. If winds are from behind, the groundspeed is greater than the true airspeed, and vice versa if flying into a headwind. A flight computer is used to calculate true heading (where to steer to eliminate drift) and groundspeed (actual speed over ground). To use a flight computer, enter TAS, true track (measure off map with protractor), wind speed and direction (from weather information). Always ask yourself "do the numbers make sense". EXAMPLE: If the desired track is 090°T, and the winds are from 090°T, will the groundspeed be greater, less or equal to the true airspeed?

ANSWER: Groundspeed will be less than true airspeed.

6.02 VFR Charts

Because the earth is a sphere, and maps are flat, there are various inaccuracies with maps and charts. Map construction is based on a PROJECTION. Most aeronautical charts are LAMBERT CONFORMAL CONIC projections, which means that a straight line drawn between two points on the map will be a great circle, and longitude or meridians will converge towards north. Another type of chart is based on a MERCATOR projection, where longitude lines are vertical and latitude is parallel. A straight line on mercator chart is a rhumb line.

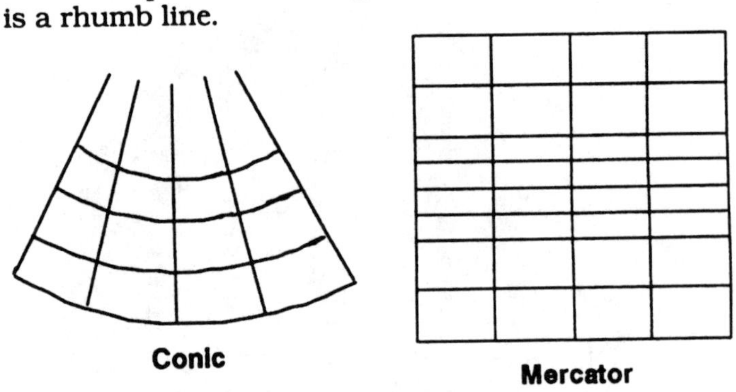

Fig. 6-5: Conic and Mercator Projections.

VFR Navigation Charts (VNC) use a conic projection and have a scale of 1:500,000. 1:250,000 VTA charts are used for flight within the airspace surrounding the more active airports such as Vancouver and Toronto. World Aeronautical Charts (WAC) are based on a conic projection and have a scale of 1:1,000,000. Canadian Pilotage Charts (CPC) are based on a TRANSVERSE MERCATOR PROJECTION and have a scale of 1:500,000.

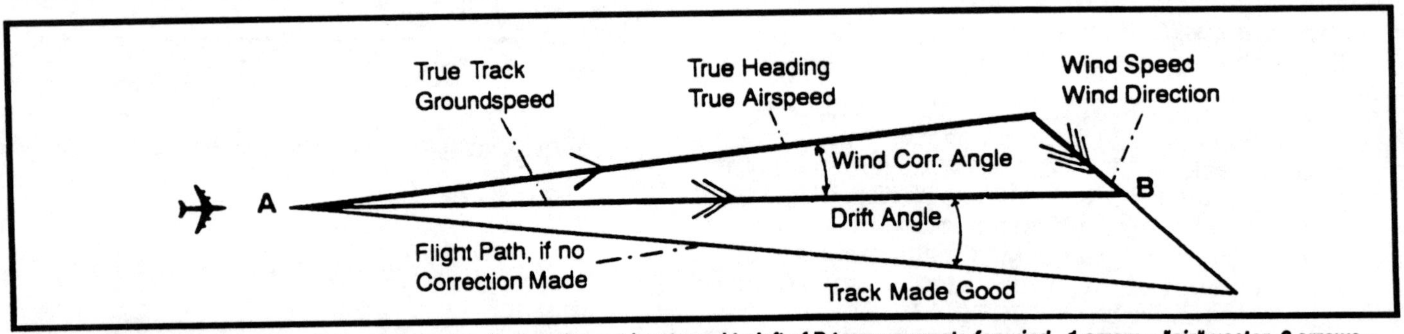

Fig. 6-4: Triangle of Velocities. To track from A to B, aircraft must be steered to left of B to compensate for wind. 1 arrow = "air" vector, 2 arrows = "ground" vector, 3 arrows = "wind" vector.

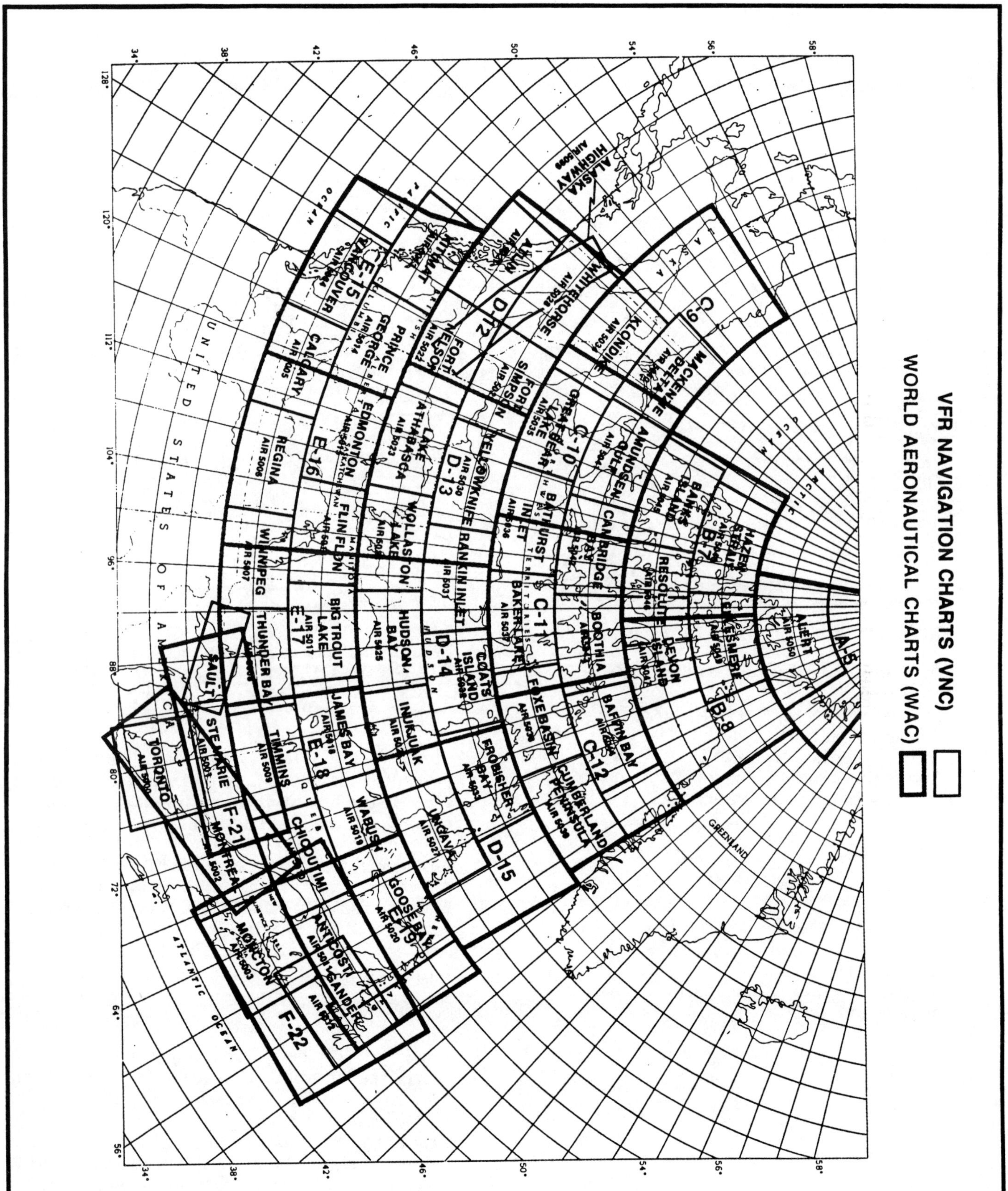

Fig. 6-6: VFR Charts in Canada.

6.03 VFR Navigation Charts (VNC)

Fig. 6-7 on the next page has a complete VNC chart legend. Review all symbols used on the legend, then complete the following exercises.

Cranbrook, B.C.

EXERCISE: Cranbrook, B.C. VNC excerpt.

1. How many hard surface runways are at Cranbrook?
2. Are Customs facilities available?
3. What is the airport elevation?
4. Is airport lighting available?
5. What is the length of the longest runway?
6. Is there a heliport at Cranbrook?

ANSWERS:
1. One.
2. Yes.
3. 3082 feet ASL.
4. Lighting available 24 hrs.
5. 6000 feet.
6. There are 2 heliports to the south of the airport. One is a hospital heliport, and the other is a conventional heliport.

Penticton, B.C.

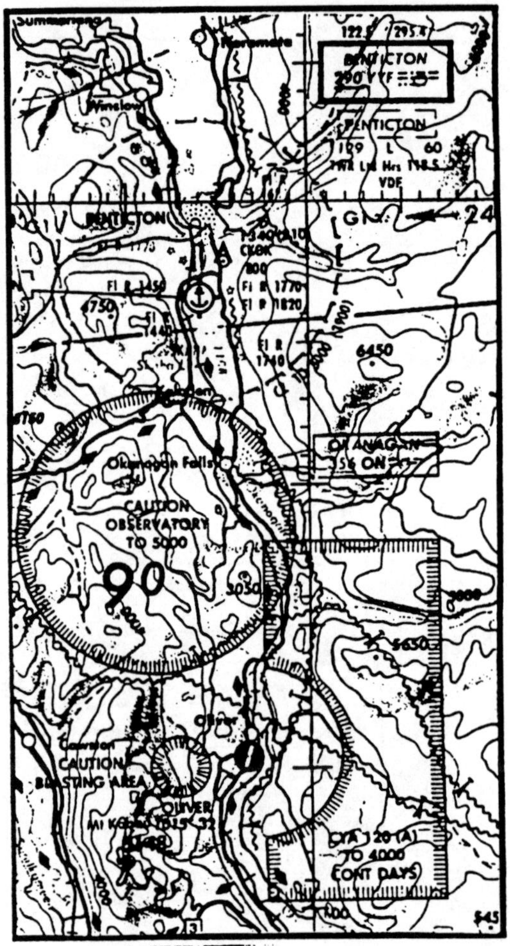

EXERCISE: Penticton, B.C.

1. Is Penticton a Class C or D control zone?
2. What is the height of the radio station tower CKOK? What is its height AGL?
3. What is CYA 120?
4. What height is the nearest MEF?
5. Are there any transmission lines in the area?
6. Is the terrain generally flat or steep to the east and west of the airport?
7. Are there any suggested VFR routes?
8. What is the highest terrain within the area?

ANSWERS:
1. Class C.
2. Tower is 1340 feet ASL, 210 feet AGL.
3. Alert Area (acrobatics) to 4,000 feet ASL, having continuous operations during days.
4. 9,000 feet.
5. Yes, to the SE and NE of the airport.
6. Contours show steep, rising terrain to E and W.
7. Suggested VFR routes (diamonds) along valleys.
8. 6450 feet ASL.

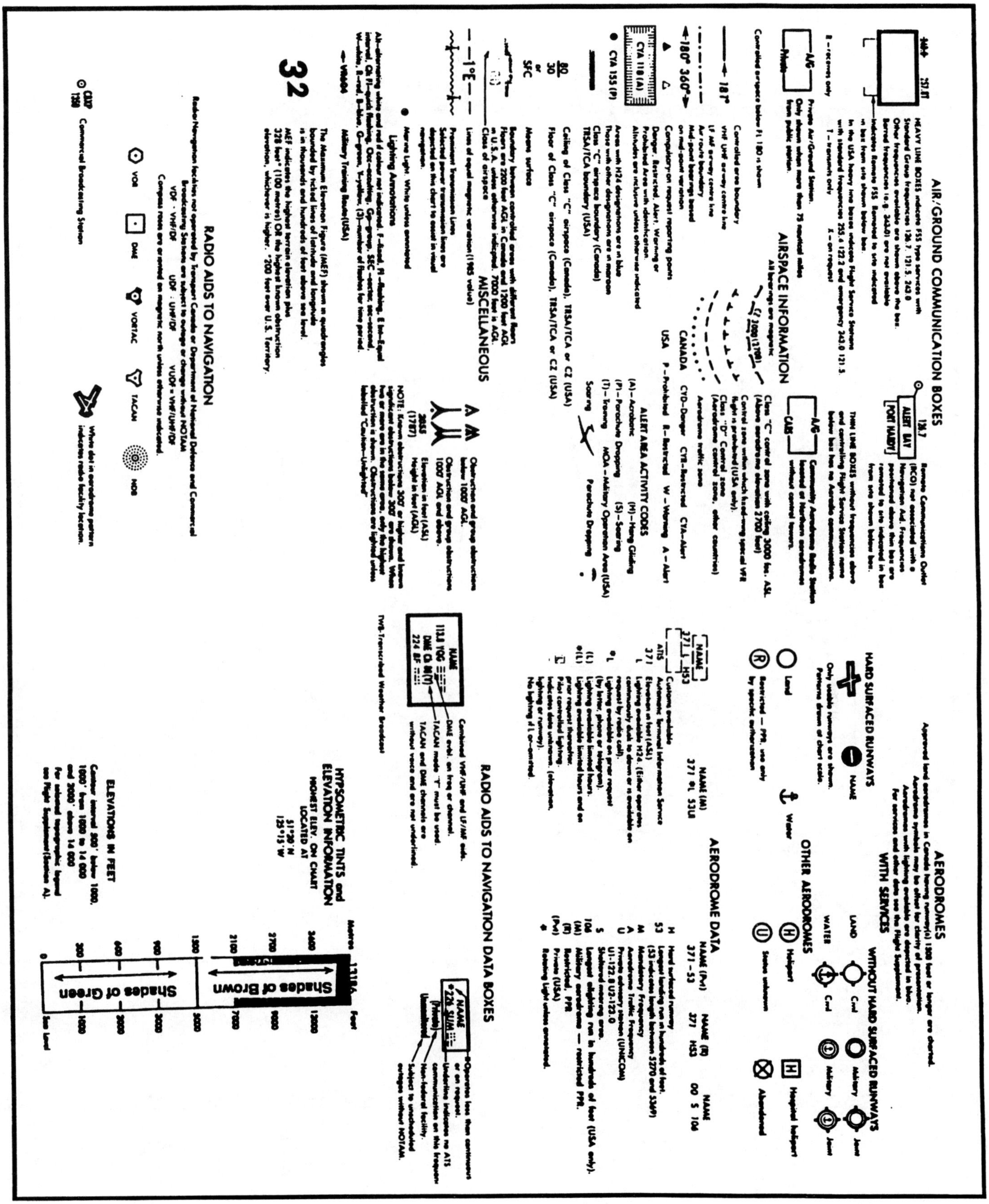

Fig. 6-7: VNC Charts legend and symbols.

6.04 Time and Longitude

Because there are 360 meridians of longitude, and the earth takes 24 hours to complete 1 rotation, each hour or time zone represents a change in longitude of 15°.

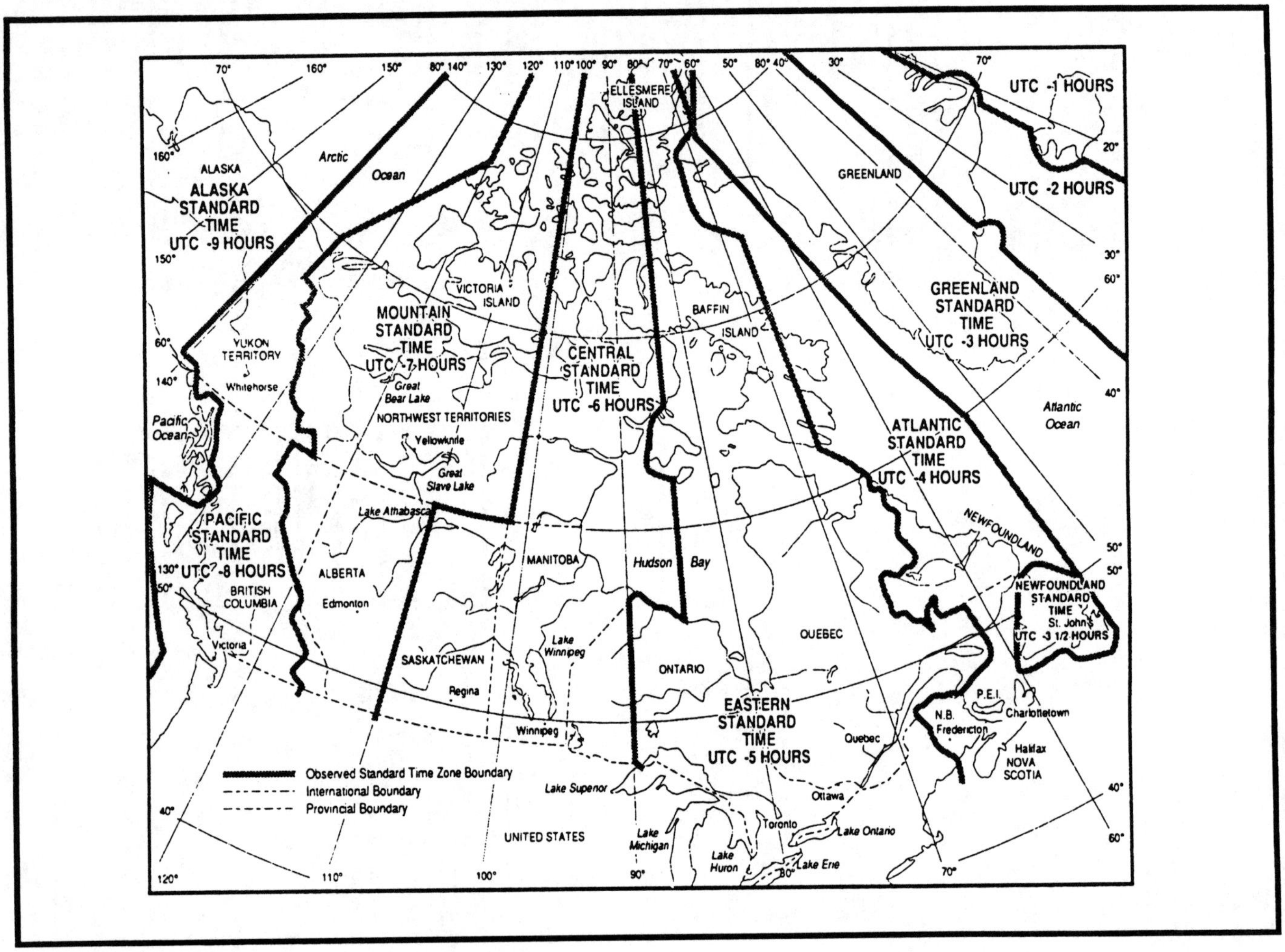

Fig. 6-9: Time Zones in Canada.

EXERCISE: Complete the following.

1. 1600 PST = ____ Z
2. 1000 AST = ____ Z
3. 0000 MDT = ____ Z
4. ____ CST = 0900 Z
5. ____ EDT = 2300 Z
6. ____ PST = 1430 Z

ANSWERS:

1. 2400Z
2. 1400Z
3. 0600Z
4. 0300 CST
5. 1900 EDT
6. 0630 PST

Note: UTC or Z refers to Universal Time Coordinated or GMT (Greenwich Mean Time). ST refers to Standard Time. DT refers to Daylight Saving Time (subtract 1 hour from UTC conversion).

6.05 Determining Position Using Latitude and Longitude

Parallels of latitude are measured north or south of the equator, in degrees and minutes. There are 60 minutes to every degree. Lines of longitude are measured east or west of the prime meridian, in degrees and minutes. [The vertical lines on a map are longitude, but the vertical scale represents latitude. The horizontal lines are latitude, but the scale on the horizontal lines is longitude.]

EXAMPLE 1: Determine the latitude and longitude of Princess airport and Brooks airport.

EXAMPLE 2: Determine what is located at grid coordinates 50° 42'N, 111° 40'W.

ANSWER 1: The position of Princess airport is latitude 50° 41'N, longitude 111°32'W. Brooks airport: 50° 38'N, 111° 55'W.

ANSWER 2: The town of Patricia.

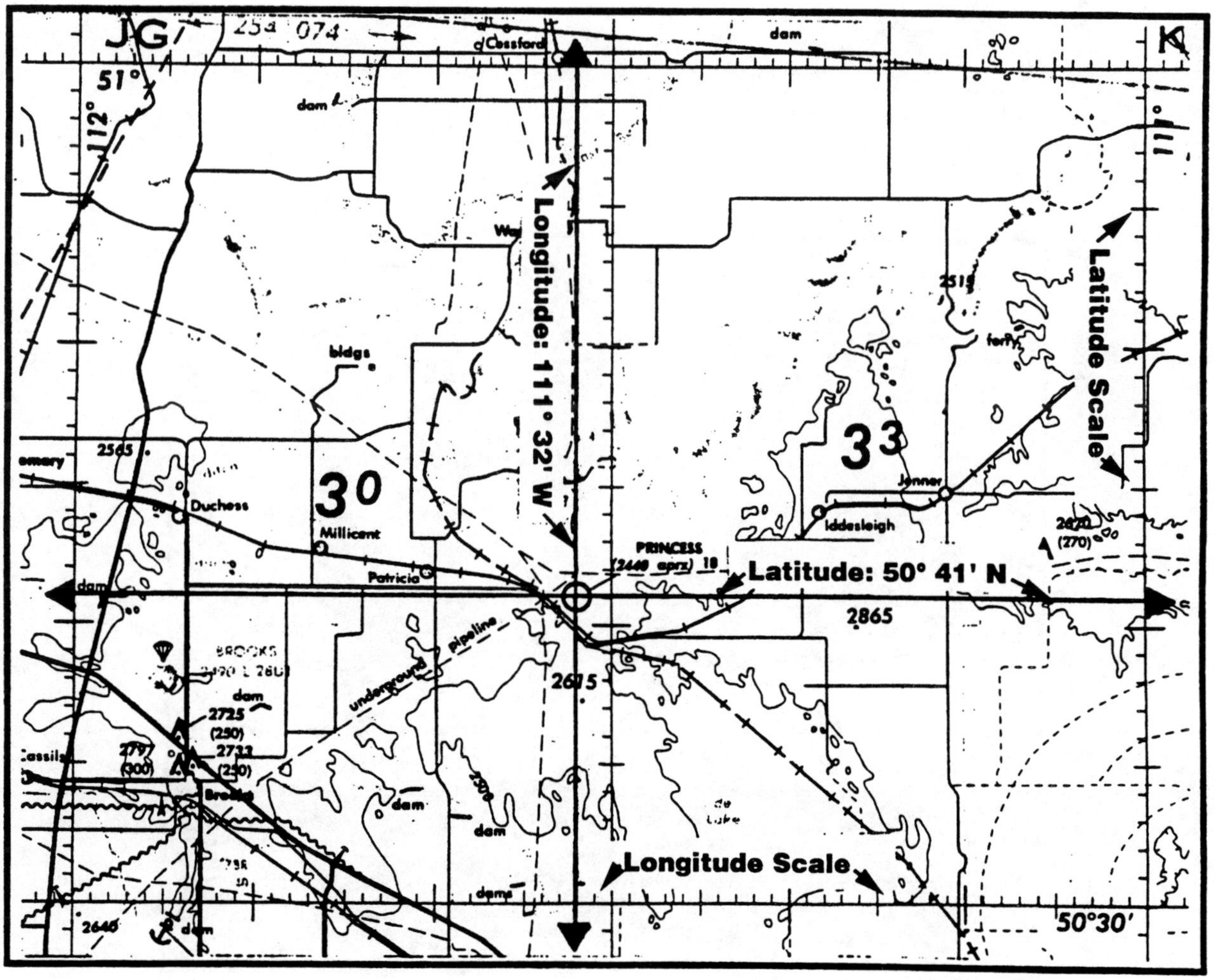

6.06 Measurement of Track and Distance

To measure true track, first draw a line on the map connecting the two airports. Next, place a protractor over the approximate mid-point (so as to average out the errors caused by the convergence of meridians towards the top of the map). Then, line up the north mark of the protractor so that it lies parallel to the nearest meridian. Measure clockwise on the protractor to determine the true track.

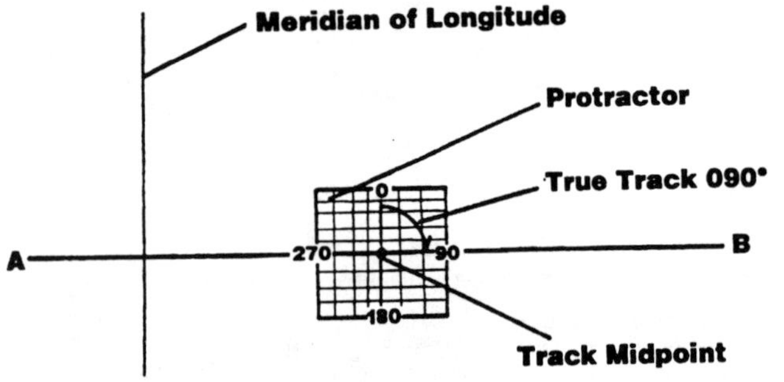

Fig. 6-10: Measurement of track and distance. The square Douglas type protractor is recommended.

To measure distance, remember that every minute of latitude is equivalent to 1 nautical mile. Take some scrap paper, make markings over the two locations, then line up with latitude scale (running vertically on map) for nautical miles (do not use longitude scale to measure distance).

EXAMPLE: What is the distance and true track from Medicine Hat airport to Suffield airport?

ANSWER: 23 NM (measure from the center of the runway complex). True track is 309.

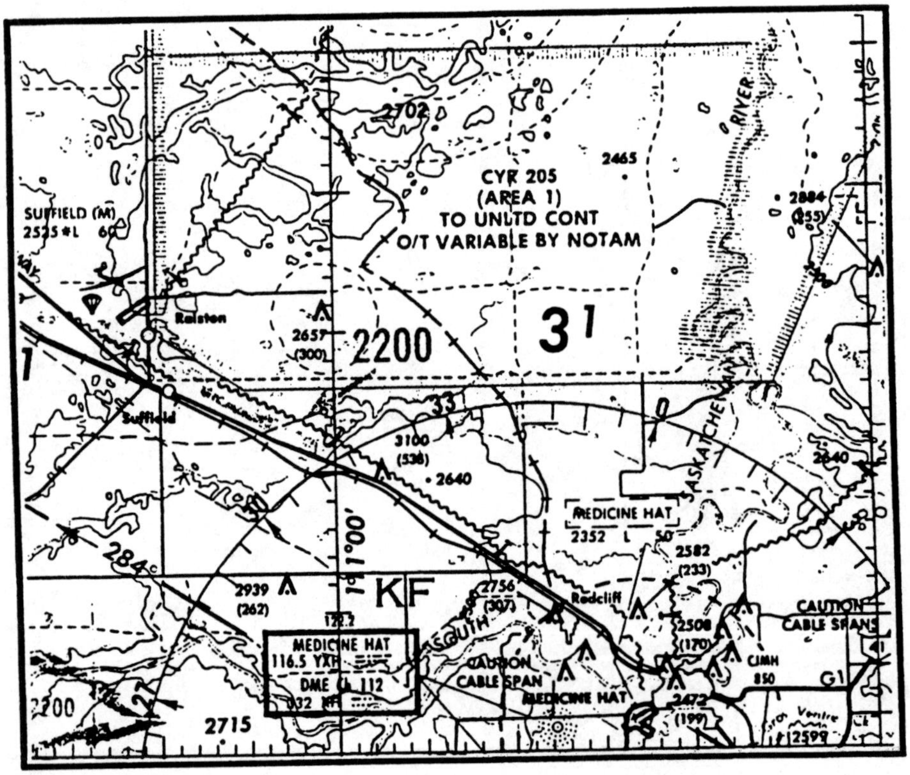

To find obstructions within a specified distance either side of track, such as 5 NM, take a scrap piece of paper and make two 5 NM markings either side of a center mark. Then, run the scrap paper along the track, and look on the map for obstructions either side of the 5 NM marks.

EXAMPLE: Refer to the previous chart. What is the highest obstruction within 5 NM of track from Medicine Hat airport to Suffield airport?

ANSWER: 3,100 feet ASL. It is a single obstruction, 538 feet AGL.

6.07 Setting Heading

The point to commence navigation need not be directly over the departure airport. Due to traffic, departure procedures or other restrictions it may better to commence navigation from a prominent position near the airport. This location is referred to as the "set heading" position.

6.08 Map Preparation

To prepare a map for VFR navigation follow these steps:

a) Draw track.
b) Mark distance every 10 NM.
c) Use protractor to draw 10° drift lines.
d) Measure distance, record in navigation log.
e) Measure true track, record in navigation log.

The following map excerpt shows the map preparation for a VFR flight from Castor to Consort:

It may be necessary to draw a track between airports on opposite sides of the chart. Take a clean piece of paper and line up the bottom edge of the paper with the outer edge of the map, draw a box around one of the airports and mark in the latitude scale, longitude scale and the airport in 5° increments. Then take the paper and line it up on the side of map for the other airport. You will see a thin area of duplication on both sides of the map. Line up your paper using the latitude and longitude markings. Then draw your track so it extends from the map and onto your "new map". Once your track is drawn, measure it from the map, find a common landmark or grid reference at the edge of the map. Go to opposite side of map, draw the track from the common landmark to the other airport, then measure it to verify it is the same. An example is a trip from Gimli, Man. (one side) to Morden Man. (other side). You will see with this method that St. Francis Xavier airport is a common landmark on track 025/205 in the overlap area.

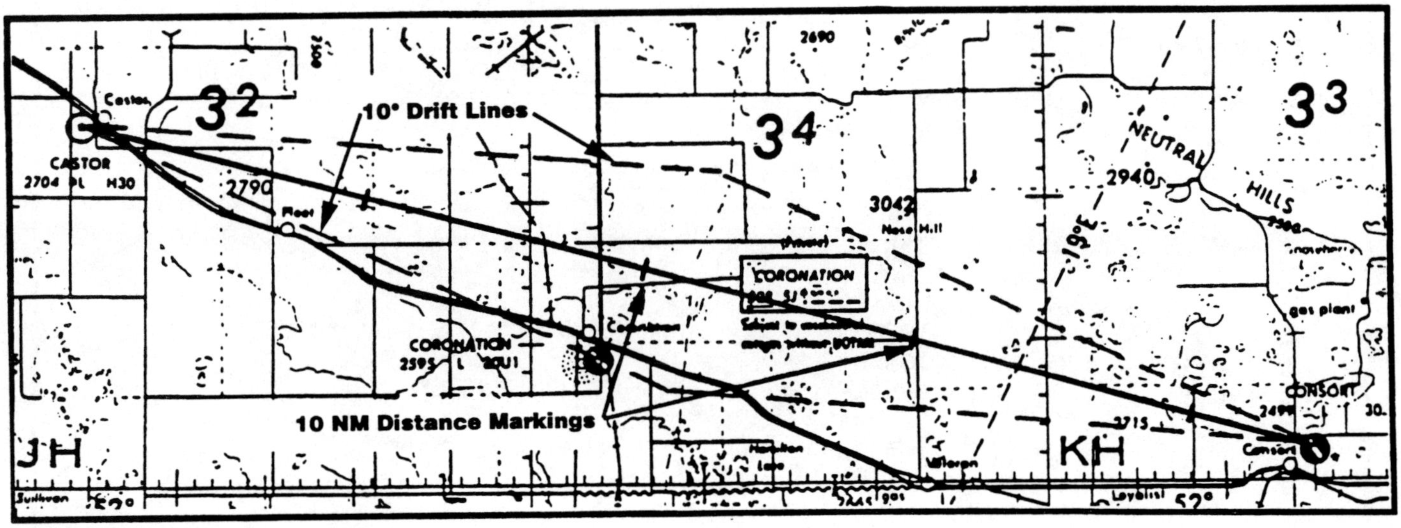

Fig. 6-11: Map Preparation for VFR trip.

6.09 Calculation of True Heading and Groundspeed

After drawing and measuring true track, use a flight computer to calculate true heading and groundspeed. Follow these steps:

a) Refer to the aircraft operating handbook to determine true airspeed for desired cruising altitude. Enter TAS and fuel consumption in navigation log.
b) Record wind speed and wind direction in navigation log.
c) Enter wind speed, wind direction, TAS and true track in flight computer. The computer is used to calculate wind correction angle, true heading and groundspeed. Use either an electronic or manual flight computer. To operate, follow the instructions on the computer.
d) Record wind correction angle (WCA), true heading (TRUE HDG) and groundspeed (G/S) in navigation log.

The purpose of calculating true heading is to know what heading to steer to overcome the effects of wind drift. The purpose of calculating groundspeed is to determine how long it will take to get to the destination airport (estimated time enroute).

Complete the exercise below using either an E6B or electronic flight computer.

EXERCISE: Using a flight computer, determine true heading and groundspeed.

	Reported Wind	True Track	True Airspeed	WCA	True Heading	Ground-Speed
1.	040/25 kts	360	160 kts	___	_____	_____
2.	240/30 kts	290	130 kts	___	_____	_____
3.	190/20 kts	050	150 kts	___	_____	_____
4.	320/30 kts	090	145 kts	___	_____	_____
5.	120/25 kts	120	170 kts	___	_____	_____
6.	0000	010	165 kts	___	_____	_____

ANSWERS:

	WCA	True Heading	Ground-Speed
1.	6° R	006	140 kts
2.	10°L	280	109 kts
3.	5° R	055	165 kts
4.	9° L	081	163 kts
5.	0°	120	145 kts
6.	0°	010	165 kts

6.10 Calculation of Estimated Time Enroute and ETA

With a known groundspeed and distance, the estimated time enroute (ETE) can be calculated with a flight computer. To calculate ETE, follow the instructions on/in the flight computer. Once the ETE is known, add the ETE to the departure time to calculate the estimated time of arrival (ETA).

EXERCISE: Calculate ETE and ETA.

	Ground Speed	Dist.	ETE	Dep. Time	ETA
1.	125 kts	63 nm	___	10:00Z	___
2.	110 kts	307 nm	___	09:30Z	___
3.	106 kts	29 nm	___	23:58Z	___
4.	119 kts	50 nm	___	16:10Z	___
5.	164 kts	10 nm	___	09:32Z	___
6.	210 kts	502 nm	___	13:20Z	___

ANSWERS:

	ETE	ETA
1.	:30	10:30Z
2.	2:48	12:18Z
3.	:16	00:14Z
4.	:25	16:35Z
5.	:04	09:36Z
6.	2:24	15:44Z

Groundspeed can be determined at any point during a flight, by noting the distance flown and time taken, then using the flight computer to calculate groundspeed. It is good pilot procedure to calculate groundspeed while enroute, and revise the estimated time of arrival if necessary.

EXERCISE: Using a flight computer, calculate revised groundspeed and time remaining to destination.

	Dist. Flown	Leg Time	Rev. G/S	Dist. Rem	Time Rem
1.	36 nm	:13	_____	12 nm	_____
2.	49 nm	:21	_____	91 nm	_____
3.	102 nm	:36	_____	106 nm	_____
4.	591 nm	2:13	_____	52 nm	_____
5.	290 nm	1:49	_____	196 nm	_____

ANSWERS:

	Ground-Speed	Time Remaining
1.	167 kts	:04
2.	140 kts	:39
3.	170 kts	:37
4.	268 kts	:11
5.	160 kts	1:14

6.11 Fuel Consumption

The amount of fuel burned per hour is found in the aircraft operating handbook. Refer to the "Performance" section.

To calculate the amount of fuel consumed for a given time, use the flight computer. Remember to include the required minimum reserve fuel of 30 minutes (day) or 45 minutes (night) based on consumption at normal cruise power.

EXERCISE: Calculate the minimum VFR fuel (not including climb, take-off and start).

	Rate	Time	Fuel	Reserve	Total
1.	8 GPH	:30	____	____	____
2.	6 GPH	1:45	____	____	____
3.	100 PPH	2:06	____	____	____

ANSWERS:

	Total Fuel
1.	8.0 Gals.
2.	13.5 Gals.
3.	260 Pounds.

The flight computer can be used for conversions such as gallons to liters. Refer s. 7.21 in this manual.

6.12 IAS - CAS - TAS Calculations

IAS is read directly off the face of the airspeed indicator. To convert to Calibrated Airspeed (CAS), refer to the aircraft operating manual, where there will be a conversion table such as the following:

AIRSPEED CALIBRATION

KIAS	KCAS
40	43
50	51
60	59
70	68
80	77
90	87
100	98
110	108
120	118
130	129
140	140

True Airspeed (TAS) is used for navigation calculations since Indicated Airspeed (IAS) varies with temperature, pressure and altitude. The flight computer and airspeed calibration table are used to convert from IAS to TAS and vice versa. To find TAS, convert IAS to CAS and enter outside air temperature for the flying altitude, pressure altitude and CAS.

EXERCISE: Use a flight computer and the above airspeed calibration table to complete the following.

	OAT °C	Pressure Altitude	TAS	CAS	IAS
1.	-10°C	8,000 ft	120 kts	____	____
2.	+10°C	9,000 ft	110 kts	____	____
3.	+30°C	5,000 ft	____	120 kts	____
4.	-20°C	7,000 ft	____	108 kts	____
5.	-10°C	12,000 ft	____	____	94 kts
6.	+10°C	4,000 ft	____	____	100 kts

ANSWERS:

	TAS	CAS	IAS
1.	-	108 kts	110 kts
2.	-	94 kts	96 kts
3.	135 kts	-	122 kts
4.	115 kts	-	110 kts
5.	109 kts	91 kts	-
6.	105 kts	98 kts	-

6.13 Pre-Flight Prep

Once you have determined that the weather for a flight will be satisfactory, all relevant factors must be analyzed and a route decided upon. The most direct route may not be the most desirable; it could be largely over water beyond gliding distance to land, or over a prohibited or danger area.

In selecting routes, try to avoid those which include extended periods of flight over areas where map reading is difficult. The ability to read maps is a very important part of pilot navigation. The most important objective, however, is to get there safely.

A CHECK-POINT, or pin-point is a prominent feature or group of features along, or close to the route, used to establish the aircraft position. They are also used for other purposes such as in-flight calculation of groundspeed and revisions to the ETA.

Main highways, rivers, railroads, and lakes are excellent and readily identifiable check points for pilot navigation. Take full advantage of them and choose a route, if possible, which utilizes all good landmarks. A diversion, if necessary, will be much easier if you know where you are.

The aeronautical chart used should be UP TO DATE: all kinds of things can be different on an old chart. It is also a good practise to supplement aeronautical charts with road maps, since they have much better secondary road detail. Don't rely on road maps for primary navigation because the scales are not accurate.

Be sure that any relevant NOTAM's have been checked with respect to the route selected.

The map should be folded neatly with the set heading point, track, drift lines and distances marked out. You may wish to use a highlighter pen to give prominence to the track.

Weather should be suitable for the trip and an alternate airport should be available and circled on the chart.

A flight log should be prepared: greater accuracy here will pay off in an easier flight.

A flight plan, notification or itinerary should be filed as appropriate.

A low level flight will require careful planning: know the location and height of all obstructions such as towers, cable crossings, spot elevations, terrain etc.

6.14 Navigation Exercise

This exercise is a trip from Consort to Castor. Everything needed to complete the exercise is provided on this page. Fill out the navigation log as you answer the questions.

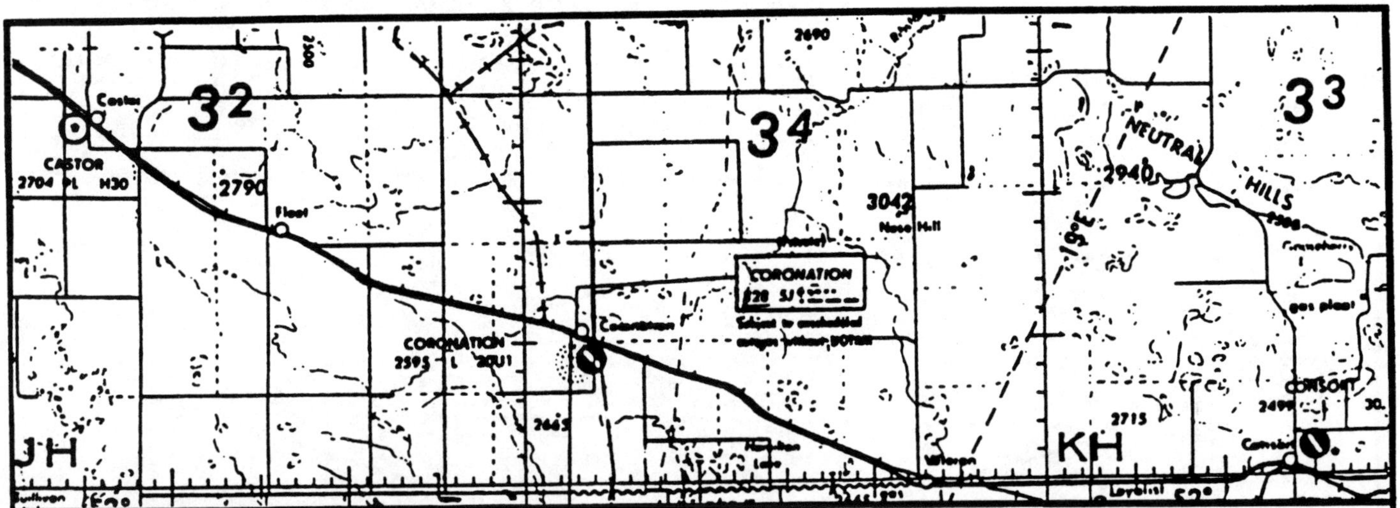

Compass Correction Card												
For (magnetic)	000	030	060	090	120	150	180	210	240	270	300	330
Steer (compass)	002	032	061	090	118	147	178	208	240	271	303	334

Navigation Log

FROM TO	TRUE TRK	ALT	RPM	OAT °C	TAS IAS	WIND	WCA	TRUE HDG	VAR	MAG HDG	DEV	COM HDG	G/S	DIST	ETE	ETA	REV G/S	FUEL CONS	FUEL REQD

QUESTIONS:

1. What is the true track? What is the variation? What is the distance?
2. Assume a cruising altitude of 4500 ft is selected. Is this an appropriate cruising altitude? Why?
3. Assume TAS and IAS are both 91 knots, and winds are 160°T at 10 knots. What is the true heading and groundspeed?
4. With reference to the compass correction chart, what is the required compass heading?
5. Determine estimated time enroute. If the departure time is 10:00 Z, what is the ETA?
6. Assume average fuel consumption of 8 U.S. GPH. What is the minimum DAY VFR fuel required for the proposed trip (adding 4 gallons for start, taxi, climb and delays)?
7. At 10:15Z, you are over track with Coronation to the left. Calculate revised G/S and new ETA.

ANSWERS:

1. True track from Consort to Castor = 285. Variation = 19°E (see isogonic line). Distance = 45 nm (use latitude scale to measure distance).
2. Yes, because the magnetic track is 266 (285 - 19), which is from 180 - 359, therefore cruising altitude should be even thousands plus 500 feet. The highest obstruction is 2790, and there will be adequate terrain clearance.
3. Groundspeed = 96 kts, true heading = 279. Wind correction angle = 6°L.
4. Compass heading = 261 (1° deviation).
5. Estimated time enroute = 28 minutes. Estimated time of arrival 10:28Z.
6. 11.7 U.S. Gals. Cruise :28 + :30 (reserve) = :58 @ 8 GPH = 7.7, add 4.0 = 11.7.
7. Revised G/S = 100 kts. (25 nm in 15 minutes = 100 kts.) Revised ETE = 12 minutes for remaining 20 nm to Castor. ETA = 12 minutes + 10:15Z = 10:27Z.

6.15 Drift Correction Procedures

There are 4 methods to correct for drift, when an aircraft drifts away from track.

Visual Method.

Note off-course position, fly back to track, apply a correction into wind to prevent further drift.

Double Track Error

This method requires a flight envelope with 10° drift lines. Note off-course position, and determine drift angle. Double the drift angle, and steer this amount towards track, for the same amount of time it took to reach the off-course position. Once over track, steer into wind by the drift angle to avoid further drift. Note: This method can only be used where off-course position is before mid point of track.

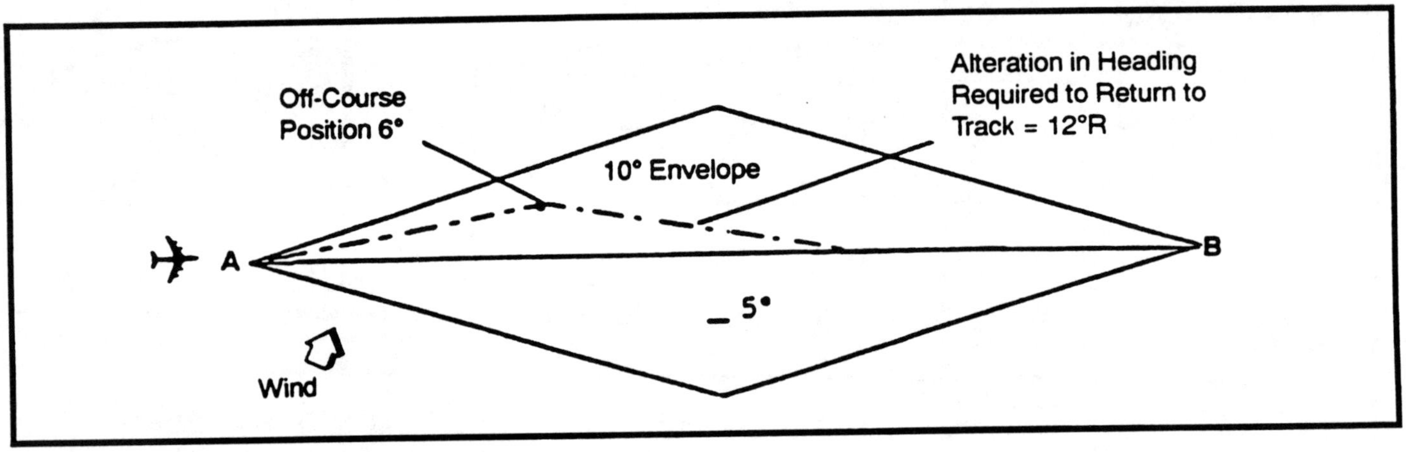

Fig. 6-12: Double Track Error Method of Drift Correction.

Opening - Closing Angle Method.

The advantage of this method is that it can be used at any point along track. To use this method, add the angle made between the off-course position and departure airport (opening angle) to the angle made between the off-course position and destination airport (closing angle). Steer towards the track by this amount to fly direct to destination.

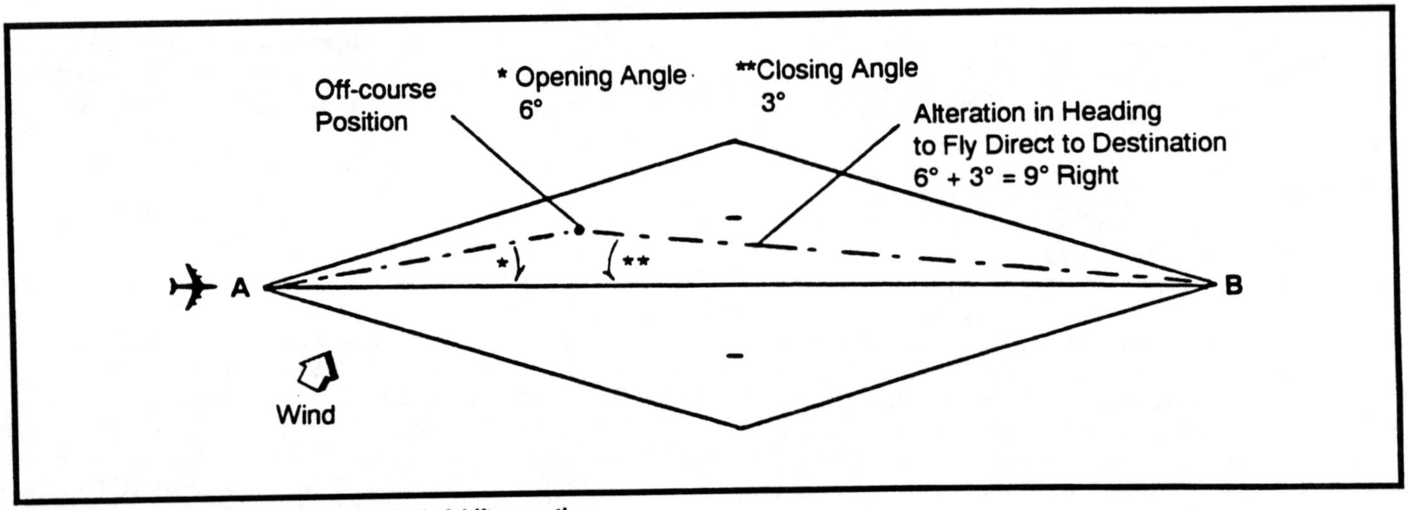

Fig. 6–13: Opening - Closing Angle method of drift correction.

1:60 Rule Method.

This method of drift correction is used where a flight envelope has not been drawn. The 1:60 rule states that every error in track of 1 nm in 60 nm is equivalent to 1° in 60°, therefore the opening angle and closing angle can be estimated using these formulas:

$$\text{opening angle} = \frac{\text{dist off course}}{\text{dist flown}} \times 60$$

$$\text{closing angle} = \frac{\text{dist off course}}{\text{Dist remaining}} \times 60$$

Using this method will also give and estimate of DRIFT: the opening angle = drift.

To determine the required correction to destination, add the opening angle to the closing angle.

EXAMPLE: An aircraft has flown 15 miles of a 75 miles track and is 5 miles off track. Determine the drift and required correction to proceed direct to destination.

ANSWER: Drift (opening angle) = 20°. Closing angle = 5°. Total correction = 25°.

6.16 Diversion to Alternate

Various events such as bad weather, darkness etc. can require a change in the planned destination and a diversion to an alternate aerodrome.

Because of space and time limitations, and operating an aircraft in flight, the level of accuracy of pre-flight planning can not be realized in a diversion procedure and short cuts are necessary.

Suggested procedure:

a) note present position, circle on chart.
b) note destination, circle on chart.
c) draw rough straight line track, estimate direction of track using nearby VOR rose or true estimate followed by adding or subtracting variation.
d) Turn to track, note time, keep over track using landmarks.
e) Divide track into quarters. After you have flown 1/4 of distance, note time taken. Time to destination will be 3 more units of the time for the first quarter.
f) Maintain track, double-check timing with map reading, radio FSS/brief passengers as appropriate.

6.17 Reciprocal Track

To turn around and return on the same track, simply turn 180°. Look to the heading indicator: at the bottom will be the reciprocal heading to turn to. Based on a standard, 2-minute, rate 1 turn, it will take 60 seconds to complete the 180° turn.

If you were correcting for wind to the right outbound, then inbound on the reciprocal track, you will correct to the left.

6.18 Procedures when Lost

If you become uncertain of your position, a climb will often assist in map reading. A circle of uncertainty can be drawn: draw a circle from your last known position plus an allowance for the distance covered from last know position. You should be within the circle somewhere.

Basically, when lost, the key to success is careful map reading and identification of prominent landmarks.

You may wish to land at a suitable aerodrome to preserve fuel and verify position: if you do so, however, don't take off again until you know where you are.

If unable to determine position, if possible, contact a local airport for DF steer (see s. 6. 21), if the airport has the necessary equipment. If the situation becomes serious, consider notification to FSS or in an emergency complete a MAYDAY call. You may also need to consider completing an off airport procedure and land on a suitable surface.

6.19 VOR Procedures

VOR means "Very High Frequency Omni Directional Radio Range" and is a navigation system based on VHF transmissions. VHF signals are "line of sight", which means they will be blocked by mountains and high terrain. The VOR frequency range is from 112.1 MHz to 117.9 MHz. The aircraft equipment consists of a radio tuner and navigation indicator. Before using any VOR station, the proper frequency should always be verified by listening to the morse code identifier. Some VOR stations have a voice feature which means ATC can send voice messages over the VOR frequency.

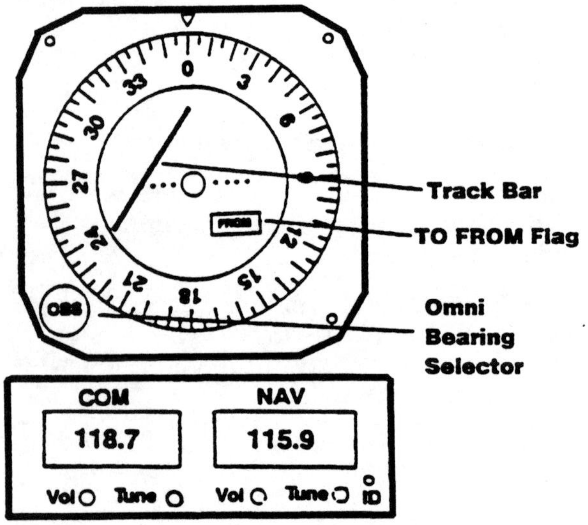

Fig. 6-14: VOR equipment: navigation indicator and radio tuner.

VOR signals radiate from a ground station. There are 360 radials sent out from the station, like spokes on a wheel. With a VOR, the pilot can verify position and track to or from a VOR station on a given radial.

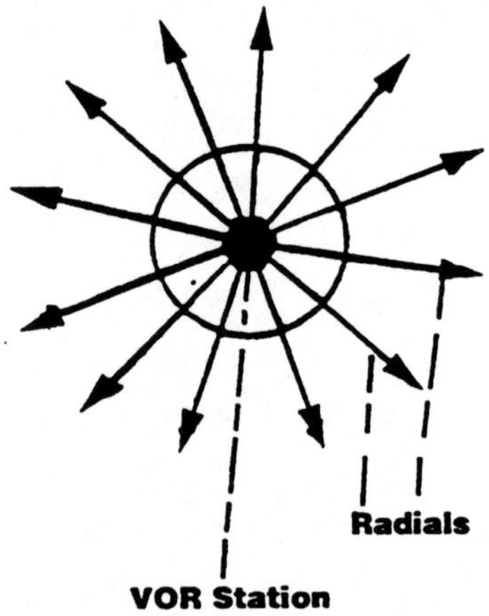

Fig. 6-15: Radials are sent out from a VOR station like spokes on a wheel. Some radials are designated as Victor airways.

On aeronautical charts, VOR stations have a surrounding COMPASS ROSE, to help identify a radial. The 360 radial is aligned with magnetic north. Certain radials are designed as airways and numbered with a "Victor" designation, such as "V-308".

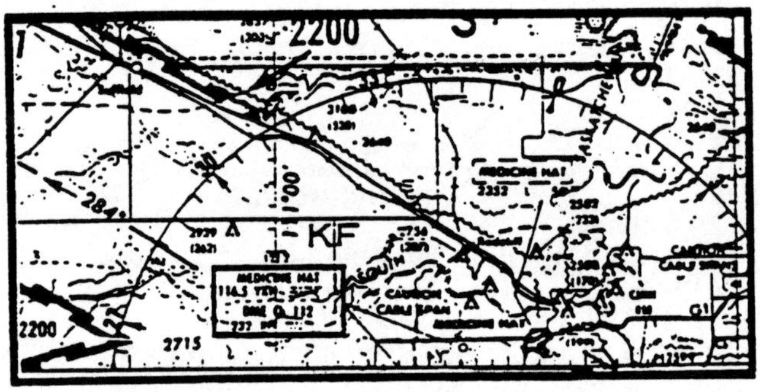

Fig. 6-16: Compass Rose and VOR station as represented on a VNC chart.

To use a VOR, remember that, for a selected radial, there are various quadrants or envelopes. The envelopes are the "TO-FROM" envelope, and the "TRACK BAR" envelope (needle left, needle right). The envelopes are always oriented on the basis of the selected radial. When over the selected radial, the track bar needle is centered, with a FROM flag. When over the reciprocal radial to the selected radial, the track bar needle is centered with a TO flag. 90° either side of the selected radial is the CONE OF AMBIGUITY, where the TO-FROM flag indicates OFF. The OFF indication means the VOR indication should not be relied upon (the TO-FROM sensor cannot properly sense whether the aircraft is in the TO or FROM area).

The TRACK BAR needle senses whether the aircraft is within 10° of the selected radial. Each dot on the face of the instrument represents 2°. Full scale deflection indicates that the aircraft is more than 10° away from the selected radial.

The VOR indications depend upon the location of the aircraft in relation to the radial selected (not in relation to the aircraft's heading). The key to understanding the VOR is that the VOR indications are based upon orienting the various envelopes to the selected radial. If you can't figure out a VOR indication, visualize the selected radial, then reconstruct the TO-FROM and track bar envelopes. Study Fig. 6-17 on the following page.

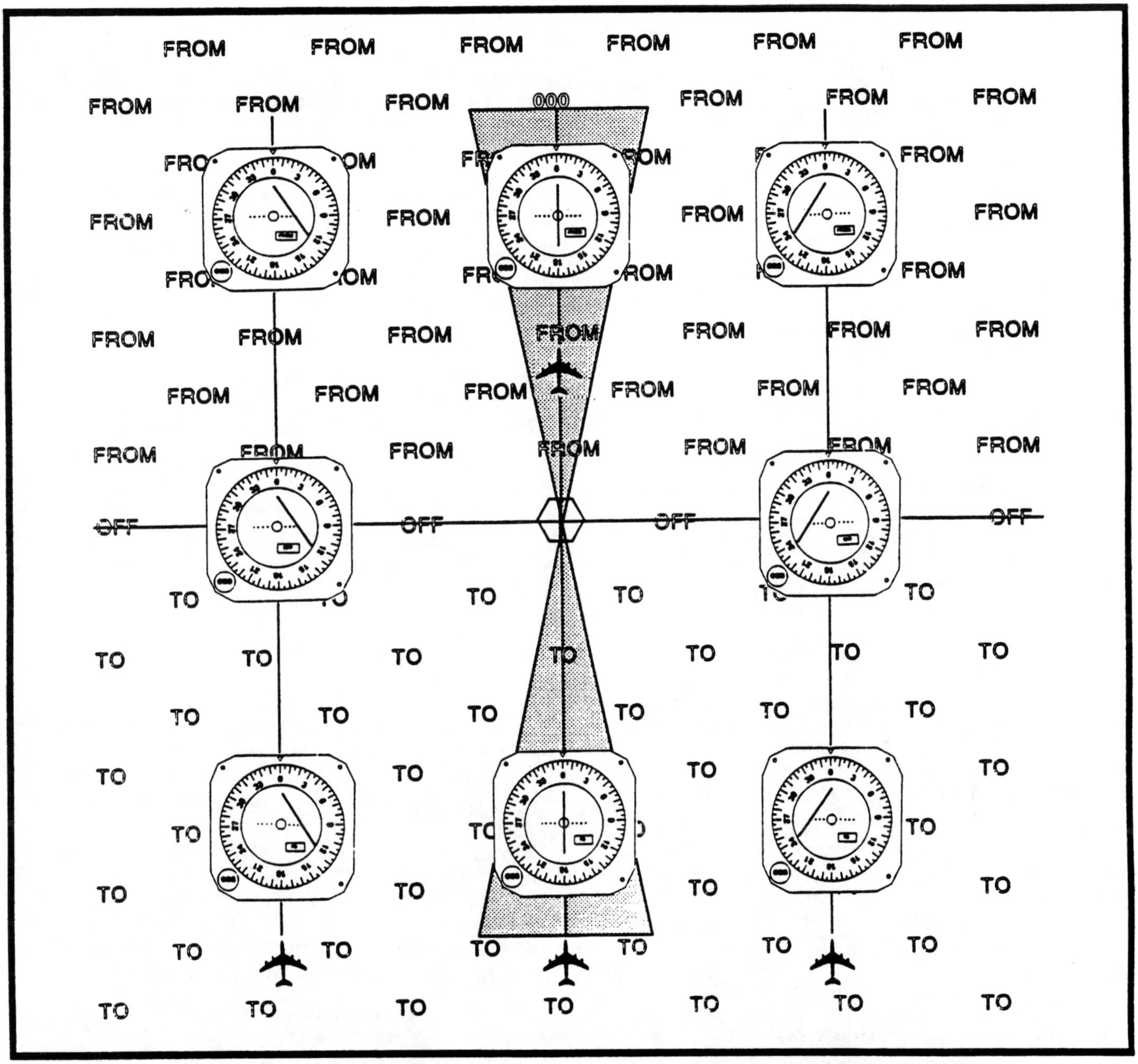

Fig. 6-17: VOR envelopes for 000 Radial. Note TO-FROM areas: FROM is indicated when on same side of station as selected radial (000) and TO is indicated when on opposite side of station. Cone of Ambiguity is 90° either side of radial where VOR indicates OFF. Track Bar indicates left when on right side of radial, and right when on left side of radial. When other radials are selected, orient the above figure to the selected radial. If you remember these fundamentals, you will never be confused by the VOR.

EXERCISE: Based on aircraft A - G, determine VOR indications for selected radials.

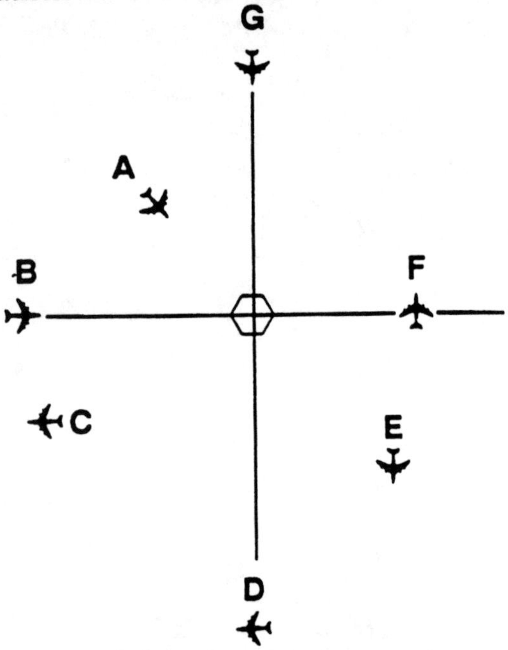

Position		Selected Radial	Track Bar	TO-FROM
1.	A	360	______	______
2.	B	270	______	______
3.	C	180	______	______
4.	D	360	______	______
5.	E	270	______	______
6.	F	______	Right	OFF
7.	B	______	Centered	FROM
8.	G	______	Centered	TO
9.	F	______	Centered	FROM
10.	D	______	Left	OFF

ANSWERS:
1. Right, FROM
2. Centered, FROM
3. Left, FROM
4. Centered, TO
5. Right, TO
6. 180
7. 270
8. 180
9. 090
10. 090

One useful application of VOR principals is taking a POSITION FIX. A position can be verified by finding the intersection point of 2 LINES OF POSITION (LOP). An LOP is also known as a BEARING. A VOR radial represents a magnetic track (or magnetic bearing) away from a station. When the track bar is centered, with a FROM flag, the selected radial will be a line of position from the VOR station. To take a fix, take 2 VOR indications with a centered Track Bar and FROM indication, and draw in the identified radials. Use the VOR Compass Rose as a reference. Where the two radials intersect will be your position.

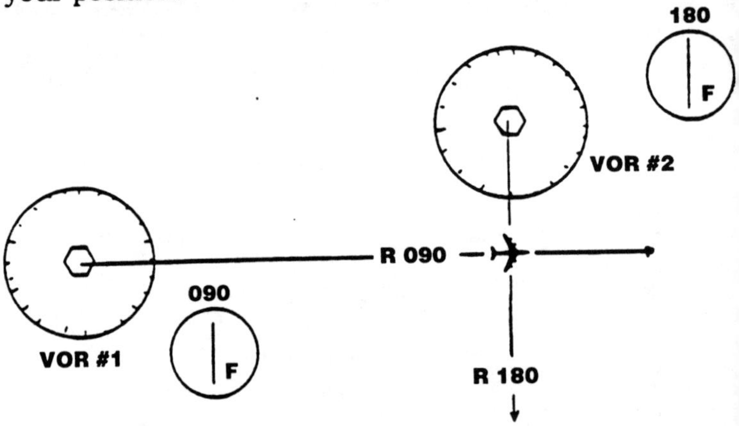

Fig. 6-18: Position Fix using 2 VOR stations. Centered needle with FROM flag indicates radial from each station.

VOR Receiver Checks - VOT: OMNITEST

2 omnitest transmitters send out low power signals to test the accuracy of VOR equipment at selected airports. To test the VOR, first tune in the VOT frequency (usually 114.8). You should hear the ATIS broadcast over this frequency. Next, move the OBS to confirm the following indications:

	OBS	CDI Needle	TO-FROM	Max Error
1.	000	Centered	FROM	+/- 4°
2.	090	Full Right	OFF	N/A
3.	180	Centered	TO	+/- 4°
4.	270	Full Left	OFF	N/A

Take advantage of the fact that, while doing the test, you will be listening to the ATIS, so copy it at this stage, instead of selecting the ATIS a second time prior to taxi. If you have difficulty remembering the various checks, think of an aircraft sitting on the 360 radial, and ask yourself what indications would be expected as you change the OBS.

6.20 Automatic Direction Finder (ADF)

The ADF is a navigation aid which is based on signals in the low frequency range. The system allows navigation over a given track in relation to a ground station termed a non-directional beacon (NDB). The aircraft equipment consists of a receiver and ADF compass. The ADF frequency range is 190 to 1750 KHz.

The ADF is easy to interpret: the 0 mark on the top of the instrument is aligned with the longitudinal axis of the aircraft. The ADF needle points to the NDB station. The angle between the pointer and the 0 mark is the RELATIVE BEARING. The ADF on the below indicates a relative bearing of 120.

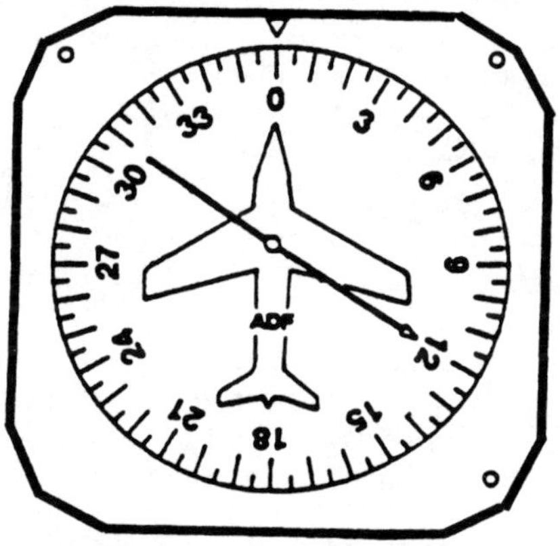

Fig. 6-20: Face of ADF.

Homing

To home an aircraft to a station, maintain a heading that will keep the ADF needle aligned on the "0". Eventually the aircraft will reache the station. Wind causes drift, resulting in a curved path.

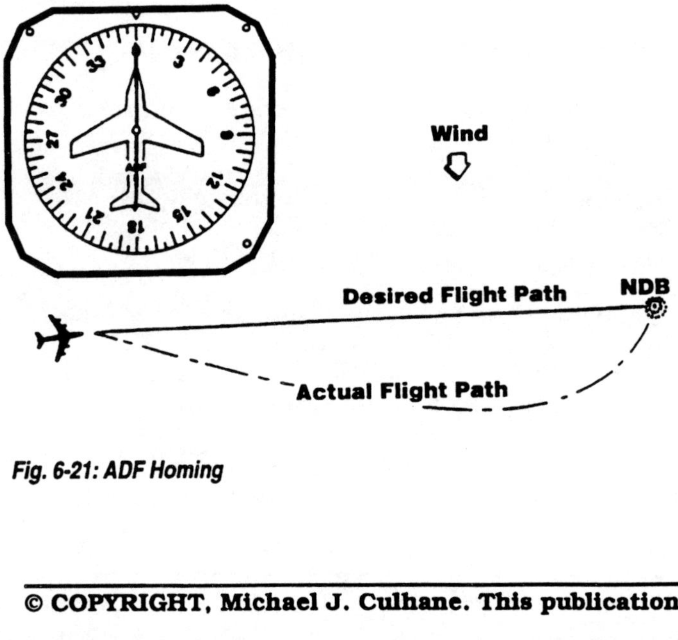

Fig. 6-21: ADF Homing

The main advantage of the ADF over the VOR is that low or medium frequencies are not affected by line of sight. The signals follow the curvature of the earth, therefore if the aircraft is within the power range of the station, the signals can be received regardless of altitude.

When using the ADF, special care should be taken to verify the frequency selected is identified correctly by cross checking the morse code identifier indicated on the chart. NDB stations have a voice feature, and ATC or FSS can transmit over the frequency. Recall from Section 5 that TWB weather information is transmitted over some NDB's.

6.21 DF Steer/VHF Direction Finder

A VHF direction finder is a ground facility used to locate lost aircraft and provide pilots with a heading to reach the airport. VHF direction finding equipment is based on "VHF" radio signals and is normally mounted on top of control towers. To obtain a DF steer, the pilot should contact the controller and request a DF steer. The controller provides the pilot with a heading to reach the airport.

Note: A DF steer will not be provided to special VFR aircraft.

6.22 Long Range Navigation - LORAN

LORAN-C is a long range navigation system operating in the 90 - 100 kHz frequency band which is used primarily for marine navigation. The system is becoming popular for air navigation. It is based upon the measurement of the time difference in the arrival of signal pulses from a group or chain of stations. The chain of ground stations consists of a Master station linked with 4 secondary stations. To provide a position fix, the LORAN-C measures the time difference between the Master and at least 2 of the secondary stations.

LORAN-C stations send ground waves and sky waves. Ground waves are used for navigation. Signal coverage ranges from 120 nm to 900 nm, depending upon ground signal strength. Accuracy is generally +/- 1/2 nm. Currently, testing is underway to develop non-precision approaches utilizing LORAN-C. Errors affecting LORAN occur if the receiver locks on to sky waves or other erroneous signals. Further, there are several "dead zones" where signal reception is impossible due to lack of stations.

6.23 RADAR Systems

RADAR means "Radio Direction and Ranging" and is based upon the time difference between transmission of a radio signal and its reflection by an object. The reflection is illuminated on a cathode ray tube. The relectivity of a radar echo decreases with the distance of the aircraft. The lag or time delay between the radar signal and its reflection provides distance information.

Radar increases airspace utilization by allowing ATC to reduce separation between aircraft. It also permits more flight information services such as traffic information and navigation assistance. Radar is also used by weather stations to locate storms and to obtain upper weather information.

PRIMARY SURVEILLANCE RADAR (PSR) determines the position (range and direction) of contacts (aircraft and weather) by measuring and displaying reflected radio frequency signals from the contacts. It does not require any information from the aircraft.

SECONDARY SURVEILLANCE RADAR (SSR) relies on measurement of the time interval between the INTERROGATION and REPLY by an airborne transponder to determine aircraft range. The instantaneous direction of the antenna determines contact direction. By itself, SSR will not provide a position for an aircraft without an operating transponder, and it will not locate weather. The main advantage of SSR is that it allows positive identification of particular aircraft.

WEATHER RADAR is based on primary radar. Aircraft equipped with airborne weather radar systems will have the precise location of severe weather displayed on the scope.

AIRPORT SURVEILLANCE RADAR (ASR) refers to primary radar systems in use by ATC to control traffic. The following types of ASR are used:

a) AASR-1. A medium to long range primary radar for distances of 180 to 200 NM.
b) ASR-5. A short range primary radar for distances of up to 60 NM.
c) ASR-803. A short to medium range radar for up to 80 NM.

AIRPORT SURFACE DETECTION EQUIPMENT (ASDE) provides radar surveillance of traffic at certain airports. It is a high definition, primary surveillance radar and is used by tower controllers to monitor the position of aircraft and vehicles on the maneuvering areas of the airport, particularly in conditions of reduced visibility.

6.24 Radio Theory

Radio waves in aviation range from VLF to UHF. Aviation transmissions are expressed in Kilo Hertz (KHz - 1,000 Hz) or Mega Hertz (MHz - 1,000,000 Hz). The following summarizes the frequency ranges and corresponding uses:

a) Very Low Frequency (VLF). Frequency range: 3 - 30 KHz. Used for long range navigation. Produces strong surface waves.
b) Low Frequency (LF). Frequency range: 30 - 300 KHz. NDB stations. Strong surface waves.
c) Medium Frequency (MF). Frequency range: 300 - 3,000 KHz. Long range navigation systems. Static problems may be encountered.
d) High Frequency (HF). Frequency range: 3,000 - 30,000 KHz. Very long range transmission. Used for trans oceanic radio communications for enroute weather, position reports. Static difficulties common.
e) Very High Frequency (VHF). Frequency range: 30 - 300 MHz. Short range, used in aircraft communications radio systems and VOR. Line of sight limitations.
f) Ultra High Frequency (UHF). Frequency range: 300 - 3,000 MHz. Pulse signals. Used in DME navigation systems, Tactical Air Navigation systems (military), and radar systems.

Common VFR communications frequencies:

121.5 MHz	Emergency frequency.
122.2 MHz	Often used at FSS stations.
122.8 MHz	First choice UNICOM.
123.0 MHz	Second choice UNICOM.
123.2 MHz	Uncontrolled aerodrome where no UNICOM.
126.7 MHz	Standard frequency for uncontrolled airspace, FSS stations.

6.25 Global Navigation Satellite System

Navigation systems will in future be based primarily on satellite technology. Eventually, a Global Navigation Satellite System (GNSS) will be completed. There are two satellite navigation systems being tested: the U.S. Global Positioning System (GPS) and the Russian GLONASS.

GPS navigation is based on a number of satellites that transmit a PRECISE NAVIGATION SIGNAL (P code) and a COARSE ACQUISITION SIGNAL (C/A code). Using this information, aircraft GPS equipment can provide precise 3-D navigation information (at least four operational GPS satellites are needed to determine 3-D position).

Many aircraft are now equipped with GPS units, and GPS makes VFR navigation far easier. With GPS, direct navigation is possible between locations without any ground based navaids - the grid coordinates of the departure and destination aerodromes are simply programmed into the unit. As well, WAYPOINTS can be entered to define a particular route.

Several airports in North America now have non-precision GPS approaches.

VFR pilots navigating from point to point using GPS in mountainous regions should not become slack on basic flight planning requirements: many accidents have resulted from VFR pilots coming into contact with terrain while using GPS to track direct to destination.

NOTES

NOTES

7 Flight Operations

7.01 Winter Operations

Critical Surface Contamination on Ground

The air regulations prohibit takeoff when frost, ice or snow is adhering to any critical surface of the aircraft. This is referred to as the CLEAN AIRCRAFT CONCEPT. Whenever frost, ice or snow may reasonably be expected to adhere to the aircraft, an inspection of the aircraft is legally required before takeoff.

Frozen Contaminants

An aircraft needs a smooth, uninterrupted flow of air over the wings and control surfaces. The slightest accretion of frost can seriously affect performance capabilities, because frost increases the stalling speed. Ice contamination having the thickness and roughness of medium to coarse sandpaper can decrease lift by as much as 30% and increase drag by 40%. Ice or frost can be removed with special de-icing solutions or by putting the aircraft in a heated hangar (ensure the aircraft is completely dry before taking outside).

The Cold Soaking Phenomenon

Since fuel tanks are normally located in the wings, after a flight, the aircraft and fuel temperature can be much colder than ambient temperatures (fuel temperature drops of 12° to 18°C have been recorded after a flight of two hours). An aircraft COLD SOAKED WINGS will conduct heat away from precipitation so that a thin sheet of clear ice can easily form on the wing surface that is hard to detect. Cold soaking can also cause frost formation at airports much warmer than freezing where there is high relative humidity.

In-Flight Airframe Ice Contamination

Airframe icing can be a serious weather hazard to aircraft in flight. Icing will result in a loss of performance in the following areas:

a) ice accretions on lifting surfaces will change their aerodynamic properties resulting in a reduction in lift, increase in drag and weight with a resultant INCREASE in stalling speed and a reduction in the stalling angle of attack. Therefore, an aerodynamic stall can occur before the stall warning systems activate;
b) ice adhering to propellers will drastically affect their efficiency and may cause an imbalance with resultant vibration;
c) ice on the windshield or canopy will reduce or block vision;
d) carburetor icing can affect engine output.

The three types of ice to contend with are clear ice, rime ice and frost. Review s. 5.11 in this manual for the characteristics of airframe ice. Frost may form on an aircraft in flight when descent is made from freezing conditions to a layer of warm moist air. In these circumstances vision may be restricted as frost forms on the windshield or canopy.

Due to these hazards, pilots should insure they keep away from areas of probable icing.

Landing Wheel Aircraft on Snow Surfaces

Each winter, accidents occur due to pilots attempting to land wheel-equipped aircraft on surfaces covered with deep snow. This usually results in the aircraft nosing-over. Light aircraft should not be landed on surfaces covered with snow unless it has previously been determined that the amount of snow will not constitute a hazard. Some aircraft are fitted with skis for winter operations.

Whiteout

Whiteout is defined as:

> An atmospheric optical phenomenon of the polar regions in which the observer appears to be engulfed in a uniformly white glow. Neither shadows, horizon, nor clouds are discernible; sense of depth and orientation is lost; only very dark, nearby objects can be seen. Whiteout occurs over an unbroken snow cover and beneath a uniformly overcast sky, when with the aid of the snowblink effect, the light from the sky is about equal to that from the snow surface.

Blowing snow may be an additional cause. Due to the lighting conditions encountered in whiteout, it becomes impossible to see the horizon. The danger is that the pilot will become disoriented. There have been numerous accidents where pilots have flown their aircraft directly into the ground, in whiteout conditions. If whiteout conditions are encountered, or suspected, the pilot should immediately climb if at low level, or level off and turn towards an area where sharp terrain features exist. The following phenomena cause whiteout and should be avoided:

a) WATER FOG WHITEOUT results from thin clouds of super-cooled water droplets contacting cold snow surface. Depending on the size and distribution of the water droplets, visibility may be minimal or zero.
b) BLOWING SNOW WHITEOUT results from fine snow being plucked from the surface by winds of 20 knots or more. Sunlight is reflected and diffused resulting in zero visibility.
c) PRECIPITATION WHITEOUT results from small wind-driven snow crystals falling from low clouds above which the sun is shining. Light reflection complicated by spectral reflection from the snow flakes and obscuration of landmarks by falling snow may reduce visibility and depth perception to zero.

7.02 Flight In Rain

It is hazardous to fly low in rain, due to illusions from looking through a windscreen covered in water droplets. The error is up to about 5° in angle. For example, a hilltop 1/2 nautical mile forward can appear to be 260 ft lower than its actual elevation.

Be alert for this hazard when flying in conditions of low visibility in rain and maintain sufficient altitude and take other precautions as necessary to allow for this error. Applying a rain repellent coating to the windshield can reduce the effect.

7.03 Flight Near Thunderstorms

Thunderstorms can contain nearly all weather hazards known to aviation, including tornadoes, turbulence, squall lines, microbursts, heavy updrafts and downdrafts, icing, hail, lightning, precipitation static, heavy precipitation, low ceiling and low visibility. Severe turbulence may extend up to 20 NM from a thunderstorm and there is no way to be sure by looking at a thunderstorm how dangerous it will be. Turbulence beneath a thunderstorm, in clear air, can be very dangerous. The probability of lighting strikes occurring to aircraft is greatest when operating at altitudes where temperatures are between -5°C and 5°C.

Lightning can occur in clear air and can puncture the skin of an aircraft, damage electronic equipment, cause engine failure and induce permanent error in magnetic compass.

Airborne and ground based radar picks up areas of precipitation. Generally, areas of high water content indicate the degree of turbulence to expect. These precautions should be followed:

a) don't land or take-off when a thunderstorm is approaching. The sudden wind shift of the gust front creating by the advancing cell or low level turbulence could result in a loss of control;
b) don't fly under a thunderstorm;
c) avoid any area where thunderstorms cover 6/10 or more of that area;
d) always review weather forecasts, reports and PIREPS to avoid thunderstorm activity.

7.04 Low Level Wind Shear

Review section 5.08 in this manual and the factors that cause wind shear.

Wind shear is most hazardous for aircraft within 1,000 ft AGL, primarily during approach for landing and during takeoff. On take-off, in areas of strong wind shear, and aircraft may encounter an area of strong increased performance wind shear as a headwind is entered, followed by sudden decreased performance wind shear as a downdraft and tailwind are encountered.

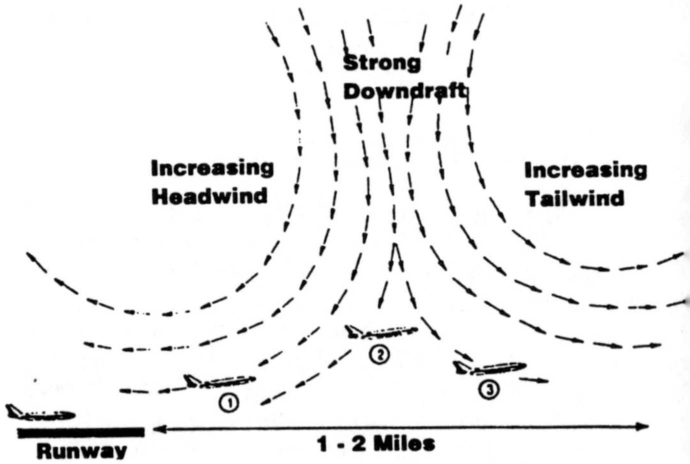

Fig. 7-1: Wind shear on departure.

The best defence to wind shear is to avoid it, and to know when it will exist based on reviewing weather information and PIREPS. However, should you recognize a wind shear encounter, prompt action is required. In all aircraft, the recovery could require full power and a pitch attitude consistent with the maximum angle of attack for your aircraft.

If increased performance wind shear is encountered, during descent, power will have to be initially reduced, then increased to a higher setting than before the shear was encountered to maintain a constant ground speed and approach angle. If decreased performance wind shear is encountered, during descent, power will have to be initially increased, then decreased to a lower setting than before the shear was encountered to maintain a constant ground speed and approach angle. If a downburst or vertical wind shear is encountered, the angle of attack will have to be INCREASED, because the relative wind is from a more vertical direction, lowering the normal in-flight angle of attack.

7.05 Collision Avoidance

To prevent a mid-air-collision, keep a constant look-out for other traffic, especially when taking off or landing, and in busy airspace. Scanning the sky for other aircraft is extremely important. To scan effectively, make a series of short, regularly spaced eye movements that bring successive areas of the sky into the central visual field. Each movement should not exceed 10 degrees, and each area should be observed for at least 1 second to enable detection. Although horizontal back-and-forth eye movements seem easiest, it depends on the individual and once you have developed a method of scanning, you should adhere to it. If you have a right seat passenger, have them scan for traffic as well.

Use the landing light frequently, especially when below 2,000 feet within terminal areas and aerodrome traffic patterns, and when taking off and landing. Birds also can see the landing light and avoid the aircraft. When scanning for traffic, if an object shows no movement on the windscreen, if the object is an aircraft, it is either coming directly towards you or moving directly away. Keep the windshield clean - dirt and bugs can look like distant aircraft and are an unnecessary distraction.

STROBE LIGHTS also help to make aircraft seen, but should not be used while taxiing, or when there is an adverse reflection from clouds while in flight.

7.06 Pilot-in-Command Responsibilities

The PIC of an aircraft, for every flight, should:

a) insure the aircraft is airworthy, both legally in terms of required documentation and that the aircraft is in a fit and safe condition for flight by virtue of thorough pre-flight inspection;
b) know all relevant air regulations, charts, Canada Flight Supplement, NOTAM, PIREPS, special procedures in effect etc. Keep up to date on current regulations by having up to date AIP. Dedicate yourself to ongoing re-currency and refresher flight training and ground school;
c) know the forecast and reported weather and complete required flight planning including accurate weight and balance computation;
d) be current and competent on the aircraft for intended flight;
e) be physically, mentally and medically fit for flight;
f) insure passengers have been given complete safety briefing including the location, operation and use of seat belts, shoulder harnesses, ELT, life jackets, first aid kit, fire extinguisher, exit procedures, and have safety belts fastened;
g) in flight, keep active lookout for other traffic at all times and comply with right of way rules, fly at proper cruising altitudes;
h) insure proper ATC clearance has been obtained prior to entering Class A, B, C, D, or E airspace, and that such clearance is adhered to;
i) maintain awareness and avoidance of restricted airspace such as Class F;
j) use proper communications procedure when in Class G airspace, including use of MF or ATF;
k) know all relevant numbers for the aircraft flown and know emergency procedures, and have POH on board and know it thoroughly;
l) have all appropriate, up to date charts on board aircraft;
m) have required safety equipment on board aircraft (lifejackets, ELT, survival equipment etc.);
n) insure that all ATC clearances and instructions can be complied with safely - if not, request alternate instructions;
o) always use vital actions checklists;
p) remain alert at all times and use good judgment.

7.07 James Brake Index

JAMES BRAKE INDEX (JBI) is a measurement used to indicate how slippery a runway is, due to conditions such as a wet or ice covered runway. A JBI of 1.0 indicates maximum theoretical friction coefficient for a bare and dry runway. In heavy rain or on an ice covered runway, the JBI may be in the 0.1 to 0.3 range, indicating slippery conditions. With a low JBI, the aircraft will take a greater distance to stop and will be more prone to drift in a crosswind. The greater the crosswind, the greater the required minimum safe JBI - for a 10 kt crosswind, the minimum JBI is .3, for a 15 kt crosswind, the minimum JBI is .4. See crosswind limits, s. 7.18 in this manual.

EXAMPLE JBI Report:

CYRB JBI 17/35 -10 .30 9801191055

Meaning: James Brake Index report Resolute NWT for runways 17 and 35, temperature -10°C, JBI .30 January 19, 1998 at 1055 UTC.

A RUNWAY SURFACE CONDITION (RSC) report indicates when conditions yield a slippery runway.

EXAMPLE RSC Report:

CYRB RSC ALL RWYS COVERED 1 IN LIGHT SNOW 9801190630

Meaning: Runway Surface Condition Report Resolute NWT all runways covered in 1 inch light snow January 19, 1998 at 0630 UTC.

7.08 Runways and Runway Markings

Runway numbering is based on runway direction. The number is the whole number nearest one-tenth the magnetic direction of the centerline of the runway, measured from magnetic north. For example, if a runway direction is 098° M, the runway would be numbered 10. In the northern domestic airspace, runways are numbered to the nearest degree in ° T (Example: Baker Lake NWT, Rwy 341T).

The take-off or landing area boundaries of aerodromes without runways are indicated by PYRAMIDAL or CONICAL type boundary markers, which are coulored international orange and white (airports) or solid international orange (aerodromes).

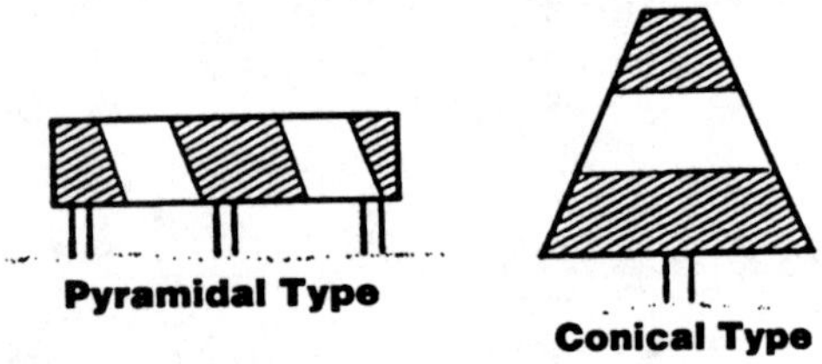

Fig. 7-2: Boundary Markers.

Airplane runway markings for non instrument runways include THRESHOLD MARKINGS, FIXED DISTANCE MARKINGS, and CENTRE-LINE MARKINGS.

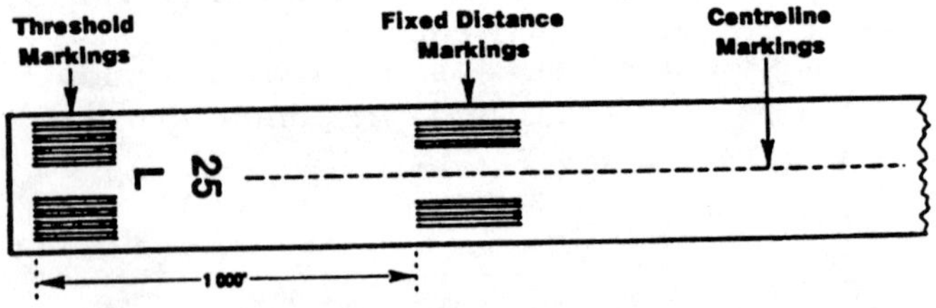

Fig. 7-3: Runway markings, non-instrument runway, over 5,000 ft.

Instrument runways will also have TOUCH DOWN ZONE MARKINGS.

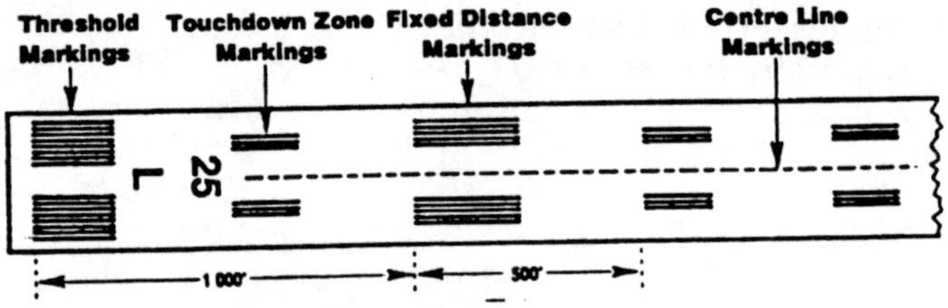

Fig. 7-4: Instrument runway markings.

Taxiways are marked with a centerline painted yellow. The taxiway edge has 2 yellow, continuous lines. Taxiway HOLDING LINES consist of a single yellow line and a dashed line (non-instrument runway) or double single lines and double dashed lines (instrument runway). Aircraft on the solid line side are required to hold short of the line until further clearance is given. Where no holding position lines are marked, aircraft must remain at least 200 feet from the runway edge.

Fig. 7-5: Taxiway holding markings.

7.09 Mountain Flying Operations

Before flying in a mountainous area, you should first receive a "mountain check-out" from a flying school. Such courses are generally 5-10 hours of ground school, followed by about 5 hours of flying. A mountain checkout is not a formal requirement in relation to aviation licensing, but is highly recommended for pilots who plan mountain flying. There are many factors involved in flying safely in the mountainous areas of Canada, especially areas such as B.C.

In B.C., the combined effect of mountain terrain and pacific weather systems leads to very changeable and often unpredictable weather. Other factors to consider are airport elevation, temperature, pressure, turbulence and localized mountain weather.

Some recommended mountain flying techniques and practises are:

a) File a flight plan.
b) Often, downdrafts are encountered in the shade areas of mountains, so the aircraft should be flown on the "sunny" side which will be more likely to have updrafts.
c) When crossing a ridge, never cross it direct in case you enter a downdraft - always cross a ridge at a 45° angle because this will allow you to turn away from the ridge if you will not have enough altitude to clear it.
d) Due to the remoteness of the terrain below, flying a perfectly straight track is often not the wisest route to take. The selected route should avoid topography that would prevent a safe forced landing. The route should be over populated areas and well known mountain passes. Sufficient altitude should be maintained to permit gliding to a safe landing in the event of an engine failure.
e) Wear appropriate clothing and bring appropriate survival gear (take a wilderness survival course).
f) Approach mountain passes with as much altitude as possible - downdrafts of 2,000 FPM on the leeward side are not uncommon.
g) Don't fly a light aircraft if the winds are forecast to be 35 kts or more at your chosen altitude.
h) Know and calculate the effects of density altitude (review s. 7.17 in this manual).
i) You should be able to complete a "maximum rate, minimum radius turn", which is generally achieved by slowing the aircraft, adding 10° - 20° flap and turning with full power at 30° - 45° bank.
j) Before entering the mountains after departure, climb overhead the departure airport until you have reached a safe cruising altitude, and only then should you turn en-route (often the terrain can climb faster than the aircraft).
k) VFR routes in western Canada are marked on VFR charts by a series of diamonds. These routes should generally be followed as they will usually be above roads and are well travelled by other VFR aircraft. However, much more is involved in choosing a route through mountainous areas and following the suggested routes does not imply any special level of facilities. The diamond routes are not always wisest in the particular circumstances.

Many other factors are involved in mountain flying and these are only a few examples of some specialized techniques that would likely be taught in a mountain check-out.

7.10 Wake Turbulence, Wing Tip Vortices

Wake turbulence is caused by wing tip vortices. The most powerful vortices are caused by heavy, slow moving aircraft in CLEAN configuration. Vortices are most likely encountered behind descending or climbing aircraft.

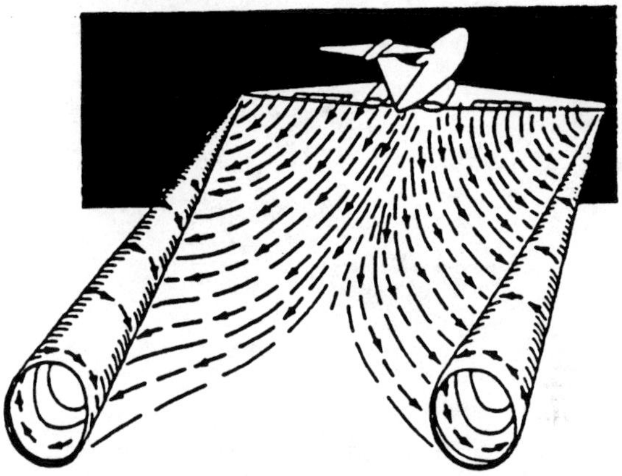

Fig. 7-6: Vortex generation.

Vortices persist for a minimum of 2 minutes. To avoid wake turbulence upon arrival, plan an approach path that is higher than the path of the generating aircraft, and plan to touchdown beyond the touchdown point of the generating aircraft. To avoid wake turbulence on departure, plan to lift-off well in advance of the lift-off point of the generating aircraft, and do not fly through its flight path.

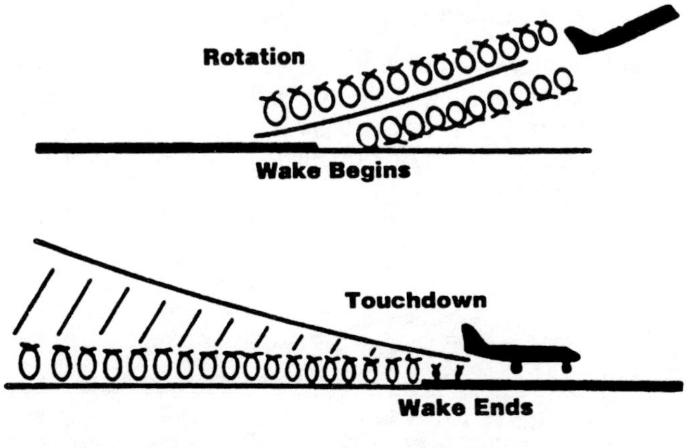

Fig. 7-7: Where vortices begin and end. Visualize and avoid these areas in take-off and landing.

Flight into a vortex core can produce roll rates of 80 degrees per second, and downdrafts of 1,500 feet per minute. It is extremely important to visualize the location of vortices and avoid flying into these areas.

Fig. 7-8: Vortex Roll.

When the vortices of a large aircraft sink to within about 200 feet of the ground, they tend to move laterally at about 5 kts. In a 5 - 10 kt crosswind, the upwind vortex remains stationary with the runway.

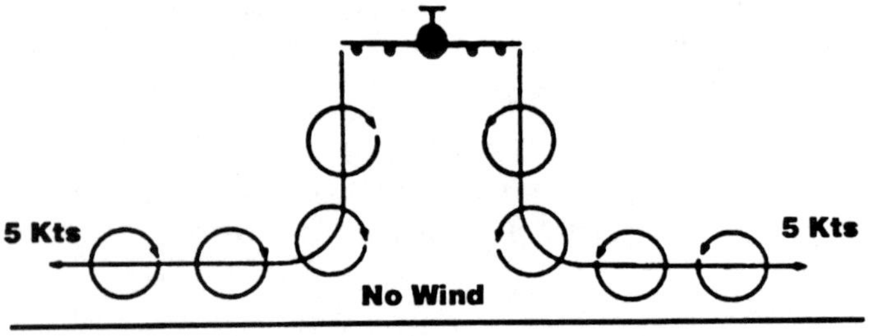

Fig. 7-9: Vortex movement in ground effect, no wind.

A tailwind can push the vortices back on the runway following landing of a heavy aircraft. The worst ground conditions would be a light, quartering tailwind, because the vortices will creep forward and remain over the runway.

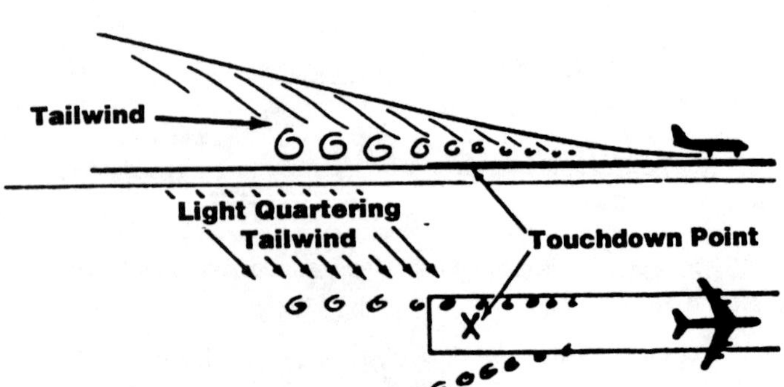

Fig. 7-10: Vortex movement in wind.

In forward flight, HELICOPTERS generate a pair of trailing vortices similar to wing-tip vortices of fixed wing aircraft. Helicopter vortices are highly concentrated and the size of the helicopter is not a significant factor in vortex strength. Pilots of small aircraft should avoid the vortices as well as the downwash.

Pilots should also know that when operating in an ATC environment, the receipt of any traffic information, instructions to follow another aircraft, or the acceptance of approach clearance does not make ATC responsible for wake turbulence. THE PILOT IS ALWAYS RESPONSIBLE TO INSURE COMPLIANCE WITH ATC WILL PERMIT A SAFE APPROACH AND LANDING. If in doubt, advise ATC and request alternate instructions.

7.11 Wheelbarrowing

When a landing aircraft has an excessive approach speed, the wings carry too much lift for a proper flare. The result, on nose-wheeled aircraft, is that the main gear is lifted off the ground while the nosewheel remains on the ground, and a "wheelbarrow" effect because the aircraft pivots about the nosewheel. To recover from wheelbarrowing, reduce speed to put more weight on the main landing gear. Another recovery technique is to raise the nose wheel, apply full power and overshoot. Refer to the aircraft manual for the specific recovery technique for your aircraft. Low wing, steerable nose gear are more susceptible to wheelbarrowing.

EXAMPLE: An aircraft on take-off roll goes up on its nosewheel and begins to wheelbarrow. What should the pilot do?

ANSWER: Apply back pressure to the control column so as lift the nosewheel off the ground.

7.12 Ground Effect

Close to the ground, ground effect reduces induced drag. Once above about 18 feet however, there will be a sharp increase in induced drag as ground effect stops. The result is that an aircraft can become airborne quickly, but without sufficient airspeed, and the aircraft is in danger of stalling. To compensate for the loss of ground effect, ensure the aircraft is at a safe climb speed prior to commencing the climb out, and avoid an excessively high approach speed.

Ground effect is useful for a rough field take-off, to get the aircraft airborne quickly to minimize the stress on the landing gear. Once airborne, accelerate to the appropriate climb speed.

7.13 Spiral Dives

A spiral dive is a steep, descending turn with the aircraft in an excessively nose-down attitude. To confirm a spiral dive, note the airspeed indicator: it will show a rapid build up and be well within the caution range. Spiral dives are hazardous because in the pull up, the forces on the aircraft can cause structural damage. The recovery technique for a spiral dive is as follows:

1. Close the throttle.
2. Level the wings. (Actions 1 and 2 almost simultaneously.)
3. Keep straight.
4. Ease out of the dive.
5. Apply power as required to maintain height.

7.14 Spins

A spin is a condition of stalled flight where the aircraft is simultaneously rolling, yawing and pitching about a vertical spin axis. All that is needed to enter a spin is a stall and yaw. The first stage of the spin is the INCIPIENT STAGE. After about 2 turns, the FULLY DEVELOPED STAGE occurs. To confirm a spin, note the airspeed indicator: the airspeed is about the stall speed, the vertical speed shows a high rate of descent, and the altimeter shows a rapid loss of height. The turn indicator shows a turn in the direction of the spin. Use the recovery technique specified by the manufacturer. Generally, the recovery technique is as follows:

1. Power to idle, neutralize ailerons.
2. Insure flaps are retracted.
3. Apply and hold full opposite rudder to the direction of rotation.
4. Just after the rudder reaches the stop, move the control column forward far enough to break the stall.
5. Hold these control inputs until the rotation stops.
6. As rotation stops, neutralize rudder, level the wings and recover smoothly from the dive.

All practice spin recoveries must be completed not less than 2,000 feet AGL, or at a height recommended by the manufacturer, whichever is greater.

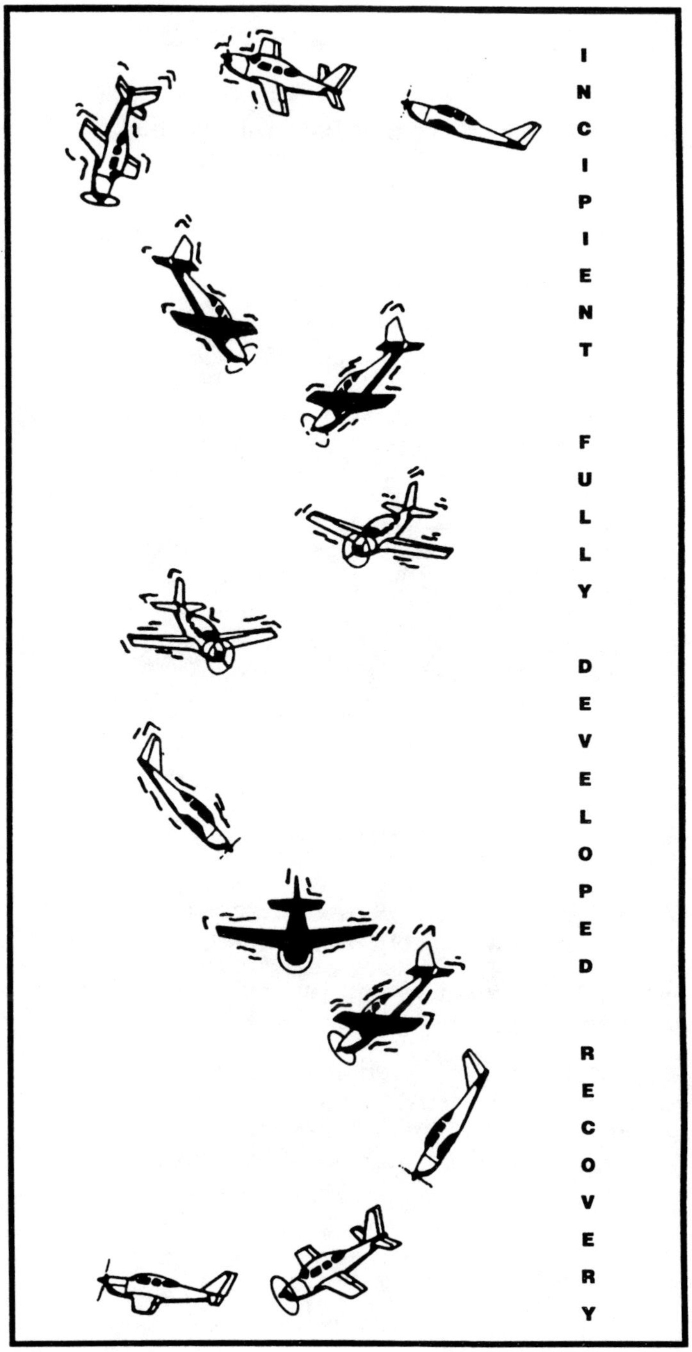

Fig. 7-11: 2-turn spin.

7.15 Position of Flight Controls While Taxiing in Wind

Fig. 7-12 below shows how the controls should be positioned where the wind is from various directions for a nose-wheel, tricycle type airplane.

Fig. 7-12: How to hold controls while taxiing in wind: nose wheeled aircraft.

EXAMPLE: What control inputs are necessary when taxiing with a wind from 8:00 O'clock?

ANSWER: Down aileron on left wing, down elevator. [ie: move control column forward and right]

For taxiing a tail-wheeled airplanes in a tailwind, move controls forward to maintain steering on tailwheel. When taxing a tail-wheeled airplane in a headwind, move the controls aft to hold the tail down.

7.16 Weight and Balance

General

The purpose of a weight and balance calculation is to ensure the aircraft is safely loaded. Each aircraft has a maximum permissible weight, referred to as the MAXIMUM GROSS WEIGHT. In addition, the aircraft should be loaded so that it remains balanced. The aircraft manual specifies the appropriate range of positions for the center of gravity. If the aircraft's center of gravity is outside its limitations, the aircraft must not be flown.

Pilots should also remember that an overloaded or improperly loaded aircraft invalidates the Certificate of Airworthiness and flying the aircraft in such a condition is a violation of the Air Regulations. Further, most insurance policies will become invalidated by improper aircraft loading, and should an accident occur, the insurer will likely deny coverage.

Effects of Weight

An overloaded aircraft may not be able to leave the ground, or if it does become airborne, it may show unexpected and unusually poor flight characteristics. Excessive weight will also result in:

a) higher take-off speed;
b) longer take-off run;
c) reduced rate and angle of climb;
d) lower maximum altitude;
e) shorter range;
f) reduced maneuverability;
g) higher stalling speed;
h) higher landing speed;
i) longer landing roll;
j) excessive weight on the nosewheel.

Weight Changes

The weight of an airplane can be changed by altering the fuel load. Fuel has considerable weight - 30 U.S. gallons may weigh more than one passenger. But if weight is reduced by lowering the fuel carried, the airplane range will be decreased.

During flight, fuel burn will result in a weight reduction, improving aircraft performance.

Changes of fixed equipment have a major effect upon the weight of the airplane. An airplane can be overloaded by the installation of extra radios or instruments. Repairs or modifications usually add to the weight of the airplane. When an airplane ages, it tends to put on weight. The total effect of this gain is known as SERVICE WEIGHT PICKUP. Most service weight pickup is from the weight of new and replacement parts installed in repair, overall, and modification. In addition, an unknown weight pickup results from the collection of trash and hardware, moisture absorption in soundproofing, and from the accumulation of dirt and grease. This pickup can only be determined by weighing the airplane.

Balance, Stability, and Center of Gravity

The positioning of loads in an aircraft has a major effect on how it flies. Balance refers to the CG position. The prime concern of the CG position is whether it is longitudinally correct, but the LATERAL CG can also be significant. A lateral unbalance condition can occur when the fuel load is mismanaged by supplying the engine(s) unevenly from tanks on one side of the airplane. The pilot can compensate for the resulting wing-heavy condition by adjusting the aileron trim tab or by holding a constant aileron control pressure. However, this places the airplane controls in an out-of-streamline condition, increases drag, and results in decreased operating efficiency.

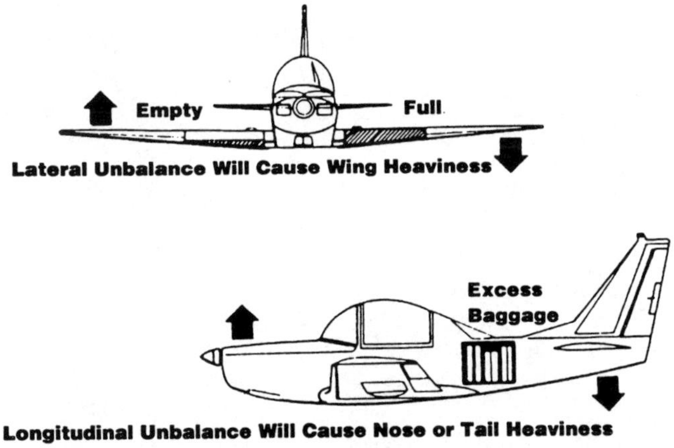

Fig 7-13: Lateral and longitudinal imbalance.

Since lateral balance is relatively easy to control and longitudinal balance is more critical, further reference to weight and balance in this book will be in reference to longitudinal CG.

An aircraft with a forward CG is "heavier" in flight and will be slower than the same aircraft with an aft CG. The reason for this is that with forward loading, nose up trim is required in most airplanes to maintain level cruising flight. Nose-up trim involves setting the tail surfaces to produce a greater down load on the aft portion of the fuselage, which adds to the wing loading and the total lift required from the wing if altitude is to be maintained. This requires a higher angle of attack of the wing, which results in more drag and, in turn, produces a higher stalling speed.

With aft loading and nose-down trim, the tail surfaces will exert less download, relieving the wing of that much wing loading and lift required to maintain altitude. The required angle of attack of the wing is less, so the drag is less, allowing for a faster cruise speed.

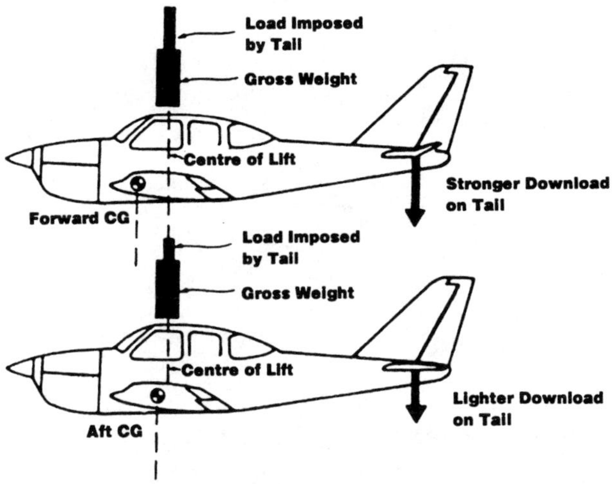

Fig. 7-14: Load distribution affects performance.

If the center of gravity is too far aft, the aircraft will be extremely unstable when in slow flight and when approaching the stall. Where an aircraft stalls with an excessively aft center of gravity, the stall will be more severe, leading to a spin if yaw is uncorrected. If the center of gravity is too far forward, the aircraft will have a nose down tendency and there may not be enough back elevator to complete a safe landing flare. For these reasons, aircraft must be properly loaded with the center of gravity in the proper range.

Know the following terms:

a) Reference Datum. An imaginary vertical plane from which all distances are measured. Positioned at firewall (Cessna) or at tip of propeller spinner (Piper).
b) Arm. The distance from the reference datum to the center of an item, such as fuel tanks, front seats etc. Measured in inches.
c) Center of Gravity. The balance point of the aircraft.
d) Center of Gravity Limits. Permissible range of center of gravity.
e) Basic Empty Weight. Empty aircraft, unusable fuel, residual oil (full oil in some aircraft) and optional equipment (radios etc.). Also known as Licensed Empty Weight (LEW).
f) Moment Index. A moment divided by a constant such as 100, or 1,000. The purpose of the moment index is to simplify calculations where moment numbers are too large.
g) Station. Refers to specific location as a number equivalent to the arm. An item located at station +50 would have an arm of 50 inches aft.
h) Useful Load. Refers to the weight of the pilot, copilot, passengers, baggage, usable fuel, and drainable oil. It is the empty weight subtracted from the maximum allowable gross weight.

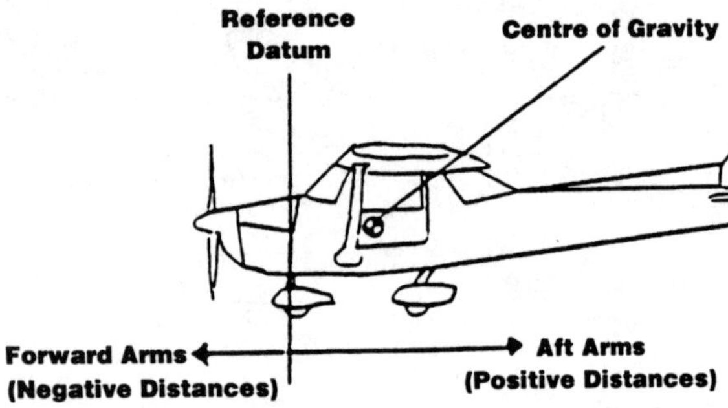

Fig. 7-15: Basic terminology: Weight and Balance.

The first step in completing a weight and balance calculation is to determine the weights. For example, if the aircraft has 20 U.S. gallons of fuel, the fuel weighs 20 X 6.0 lbs/U.S. gallon = 120 lbs.

With respect to passengers, ACTUAL WEIGHTS are best determined by weighing each passenger, including exterior clothing and articles of carry-on baggage. Where weighing scales are not available, standard weights may be used as follows:

	summer	winter
Males (12 yrs up)	182 lbs	188 lbs
Females (12 yrs up)	135 lbs	141 lbs
Children (2 - 11 yrs)	75 lbs	75 lbs
Infants (less than 2)	30 lbs	30 lbs
Large athletes	215 lbs	215 lbs

If no carry-on baggage is involved, reduce the weights for males and females by 8 pounds. If the standard weights are inaccurate, ask the passenger for their weight and add an 8 pound allowance for clothing in summer, and 14 pounds in winter.

Once a particular weight is known, the next step is to multiply it by the arm, to obtain the moment. A loading graph can also be utilized to determine the moments.

Finally, add up all the moments and divide by the total weight to determine the location of the center of gravity. Refer to the aircraft moment envelope to ensure the aircraft if safely loaded.

Memorize these formulas:

WEIGHT (lbs) X ARM (") = MOMENT ("-lbs)

$$\frac{\text{MOMENT}}{\text{WEIGHT}} = \text{ARM}$$

$$\frac{\text{MOMENT}}{\text{ARM}} = \text{WEIGHT}$$

EXERCISE:

	Weight	Arm	Moment
1.	1600 lbs	31"	______
2.	_____	16"	32,000 "-lbs
3.	340 lbs	___	12,500 "-lbs

ANSWERS:
1. 49,600 "-lbs
2. 2,000 lbs
3. 36.8"

On older aircraft, the manufacturer may not provide a moment envelope to confirm whether a loading condition is safe. Instead, they provide a range of permissible locations for the center of gravity. To determine the loaded center of gravity, multiply the weights by the arms to determine each moment, then divide the total moment by the total weight.

EXAMPLE: Complete the following weight and balance. The manufacturer indicates that the aircraft is safe to fly for a loaded gross weight under 3900 lbs, and a center of gravity between 28" and 34". Is the aircraft safely loaded?

	Weight (lbs)	Arm (")	Moment ("-lbs)
Empty Aircraft	3,000		94,000
Pilot, Pass.	360	18"	______
Fuel	200	20"	______
Oil	10	-10"	______
Baggage	100	23"	______
TOTALS	______		______

$$\text{CG} = \frac{\text{TOTAL MOMENT}}{\text{TOTAL WEIGHT}} = \_\_\_\_\_\_ = \_\_\_\_"$$

ANSWER: Total moment = 106,680"-lbs (subtract -100 moment for oil). Total weight = 3670 lbs. Center of Gravity is 29.1" aft of datum. Aircraft is loaded below maximum gross weight, and center of gravity is within limitations, therefore aircraft is safely loaded.

A weight and balance calculation is simplified by using a LOADING GRAPH to determine the moments. Line up the weight for an item and read off the resulting moment (remember that all moments have been divided by 1000). The advantage of using a loading graph is that it isn't necessary to multiply all the arms by the weights to determine the moments.

A CENTER OF GRAVITY MOMENT ENVELOPE is used to determine whether the aircraft is within acceptable limitations. If the intersection point of the total weight and total moment is within the envelope, the aircraft is safely loaded.

EXAMPLE: Complete the following weight and balance using the loading graph and moment envelopes at Fig. 7-16 and 7-17.

Item	Weight (lbs)	Moment ("-lbs)
Empty Aircraft	1100	35,900
Fuel	135	______
Front Seats	340	______
Baggage (area # 1)	25	______
TOTALS	______	______

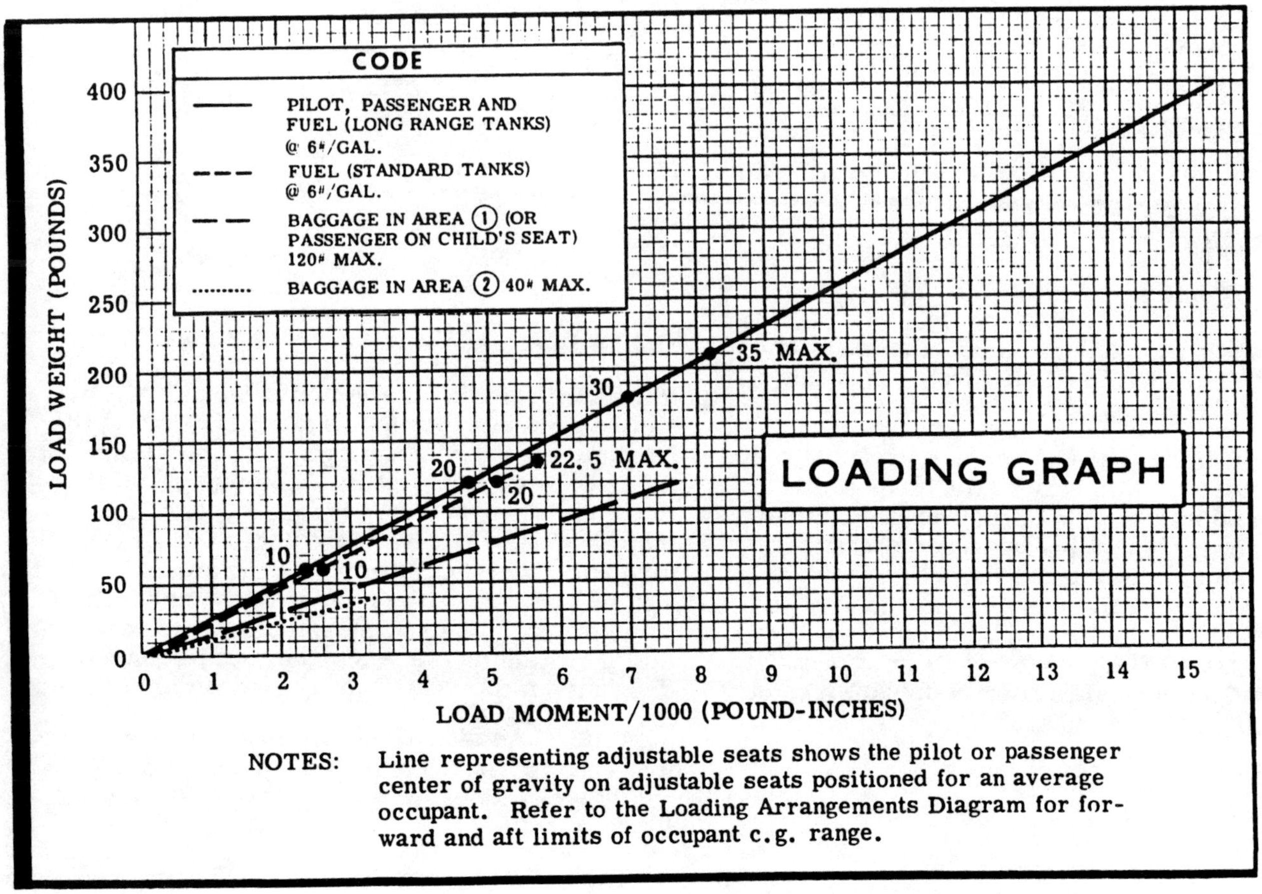

Fig. 7-16: Loading Graph.

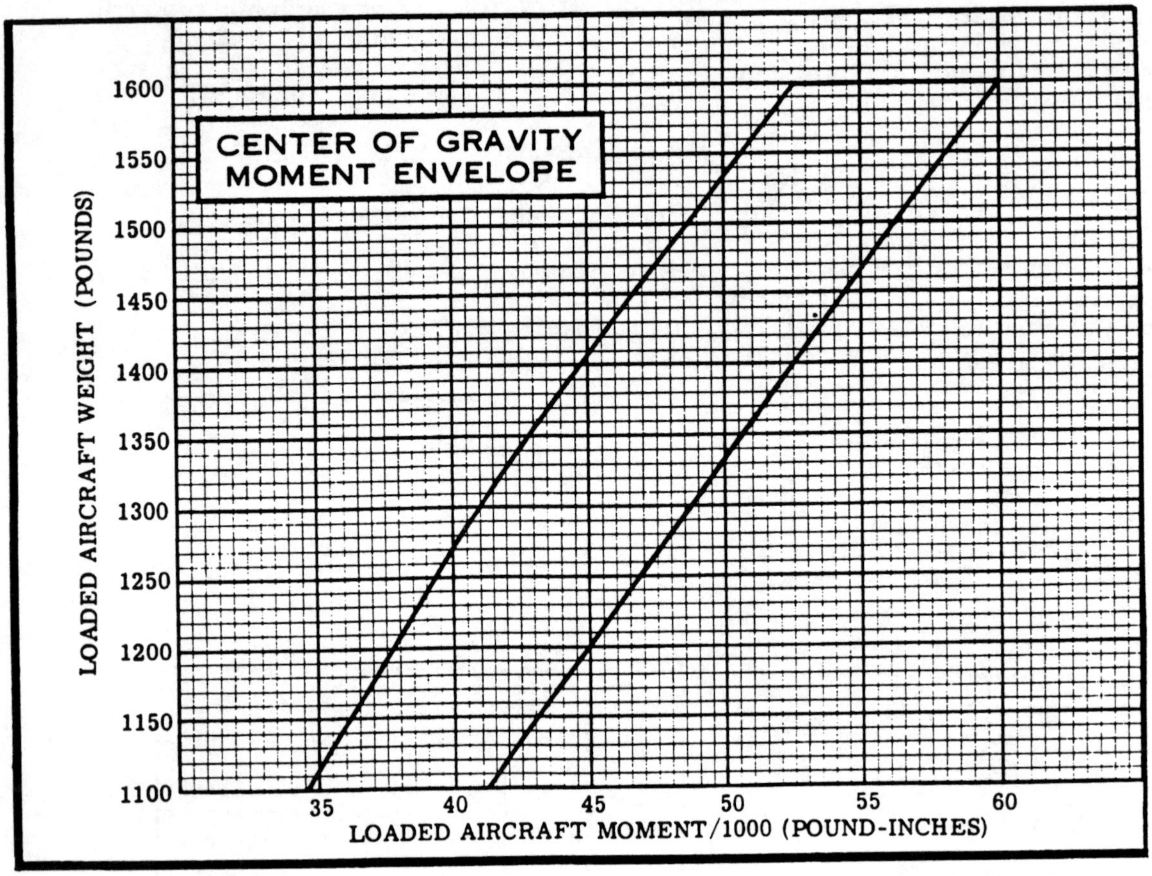

Fig. 7-17: Moment Envelope.

ANSWER: Total moment = 56,500 "-lbs. Total weight = 1600 lbs. C of G = 35.3" aft. Aircraft is within envelope and safely loaded.

For some airplanes, the CG limits, both forward and aft, may be specified to vary as the gross weight changes in flight with fuel burn. The CG limits may also be changed depending upon the aircraft operations such as acrobatic flight. The term UTILITY CATEGORY in such aircraft will refer to the loading and CG limits for aerobatic operation, while the NORMAL CATEGORY will refer to non-aerobatic operations. The normal category will allow greater weight and wider CG limits.

Load Adjustment

If the CG is outside of the limits, weight can be shifted from one location to another. Moving the location of a particular weight will not change the total aircraft weight, but will change the total moment. When weight is moved forward, total moments decrease; when weight is moved aft, total moments increase. The greater the movement of the weight, the greater the change in moment. These principles can be used to re-locate various weights and then calculate a new weight an balance to determine if the CG is brought within limits.

EXAMPLE: An aircraft loaded moment is 57,000 "-lbs. The aircraft maximum moment is 55,000. What would be the effect of moving 100 lbs of baggage forward by 20"?

ANSWER: The new moment for the baggage would be 2,000 "-lbs less, bringing the aircraft total moment to 55,000 which would be within limits.

7.17 Airplane Performance

Density Altitude

Many accidents have occurred because pilots did not understand the effect of varying conditions of airplane performance. In addition to the effects of weight and balance previously discussed, other factors such as density altitude, humidity, winds, runway gradient, and runway surface conditions all have a large impact in changing aircraft performance.

The factor which most significantly affects aircraft performance is DENSITY ALTITUDE. Air density is perhaps the single most important factor in performance: it directly affects the power output of the engine, efficiency of the propeller, and the lift generated by the wings. When the air temperature increases, the air density decreases. As well, air density decreases with an increase in altitude. This is why the term used to discuss the effects of air density is density altitude. Remember that the HIGHER the density altitude, the LOWER the aircraft performance. Think of density altitude as the altitude the aircraft "thinks" it is operating at.

To obtain an approximate density altitude using a flight computer, enter pressure altitude and outside air temperature.

EXAMPLE: Using a flight computer, what is the approximate density altitude where the pressure altitude is 10,000 feet, and the outside air temperature is 20° C?

ANSWER: 13,000 feet density altitude.

Humidity

Because of evaporation, the atmosphere always contains some moisture in the form of water vapour. This water vapour replaces molecules of dry air and because water vapour weighs less than dry air, any given volume of warm air will weigh less and will be less dense than the same volume of dry air. This means that humidity will decrease aircraft performance.

Density Altitude Effect on Engine Power and Propeller Efficiency

At high density altitude, and in high humidity, the engine delivers less power. This is because the engine delivers power in proportion to air density, therefore, as density decreases, power output decreases. This is true for all normally aspirated, non-supercharged, non-turbocharged engines.

The propeller produces thrust in proportion to the mass of air being accelerated through rotating blades. If the air is less dense, propeller efficiency will be decreased.

The problem of high density altitude operations is made worse by the fact that when the air is less dense, more engine power and increased propeller efficiency are needed to overcome the decreased lift efficiency of the wing. This additional power and propeller efficiency are not available under high density altitude, and as a result, performance will decrease.

Effect of Wind on Airplane Performance

During take-off, a headwind will shorten the takeoff run and increase the angle of climb. A headwind in flight, will decrease groundspeed and increase fuel consumption for the distance covered.

A tailwind will increase the take-off run and decrease the angle of climb. A tailwind in flight will increase the groundspeed and reduce fuel consumption.

During landing, a headwind will steepen the approach angle and shorten the landing roll, while a tailwind will decrease the approach angle and increase the landing roll. Downwind operations should be avoided if at all possible, as the speed at which the aircraft contacts the ground will be significantly greater.

Runway Surface Condition and Gradient

The take-off distance is affected by the surface condition of the runway. If the runway is muddy, wet, soft, rough, or covered with grass, there will be increased friction and the take-off distance will be increased.

Runway conditions which will increase the landing roll are ice, snow or standing water covering the landing surface.

The RUNWAY GRADIENT or slope is important when runway length is limited. UPSLOPE runways will provide a retarding force which slows acceleration, increasing the take-off run. A DOWNSLOPE runway will decrease the take-off distance but increase the landing roll. Due to terrain, some runways may be one way, and special precautions must be taken.

Performance Charts

Most aircraft manufacturers provide adequate information from which the pilot can determine aircraft performance. Such information can be found in the Pilot's Operating Handbook for the aircraft used. Always used the proper make, model and year Handbook for the particular aircraft: for example, never use a C-150 manual for a C-152, and never use a 1969 C-172 manual for a 1979 C-172 etc.

Remember that any performance chart is based on flight tests conducted under normal operating conditions, using average piloting skills, with the airplane and engine in good operating condition. Any deviation from these conditions will affect performance. You should maintain the assumption that even after correctly using a performance chart, your aircraft will probably not perform as well as stated in the POH.

Standard atmospheric conditions are used in producing performance charts: MSL temperature 15° C (59° F), zero relative humidity, 29.92 " Hg MSL Pressure. This provides a base from which to evaluate performance when conditions change.

When reading any performance chart, first review the conditions and notes. Often, it is necessary to vary the performance figure obtained due to some condition.

Cruise Performance Charts

Cruise charts indicate true airspeed and fuel consumption for various conditions.

EXAMPLE: Refer to the cruise performance chart at Fig. 7-18 to determine how many hours of fuel remain and what is the TAS under the following conditions:

Altitude . . . . 5,000 ft
Propeller RPM 2,300
Manifold Pressure . . . . 22" Hg
Mixture . . . . Lean
Fuel Remaining . . . . 40 Gal

ANSWER: 3 hrs 10 minutes fuel remaining, TAS 155 MPH. Locate 5,000 ft altitude in first column, move down and right to 2,300 RPM, then down and right to 22 inches MP. Brake Horsepower - 158, % power 69, TAS 155, fuel consumption 12.6 GPH. With 40 gallons fuel at 12.6 GPH, there will be 3.17 hours flying time or 3 hours 10 minutes.

CRUISE PERFORMANCE CHART

Altitude	RPM	MP	BHP	% BHP	TAS/mph	Gal/Hr
2500	2450	23	175	76	158	14.2
		22	166	72	154	13.4
		21	157	68	151	12.7
		20	148	63	148	12.0
	2300	23	164	71	154	13.1
		22	153	67	149	12.2
		21	143	62	145	11.5
		20	135	59	142	11.0
	2200	23	153	67	149	12.1
		22	144	63	146	11.4
		21	135	59	142	10.8
		20	126	55	138	10.2
Maximum Range Settings	*2000*	*20*	*107*	*47*	*126*	*8.7*
		19	*99*	*43*	*121*	*8.2*
		18	*89*	*39*	*113*	*7.5*
		17	*81*	*35*	*105*	*7.0*
5000	2450	23	179	78	163	14.5
		22	169	73	159	13.6
		21	161	70	156	13.0
		20	150	65	151	12.2
	2300	23	167	73	158	13.4
		22	158	69	155	12.6
		21	148	64	151	11.9
		20	139	60	146	11.2
	2200	23	157	68	155	12.4
		22	148	64	151	11.7
		21	138	60	146	11.0
		20	131	57	143	10.5
Maximum Range Settings	*2000*	*19*	*103*	*45*	*126*	*8.5*
		18	*94*	*41*	*118*	*7.9*
		17	*86*	*37*	*111*	*7.3*
		16	*79*	*34*	*103*	*6.8*
7500	2450	21	163	71	161	13.1
		20	153	67	157	12.4
		19	143	62	152	11.7
		18	133	58	147	11.0
	2300	21	151	66	156	12.2
		20	142	62	151	11.6
		19	133	58	147	11.0
		18	125	54	142	10.5
	2200	21	143	62	152	11.4
		20	134	58	148	10.7
		19	126	54	143	10.2
		18	118	51	138	9.7
Maximum Range Settings	*2000*	*19*	*107*	*47*	*131*	*8.7*
		18	*98*	*43*	*123*	*8.1*
		17	*90*	*39*	*116*	*7.6*
		16	*82*	*36*	*107*	*7.0*

Data based on lean mixture, ISA conditions, maximum gross weight.

Fig. 7-18: Cruise Performance Chart.

Take-Off Distance, Landing Distance Charts

Take-off and landing distance charts can be used to determine the expected take-off or landing distance over a 50 foot obstruction and ground roll under various conditions. Always read any notes provided on the chart (for example, distances have to be increased for a tailwind). INTERPOLATION means calculating a value between two other values. For example, using the distances for 10° and 20°, you could calculate the take-off distance at 14°C:

Temperature (°C)	10°	20°
Take-off Distance (feet)	1400	1600

Notice that no distance is provided for a temperature of 14° C. To interpolate for 14° C, take the spread from 10° C to 20° C, and convert the 14° as a percentage between 10° and 20°. The spread is 4°/10° = 40%. Next, multiply the percentage by the spread for the take-off distance. 40% X 200 feet = 80 feet. Add the 80 feet to 1,400 feet = 1,480 feet take-off distance at 14° C.

EXERCISE (use Fig. 7-19 below):

1. Find the total take-off distance to clear a 50 ft obstruction, at sea level and 10° C.
2. Find the total take-off distance to clear a 50 ft obstacle at 1,000 ft pressure altitude and 15° C.
3. Find the total take-off distance to clear a 50 ft obstacle, at 3,000 ft pressure altitude and 18° C.
4. Find the ground roll at 1,600 ft pressure altitude and 20° C.
5. Find the total take-off distance to clear a 50 ft obstacle, given 4,000 ft pressure altitude, 26° C with a 9 kt headwind.
6. Find the total take-off distance to clear a 50 ft obstruction, given 3,500 ft pressure altitude, 30° C and a dry, grassy runway.
7. Find the total take-off distance to clear a 50 ft obstacle, given 6,000 ft pressure altitude, 27° C and a tailwind of 6 kts.
8. Find the total take-off distance to clear a 50 ft obstacle, given airport elevation 2,600 ft ASL, airport altimeter setting 28.92, and 24°C.
9. Find the total take-off distance to clear a 50 ft obstacle, given altimeter setting 28.63, 23° C, airport elevation 5,286 ft ASL, and a dry, grassy runway.
10. Would you attempt take-off where the pressure altitude was 7,000 feet at 40° C?

TAKEOFF DISTANCE

CONDITIONS:
Flaps Up
Full Throttle Prior to Brake Release
Paved, Level, Dry Runway
Zero Wind

NOTES:
1. Maximum performance technique as specified in Section 4.
2. Prior to takeoff from fields above 5000 feet elevation, the mixture should be leaned to give maximum RPM in a full throttle, static runup.
3. Decrease distances 10% for each 9 knots headwind. For operation with tailwinds up to 10 knots, increase distances by 10% for each 2 knots.
4. Where distance value has been deleted, climb performance after lift-off is less than 150 fpm at takeoff speed.
5. For operation on a dry, grass runway, increase distances by 15% of the "ground roll" figure.

WEIGHT LBS	TAKEOFF SPEED KIAS		PRESS ALT FT	0°C		10°C		20°C		30°C		40°C	
	LIFT OFF	AT 50 FT		GRND ROLL	TOTAL TO CLEAR 50 FT OBS	GRND ROLL	TOTAL TO CLEAR 50 FT OBS	GRND ROLL	TOTAL TO CLEAR 50 FT OBS	GRND ROLL	TOTAL TO CLEAR 50 FT OBS	GRND ROLL	TOTAL TO CLEAR 50 FT OBS
1600	53	60	S.L.	655	1245	710	1335	765	1435	820	1540	880	1650
			1000	720	1365	775	1465	835	1575	900	1690	970	1815
			2000	790	1500	855	1615	920	1735	990	1865	1065	2005
			3000	870	1650	935	1780	1010	1915	1090	2065	1170	2225
			4000	955	1820	1030	1965	1115	2125	1200	2290	1290	2475
			5000	1050	2015	1140	2185	1230	2360	1325	2555	1430	2770
			6000	1160	2245	1255	2435	1360	2640	1465	2870	1580	3120
			7000	1285	2510	1390	2730	1505	2970	1625	3240	---	---
			8000	1420	2820	1540	3080	1670	3370	---	---	---	---

Fig. 7-19: Take-Off distance Chart

ANSWERS:

1. 1,335 ft.
2. 1,520 ft.
3. 1,888 ft.
4. 886 ft.
5. 2,002 ft (decrease distance by 10% for 9 kts headwind).
6. 2,349 ft (increase distance by 15% of "ground roll" figure).
7. 3,641 ft (increase distance by 10% for each 2 kts tailwind).
8. 2,105 ft (first, calculate pressure altitude: 29.92 - 28.92 = 1" Hg. 1,000 ft + 2,600 ft = 3,600 ft PA. Then, interpolate distance for 20° C = 2041 ft and 30° C = 2,200 ft. Interpolate for 24°C, total distance = 2105 ft).
9. 3,128 feet (first calculate pressure altitude: 29.92 - 28.63 = 1.29" Hg. 1,290 ft + 5286 ft = 6,576 ft PA. Then, interpolate total distance for 20° C = 2,830 ft and 30° C = 3,083 ft. Interpolate for 23° C = 2,906 ft total distance. Next, interpolate as above for ground roll because of dry grass runway = 1,480 ft. Finally, add 15% of ground roll figure to total = 222 ft + 2,906 ft = 3,128 ft).
10. NO! Climb performance after lift-off will be less than 150 feet per minute, so take-off should not be attempted.

LANDING DISTANCE charts involve exactly the same principles: remember to read carefully all associated conditions.

Gliding for Range

To achieve a flat gliding angle, the aircraft should be flown at the angle of attack that gives the best LIFT/DRAG RATIO (L/D). The L/D is a measure of the gliding efficiency or aerodynamic cleanness of the airplane. If the L/D is 10/1, it means that lift is 11 times greater than drag.

Flying the airplane at the maximum angle of attack will not give the best L/D, because drag is large at high angles of attack. By reducing the angle of attack below the maximum, the airspeed increases and the induced drag decreases. At some point the best L/D will be achieved, and if airspeed is increased further, parasite drag begins to rise rapidly and the aircraft will begin to lose more altitude per forward distance travelled.

Higher gross weight does not decrease the glide angle. A fully loaded airplane will sink faster, but at a greater forward speed, with the same gliding range as a lightly loaded airplane. However, a heavier loaded aircraft will glide to the ground in less time.

The approximate speed to obtain the best L/D can be obtained from the POH.

Gliding range will be increased when gliding with a tailwind.

Fig. 7-20 below shows a typical maximum glide chart. Note the conditions specified: speed - 80 MPH, windmilling propeller, flaps up with zero wind. It is important to be familiar with such charts so that gliding range can be estimated in case of engine failure.

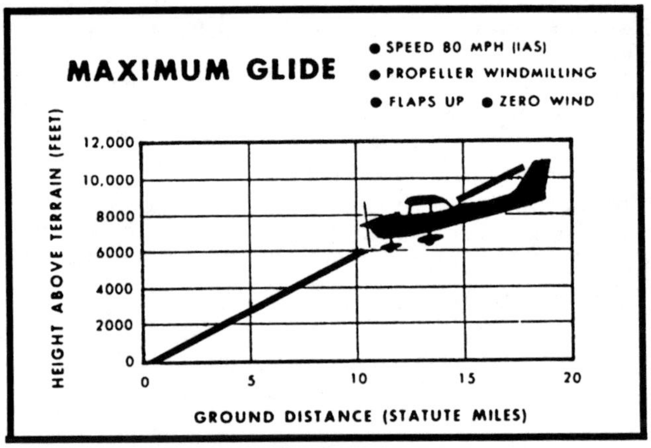

Fig. 7-20: Typical Glide Range Chart.

Flying for Range and Endurance

The POWER CURVE below shows power required, power available and airspeed. Note that the difference between horsepower required and available is greater with higher speed, meaning less power is available for climb. At the aircraft maximum speed, the power required and power available meet.

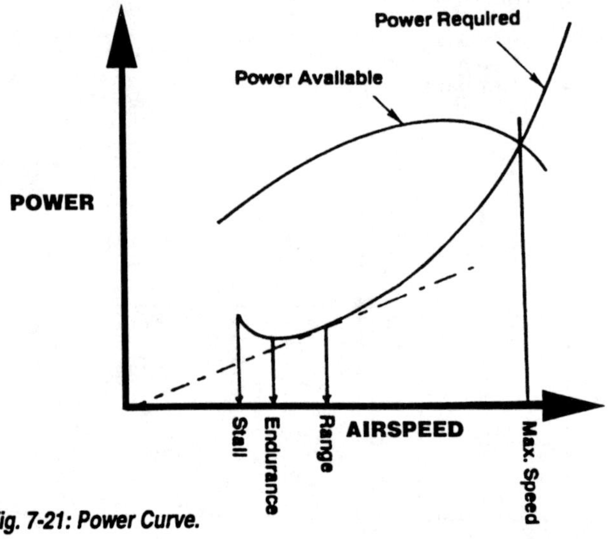

Fig. 7-21: Power Curve.

The power curve also illustrates the speed required to obtain maximum range. This point is the airspeed indicated at the intersection of a straight line drawn from the origin to the power required curve. This corresponds to the best L/D. Accordingly, to obtain maximum range, or greatest forward distance per unit fuel consumed, the power setting required is that for the airspeed for best L/D.

A RANGE CHART provides a good indication of the forward distance an aircraft can travel for a given amount fuel.

EXAMPLE: Using the range chart below, what is the approximate range based on 2,400 lbs gross weight, zero wind, 60% power at 6,000 feet density altitude?

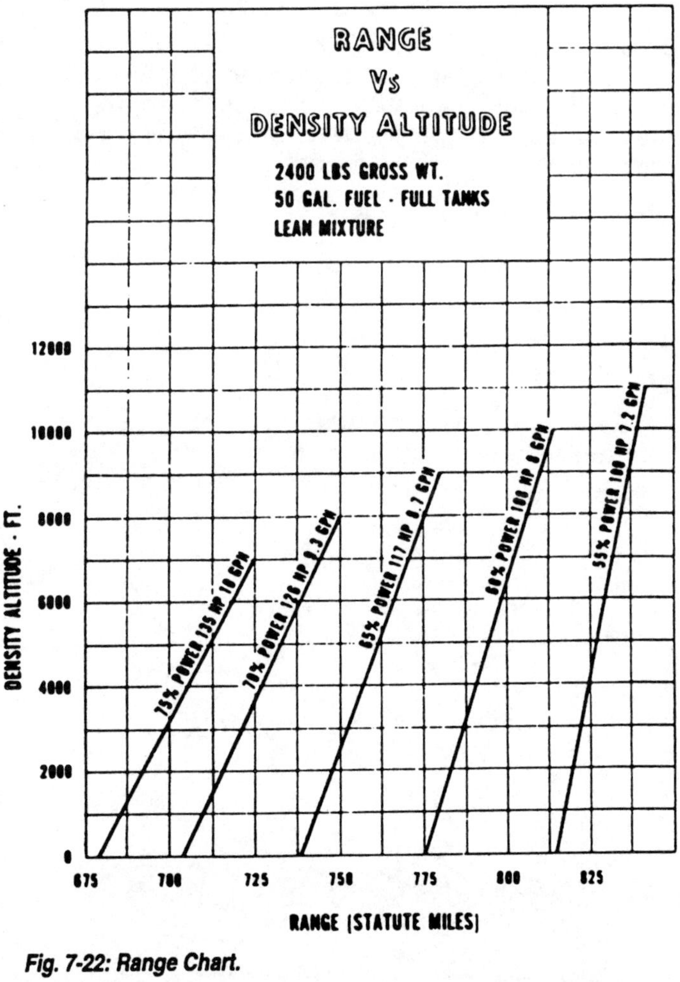

Fig. 7-22: Range Chart.

ANSWER: 795 statute miles. Note that range could be improved by reducing power to 55%.

At higher altitude, the speed and power required for best range increases because of the greater true airspeed. This is because the drag of the airplane at altitude is the same as the drag at sea level but the higher true airspeed causes a proportionately greater power required.

At higher gross weight, greater lift is required to obtain the best L/D, more power is required for best range, and the airspeed will be higher. While cruising for range, as fuel burns, the gross weight decreases, so the power for best range will decrease.

To improve range performance, the aircraft should be flown at the best lean mixture setting to maximize engine performance. Additionally, the aircraft should be flown at a cruising altitude having the most favorable winds.

To obtain MAXIMUM ENDURANCE, the aircraft is flown to achieve the greatest time aloft for the minimum fuel consumed. Flying an aircraft at the endurance power setting will be necessary when it becomes impossible to land due to snow covering an airport, for example. On the power curve, the lowest point of the power required represents the power setting for endurance. For normally aspirated engines, best endurance will be found at sea level.

SLOW FLIGHT is defined as flight in the region of airspeeds from the stall to endurance. Reviewing the power curve, you will see that power settings in slow flight will be greater than for endurance, and that for a slow flight power setting, that same power would provide an airspeed usable for cruise. This is why flight in slow flight is called flying "behind the power curve" or flight in the region of "reversed command", and the aircraft will be sluggish and easily stalled. Control of an aircraft in slow flight is important: every landing requires flight into slow flight.

Take-off, Landing Performance

For any take-off or landing, all procedures recommended in the aircraft manual should be adhered to.

A NORMAL TAKEOFF technique is used in light winds on a paved surface, into wind.

Different conditions require specialized techniques. In a CROSSWIND TAKEOFF, generally, the ailerons are turned into wind and directional control is maintained with rudder. The aircraft is lifted off positively so it won't settle back to the runway. In a ROUGH or SOFT FIELD TAKEOFF, more distance will be required, taxiing to position should be in a continuous movement, and a nose high attitude should be maintained during the take-off roll to lift the aircraft off as soon as possible. The aircraft should then be accelerated in ground effect before climbing. If runway length is minimal, a SHORT

FIELD TAKEOFF is required. This may involve using partial flaps. Maximum acceleration is obtained by holding brakes while full power is applied, and the aircraft is lifted off in a minimum drag attitude.

Landing technique is also dictated by existing conditions and the aircraft manual's procedures.

A NORMAL LANDING is made in light winds on a dry, paved surface. For a CROSSWIND LANDING, generally, the aircraft is put in a sideslip toward wind, and the rudder is used to keep the aircraft aligned with the runway. The aircraft should touchdown slightly on the upwind wheel and should roll straight with the ailerons into wind. In a SHORT FIELD LANDING, generally, flaps are required and a slower than normal approach speed is maintained. The aircraft should be touched down on the main wheels at the lowest possible speed, and flaps should be retracted on touchdown. An OBSTACLE CLEARANCE LANDING will require the use of power and flaps to acquire a steep approach. In GUSTY WINDS, it is advisable to add half the gust factor to a maximum of 10 to the approach speed. For a SOFT or ROUGH FIELD LANDING, the aircraft should be touched down at the lowest possible speed consistent with safety. The touchdown attitude should be a high nose up attitude, and the nose wheel should be held clear with minimum use of brakes; power may be necessary to keep the aircraft from getting stuck.

Performance Speeds

Airplane performance speeds are also referred to as V-SPEEDS. Know the following V-speeds:

Va — Design maneuvering speed. Maximum speed at which the limit load factor can be imposed (either by gusts or full deflection of the controls) without causing structural damage.

Vfe — Maximum airspeed permissible with wing flaps in a given position. Exceeding Vfe with flaps can impose damaging loads on flaps.

Vle — Maximum airspeed at which the airplane can have landing gear extended.

Vlo — Maximum airspeed at which the landing gear can be safely extended or retracted.

Vne — The airspeed which should never be exceeded. Flight above this speed can result in structural damage/failure.

Vno — Maximum airspeed for normal operation or maximum structural cruising speed. Beyond this speed, load factor limit can be exceeded.

VsO — Calibrated power-off stall speed or minimum steady flight speed at which the airplane is controllable in the LANDING configuration.

Vs1 — Calibrated power-off stall speed or minimum steady flight speed at which the airplane is controllable in a SPECIFIED configuration.

Vx — Speed at which the airplane will obtain the highest altitude in a given horizontal distance. Also called best angle of climb speed.

Vy — Speed at which the airplane will obtain the maximum increase in altitude per unit of time. Vy decreases slightly with altitude.

7.18 Crosswind Limitations

U.S. manufactured aircraft are designed to withstand a maximum 90° crosswind of 20% of the stall speed. This means that take-off or landing should not be attempted where the existing winds exceed the maximum crosswind.

Example maximum crosswinds:

Aircraft Stall Speed	Maximum 90° Crosswind
60 kts	12 kts
50 kts	10 kts
45 kts	9 kts

Note: Refer to the owner's manual for the specific maximum crosswind for a particular aircraft.

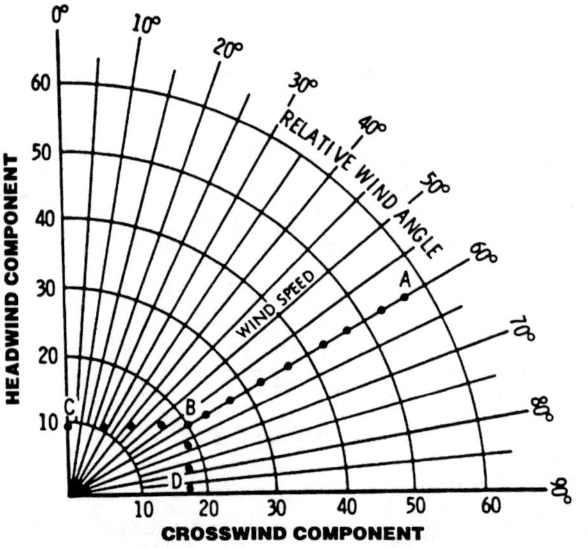

Fig. 7-23: Crosswind Chart.

The purpose of a crosswind chart is to calculate the crosswind component for various wind conditions. To use a crosswind chart, follow these steps:

1. Determine aircraft maximum 90° crosswind (20% X Vso or from A/C manual).
2. Determine angle between wind and runway.
3. Determine wind speed.
4. Mark on crosswind chart where wind angle and wind speed intersect.
5. Draw a straight line from mark to bottom of chart. Where this line crosses horizontal axis is existing crosswind. If this amount exceeds the aircraft maximum 90° crosswind, then crosswind limit is exceeded.
6. Headwind component is found by drawing a straight line to vertical line. Read off wind component parallel to runway.

EXAMPLE 1: What is the crosswind component where runway direction is 360° M, and winds are from 060° M at 20 kts?

ANSWER 1: 18 kts. Put mark where 60° off runway meets curved wind arc of 20 kts. Crosswind of 18 kts is on crosswind axis, effective headwind is 10 kts on headwind axis.

EXAMPLE 2: What is the maximum wind speed from 45° off runway, for 15 kts max 90° crosswind?

ANSWER 2: 22 kts. Draw straight line up from 15 kts on crosswind scale. Where this line meets 45° = 22 kts.

The crosswind chart can be used to find the maximum permissible wind speed for a given wind direction (see example 2 above). To find maximum permissible winds, take the maximum 90° crosswind, and draw a vertical line upwards on the crosswind chart: where the line crosses the speed lines and wind degrees off runway is the maximum permissible wind speed.

EXERCISE 1: (refer to the crosswind chart).

	Aircraft Stall Speed	Maximum 90° Crosswind	Runway Direct-tion	Wind Degrees magnetic	Cross-wind Comp.	Head-wind Comp.	Limits Exceeded (yes/no)
1.	60 kts	_______	030°M	060/10 kts	____	____	______
2.	80 kts	_______	260°M	240/50 kts	____	____	______
3.	75 kts	_______	180°M	260/15 kts	____	____	______

ANSWERS:

	Maximum 90° Crosswind	Crosswind Component	Headwind Component	Limits Exceeded
1.	12 kts	5 kts	8 kts	No
2.	16 kts	17 kts	48 kts	Yes
3.	15 kts	15 kts	4 kts	Yes

EXERCISE 2: (refer to the crosswind chart).

	Maximum 90° Crosswind Component	Wind Degrees Off Runway	Maximum Permissible Wind Speed
1.	10 kts	30°	______
2.	15 kts	45°	______
3.	12 kts	80°	______

ANSWERS:
1. 18 kts.
2. 22 kts.
3. 13 kts.

7.19 Overshoot Technique

An overshoot may be necessary due to a poor approach or instructions from ATC. Once the decision has been made to overshoot, apply full power, accelerate to a safe climb speed in level flight, reduce flap extension as required according to type, and raise the nose to the climbing attitude.

7.20 Open Fuel Cap In Flight

If the fuel cap is open in flight, fuel escapes rapidly into the low pressure area above the wing. Accordingly, the fuel caps should be carefully checked to insure they are securely closed, especially after refuelling.

7.21 Conversion Factors

Many conversions have to made when operating an aircraft. Standard E6B slide type computers are set up to do many of the basic conversions. Use the reverse side of the computer (calculator side) where the various conversions are marked with reference lines. For example, to convert statute miles to nautical miles, there will be a small reference line for "stat", and a small reference line for "naut" - simply line up the number sought to be converted, and under the other line will appear the desired number of other units.

An electronic aviation computer can also be utilized for conversion factors. Generally, enter the number of units sought to be converted, then press "CONV", then press the appropriate button having the indicated unit conversion (ie nm - sm) and the desired number of units will appear in the display.

Another way to do conversions is to simply memorize conversion ratios, or to make up a handy reference table and attach it to your lapboard. Some of the more common conversions ratios are:

To Convert	Into	Multiply by
Imperial Gallons	U.S. Gallons	1.2
U.S. Gallons	Imp. Gals	0.8
U.S. Gallons	Litres	3.8
Imperial Gallons	Litres	4.5
Litres	U.S. Gallons	.26
Kilograms	Pounds	2.2
Nautical Miles	Statute Miles	1.2
Statute Miles	Nautical Miles	0.9
Statute Miles	Kilometers	1.6
Litres Fuel	Pounds	1.6
U.S. Gallons Oil	Pounds	7.5

7.22 Canada Flight Supplement

The Canada Flight Supplement (CFS) provides detailed information on aerodromes in Canada and is a reference for the planning and safe conduct of air operations. For all flights you should review the appropriate aerodrome information.

Remember that the Air Regulations require that the Pilot in Command familiarize themselves with all relevant information appropriate to an intended flight. Accordingly, the UP-TO-DATE CFS extract for the aerodrome(s) used should be reviewed thoroughly. A pilot who is not familiar with an aerodrome and unaware of any special procedures in effect will be unsafe and risk enforcement action being taken against them (including license suspension). Get a yearly subscription to the CFS - see the AIP in the MAP section for ordering and price information.

Review the table at Fig. 7-24: it provides most of the information presented in the CFS that is relevant to VFR, private operations. After reviewing the table, complete the exercise at p.142.

TOPIC	Information Provided
REF	Reference. Latitude and longitude of aerodrome. Distance to nearest community. Variation. Time zone conversion. Aeronuatical charts in use. Elevation.
OPR	Operator of Aerodrome. Public License (Pub Lic). Private License (Pvt Lic, owner's permission required in advance). Unlicensed (Unl). Military (Mil). Transport Canada (TC).
PF	Public Facilities. A Facilities available in terminal building. B Facilities on aerodrome. C These facilities exist in adjacent town. D These facilities exist within 30 nm of aerodrome. E These facilities are scarce. F No facilities exist at aerodrome or within 30 nm. X Facilities are unknown. 1 Telephone. 2 Food. 3 Taxi. 4 Medical Facilities. 5 Accommodation (rental). 6 Car Rental.
CUST	Customs. AOE - A. Authorized airport of entry. AOE/RA/CAPACITY. Authorized for limited weights or number of passengers. AOE-XA. Authorized for general aircraft only and small charter with less than 15 passenger. If no suffix "A"appears, direct contact must be made in advance with customs.
FLT PLN	Flight Planning. TWB = Transcribed Weather Broadcast. DUAT = Direct User Access Terminal. CARS. Community Aerodrome Radio Station. NOTAM FILE = location where NOTAM info can be found. ACC= Area Control Center Weather: WD1 = Military Wx office (with forecaster). WD2 = Military Wx office. W1 = Environment Canada Wx office. W2= Weather at FSS including pictorial display. W3 = Weather at FSS. W4= limited service such as self service.
SERVICES	1 Fuel grades available. 2 Oil grades available. 3 Servicing (S). 0 = no storage. 1= storage avail. 2 = servicing avail. 3 = minor repairs. 4 = major repairs. 5 = parking (extended) 6 = tie downs. 7 = plug in facilities. 4 National Crash Fire Fighting and Rescue Services (CFR). Airport categories 1 - 9 establish increasing capability. 5 Supporting Fluids, Systems and Oxygen. Examples: D-Ice, OXRB (oxygen replacement bottles) etc. 6 Jet Aircraft Starting Units (JASU). Specifications of units avail. 7 Deceleration Parachutes (DEC P). Aircraft types specified for which parachutes avail. 8 Military Advisory (MIL ADV) 9 Private Advisory (PVT ADV)
RWY DATA	Runway Data. Runway dimensions, magnetic bearing, weight limitations. JBI = James Brake Index (braking deceleration information). OPR = Operator. RCR = Runway Condition Report. PLR = Pavement Loading Rating. PCN = Pavement Classification Number. CABLE = Arrestor Cable.
LIGHTING	Runway lighting available. ARCAL = Aircraft Radio Control of Aerodrome Lighting, Type J = key microphone 5 times in 5 seconds, Type K = key mike 7 times initially. AS = runway identification strobes. SF = Sequenced Flashing. TE = Threshold and Runway End. A = Centerline and Rapid Exit from Taxiway. TDZL = Touchdown Zone and Runway Centerline Lighting. V1 = 2-Bar VASIS. V2 = 2 Bar VASIS (DC 8 and smaller). V3 = 3-bar Vasis. AV = AVASIS (small aircraft). PAPI = Precision Approach Path Indicator. Vt = T-VASIS .
COMM	Communications. X = frequency can be requested. T = transmit only. R = receive only. MF = Mandatory Frequency. ATF = Aerodrome Traffic Frequency. Other frequencies self explanatory (TWR, GND, ATIS etc.)
NAV	Navigation Services. VOT = Omnitest. ILS = Instrument Landing System. NDB = Non directional Beacon. TACAN = Tactical Navigation Equipment (military). VORTAC, VOR/DME. PAR = Precision Radar Approach.
PRO	Procedures. Special procedures such as right hand circuits , taxiway restrictions, arrival procedures etc.

Fig. 7-25: Summary of information provided in Canada Flight Supplement.

CALGARY/SPRINGBANK ALTA YBW

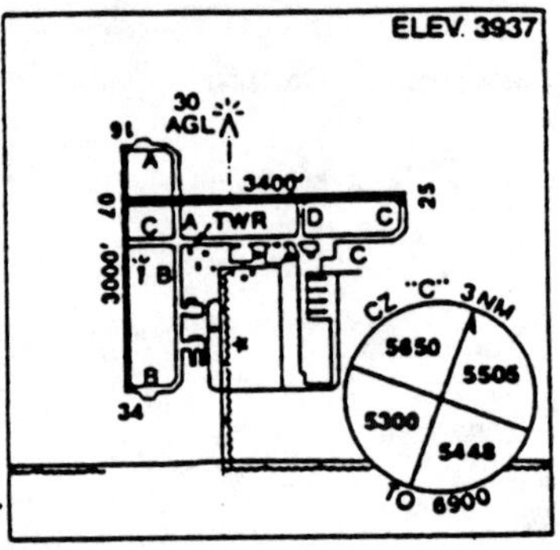

REF	N51 06 11 W114 22 24 8W 20°E UTC-7(6) Elev 3937' A5005 E-15 E-16 LE4 LE5 LE11 T1
OPR	TC 403-288-9476 Pub Lic
PF	A-1.2 avbl H24 D-2.3.5
FLT PLN FSS WX TWR	 Calgary 403-250-0337 W4 403-247-2739 1345-2315Z (DT 1245-2215Z) Calgary Intl area 403-274-7542
SERVICES FUEL OIL S PVT ADV	 100, JB 15-24Z (DT 14-23Z) Mon-Fri All 1(ltd).2.3.4.5.6.7 North American Air Training College 123.4 14-06Z (DT 13-05Z)
RWY DATA RCR	Rwy 07(070°)/25(250°) 3400x100 asphalt Rwy 16(160°)/34(340°) 3000x75 asphalt Opr. 1430-23Z (DT 1330-22Z) Mon - Fri. Aprt ltd to propeller driven acft up to 12,500 lbs gross tkof wt. PLR/PCN
LIGHTING	07-(TE ME). 16-(TE ME). 25-(TE ME). 34-(TE ME) 14-06Z (DT 13-06Z) O/T ARCAL - 118.2 type K.
COMM ATIS GND TWR ATF VDF	 125.5 121.7 14-06Z (DT 13-06Z) 118.2 120.7 (V) 14-06Z (DT 13-06Z) tfc 118.2 06-14Z (DT 06-13Z) 3NM 6900 ASL 118.2 120.7 121.7 (V)
NAV VOR	 YBW 108.6 N51 06 24 W114 22 19 at A/D on test Unmonitored
PRO	Rgt hand circuits rwys 16 & 25. Twy C to "T" hgs not visible to twr. For VFR routes see Calgary VTPC or as directed by ATC. Aerobatics may be in progress over aprt. 7500 ASL & below.

NOT FOR NAVIGATION

Fig. 7-25: CFS Extract - Springbank, Alta.

QUESTIONS BASED ON CFS EXTRACT ABOVE:

1. What is the location identifier for the airport?
2. Who is the operator? Is prior permission required to use the airport?
3. What is the elevation? Length of longest runway? What is the magnetic variation?
4. Is there a phone, taxi or restaurant at the airport? Car rental? Medical Facilities?
5. What octanes of fuel are available? Can major repairs be carried out?
6. Are the runways hard surfaced? What is the width of the widest runway?
7. How could one get a runway condition report? Are there limitations for heavy A/C?
8. How would lighting be activated for arrival at night when tower is closed?
9. Is a DF steer service available? What are the control tower hours of operation?
10. Which runways have a non-standard circuit pattern? Are taxiing aircraft visible to tower?

ANSWERS:

1. YBW.
2. Transport Canada. No prior permission required because it is a public licensed airport.
3. Elev. 3937 ft ASL. 3,400 ft. Mag. variation 20° E.
4. Telephone and food at airport 24 hours. Food, taxi, rental accommodation within 30 nm. No medical facilities at airport.
5. Fuel grades 100, Jet B from 15Z to 24Z UTC, Monday to Friday. Major repairs available.
6. Runways are all hard surfaced asphalt, widest runway 100 ft.
7. For runway condition report (RCR), contact Transport Canada, 403-288-9476. Airport is limited to propeller driven aircraft up to 12,500 lbs gross take-off weight.
8. Key mike over 118.2 MHz 7 times.
9. DF steer available over 118.2. Tower hours of operations 1400 - 0600 UTC.
10. Right hand circuit runways 16 /25. From taxiway C to T-hangars aircraft not visible to tower.

7.23 Rate and Radius of Turn

For the same angle of bank, slower aircraft have a greater rate of turn and smaller radius of turn, and faster aircraft have a smaller rate of turn and larger radius of turn.

EXAMPLE: Three aircraft are all turning at an angle of bank of 20°. Aircraft A has a speed of 130 kts TAS, aircraft B 195 kts TAS, and aircraft C 260 kts TAS. Which airplane has the greatest rate of turn?

ANSWER: Aircraft A, having the slowest airspeed and consequently the shortest radius of turn, has the greatest rate of turn.

7.24 Medical Facts for Pilots

Pilots should commit themselves to a high standard of PERSONAL HEALTH. Regular exercise is a must, and pilots should eat wisely to insure proper nutrition and avoidance of obesity and other risk factors.

HYPOXIA is a shortage of oxygen in the blood. It is a dangerous condition, because it severely impairs one's ability to operate an aircraft. An early symptom of hypoxia is a sensation of well-being (referred to as euphoria). To avoid hypoxia, do not fly above 10,000 ft without supplemental oxygen.

HYPERVENTILATION is overbreathing caused by anxiety or panic. The result is a lower percentage of blood carbon dioxide. Sensations of hyperventilation are a tingly feeling and shortness of breath, followed by dizziness and eventual unconsciousness. Breathing into paper bag, reducing breathing rate to 10 - 12 breaths per minute, or talking aloud raises carbon dioxide levels and stops the condition.

FATIGUE slows reaction times, reduces concentration and leads to attention errors. Fatigue may be caused or aggravated by lack of sleep, insufficient rest, business pressures or other stresses. Boredom and fatigue worsen each other. To overcome boredom, make frequent ground speed checks, weather updates, fuel consumption calculations, flight planning for diversions etc. as needed to keep alert.

HEARING can be damaged from aircraft noise and vibration. Using noise protection with ear plugs, headsets or helmets is a good practice.

HYPOTHERMIA is severe accidental cooling of the body. Prolonged exposure to water or cold air temperatures following an aircraft accident or unplanned landing without proper clothing typically lead to this condition. Basic first aid treatment is to provide blankets and lukewarm, sweet drinks to raise lowered blood sugar level. Do not use hot water bottles or other direct heating as this will cause sudden opening of superficial blood vessels, taking away blood from deep tissues and essential organs.

Do not fly under the influence of alcohol. Alcohol decreases altitude tolerance - 1 drink at 6,000 feet has the same effect as 2 at sea level. Although the Air Regulations require a minimum of 8 hours lapse before flying, following alcohol consumption, an excellent rule is to allow 24 hours. ALCOHOL ABUSE AND DEPENDENCY is a dangerous condition for pilots and must be avoided or treated.

Do not fly while under medication, except on the advice of an Aviation Medical Examiner. Drugs associated with aviation accidents are: antihistamines, tranquilizers, appetite suppressing drugs, barbiturates, nerve tonics and digestive pills. Where anesthetics have been given following serious operations, do not fly until advised it is safe by a medical practitioner. Following dental procedures using local anesthetics, wait at least 24 hours before flying. DRUG ABUSE can occur with illicit drugs such as cannabis, codeine and cocaine. The effect of drug dependency will be lower flight performance.

In Canada, PREGNANT pilots may continue to fly until the 30th week if approved by their doctor. Flight beyond the 30th week of pregnancy is not advisable due to possible risks to the unborn child (from seat belt injury and cosmic radiation), and the danger of pre-mature delivery. Generally, 4 - 6 weeks after birth should be waited before flying again.

VERTIGO is a sensation of disorientation which results from various causes, such as the absence of visual references. FLICKER VERTIGO results from the strobe effect caused by looking through a windmilling propeller into the sun. To avoid flicker vertigo, increase the propeller RPM or chose a runway that does not line up into the sun.

G-FORCE affects all persons in aircraft. It refers to the acceleration forces acting on an aircraft - 1G is one times the force of gravity. NEGATIVE G FORCE is when the forces are upwards, such as when pushing forward on the controls. The most serious effect of positive G force is to draw blood away from the head, leading to hypoxia and a loss of vision called GREY OUT. Continued high positive G force can eventually lead to black out and unconsciousness. Greater positive G forces can be tolerated with the VALSALVA MANOEUVRE - which is accomplished by pushing against a closed throat like a child just about to throw a temper tantrum. Pressure suits used by military pilots also increase positive G tolerance by putting pressure on the legs, minimizing downward blood flow.

The human body can only tolerate small sustained negative G force. During high negative G, the blood rushes to the head, leading to RED OUT when there is a red haze in vision as the blood vessels in the eye expand.

CARBON MONOXIDE is an odorless, colorless gas that results from leaks in the exhaust systems. Symptoms of carbon monoxide poisoning are blurred thinking or headache, and cherry-red lips. If carbon monoxide poisoning is suspected, immediately shut off the heater, open the air ventilators, descend to lower altitudes, and land at the nearest airfield. Consult an Aviation Medical Examiner. It may take several days to fully recover. Carbon monoxide also enters the blood from SMOKING: individuals who smoke a pack of cigarettes per day will have 5 - 10% of their blood hemoglobin poisoned and will experience hypoxia at significantly lower altitudes than non smokers.

Do not fly an aircraft immediately following SCUBA DIVING: due to changes in nitrogen gas pressure, bubbles can form in the bloodstream causing difficulty in breathing or death. Wait at least 24 hours prior to flying after scuba diving.

Following BLOOD DONATION, pilots should wait at least 48 hours before flying.
EAR AND SINUS PAIN may be caused by changes in air pressure following climb or descent, especially at high rates of descent. Pressure equalization can generally be accomplished by yawning or swallowing. However, an upper respiratory infection such as a cold or sore throat, or a nasal allergic condition can make equalization difficult. For the same causes, a block may be experienced in the frontal sinuses, above each eyebrow, or in the maxillary sinuses, in each upper cheek. In a sinus block, there is often excruciating pain in the sinus area, and sometimes bloody mucus may discharge from the nasal passages. A sinus block or ear block can be prevented by not flying with an upper respiratory infection or allergy. If an ear or sinus block does not clear after landing, a doctor should be consulted.

The CIRCADIAN RHYTHM refers to normal sleep - wake cycles, which can be disturbed by shift work and JETLAG. To minimize jet lag, try to get into the new time rhythms and do not sleep at day. It is estimated that it takes 1 day per time zone crossed to re-establish circadian rhythm.

STRESS from the pressures of everyday living can impair pilot performance, often in very subtle ways. Difficulties, particularly at work can occupy thought processes enough to markedly decrease alertness. Distraction can so interfere with judgment so that unwarranted risks are taken, such as flying into deteriorating weather conditions to keep on schedule. Stress and fatigue are an extremely hazardous combination.

Certain upsetting events, including a job loss, divorce, loss of family member or financial difficulties can make EMOTION a factor in a pilot's ability to fly safely. The emotions of anger, depression, and anxiety from such events not only decrease alertness but also may lead to taking risks that border on self-destruction. Any pilot who experiences an emotionally upsetting event should not fly until satisfactorily recovered from it.

Aircraft accidents show that pilots should be conducting a PERSONAL CHECKLIST of themselves as well as their aircraft, because pilot impairment contributes to many more accidents that failures of aircraft systems. A personal checklist should cover all items of general health mentioned above.

7.25 VASIS

VASIS means "Visual Approach Slope Indicator System". It consists of lights on the edge of the runway which indicate if the aircraft is descending on the proper approach slope to the runway (usually 3°). VASIS systems provide safe wheel clearance over the runway threshold and at the same time ensure that available runway length is fully utilized.

A 2-BAR VASIS (V1 and V2) consists of 4 light units on the left side of the runway in the form of 2 pairs of wing bars referred to as the upwind and downwind wing bars. The wing bars project a beam of light having a white colour in the upper part and a red colour in the lower part. The indications provided are:

a) when on the approach slope, the upwind bar will show red and the downwind bar will show white;
b) When above the approach slope, both upwind and downwind bars will show white;
c) when below the approach slope, both upwind and downwind bars will show red;
d) when well below the approach slope, the lights of the 2 wing bars will merge into a red signal.

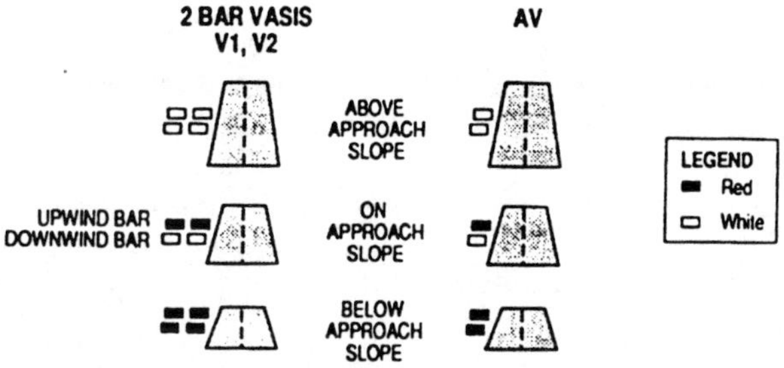

Fig. 7-26: 2 Bar VASIS System. V1 is for aircraft having an Eye-to-Wheel (EWH) height of up to 10 feet, DC 3 or smaller. V2 is for aircraft with EWH up to 25 feet, DC 8 or smaller.

A 3-Bar VASIS is basically a 2-bar VASIS (V2) with one light unit added to form an additional upwind bar. This provides a greater threshold wheel clearance for aircraft with large EWH (wide body). Wide bodied aircraft use the upwind and middle bars to provide safe wheel clearance and conventional aircraft (up to 25 ft EWH) use the middle and downwind bars as with V2.

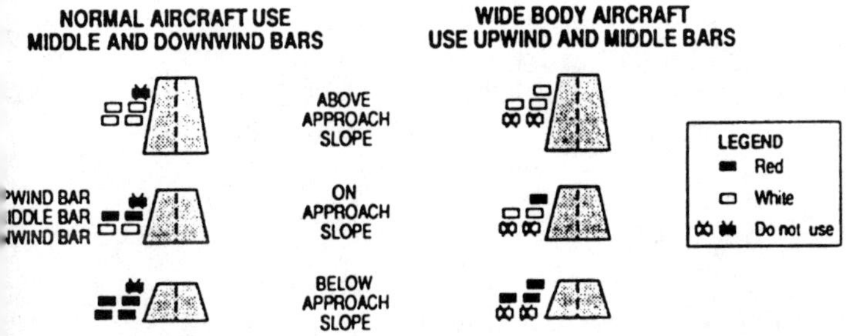

Fig. 7-27: 3 Bar VASIS.

7.26 Precision Approach Path Indicator (PAPI)

PAPI consists of 4 light units sited on the left side of the runway in the form of a wing bar. The indications are as follows:

a) When on the approach slope, the 2 units nearest the runway show red and the 2 units furthest from the runway show white.

b) When slightly above the approach slope the 1 unit nearest the runway shows red and the other 3 show white; when further above the approach slope, all 4 units show white.

c) When slightly below the approach slope, the 3 units nearest the runway show red and the other white; when well below the approach path, all 4 units show red.

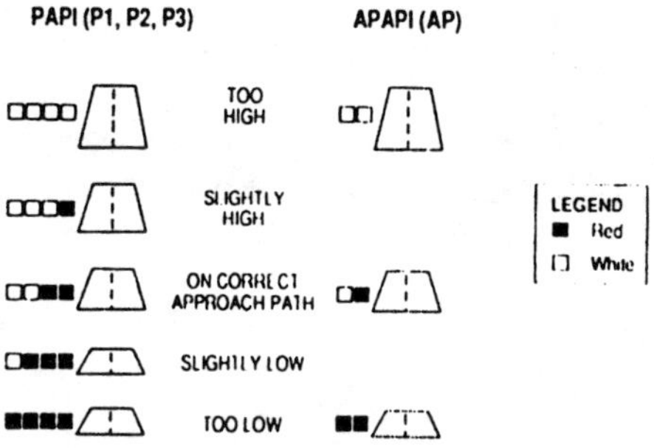

Fig. 7-28: PAPI systems.

Although the single wing bar configuration remains the same for all PAPI systems, it is possible to provide for safe wheel clearance over the threshold for aircraft with different EWH by varying the distance of the wing bar from the runway threshold (P1, P2, P3).

7.27 Search and Rescue (SAR)

SAR in Canada is provided through 4 rescue coordination centers, in Victoria, Edmonton, Trenton and Halifax. SAR service is available throughout Canada and the oceanic regions in Canadian territorial waters.

When an aircraft is overdue operators or owner's should immediately notify any air traffic services unit. When an aircraft is reported missing, a Missing Aircraft Notice (MANOT) is issued. Pilots flying in the search area will be notified of the MANOT. Pilots receiving MANOT's should monitor 121.5 MHz and maintain a visual lookout for the missing aircraft.

If you spot an aircraft in distress, keep the aircraft in sight until your presence is no longer necessary and contact ATS. If possible, monitor 121.5 MHz for any ELT signals when in flight.

If you have crashed, turn on your ELT, and keep it on. If it is cycled through "on" and "off", the battery life will be too short. In a non-emergency situation, SAR will commence 1 hour past the ETA (flight plan) or 24 hours past the ETA (flight itinerary, unless a specific SAR time has been specified).

Survival after an aircraft accident in a remote area will depend on your preparation - if you have proper dress and useful survival gear and basic survival skills knowledge, your chances of survival are much enhanced. I would recommend that any Canadian pilot complete a basic wilderness survival course. A useful survival skills manual that I would highly recommend reading is *Down But Not Out*, by the Canadian Forces and available at pilot supply shops.

Some really smart "high tech" equipment to bring along on all flights would be a cellular phone, portable 2-way VHF radio and a portable GPS unit. Recently, a pilot ferrying a light twin across the Atlantic to Canada had a complete electrical failure at night, over the mid-Atlantic ocean! With all electrics out, including all radios, navaids and lights, the well prepared pilot made it to destination successfully relying on a hand held GPS, portable and flashlight (while taxiing to the ramp, one of the engines failed)!

7.28 Forced Approach Procedures

If an engine fails in flight, the pilot should complete the following:

1. establish attitude for best glide;
2. confirm wind direction;
3. select appropriate landing surface;
4. complete cause/restart checks;
5. declare emergency;
6. complete emergency landing in accordance with procedure in owner's handbook;
7. provide passenger briefing as required.

7.29 Effect of Changing Winds on Approach Angle

When descending at a constant angle, changes in winds affect the groundspeed.

If descending into a HEADWIND, the angle of approach will steepen, and the groundspeed will be reduced.

If descending into a TAILWIND, the angle of approach will flatten, and the groundspeed will be increased.

EXAMPLE: A pilot of an aircraft on a long final is maintaining a constant descent rate of 500 FPM and a ground speed of 100 knots. While on the descent, the winds shear from a 10 kt headwind to a 10 kt tail wind. To maintain the 500 FPM descent rate, and to maintain the ground speed, what power change will be required?

ANSWER: Power reduction is required.

7.30 Effect of Winds on Angle of Climb and Rate of Climb

When climbing into a headwind, the ANGLE of climb (angle between ground and flight path) steepens. Winds do not have any effect on the RATE of climb because rate of climb (feet per minute) is determined by engine power (it takes the same time to climb regardless of whether nose is pointed into wind).

7.31 Visual Illusions

Spatial Disorientation

ORIENTATION is knowing which way is up and is provided by vision, kinesthetic sensors (muscle - bone - joints) and the vestibular organs in the inner ear. Spatial disorientation can occur in conditions of reduced visibility, such as in whiteout conditions, in night flying, or when flying under IFR. One example is the FALSE HORIZON effect, where sloping cloud formations, a dark scene spread with ground lights and stars, and certain geometric patterns of ground lights can create illusions of not being aligned correctly with the actual horizon. The disoriented pilot may place the aircraft in a dangerous attitude. Spatial disorientation from these illusions can be prevented only by visual reference to reliable, fixed points on the ground or to flight instruments.

Illusions on Approach

Various surface features and atmospheric conditions encountered in landing can create illusions of incorrect height above and distance from the runway threshold. Landing error from these illusions can be prevented by anticipating them during approaches, precautionary inspection of unfamiliar airports before landing, using VASIS systems when available, and remaining proficient in landing procedures.

Approaching a DOWNSLOPE runway can lead to a high approach, because the effect of approaching the downsloping runway is to give an illusion of undershooting or being low. Approaching an UPSLOPE runway can lead to a low approach, because the effect of approaching the upsloping runway is to give an illusion of overshooting or being high.

Rain causes a refraction error and can lead to a low approach, because the runway appears closer than it actually is. The rain on the windshield acts as a kind of a "lens" shifting the runway image closer.

A NARROW RUNWAY can lead to a high flare by creating the illusion that the aircraft is at a greater height.

GROUND LIGHTING illusions are caused by mistaking ground lights such as street lights or a moving train for a runway and approach lights. Bright approach lighting can lead to a high approach, because it appears the runway is closer.

7.32 Aircraft Tie Downs

When tying down an aircraft, tension should be tight for nylon rope (to allow for stretching), and loose if using manila rope (moisture tends to tighten manila rope).

7.33 Marshalling Signals

Marshalling Signals are used by ground attendants to provide instructions for aircraft entering, departing or maneuvering on the surface of an aerodrome. The ground marshaller may use flashlight-lamps to enhance visibility. The marshaller normally stands forward of the left wing tip to be in full view of the pilot. If you arrive at an unfamiliar aerodrome, be able to comply with the marshalling signals below.

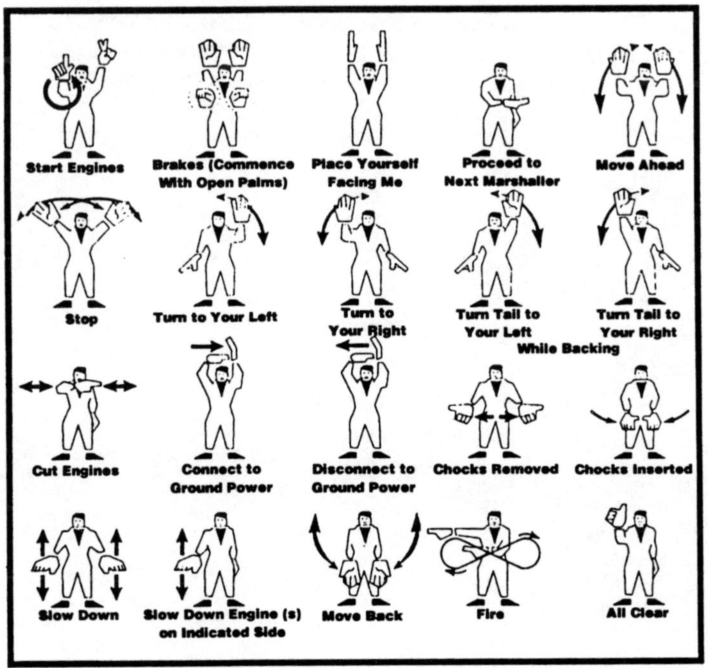

Fig. 7-29: Marshalling Signals.

7.34 Pilot Decision Making and Psychology

Pilots make better decisions when they understand the factors and process of decision making. An awareness of decision making will help the pilot make safer choices both for in flight situations and for go/no-go decisions before flight.

The Decision Making Process

To make safe decisions, the pilot must have SITUATIONAL AWARENESS, by finding out all they can about important information in a flying situation. Next, the pilot should identify as many OPTIONS as possible to choose from. Once the options are worked out, the pilot must CHOOSE the best option for the situation at hand. Finally, once an option is chosen, the pilot must ACT and EVALUATE the results to confirm the best option has been chosen. If the best option has not been chosen, the decision making process will have to be repeated.

Many factors affect the decision making process, such as job demands, stress, monetary gain, skill and level of training.

Factors that Influence Decision-Making

Anything that gets in the way of clear thinking (such as fatigue, a bad cold etc.) can harm good decision-making. The more the background KNOWLEDGE available, such as pre-flight planning and preparation, the better. SKILLS should be practised until they become automatic reactions and must be maintained long after the flight test. EXPERIENCE usually aids decision making, although getting away with a sloppy practise in the past can lead to unsafe flying. REASONING is the process of thinking carefully to recognize early the situations that may lead to trouble. RISK ASSESSMENT is important to evaluate how serious a problem is and how much time is available to deal with it. STRESS can place the pilot into problem situations and negatively affect the pilot's ability to deal with them. A good ATTITUDE will help to avoid dangerous situations.

EXAMPLE: A pilot and his passenger removed 4 inches of snow from their Cessna 172 before flight, leaving an underlying layer of snow and ice. Following two rejected takeoffs, the pilot returned to the parking area and swept more of the residual ice and snow from the wings and tail. On the third take-off, the aircraft lifted off but did not climb well and wouldn't accelerate. As the pilot turned right for the downwind leg in the circuit, he recognized that he didn't have enough altitude to complete a circuit.

While turning directly towards the field, the aircraft stalled and commenced a spin. Fortunately, the pilot and passenger survived the crash, although they had facial injuries because the available shoulder harnesses were not worn. After the accident, investigators noted a layer of rough ice on the wings, tail and control surfaces up to one-quarter inch thick. A contributory factor was temperature conditions conducive to the formation of carburetor ice, and it is likely that the engine was not producing full power. [Excerpt from actual accident report, Issue 7/90 TP 185E Aviation Safety Letter].

ANALYSIS: The pilot did not properly ASSESS THE RISK of ice contamination and conditions leading to carb ice. The CARs state no person shall operate an aircraft if there is ice on wings/control surfaces. Pilot disobeyed air regulations either because a) he was unaware of them, indicating insufficient KNOWLEDGE or b) he knew the rules but chose to ignore = ANTI-AUTHORITY (rules are made for good reasons). Pilot showed poor SITUATIONAL AWARENESS of his ENVIRONMENT and AIRCRAFT. When pilot rejected first take-off, to continue attempted take-offs showed FALSE ASSUMPTION that everything was normal when in fact situation was very dangerous. Other factors may have been involved leading to this pilot's poor decision-making such as STRESS, GET-HOME-A-TITIS, IMPULSIVENESS or INVULNERABILITY. Failure to wear shoulder harness indicates HAZARDOUS ATTITUDE.

Pilot-Equipment Relationship

The human mind is often unreliable and pilots should be in the habit of double-checking information received from aircraft displays. Such errors can be found in instrument flying, when an instrument is incorrectly read, resulting in a control error.

Often, a pilot may completely forget a crucial check - such as fuel tank selection. It is for this reason that checklists have been developed. A checklist should be accurate and followed to the letter - it is a good operating practise to read each item out loud as the checks are being carried out. ALWAYS USE A CHECKLIST!

Errors can also result from improper map reading. Always be sure of what the various symbols indicate when map reading.

Interpersonal Relations

Relationships with other people can often lead to unsafe flying practises. For example, passengers, unaware of many risks in aviation, can bring pressure on pilots to take chances. The final authority for any flight always rests with the judgement of the pilot in command, and pilots should never bend to such pressures.

Pilots should instill confidence in their passengers and be completely in charge of their flight. When communicating with ATC, passengers, maintenance personnel and others, the pilot of even the smallest light aircraft should behave with the same seriousness, professionalism and maturity of a 747 captain.

7.35 Hydroplaning

When operating on wet runways, due to standing water, snow or slush, braking capability is reduced because water cannot be completely squeezed out between the tire and runway. DYNAMIC HYDROPLANING occurs at high speeds with standing water on the runway. Standing water on the runway also reduces an aircraft's crosswind capabilities.

Where the runway surface is very smooth, VISCOUS HYDROPLANING may occur at considerably slower speeds with only a thin film of water on the runway.

REVERTED RUBBER HYDROPLANING occurs when the pilot locks the wheel brakes for a long period, resulting in liquid-hot rubber paste decreasing braking efficiency.

7.36 Take-Off Blast Areas

Pilots should exercise caution when operating near active runways and taxiways, due to hazards from jet blast or propeller wash. The most hazardous areas are behind the aircraft, in the GROUND IDLE DANGER AREA, and in the TAKEOFF THRUST DANGER AREA.

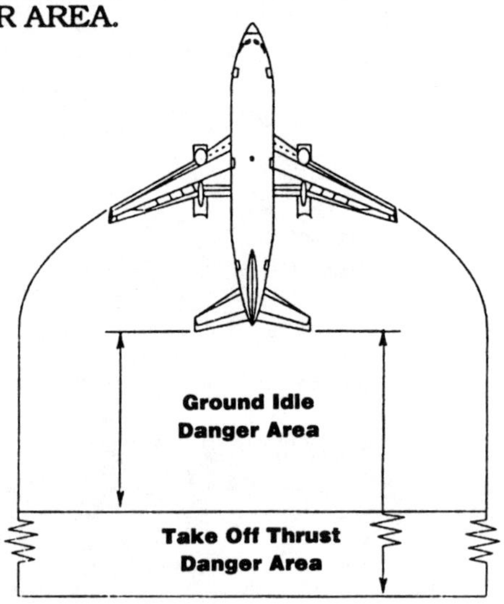

Fig. 7-30: Jet Blast Danger Areas

The danger areas extend back as follows:

	Ground Idle	Take-off Thrust
Jumbo Jet (747)	600 ft	1,600 ft
Medium Jet (737)	450 ft	1,200 ft
Executive Jet (Lear)	200 ft	500 ft

A 45 Kt blast area can be encountered 60 feet behind the propellers of a taxiing large turbo-prop aircraft.

Regardless of these distances, pilots of light aircraft should always be alert for jet blasts when in the vicinity of large aircraft and always keep well clear. Advise ATC immediately should you be put dangerously close to a large aircraft or should one move too close to you.

NOTES

8 Sample Exam

INSTRUCTIONS: The following exam is a simulated Transport Canada Private Pilot - Aeroplane exam (PPAER). This exam is also recommended for those preparing for the Recreational Pilot Permit written - (RPPAE). Do not write this exam until you have learned the previous subjects in this manual. When writing this exam, you should simulate actual exam conditions: do not refer to any notes, no "peeking" at the answers, and keep to a strict time limit of 3 hours. You may use an E6B or electronic flight computer and protractor. Any figures or charts referred to in this exam are found in the exam appendix, which starts at page 161. The answer key is at page 167. After scoring yourself, cross reference any subjects missed by using the subject references provided.

Part 1: Air Law

1. Assuming no passengers are carried, for a licenced private pilot, which of the following would meet the Transport Canada recency requirements within the previous 24 months?

a) Completion of TC private pilot written.
b) Completion of TC PSTAR written.
c) Completion of a "check out" with an instructor.
d) Provided at least 5 take-offs and landings have been completed within 6 months preceding flight, no further recurrent training is needed.

2. A flashing red light signal means:

a) Ground: Stop.
b) Air: Give way to other aircraft and continue circling.
c) Ground: Taxi back to starting point.
d) Air: Airport unsafe, do not land.

3. What is the waiting period before flight after donating blood?

a) 8 hours.
b) 24 hours.
c) 48 hours.
d) Consult an Aviation Medical Examiner.

4. Who is responsible for compliance with AD's?

a) The pilot.
b) The owner.
c) The pilot and the owner.
d) The aircraft maintenance engineer.

5. A "maneuvering area" is

a) that part of an aerodrome to be used for the surface maneuvering of aircraft, including movement areas and aprons.
b) that part of an aerodrome used for the taking off and landing of aircraft and movement of aircraft associated with take-off and landing, including aprons.
c) that part of an aerodrome used for the taking off and landing of aircraft and movement of aircraft associated with take-off and landing, excluding aprons.
d) that part of an aerodrome to be used for the surface movement of aircraft, including maneuvering areas and aprons.

6. Cruising altitude requirements apply to

a) all aircraft, regardless of altitude.
b) aircraft operating above 3,500 feet AGL.
c) aircraft with magnetic track of 000° - 179°.
d) all aircraft above 3,000 feet AGL.

7. The required documents to be carried on board an aircraft are the Certificate of Airworthiness, Certificate of Registration, Aircraft Manual, Weight & Balance, Intercept Signals, Insurance and

a) technical log, radio license, pilot license, pilot medical and pilot radio license.
b) journey log, radio license, pilot license, pilot medical and pilot radio license.
c) technical log, journey log, radio license, pilot medical and pilot license.
d) journey log, radio license, pilot license, and pilot medical.

8. ELT's should be tested only

a) during the first 5 minutes of the hour, for not longer than 5 seconds.
b) during the first 10 minutes of the hour, for not longer than 5 minutes.
c) at any time, provided it is not longer than 5 seconds.
d) during the first 5 minutes of the hour, for not longer than 10 seconds.

9. On initial contact with ATC, pilots should give the type of aircraft, followed by

a) the last four letters of the registration.
b) the last three letters of the registration.
c) the same as whatever the controller says.
d) the last four letters of the registration in phonetics.

10. "CAVOK" means there are no thunderstorms, shallow fog, low drifting snow, cumulonimbus clouds or precipitation, and

a) no cloud below 5,000 feet.
b) no cloud below 5,000 feet, a visibility of 6 sm or more, and winds of less than 10 kts.
c) no cloud below 5,000 feet and a visibility of 6 sm or more.
d) no cloud, unlimited visibility, and winds less than 10 kts.

11. A stopover is indicated on a flight plan by

a) indicating the location of the stopover in the "route" section, and including the duration in the estimated elapsed time.
b) repeating the name of the stopping point and its duration in the "route" section.
c) indicating the location and duration of the stopover in the "route" section.
d) indicating the location and stopover duration in the "other information" section.

12. Automatic Terminal Information Service (ATIS) broadcast is intended for:

a) Arriving aircraft.
b) Departing aircraft.
c) Landing and taking off aircraft.
d) Arriving and departing aircraft.

13. When operated in accordance with VFR, aircraft shall be flown with

a) visual reference to the ground or water.
b) visual reference to the ground or water at all times.
c) visual reference to ground or water, unless otherwise authorized by ATC.
d) specific authorization from ATC only.

14. With respect to right of way,

a) Where two aircraft are approaching to land, the lower aircraft shall give way to the higher aircraft.
b) Where two aircraft are approaching to land, the higher aircraft shall give way to the lower aircraft.
c) Balloons shall give way to airships.
d) Gliders shall give way to airships.

15. When approaching the Air Defence Identification Zone, the tolerances and position reporting requirements apply

a) only to VFR aircraft.
b) only to IFR aircraft.
c) only VFR aircraft that have filed a DVFR flight plan.
d) to all aircraft.

16. Which of the following is true?

a) No person shall execute any acrobatic flying unless they are the sole occupant or are receiving dual instruction.
b) Acrobatic flight is only permitted with special permission from the Minister.
c) Only flying instructors can execute acrobatic flight.
d) No person shall execute any acrobatic flight unless that person holds an acrobatic endorsement.

17. Which of the following is true?

a) No person shall drop anything from an aircraft in flight.
b) Specific permission from the Minister is required to drop objects from an aircraft.
c) No person shall drop anything from an aircraft in flight if it creates a hazard to persons on the ground.
d) No person shall drop anything from an aircraft in flight if it creates a hazard to persons or property on the ground.

18. Which of the following is true with respect to a Special VFR clearance?

a) The pilot is responsible for avoiding weather conditions beyond the capabilities of the aircraft.
b) The pilot is responsible to keep the aircraft clear of cloud and within sight of the ground at all times.
c) The pilot is responsible for avoiding weather conditions beyond their own flight capabilities.
d) All of the above.

19. The engine or engines of any aircraft

a) shall not be started unless there is a pilot seated in the pilot's seat.
b) shall not be started unless the pilot's seat is occupied by a person competent to control the aircraft.
c) may be left running without anyone occupying the pilot's seat, if the aircraft is prevented from moving forward.
d) choice b) and c).

20. Supplemental oxygen and masks are required

a) for flight crew on flights above 13,000 ft.
b) for passengers on flights above 13,000 ft.
c) for 10% of passengers for flights from 10,000 to to 13,000 for less than 30 mins.
d) Choice a) and b).

21. The CARs define "night " as the period

a) from dusk till dawn.
b) from 1/2 hour before sunrise to 1/2 hour before sunset.
c) where the center of the sun's disc is greater than 6° below the horizon.
d) any time the sun is not visible.

22. If ATC instructs you to "squawk code 1232", you should select 1232 on the transponder, and

a) select the "ON" position.
b) select the "IDENT" position.
d) select "STANDBY" and wait for ATC to confirm identification.
d) select "IDENT", then "ON" when ATC instructs.

23. After a pilot completes elementary work on his/her aircraft (such as replacing a tire), the pilot should

a) make the appropriate entries of the work completed in the technical records and journey log and release the aircraft for return to service.
b) have a licenced AME go over the work and if satisfactory, the AME will sign a maintenance release in the technical records.
c) enter maintenance details and release the aircraft for return to service in the journey log only since elementary work need not be detailed in techincal records.
d) not complete any recording of the work done since no record keeping requirements apply to elementary work.

24. When an aircraft is overdue on a flight itinerary, the responsible person should immediately

a) notify the nearest detachment of the RCMP.
b) notify the nearest ATC unit or FSS station.
c) notify an ATC unit, FSS station, CARS station or rescue coordination centre.
d) call 911 and ask for the search and rescue coordination centre.

25. Class B airspace is normally that airspace

a) within low level airways above the MEA.
b) within the Southern Control Area extending upwards from 12,500 feet ASL.
c) from 12,500 feet to 18,000 feet ASL.
d) within low level airways extending upwards from 12,500 feet ASL.

Part 2: General Aeronautical Knowledge

26. An excessively lean mixture can result in

a) detonation.
b) plug fouling.
c) engine will run too hot.
d) excessive fuel consumption.

27. Which of the following is false with respect to constant speed propeller systems?

a) Oil pressure varies propeller blade angle.
b) Select full coarse for take-off.
c) "Low pitch" refers to small blade angles.
d) "High pitch" is the same as "coarse pitch".

28. The ______ stall speed increases with altitude; the ______ stall speed remains the same with altitude.

a) CAS, IAS.
b) TAS, IAS.
c) IAS, TAS.
d) CAS, TAS.

29. An aircraft can stall at any ____________.

a) Attitude.
b) Airspeed.
c) Power setting.
d) All of the above.

30. What is the turn coordinator/turn and bank indication of a slip to the right?

a) Needle/miniature aircraft right, ball right.
b) Needle/miniature aircraft left, ball right.
c) Needle/miniature aircraft right, ball left.
d) Needle/miniature aircraft left, ball left.

31. On the ASI, the lower end of the green arc is

a) Vs0.
b) The stall speed in clean configuration.
c) Vs1.
d) Choice b or c.

32. Refer to the compass correction card in the exam appendix. For a magnetic heading of 096, the compass heading required is:

a) 095 magnetic.
b) 097 compass.
c) 097 magnetic.
d) 095 compass.

33. In extremely cold temperatures, the altimeter

a) over-reads.
b) under-reads.
c) indicates normally.
d) indicates normally if the correct altimeter setting has been applied.

34. Northerly turning error is greatest on turns from ______ to ______.

a) east to north-east.
b) west to south-west.
c) north to north-west.
d) north-west to west.

35. To avoid wake turbulence, when approaching to land behind a large, heavy aircraft, plan the approach path ______ the approach of the large aircraft, and land ______ the touchdown point of the large aircraft.

a) well below, well before.
b) well above, well before.
c) well below, well beyond.
d) well above, well beyond.

36. A collision will be more likely where another aircraft appears to

a) move up on the windshield.
b) remain stationary on the windshield.
c) move down on the windshield.
d) move to the center of the windshield.

37. An important advantage of a fuel injection system over float-type carburetor is

a) there is less engine wear.
b) the mixture control is not necessary.
c) there will not be carburetor icing.
d) engine will not be as prone to detonation.

38. The use of carburetor heat, when there is carb ice, results in:

a) decreased RPM, more rich mixture.
b) decreased manifold pressure, more lean mixture.
c) increased RPM, more lean mixture, temporary engine roughness.
d) decreased manifold pressure, more rich mixture, temporary engine roughness.

39. Which of the following is false?

a) Whiteout occurs over an unbroken snow cover beneath a uniformly overcast sky.
b) Whiteout occurs when light from sky is much brighter than from snow surface.
c) In whiteout, neither shadows, horizon, nor clouds are discernible.
d) During whiteout, sense of depth and orientation is lost.

40. The recovery technique for a fully developed spin, with partially extended flaps is to

a) raise flaps, use spin recovery technique.
b) use spin recovery technique, then raise flaps.
c) apply spin recovery technique and raise flaps at the same time.
d) lower flaps to reduce the spin.

41. A principal cause of wheelbarrowing is

a) too steep approach.
b) too fast approach.
c) too fast landing speed.
d) too much flap.

42. When flying low in rain, hilltops appear

a) 230 feet lower at 1/2 statute mile.
b) 230 feet higher at 1/2 statute mile.
c) 260 feet lower at 1/2 nautical mile.
d) choice a) or c).

43. When taxiing in strong winds from the "8:00 o'clock" position, the controls should be held:

a) Down aileron on left wing, elevator up.
b) Down aileron on left wing, elevator down.
c) Up aileron left wing, elevator up.
d) Up aileron on left wing, elevator down.

44. A missing fuel cap in flight results in

a) nothing (new cap will have to be purchased after landing).
b) fuel rapidly escapes.
c) a small amount of fuel evaporates.
d) vapour lock.

45. To overshoot, use the following sequence:

a) Apply full power, accelerate to normal climb speed, reduce flaps, raise nose to climb attitude.
b) Apply full power, accelerate to normal climb speed, reduce flaps, raise nose to climb attitude.
c) Apply full power, accelerate to safe climb speed, reduce flaps, raise nose to climb attitude.
d) Apply full power, accelerate to safe climb speed in level flight, reduce flaps, raise nose to climb attitude.

46. Which of the following is false ?

a) Ground effect is most pronounced below about 18 feet above the runway.
b) Ground effect is useful for short/soft field take-offs.
c) Ground effect causes an aircraft to float just before touchdown.
d) Ground effect increases induced drag.

47. To recover from a spiral dive,

a) power idle, level wings, ease out of dive.
b) power idle, ease out of dive, level wings.
c) power idle, ease out of dive and level wings simultaneously.
d) power idle, level wings (almost simultaneous), ease out of dive.

48. For an aircraft having an aft CG,

a) the aircraft will stall at a lower airspeed, and have unstable stall characteristics.
b) the stall speed will remain unaffected.
c) the aircraft will stall at higher airspeed, with stable stall characteristics.
d) the aircraft will stall at the same stall speed, but will have unstable stall characteristics.

49. If one magneto fails in flight on an engine equipped with dual ignition,

a) a slight loss of power would result.
b) there would be no effect on the engine.
c) the engine would stop.
d) half the cylinders would not fire.

50. The use of low octane fuel in a high compression engine may result in

a) too lean a mixture.
b) carburetor icing.
c) fouling of the spark plugs.
d) detonation.

Part 3: Meteorology

51. A close spread between the dewpoint temperature and surface temperature means

a) rain is likely.
b) fog is likely.
c) high winds will develop.
d) none of the above.

52. "MVFR" in the area forecast refers to forecast ceilings ___________ and visibilities __________.

a) below 1,000 feet, less than 3 sm.
b) above 3,000 feet greater than 3 sm.
c) from 3,000 - 5,000 feet, from 3 - 6 sm.
d) from 1,000 - 3,000 feet, from 3 - 5 sm.

53. In the area forecast, forecasted winds

a) are always provided.
b) are only provided if expected to be less than 20 kts.
c) are only provided if expected to be greater than 20 kts.
d) are only provided if expected to be greater than 10 kts.

54. With respect to sky condition,

a) overcast = 7/8 - 8/8.
b) broken = 4/8 - 8/8.
c) scattered = 1/8 - 3/8.
d) few = >0/8 - 2/8.

55. A sea breeze is most likely to occur

a) at day (wind blows from sea to land).
b) at night (wind blows from sea to land).
c) at day (wind blows from land to sea).
d) at night (wind blows from land to sea).

56. Altostratus clouds are from

a) 20,000 ft to 30,000 ft.
b) 6,000 ft to 20,000 ft.
c) 6,500 ft to 20,000 ft.
d) surface to 6,000 ft.

57. The requirements for a thunderstorm are:

a) a steep lapse rate, high winds and rain.
b) a steep lapse rate, high moisture content and lifting agent.
c) a shallow lapse rate, high moisture content and lifting agent.
d) a shallow lapse rate, high winds and condensation nuclei.

58. ___________ would not qualify as a SIGMET.

a) Mountain waves.
b) Hail.
c) Snow.
d) Thunderstorms.

59. With respect to a passing squall line,

a) wind direction changes from S or SW to N or NW, temperature drops, barometric pressure rises.
b) wind direction changes from N or NW to S or SW, temperature and barometric pressure drop.
c) wind direction changes from N or NW to S or SW, temperature increases, barometric pressure drops.
d) None of the above.

Questions 60 to 75 are based on the weather information in the exam appendix.

60. The forecast winds in the South Coast region are

a) 20 - 30 kts.
b) above 25 kts.
c) less than 10 kts.
d) less than 20 kts.

61. What is the lowest expected height of the freezing level in the Southern Mountains region?

a) 9,000 feet.
b) 5,000 feet.
c) 7,000 feet.
d) 7,000 - 8,000 feet.

62. The air mass over all regions is forecast to be

a) moist and unstable.
b) moist and fairly stable.
c) arctic maritime.
d) dry and stable

63. The lowest forecast ceiling at CYVR is

a) 4,000 feet ASL.
b) 2,500 feet ASL.
c) 1,500 feet ASL.
d) 1,500 feet AGL.

64. The forecast visibility for CYYJ after 19Z is

a) 4 sm.
b) 6 sm.
c) greater than 6 sm.
d) greater than 10 sm.

65. The validity period of the TAF at CYVR is

a) 2200Z.
b) 2200Z - 1700Z.
c) 1717Z.
d) 1700Z - 1700Z.

66. What is the expected weather at CYYJ at 21Z?

a) Light rain and fog.
b) Rain.
c) Occasional light rain.
d) Rain and light fog.

67. The forecast winds for YVR at 4,500 feet are

a) 275° True at 11.5 kts.
b) 275° Magnetic at 11.5 kts.
c) 270° True at 10 kts.
d) Unknown (no winds given for 4,500 feet).

68. The reported visibility at CYVR at 1800Z is

a) 7 sm in light drizzle.
b) 7 sm in moderate drizzle.
c) 7 sm in very light drizzle.
d) 9 sm in heavy rain showers.

69. The reported ceiling at CYHE at 1800Z is

a) 400 feet.
b) 1,200 feet.
c) 2,200 feet.
d) 3,800 feet.

70. The reported winds for CYYJ at 1800Z are

a) light and variable.
b) 120°M at 10 kts.
c) 160°T at 6 kts.
d) 160°M at 6 kts.

71. At 1800Z, which airport has the highest dewpoint temperature?

a) CYVR.
b) CYCD.
c) CYYJ.
d) CYHE.

72. The altimeter setting at CYVR at 1800Z is

a) 1032.9 hPa.
b) 30.29 hPa.
c) 30.29" Hg.
d) 1032.9" Hg.

73. The barometric pressure at CYYJ at 1800Z is

a) 24.8".
b) 1606 hPa.
c) 29.26".
d) 1024.8 hPa.

74. The 1800Z ceiling at CYHE is __________ cloud.

a) stratus fractus.
b) altocumulus.
c) cumulo fractus.
d) stratocumulus.

75. The reported weather at CYCD at 1800Z is

a) drizzle and fog.
b) light drizzle and light fog.
c) light drizzle and moderate fog.
d) very light fog and moderate drizzle.

Part 4: Navigation

76. Which of the following abbreviations represents the highest terrain plus 100 meters on a VNC chart?

a) MEA.
b) MRA.
c) MEF.
d) MOA.

77. The symbol "*L" on a VNC chart indicates

a) lighting available on prior request.
b) lighting available 24 hours.
c) lighting available limited hours.
d) lighting available limited hours and on prior request thereafter.

78. An aircraft is tracking inbound on the 300° radial to a VOR station. The OBS indicates ____, the TO-FROM flag indicates ______, and the magnetic heading is ______.

a) 300, TO, 120.
b) 300, FROM, 300.
c) 120, TO , 120.
d) 120, FROM, 300.

79. On a mercator projection chart, lines of longitude ______, and on a lambert conformal projection chart, lines of longitude ______.

a) remain parallel, converge towards north.
b) converge towards north, remain parallel.
c) converge towards north, converge towards north.
d) remain parallel, remain parallel.

80. If the magnetic variation is 5° west, a true track of 095 is a magnetic track of ______.

a) 100° True.
b) 100° Magnetic.
c) 090° True.
d) 090° Magnetic.

81. Complete the following weight and balance.

Item	Weight	Arm	Moment
Empty Aircraft	1600	36	______
Fuel	240	34	______
Pilot and copilot	____	37	12,580
Rear Passengers	340	____	22,100
Fwd Baggage	60	-15	______
TOTALS	____	____	______

a) Total weight = 2580 lbs, total moment = 100,440 "-lbs, center of gravity = 38.9".
b) Total weight = 2580 lbs, total moment = 101,610 "-lbs, center of gravity = 39.1".
c) Total weight = 2580 lbs, total moment = 99,540 "-lbs, center of gravity = 38.6".
d) Total weight = 2610 lbs, total moment = 65,210 "-lbs, center of gravity = 49.6".

82. Using the center of gravity from Q. 81, if the permissible range of the C of G is from 34.6" to 38.8", is the aircraft loaded within proper range?

a) Yes.
b) No.
c) More information is needed (moment envelope or CG range chart).
d) Yes, if the weight were repositioned.

83. Refer to the loading graph and center of gravity envelope in the exam appendix to complete the following weight and balance:

Item	Weight	Arm	Moment
Empty Aircraft	1100		35,900
Fuel	135		______
Pilot and Pass.	300		______
Baggage (area 1)	25		______
TOTALS	_____	____	______

a) Total moment 54,900, total weight 1560, outside moment envelope
b) Total moment 55,600, total weight 1560, inside moment envelope.
c) Total moment 54,900, total weight 1560, inside moment envelope.
d) Total moment 55,600, total weight 1560, outside moment envelope.

84. Use the crosswind chart in the exam appendix. Conditions: stall speed 60 kts, runway 26, winds 230° M at 25 kts. Max 90° crosswind is ______. Crosswind component ______ (will, will not) be exceeded.

a) 12 kts, will not.
b) 10 kts, will.
c) 10 kts, will not.
d) 12 kts, will.

85. What is the weight of 15 litres of aviation fuel?

a) 10 lbs.
b) 24 lbs.
c) 46 lbs.
d) 13 lbs.

86. Refer to the take-off distance chart in the exam appendix. What is the total take-off distance where the pressure altitude is 1,000 feet, air temperature is 14°C and there is a 4 knot tailwind?

a) 1,811 feet.
b) 1,509 feet.
c) 1,375 feet.
d) 1,650 feet.

87. Refer to the Flight Supplement extract for Guelph, Ontario in the exam appendix. During what hours are customs available?

a) 1400 - 2200 local, Monday to Friday.
b) 0900 - 1700 local, excluding holidays.
c) 0900 - 1700 local, including holidays.
d) 1400 - 2200 UTC, Monday to Friday, excluding holidays.

For the next series of questions, refer to the VNC chart excerpt in the exam appendix.

88. What is the length of the longest runway at Port Hardy?

a) Unknown - need Flight Supplement.
b) 6,300 feet.
c) 5,000 feet.
d) 7,100 feet.

89. What is the frequency of the Port Hardy VORTAC?

a) 122.2 MHz.
b) 242.0 KHz.
c) 236.6 KHz.
d) 112.0 MHz.

Assume you are planning a cross country trip from Port Hardy airport to Woss (an unlicensed logging strip). You plan a cruising altitude of 7,500 feet. Winds are 270°T at 20 kts. Cruising TAS is 110 kts. A navigation log is in the exam appendix.

90. The true track from Port Hardy to Woss is

a) 135.
b) 149.
c) 206.
d) 198.

91. What is the highest elevation 5 nm either side of track?

a) 4,890 feet ASL.
b) 4,632 feet ASL.
c) 5,850 feet ASL.
d) 4,688 feet ASL.

92. What is the true heading and groundspeed?

a) 140°T, 130 kts.
b) 143°T, 123 kts.
c) 143°T, 120 kts.
d) 140°T, 123 kts.

93. What is the estimated elapsed time?

a) 20 minutes.
b) 22 minutes.
c) 18 minutes.
d) 28 minutes.

94. Assuming a rate of fuel consumption of 8 gallons per hour, what is the minimum amount of DAY VFR fuel necessary for the trip [assume you have adequate fuel for start, taxi and climb]?

a) 2.7 Gallons.
b) 3.0 Gallons.
c) 6.7 Gallons.
d) 8.8 Gallons.

95. Assuming calibrated airspeed is equivalent to indicated airspeed, what is the expected IAS where the pressure altitude is 7,500 feet and the temperature at 7,500 feet is +10°C?

a) 97 kts.
b) 95 kts.
c) 110 kts.
d) 100 kts.

96. When over Nimpkish Lake, to obtain a weather update,

a) contact Port Hardy FSS, use Alert Bay RCO frequency 126.7.
b) contact Port McNeill FSS, 122.8.
c) contact Alert Bay FSS, 126.7.
d) contact Port Hardy RCO, 126.7.

97. Assume after 7 minutes you are 7° left of track. To correct for drift,

a) steer 7° right for 7 minutes to regain track.
b) steer 7° left for 14 minutes to regain track.
c) steer 14° right for 7 minutes to regain track.
d) steer 14° left for 7 minutes to regain track.

98. Which of the following is true?

a) There is a UNICOM at Alert Bay, frequency 122.8.
b) There is a UNICOM at Alert Bay, frequency 123.0.
c) There is no UNICOM at Alert Bay, but there is a DF steer at Port Hardy.
d) There is no UNICOM at Alert Bay, and there is no DF steer at Port Hardy.

99. A straight line drawn on a VNC chart most nearly represents:

a) a rhumb line.
b) a great circle.
c) an isogonic line.
d) None of the above.

100. To return from Woss to Port Hardy, the Port Hardy VOR can be used for navigational assistance. To track from Woss to Port Hardy, select ______ on the OBS for a centered track bar with a "TO" flag.

a) 292.
b) 112.
c) 030.
d) 135.

EXAM APPENDIX

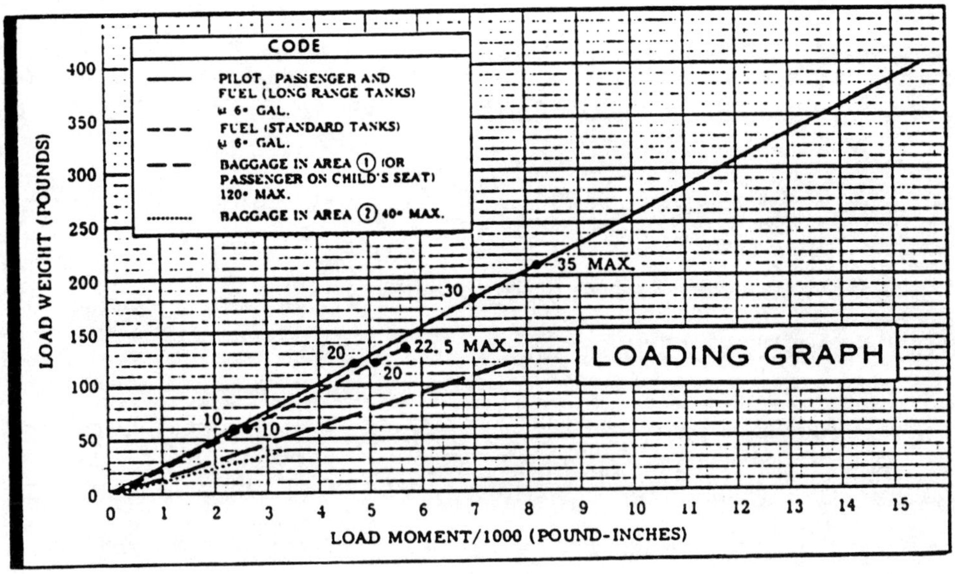

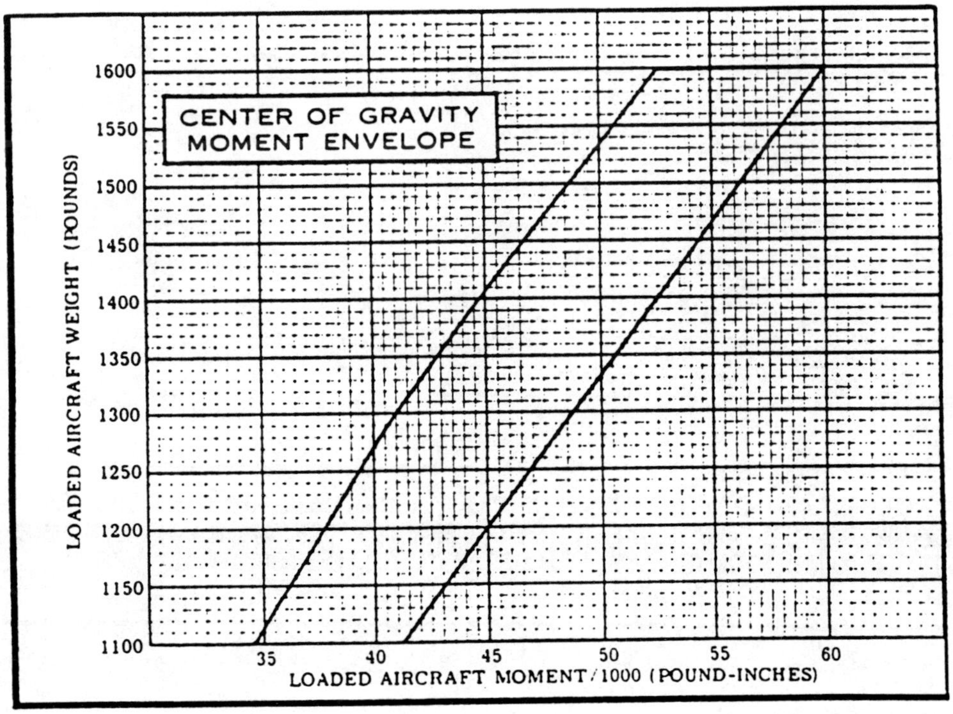

FACN1 CWVR 221730
18 - 06
0TLK 06 - 18

HTS ASL UNLESS NOTED

PROG
FNTL ZONE QSTNRY ALG 51N LAT THRU PRD. ASOCTD MRTM TROWAL BLUE RIVER-ALTA LAKE-JUST SE PORT HARDY ROTATING SEWDS TO LIE 50 N OF LN CRANBROOK-PENTICTON-COMOX BY 06Z. CMA WRM FNT 60 NE DEASE LAKE-FORT ST JOHN MOVG NEWDS AT 5-10 KTS. MDT-STG NWLY FLO ALF. MOIST FAIRLY STBL AMS OVR RGNS.

S CST RGN
CLDS AND WX. 5-15 SCT OCNL BKN 20 30 BKN V OVC 80 100-120 BKN 160 6+ OCNL -RA. VSBYS LCLY 4-6 -RA FG. LWR LYR BKN V OVC XTNSV VSBYS 1-4 -ZL FG ALG W AND N CSTS VRISL. PASSES CLOSED.

ICG. LGT-MDT RIME ICGIC. FRLVL 90-100 SLPG TO NR 70 NWRN SXNS.

OTLK. MVFR XCP IFR CIGS FG ALG W CST VRISL.

SRN MTNS RGN
CLDS AND WX. 50-70 SCT V BKN 80 90-100 OVC 150 LYRS 180 /BKN 6+ OCNL -SHRA. BCMG NW OF TROWAL...50-70 SCT V BKN SC CU 90 110 BKN LYRS 160 /SCT 6+ OCNL -SHRA. ADDNL 25-30 SCT 40 FRMG IN PCPN FQTLY BKN ALG WLY SLPS ROCKY MTNS AHD OF TROWAL. PASSES MARGINAL WRN SXNS...CLOSED THRU ROCKIES.

ICG. LGT-MDT RIME ICGIC. LGT MXD IN CU TOPS. FRLVL 70-80 LWRG TO NR 50 NW OF TROWAL.

TURBC. LGT VCNTY CU.

OTLK. MVFR THRU PASSES ROCKYS AND VCNTY TROWAL.

CYYJ TAF 221630Z 221705 00000KT 4SM -RA FG BKN008 OVC025 FM 1900Z 00000KT P6SM SCT015 BKN030 OVC100 TEMPO 1905 -RA
CYVR TAF 221630Z 221717 00000KT P6SM SCT020 BKN040 OVC100 TEMPO 1702 -RA FM0200Z 00000KT P6SM SCT030 BKN050 BKN100 PROB30 1317 BKN015

FD CN3 CWAO 221520
FCST BASED ON 221200 DATA VALID 221800 FOR USE 15-21

	3000	6000	9000	12000
YVR	2710	2813+03	2921+00	3028-03
YYF	2713	2715+02	2921+00	2927-05
YYE	9900	2708-02	2817-07	2826-12

CYYJ METAR 221800Z 16006KT 6SM FG SCT008 SCT014 OVC032 12/10 A3026 RMK SF1 SC3 SC4 SLP248
CYCD METAR 221800Z 00000KT 4SM -DZ FG SCT004 BKN010 OVC015 12/11 A3025 RMK SF1 SF3 CF5 SLP243
CYVR METAR 221800Z 10007KT 7SM -DZ SCT007 BKN013 OVC025 11/09 A3029 RMK SF1 SF4 SC3 SLP252
CYHE METAR 221800Z CCA 23002KT 7SM -SHRA SCT004 SCT012 SCT022 BKN038 OVC080 11/09 A3028 RMK SF1 CF1 CF1 SC3 AC2 SLP256

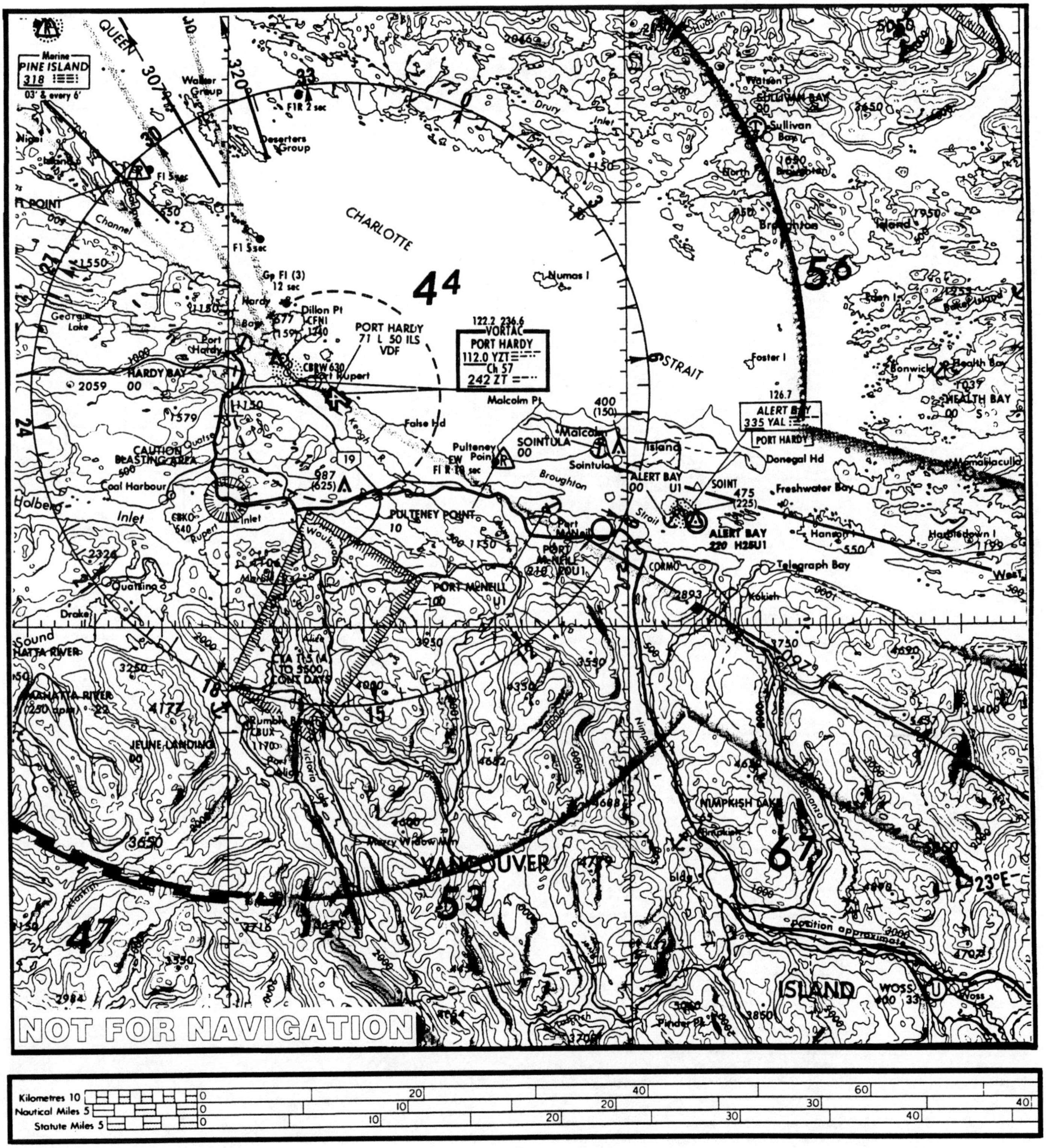
Marine
PINE ISLAND
318
03' & every 6'
QUEEN
307
320
Walker Group
Deserters Group
Drury Inlet
CHARLOTTE
44
STRAIT
Numas I
Dillon Pt
CFNI
PORT HARDY
71 L 50 ILS
VDF
122.2 236.6
VORTAC
PORT HARDY
112.0 YZT
Ch 57
242 ZT
Malcolm Pt
Port Hardy
HARDY BAY
Port Rupert
CBRW630
False Hd
Pulteney Point
SOINTULA
Sointula
Broughton
Strait
CAUTION
BLASTING AREA
Coal Harbour
Holberg
Inlet
Rupert
Quatsino
PULTENEY POINT
PORT MCNEILL
Port McNeill
ALERT BAY
335 YAL
126.7
PORT HARDY
Donegal Hd
Freshwater Bay
Telegraph Bay
Foster I
Bonwick
HEALTH BAY
Health Bay
Eden I
Sullivan Bay
SULLIVAN BAY
Watson
North
Broughton
Island
56
Harbledown I
Hanson I
Kokish
Sound
Drake
Nimpkish
NIMPKISH LAKE
Victoria Lake
Merry Widow Mt
VANCOUVER
53
ISLAND
47
67
WOSS
Woss
JEUNE LANDING
Rumble Beach
CBUX
2059
1579
2326
3250
4177
3650
3350
2984
3950
3550
4350
4662
4600
4698
4750
3750
4690
5400
5050
3650
1950
1037
2893
4707
3850
3700
23°E
position approximate
NOT FOR NAVIGATION
Kilometres 10
Nautical Miles 5
Statute Miles 5
0
20
40
60
10
30

VFR Nav Log

From To	True Track	ALT	OAT °C	TAS IAS	Wind	True HDG	VAR	MAG HDG	DEV	COM HDG	G/S	Dist	ETE	ETA	Fuel Burn	Fuel Req'd

Remarks	Stopovers		Reserve Start/Taxi	
	Total Elapsed Time		Total Fuel	

COMPASS CORRECTION CARD

For (magnetic)	000	030	060	090	120	150	180	210	240	270	300	330
Steer (compass)	001	031	061	091	120	150	179	208	238	267	299	327

GUELPH ONT NC4

REF	43 34N 80 12W 1.3NE 8°W GMT-5(4) Elev 1100' VTA A5000 F-21 LE7
OPR	Guelph Air Park 519-824-2660 Pub Lic* aprt attended ltd hrs
PF	B-1,2 C-3,5,6
CUST	AOE-X 519-822-7041 14-22Z Mon-Fri, exc hols
FLT PLN FSS	London 800-265-1031

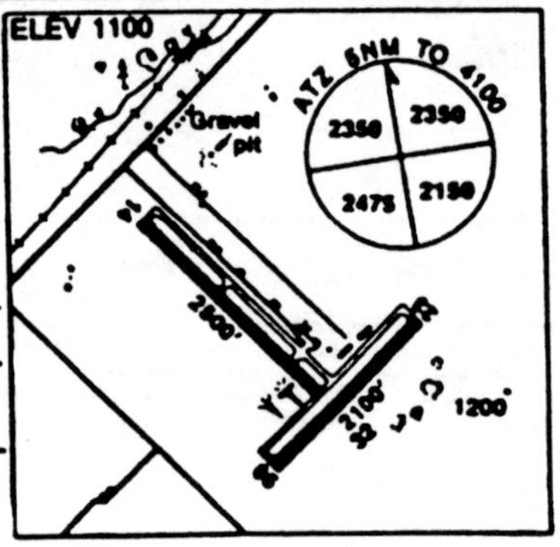

TAKEOFF DISTANCE

CONDITIONS:
Flaps Up
Full Throttle Prior to Brake Release
Paved, Level, Dry Runway
Zero Wind

NOTES:
1. Maximum performance technique as specified in Section 4.
2. Prior to takeoff from fields above 5000 feet elevation, the mixture should be leaned to give maximum RPM in a full throttle, static runup.
3. Decrease distances 10% for each 9 knots headwind. For operation with tailwinds up to 10 knots, increase distances by 10% for each 2 knots.
4. Where distance value has been deleted, climb performance after lift-off is less than 150 fpm at takeoff speed.
5. For operation on a dry, grass runway, increase distances by 15% of the "ground roll" figure.

WEIGHT LBS	TAKEOFF SPEED KIAS		PRESS ALT FT	0°C		10°C		20°C		30°C		40°C	
	LIFT OFF	AT 50 FT		GRND ROLL	TOTAL TO CLEAR 50 FT OBS	GRND ROLL	TOTAL TO CLEAR 50 FT OBS	GRND ROLL	TOTAL TO CLEAR 50 FT OBS	GRND ROLL	TOTAL TO CLEAR 50 FT OBS	GRND ROLL	TOTAL TO CLEAR 50 FT OBS
1600	53	60	S.L.	655	1245	710	1335	765	1435	820	1540	880	1650
			1000	720	1365	775	1465	835	1575	900	1690	970	1815
			2000	790	1500	855	1615	920	1735	990	1865	1065	2005
			3000	870	1650	935	1780	1010	1915	1090	2065	1170	2225
			4000	955	1820	1030	1965	1115	2125	1200	2290	1290	2475
			5000	1050	2015	1140	2185	1230	2360	1325	2555	1430	2770
			6000	1160	2245	1255	2435	1360	2640	1465	2870	1580	3120
			7000	1285	2510	1390	2730	1505	2970	1625	3240	---	---
			8000	1420	2820	1540	3080	1670	3370	---	---	---	---

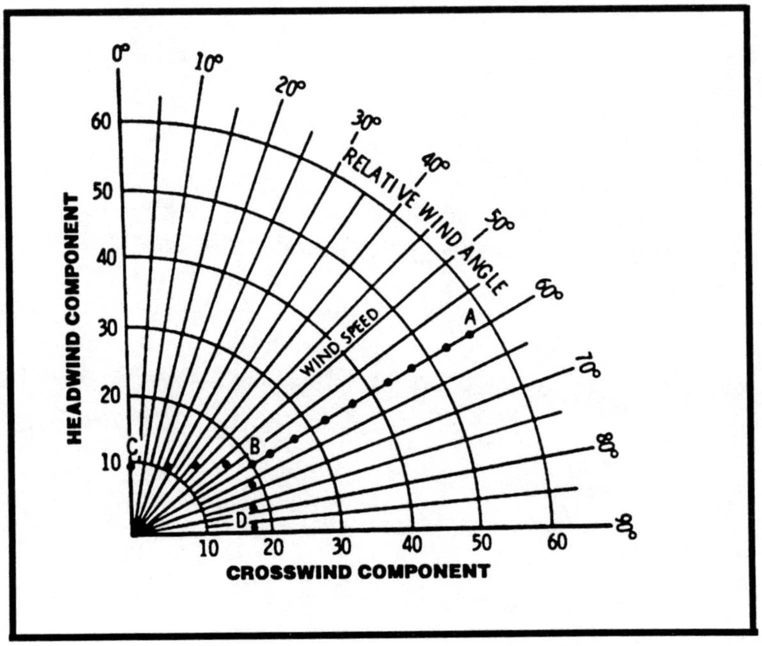

NOTES

ANSWER KEY: SAMPLE EXAM

INSTRUCTIONS: The correct choice for each question is indicated with a subject reference number. A pass is 60 out of 100 OVERALL, and at least 60% PER SECTION (15 correct for each set of 25). After scoring yourself, review any questions missed by using the subject reference number to cross reference to the supporting text in this course.

To enhance your scoring potential on the Transport Canada written, and for further exam practice, refer to **Private Pilot and Recreational Pilot Written Test Book, 1999 Edition.**

Part 1: Air Law

Ques.	Ans.	Sub Ref
1.	A	1.11
2.	D	1.31
3.	C	7.24
4.	B	1.14
5.	C	1.01
6.	D	1.37
7.	B	1.40
8.	A	1.29
9.	D	1.31
10.	C	5.24
11.	B	1.20
12.	D	5.16
13.	B	1.15
14.	B	1.13
15.	D	1.36
16.	A	1.13
17.	D	1.13
18.	D	1.15
19.	B	1.13
20.	D	1.27
21.	C	1.15
22.	A	1.19
23.	A	1.42
24.	C	1.24
25.	D	1.16

SCORE: ____/25
❑ Pass ❑ Fail

Part 2: General

Ques.	Ans.	Sub. Ref
26.	C	2.04
27.	B	3.06
28.	B	3.04
29.	D	3.04
30.	A	4.09
31.	D	4.02
32.	B	4.05
33.	A	4.04
34.	C	4.05
35.	D	7.10
36.	B	7.05
37.	C	2.09
38.	D	2.09
39.	B	7.01
40.	A	7.14
41.	C	7.11
42.	D	7.02
43.	B	7.15
44.	B	7.20
45.	D	7.19
46.	D	7.12
47.	D	7.13
48.	A	7.16
49.	A	2.07
50.	D	2.04

SCORE: ____/25
❑ Pass ❑ Fail

Part 3: Meteorology

Ques.	Ans.	Sub Ref
51.	B	5.04
52.	D	5.17
53.	C	5.17
54.	D	5.17
55.	A	5.08
56.	C	5.06
57.	B	5.12
58.	C	5.21
59.	A	5.13
60.	D	5.17
61.	B	5.17
62.	B	5.17
63.	D	5.18
64.	C	5.18
65.	D	5.18
66.	C	5.18
67.	A	5.19
68.	A	5.20
69.	D	5.20
70.	C	5.20
71.	B	5.20
72.	C	5.20
73.	D	5.20
74.	D	5.20
75.	C	5.20

SCORE: ____/25
❑ Pass ❑ Fail

Part 4: Navigation

Ques.	Ans.	Sub. Ref
76.	C	6.03
77.	A	6.03
78.	C	6.20
79.	A	6.02
80.	B	6.01
81.	C	7.16
82.	A	7.16
83.	C	7.16
84.	D	7.18
85.	B	7.21
86.	A	7.17
87.	D	7.22
88.	C	6.03
89.	D	6.03
90.	A	6.07
91.	C	6.07
92.	B	6.10
93.	A	6.11
94.	C	6.12
95.	A	6.13
96.	A	6.03
97.	C	6.16
98.	A	6.03
99.	B	6.02
100.	A	6.18

SCORE: ____/25
❑ Pass ❑ Fail

OVERALL SCORE: ____/100 ❑ Pass ❑ Fail

Subjects for review:

____

____

____

____

NOTES

TRANSPORT CANADA SUBJECT CODES: PRIVATE PILOT

INSTRUCTIONS: Transport Canada indicates the subject material for study and review in preparation for the PPL in the publication *Study and Reference Guide Private Pilot Licence - Aeroplane, Second Edition, February 1999 - TP 12880E.* Following your completion of the PPAER written, your scoring results sheet will list any subjects scored incorrectly utilizing the following Transport Canada subject codes. The number in brackets following the Transport Canada number is the equivalent subject number for this course. To cross reference from a Transport Canada subject number to the same subject in this manual, first locate the Transport Canada subject and code from the numbers below, note our corresponding subject reference number (in brackets), then look up the subject in this manual using our table of contents. EXAMPLE: To cross reference Transport Canada number **1.3.3.2** (Issuance, Renewal and Validity Period of Medical Certificate), refer to subject **1.02** in this course. Subjects with an asterisk (*) are essential knowledge areas that will receive particular emphasis on the Transport Canada written.

1. Air Law and Procedures

1.1. CARs PART I - GENERAL PROVISIONS

1.1.1. 101 - Interpretation
1.1.1.1 101.01 - Definitions [1.01]

1.1.2. 103 - Administration and Compliance
1.1.2.1 103.02 - Inspection of Aircraft, Requests for Production of Documents and Prohibitions [1.01]
1.1.2.2 103.03 - Return of Canadian Aviation Documents [1.01]
1.1.2.3 103.04 - Record Keeping [1.01]

1.2. CARs PART III - AERODROMES AND AIRPORTS

1.2.1. 300 - Interpretation
1.2.1.1 300.01 - Definitions [1.38]

1.2.2. 301 - Aerodromes
1.2.2.1 301.01 - Application [1.38]
1.2.2.2 *301.04 - Markers and Markings [1.38]
1.2.2.3 301.07 - Lighting [1.38]
1.2.2.4 301.08 - Prohibitions [1.38]
1.2.2.5 301.09 - Fire Prevention [1.38]

1.2.3. 302 - Airports
1.2.3.1 302.10 - Prohibitions [1.38]
1.2.3.2 302.11 - Fire Prevention [1.38]

1.3. CARs PART IV - PERSONNEL LICENSING AND TRAINING

1.3.1. 400 - General
1.3.1.1 400.01 - Definitions [1.01]

1.3.2. 401 - Flight Crew Permits, Licences and Ratings
1.3.2.1 401.03 - Requirements to Hold a Flight Crew Permit, Licence or Rating [1.11]
1.3.2.2 401.04 - Flight Crew Members of Aircraft Registered in Contracting States other than Canada [1.11]
1.3.2.3 *401.05 - Recency Requirements [1.11]
1.3.2.4 401.08 - Personal Logs [1.11]
1.3.2.5 *401.26 - Private Pilot Licence - Aeroplane - Privileges [1.06]
1.3.2.6 401.28 - Reimbursement of Costs Incurred in Respect of a Flight [1.12]
1.3.2.7 401.45 - Visual Flight Rules (VFR) Over-the-Top Privileges [1.08]

1.3.3. 404 - Medical Requirements
1.3.3.1 404.03 - Requirement to Hold a Medical Certificate (MC) [1.02]
1.3.3.2 404.04 - Issuance, Renewal and Validity Period of MC [1.02]
1.3.3.3 *404.06 - Prohibitions Regarding Exercise of Privileges [1.02]

1.4. CARs PART VI - GENERAL OPERATING AND FLIGHT RULES

1.4.1. 600 - Interpretation
1.4.1.1 600.01 - Definitions [1.13]

1.4.2. 601 - Airspace Structure, Classification and Use
1.4.2.1 601.01 - Airspace Structure [1.16]
1.4.2.2 601.02 - Airspace Classification [1.16]
1.4.2.3 *601.03 - Transponder Airspace [1.19]
1.4.2.4 *601.04 - IFR or VFR Flight in Class F Special Use Restricted Airspace or Class F Special Use Advisory Airspace [1.16]
1.4.2.5 601.06 - VFR Flight in Class A Airspace [1.16]
1.4.2.6 601.07 - VFR Flight in Class B Airspace [1.16, 1.33]
1.4.2.7 *601.08 - VFR Flight in Class C Airspace [1.16, 1.33]
1.4.2.8 *601.09 - VFR Flight in Class D Airspace [1.16]
1.4.2.9 601.15 - Forest Fire Aircraft Operating Restrictions [1.39]
1.4.2.10 601.16 - Issuance of NOTAM for forest Fire Aircraft Operating Limitations [1.39]

1.4.3. 602 - Operating and Flight Rules
GENERAL
1.4.3.1 602.01 - Reckless or Negligent Operation of Aircraft [1.13]
1.4.3.2 602.02 - Fitness of Flight Crew Members [1.13]
1.4.3.3 602.03 - Alcohol or Drugs - Crew Members [1.13]
1.4.3.4 602.04 - Alcohol or Drugs - Passengers [1.13]
1.4.3.5 602.05 - Compliance with Instructions [1.13]
1.4.3.6 602.06 - Smoking [1.13]
1.4.3.7 602.07 - Aircraft Operating Limitations [1.13]
1.4.3.8 602.08 - Portable Electronic Devices [1.13]
1.4.3.9 602.09 - Fuelling with Engines Running [1.13]
1.4.3.10 *602.10 - Starting and Ground Running of Aircraft Engines [1.13]
1.4.3.11 *602.11 - Aircraft Icing [1.13]
1.4.3.12 *602.12 - Overflight of Built-up Areas or Open-Air Assemblies of Persons during Take-offs, Approaches and Landings [1.13]
1.4.3.13 *602.13 - Take-offs, Approaches and Landings within Built-up Areas of Cities and Towns [1.13]
1.4.3.14 *602.14 - Minimum Altitudes and Distances [1.13]
1.4.3.15 602.15 - Permissible Low Altitude Flight [1.13]
1.4.3.16 *602.19 - Right of Way - General [1.13]
1.4.3.17 602.20 - Right of Way - Aircraft Maneuvering on Water [1.13]
1.4.3.18 *602.21 - Avoidance of Collision [1.13]
1.4.3.19 602.22 - Towing [1.13]
1.4.3.20 602.23 - Dropping of Objects [1.13]
1.4.3.21 602.24 - Formation Flight [1.13]
1.4.3.22 602.25 - Entering or Leaving and Aircraft in Flight [1.13]
1.4.3.23 602.26 - Parachute Descents [1.13]
1.4.3.24 *602.27 - Aerobatic Manoeuvres - Prohibited Areas and Flight Conditions [1.13]
1.4.2.25 602.28 - Aerobatic Maneuvers with Passengers [1.13]
1.4.2.26 *602.31 - Compliance with Air Traffic Control Instructions and Clearances [1.13]

1.4.2.27 *602.34 - Cruising Altitudes and Cruising Flight levels [1.37]
1.4.2.28 *602.35 - Altimeter Setting and Operating Procedures in the Altimeter-Setting Region [1.17]
1.4.2.29 *602.36 - Altimeter Setting and Operating Procedures in the Standard Pressure Region [1.17]
1.4.2.30 602.37 - Altimeter Setting and Operating Procedures in Transition Between Regions [1.17]
1.4.2.31 602.40 - Landing at or Take-off from an Aerodrome at Night [1.38]

OPERATIONAL AND EMERGENCY EQUIPMENT REQUIREMENTS
1.4.3.32 602.58 - Prohibition [1.35]
1.4.3.33 602.59 - Equipment Standards [1.35]
1.4.3.34 *602.60 - Requirements for Power-Driven Aircraft [1.35]
1.4.3.35 602.61 - Survival Equipment - Flights over Land [1.35]
1.4.3.36 *602.62 - Life Preservers and Personal Flotation Devices [1.35]
1.4.3.37 *602.63 - Life Rafts and Survival Equipment - Flights over Water [1.35]

FLIGHT PREPARATION, FLIGHT PLANS & FLIGHT ITINERARIES
1.4.3.38 602.70 - Definitions [1.20]
1.4.3.39 602.71 - Pre-Flight Information [1.20]
1.4.3.40 602.72 - Weather Information [1.20]
1.4.3.41 *602.73 - Requirements to file a Flight Plan or Flight Itinerary [1.20]
1.4.3.42 602.74 - Contents of a Flight Plan or Fight Itinerary [1.20]
1.4.3.43 602.75 - Filing of a Flight Plan or Fight Itinerary [1.20]
1.4.3.44 *602.76 - Changes in the Flight Plan — Subsections (3) and (4) [1.20]
1.4.3.45 *602.77 - Requirement to File an Arrival Report [1.20]
1.4.3.46 602.78 - Contents of an Arrival Report [1.20]
1.4.3.47 602.79 - Overdue Aircraft Reports [1.20]

PRE-FLIGHT AND FUEL REQUIREMENTS
1.4.3.48 602.86 - Carry-on Baggage, Equipment and Cargo [1.21]
1.4.3.49 *602.88 - Fuel Requirements — Sub-sections (1), (2), (3) and (5) [1.21]
1.4.3.50 602.89 - Passenger Briefings [1.21]

OPERATIONS AT OR IN THE VICINITY OF AN AERODROME
1.4.3.51 *602.96 - General [1.32]
1.4.3.52 *602.97 - VFR and IFR Aircraft Operations at Uncontrolled Aerodromes within an MF Area [1.32]
1.4.3.53 *602.98 - General MF Reporting Procedures [1.32]
1.4.3.54 *602.99 - MF Reporting Procedures before Entering Maneuvering Area [1.32]
1.4.3.55 *602.100 - MF Reporting Procedures on Departure [1.32]
1.4.3.56 *602.101 - MF Reporting Procedures on Arrival [1.32]
1.4.3.57 *602.102 - MF Reporting Procedures when Flying Continuous Circuits [1.32]
1.4.3.58 *602.103 - Reporting Procedures when Flying through an MF Area [1.32]

VISUAL FLIGHT RULES
1.4.3.59 *602.114 - Minimum Visual Meteorological Conditions for VFR Flight in Controlled Airspace [1.15]
1.4.3.60 *602.115 - Minimum Visual Meteorological Conditions for VFR Fight in Uncontrolled Airspace [1.15]
1.4.3.61 602.116 - VFR Over-the-Top [1.15]
1.4.3.62 *602.117 - Special VFR Flight [1.15]

RADIOCOMMUNICATIONS
1.4.3.63 602.136 - Continuous Listening Watch [1.15]
1.4.3.64 *602.138 - Two-way Radiocommunication Failure in VFR Flight [1.15]

EMERGENCY COMMUNICATIONS AND SECURITY
1.4.3.65 602.143 - Emergency Radio Frequency Capability [1.28]
1.4.3.66 602.144 - Interception Signals, Interception of Aircraft and Instructions to Land [1.28]
1.4.3.67 602.145 - ADIZ [1.36]
1.4.3.68 602.146 - SCATANA Plan [1.36]

1.4.5. 605 - Aircraft Requirements

GENERAL
1.4.4.1 605.03 - Flight Authority [1.14]
1.4.4.2 605.04 - Availability of Aircraft Flight Manual [1.14]
1.4.4.3 605.05 - Markings and Placards [1.14]
1.4.4.4 605.08 - Unserviceable and Removed Equipment - General [1.14]

AIRCRAFT EQUIPMENT REQUIREMENTS
1.4.4.5 *605.14 - Power-Driven Aircraft - Day VFR [1.15]
1.4.4.6 605.15 - Power-driven Aircraft - VFR OTT [1.15]
1.4.4.7 605.16 - Power-driven Aircraft - Night VFR [1.15]
1.4.4.8 605.17 - Use of Position and Anti-collision Lights [1.15]
1.4.4.9 605.22 - Seat and Safety-Belt Requirements [1.26]
1.4.4.10 605.24 - Shoulder Harness Requirements [1.26]
1.4.4.11 *605.25 - General Use of Safety Belts and Restraint Systems — Sub-sections (1) and (2) [1.26]
1.4.4.12 605.28 - Child Restraint System — Sub-section 1(a) [1.26]
1.4.4.13 605.29 - Flight Control Locks [1.18]
1.4.4.14 605.31 - Oxygen Equipment and Supply [1.27]
1.4.4.15 *605.32 - Use of Oxygen — Sub-sections (1) and (2) [1.27]
1.4.4.16 *605.35 - Transponder and Automatic Pressure Altitude Reporting Equipment — Sub-sections (1a), (2) and (3) [1.19]
1.4.4.17 605.38 - ELT — Subsections (1) and (2) [1.29]
1.4.4.18 605.40 - ELT Activation [1.29]

AIRCRAFT MAINTENANCE REQUIREMENTS
1.4.4.19 605.84 - General [1.14]
1.4.4.20 605.85 - Maintenance Release and Elementary Work [1.14]
1.4.4.21 605.86 - Maintenance Schedule — Sub-sections (1) (a), (2) and (3) [1.14]
1.4.4.22 605.88 - Inspection after Abnormal Occurrences [1.14]

TECHNICAL RECORDS
1.4.4.23 605.92 - Requirements to Keep Technical Records — Sub-section (1) [1.14]
1.4.4.24 605.93 - Technical Records - General [1.14]
1.4.4.25 605.94 - Journey Log Requirements [1.14]
1.4.4.26 605.95 - Journey Log - Carrying on Board [1.14, 1.40]
1.4.4.27 605.97 - Transfer of Records [1.14]

1.4.5. Miscellaneous
1.4.5.1 606.02 - Liability Insurance — Sub-section (9) [1.43]

1.5. TRANSPORTATION SAFETY BOARD OF CANADA (TSB) (AIP Canada GEN 3.0) [1.24]
1.5.1 Definitions [1.24]
1.5.2 *Reporting an Aviation Occurrence [1.24]
1.5.3 *Protection of Occurrence Sites [1.24]

1.6. AIR TRAFFIC SERVICES AND PROCEDURES
1.6.1 *Air Traffic and Advisory Services [1.30]
1.6.2 *Flight Service Stations [1.30]
1.6.3 *Communications Procedures [1.30]
1.6.4 *Radar Service — Clock Position System [1.30]
1.6.5 *ATC Clearances/Instructions/Mandatory Readback Procedures [1.30]
1.6.6 *Wake Turbulence Separation
1.6.7 *Airport/Aerodrome Operations — Controlled [1.31]
1.6.8 *Airport/Aerodrome Operations — Uncontrolled [1.32]
1.6.9 Mandatory & Aerodrome Traffic Frequencies [1.32]
1.6.10 VFR Enroute Procedures [1.33]
1.6.11 VFR Holding Procedures [1.34]
1.6.12 *VFR/IFR Traffic Mix at Uncontrolled Airports/Aerodromes [1.34]
1.6.13 *Simultaneous Intersecting Runway Operations (SIRO) [1.32]
1.6.14 *Procedures for the Prevention of Runway Incursions [1.31]

2. Navigation and Radio Aids

2.1. DEFINITIONS
2.1.1 Meridian [6.01]
2.1.2 Prime Meridian [6.01]
2.1.3 Longitude [6.01]
2.1.4 Equator [6.01]
2.1.5 Latitude [6.01]
2.1.6 Rhumb Line/Great Circle [6.01]
2.1.7 Variation [6.01]
2.1.8 Isogonal [6.01]
2.1.9 Agonic Line [6.01]
2.1.10 Deviation [6.01]
2.1.11 Track [6.01]
2.1.12 Heading [6.01]
2.1.13 Airspeed [6.01]
2.1.14 Ground Speed [6.01]
2.1.15 Air Position [6.01]
2.1.16 Ground Position [6.01]
2.1.17 Bearing [6.19]
2.1.18 Wind Velocity [6.01]
2.1.19 Drift [6.01]

2.2. MAPS AND CHARTS
2.2.1 VTA — Transverse Mercator Projection [6.02]
2.2.2 VNC — Lambert Conformal Conic Projection [6.02]
2.2.3 *Topographical Symbols [6.02]
2.2.4 *Elevation and Contours (Relief) [6.02]
2.2.5 *Aeronautical Information [6.02]
2.2.6 *Scale and Units of Measurement [6.02]
2.2.7 *Locating Position by Latitude and Longitude [6.05]
2.2.8 Navigation Aids [6.03]

2.3. TIME AND LONGITUDE
2.3.1 24 Hour System [6.04]
2.3.2 Time Zones and Relation to Longitude [6.04]
2.3.3 Conversion of UTC to Local and Vice Versa [6.04]

2.4. PILOT NAVIGATION
2.4.1 Use of Aeronautical Charts [6.06, 6.08]
2.4.2 *Measurement of Track and Distance [6.06]
2.4.3 Map Reading [6.06]
2.4.4 Setting Heading — Visual Angle of Departure [6.07]
2.4.5 Check-points and Pin-points [6.13]
2.4.6 Use of Position Lines to Obtain a Fix [6.19]
2.4.7 *Ground Speed Checks and ETA Revisions [6.14]
2.4.8 *Variation/Deviation [6.01, 6.14]
2.4.9 *True Track/Magnetic Track [6.01, 6.14]
2.4.10 Determining Drift by 10° Lines [6.15]
2.4.11 Double Track Error Method to Regain Track [6.15]
2.4.12 Sum of Opening and Closing Angles to Destination [6.15]
2.4.13 Visual Alteration Method of Correcting to Track [6.15]
2.4.14 *Diversion to Alternate [6.16]
2.4.15 *Return to Departure Point (Reciprocal Track) [6.17]
2.4.16 Low Level Navigation [6.13]
2.4.17 Deduced (Dead) Reckoning (DR Navigation) [6.00]
2.4.18 In-flight Log and Mental Calculations [6.14]
2.4.19 *Procedures When Lost [6.18]
2.4.20 Air and Ground Position [6.01]
2.4.21 True/Magnetic/Compass Headings [1.14]
2.4.22 Indicated/Calibrated Airspeed (IAS, CAS) [6.12]
2.4.23 True Airspeed/Ground Speed (TAS, G/S) [6.09]
2.4.24 Compass Errors [6.14]
2.4.25 Radio Communications (as per Sec. 1.)

2.5. TRIANGLE OF VELOCITIES
2.5.1 True Airspeed and Heading [6.01]
2.5.2 Wind Velocity [6.01]
2.5.3 Ground Speed and Track [6.01]

2.6. NAVIGATION COMPUTERS
2.6.1 *Heading and True Airspeed [6.09]
2.6.2 *Applying the Wind [6.09]
2.6.3 *True Track and Ground Speed [6.09]
2.6.4 *Magnetic Heading and Magnetic Track [6.09]
2.6.5 *Pressure/Density and True Altitudes [4.04]
2.6.6 *Indicated/Calibrated/True Airspeed [4.02]
2.6.7 *Time/Ground Speed/Distance [6.10]
2.6.8 *Fuel Consumption and Conversions [6.11]

2.7. PRE-FLIGHT PREPARATION
2.7.1 *Factors Affecting Choice of Route [6.13]
2.7.2 Map Preparation [6.08, 6.13]
2.7.3 Meteorological Information [5.16 - 5.26]
2.7.4 *NOTAM [1.22, 6.13]
2.7.5 Selection of Check-points [6.13]
2.7.6 *Fuel Requirements [6.11]
2.7.7 *Weight and Balance [7.16]
2.7.8 *Use of Canada Flight Supplement [7.22]
2.7.9 Documents to be Carried in Aircraft [1.40]
2.7.10 Flight Plans/Itineraries [1.20]
2.7.11 Flight Log Forms [6.14]
2.7.12 Aircraft Serviceability [1.20

2.8. RADIO THEORY
2.8.1 Characteristics of Low/High and Very High Frequency Radio Waves [6.24]
2.8.2 Frequency Bands Used in Navigation and Communication [6.24]
2.8.3 Reception Limitations [6.24]

2.9. VHF OMNIDIRECTION RANGE (VOR)
2.9.1 Aircraft Equipment [6.19]
2.9.2 Tuning and Identifying [6.19]
2.9.3 Serviceability Checks [6.19]
2.9.4 *Interpretation/Orientation/Homing [6.19]

2.10. AUTOMATIC DIRECTION FINDER (ADF)
2.10.1 Aircraft Equipment [6.20]
2.10.2 Tuning and Identifying [6.20]
2.10.3 Serviceability Check [6.20]
2.10.4 Interpretation/Orientation/Homing [6.20]
2.10.5 Voice Feature [6.20]

2.11. LONG RANGE AREA NAVIGATION (LORAN C)
2.11.1 Basic Use/Inaccuracies/Reception Limitations [6.27]

2.12. GLOBAL NAVIGATION SATELLITE SYSTEM (GNSS/GPS)
2.12.1 Basic Principles/Use/Limitations [6.25]

2.13. OTHER RADIO AND RADAR AIDS — BASIC PRINCIPLES AND USE
2.13.1 *Transponder [1.19]
2.13.2 *Emergency Locator Transmitter (ELT) [1.29]
2.13.3 *VHF Direction Finding (DF) Assistance [6.21]
2.13.4 Airport Surveillance Radar (ASR) Primary/ Secondary [6.23]

3. Meteorology

3.1. THE EARTH'S ATMOSPHERE
3.1.1 Composition and Physical Properties [5.01]
3.1.2 Vertical Structure [5.01]
3.1.3 The Standard Atmosphere [5.01]
3.1.4 Density and Pressure [5.01]
3.1.5 Mobility [5.01]
3.1.6 Expansion and Compression [5.01]

3.2. ATMOSPHERIC PRESSURE
3.2.1 Pressure Measurements [5.01]
3.2.2 Station Pressure [5.01]
3.2.3 Sea Level Pressure [5.01]
3.2.4 Pressure Systems and their Variations [5.01]
3.2.5 Effects of Temperature [5.03]
3.2.6 Isobars [5.01]
3.2.7 *Horizontal Pressure Differences [5.01]

3.3. METEOROLOGICAL ASPECTS OF ALTIMETRY
3.3.1 *Pressure Altitude [5.02]
3.3.2 *Density Altitude [5.02]
3.3.3 *Altimeter Settings [5.02]
3.3.4 *Considerations When Flying from High to Low Pressure and Temperature Areas, and Vice Versa [5.02]

3.4. TEMPERATURE
3.4.1 Temperature Scale — Fahrenheit/Celsius [5.03]
3.4.2 *Heating/Cooling of the Atmosphere — Convection/Advection/Radiation [5.03]
3.4.3 Horizontal Differences [5.03]

3.4.4 *Temperature Variations with Altitude [5.03]
3.4.5 *Inversions [5.03]
3.4.6 Isothermal Layers [5.03]

3.5. MOISTURE
3.5.1 *Relative Humidity/Dewpoint [5.04]
3.5.2 Sublimation/Condensation [5.04]
3.5.3 Cloud Formation [5.04]
3.5.4 Precipitation [5.04]
3.5.5 Saturated/Dry Adiabatic Lapse Rate [5.04]

3.6. STABILITY AND INSTABILITY
3.6.1 Lapse Rate and Stability [5.05]
3.6.2 Modification of Stability [5.05]
3.6.3 *Characteristics of Stable/Unstable Air [5.05]
3.6.4 Surface Heating and Cooling [5.05]
3.6.5 *Lifting Processes [5.05]
3.6.6 Subsidence/Convergence [5.05]

3.7. CLOUDS
3.7.1 Classification [5.06]
3.7.2 Formation and Structure [5.06]
3.7.3 *Types and Recognition [5.06]
3.7.4 *Associated Precipitation and Turbulence [5.06]

3.8. TURBULENCE
3.8.1 Convection [5.07]
3.8.2 Mechanical [5.07]
3.8.3 Orographic [5.07]
3.8.4 *Wind Shear [5.07]
3.8.5 Reporting Criteria [5.07]

3.9. WIND
3.9.1 Definition [5.08]
3.9.2 *Pressure Gradient [5.08]
3.9.3 Deflection Caused by the Earth's Rotation [5.08]
3.9.4 *Low Level Winds — Variation in Surface Wind [5.08]
3.9.5 Friction [5.08]
3.9.6 *Veer/Back [5.08]
3.9.7 Squalls and Gusts [5.08]
3.9.8 Diurnal Effects [5.08]
3.9.9 Land/Sea Breezes [5.08]
3.9.10 Katabatic/Anabatic Effects [5.08]
3.9.11 *Topographical Effects [5.08]
3.9.12 *Wind Shear — Types and Causes [5.08, 7.04]

3.10. AIR MASSES
3.10.1 Definition and Characteristics [5.09]
3.10.2 Formation/Classification [5.09]
3.10.3 Modification [5.09]
3.10.4 Factors that Determine Weather [5.09]
3.10.5 Seasonal/Geographic Effects [5.09]
3.10.6 Air Masses Affecting North America [5.09]

3.11. FRONTS
3.11.1 Structure [5.10]
3.11.2 *Types [5.10]
3.11.3 Formation [5.10]
3.11.4 Cross-sections [5.10]

3.12. FRONTAL WEATHER
3.12.1 *Cold Front [5.10]
3.12.2 *Warm Front [5.10]
3.12.3 TROWAL and Upper Fronts [5.10]

3.13. AIRCRAFT ICING
3.13.1 In-flight — Freezing Rain [5.11]
3.13.2 Hoar Frost [5.11]

3.14. THUNDERSTORMS
3.14.1 *Requirements for Development [5.12]
3.14.2 Structure/Development [5.12]
3.14.3 Types — Air Mass/Frontal [5.12]
3.14.4 *Hazards — Updrafts/Downdrafts/Gust Fronts/Downbursts/Microbursts/Hail and Lightning [5.12]
3.14.5 Squall Lines [5.13]

3.15. HURRICANES AND TORNADOES [5.14]

3.16. SURFACE BASED LAYERS
3.16.1 *Fog Formation [5.15]
3.16.2 *Fog Types (including Mist) [5.15, 5.20]
3.16.3 Haze/Smoke [5.15]
3.16.4 Blowing Obstructions to Vision [5.15]

3.17. METEOROLOGICAL SERVICES AVAILABLE TO PILOTS
3.17.1 Aviation Weather Information Service (AWIS) [5.16]
3.17.2 Aviation Weather Briefing Service (ABWS) [5.16]
3.17.3 Flight Service Stations (FSS) [5.16]
3.17.4 Pilot's Automatic Telephone Weather Answering Service (PATWAS) [5.16]
3.17.5 Transcribed Weather Broadcasts (TWB) [5.16]
3.17.6 Automatic Terminal Information Service (ATIS) [5.16]

3.18. AVIATION WEATHER REPORTS
3.18.1 *Aviation Routine Weather Report (METAR) [5.20]
3.18.2 Automated Weather Observation Station (AWOS) [5.20]
3.18.3 Pilot Reports (PIREP)

3.19. AVIATION FORECASTS
3.19.1 *Times Issued and Period of Coverage [5.17, 5.18, 5.19]
3.19.2 *Decoding [5.17, 5.18, 5.19]
3.19.3 *Area Forecasts (FA) [5.17]
3.19.4 *Terminal Area Forecasts (TAF) [5.18]
3.19.5 *Upper Winds and Temperature Forecasts (FD) [5.19]
3.19.6 Pilot's Meteorological Advisory (AIRMET) [5.25]
3.19.7 *Significant In-flight Weather Warning Messages (SIGMET) [5.21]

3.20. WEATHER MAPS AND PROGNOSTIC CHARTS
3.20.1 Times Issued and Validity Periods [5.23]
3.20.2 Symbols and Decoding [5.23]
3.20.3 Surface Weather Map [5.23]
3.20.4 Upper Air Charts — Weather Information to 700 mb Level [5.26]
3.20.5 Prognostic Surface Chart [5.23]

4. Airframes, Engines and Systems

4.1. AIRFRAMES
4.1.1 Types of Construction [2.01]

4.2. LANDING GEAR, BRAKES AND FLAPS
4.2.1 Mechanical [2.01, 2.02]
4.2.2 Hydraulic [2.01, 2.02]
4.2.3 Electrical [2.01, 2.02]

4.3. ENGINES
4.3.1 Two/Four Stroke Cycle [2.03]
4.3.2 Methods of Cooling [2.03]
4.3.3 Principle of the Magneto [2.07]
4.3.4 *Dual Ignition [2.07]
4.3.5 Exhaust Systems [2.12]
4.3.6 *Ancillary Controls [2.09]
4.3.7 Turbo-charging [2.11]
4.3.8 Effects of Density Altitude/Humidity [7.17]
4.3.9 Limitations and Operations [7.17]
4.3.10 Instruments [2.03]

4.4. CARBURATION
4.4.1 Theory of Operation [2.09]
4.4.2 *Fuel-Air Mixture [2.09]
4.4.3 *Mixture Controls [2.09]
4.4.4 *Carburettor Icing [2.09]
4.4.5 *Use of Carb Heat and its Effects on Mixture [2.09]

4.5. FUEL INJECTION
4.5.1 Principle and Operation [2.09]
4.5.2 Icing [2.09]
4.5.3 Alternate Air [2.09]

4.6. ELECTRICAL SYSTEM
4.6.1 Generator/Alternator/Battery [2.10]
4.6.2 Lighting [2.10]
4.6.3 Master/Alternator/Generator Switches [2.10]
4.6.4 Ammeter/Load Meter [2.10]
4.6.5 Bus Bars [2.10]
4.6.6 Circuit Breakers/Fuses [2.10]
4.6.7 Grounding/Bonding [2.10]

4.7. LUBRICATING SYSTEMS AND OILS
4.7.1 *Types — Viscosity/Grades/Seasonal Use [2.08]
4.7.2 *Purposes [2.08]
4.7.3 Methods of Lubrication [2.08]
4.7.4 Venting [2.08]
4.7.5 Filters [2.08]
4.7.6 Oil Cooler [2.08]

4.8. FUEL SYSTEMS AND FUELS
4.8.1 Types — Colour/Properties [2.04]
4.8.2 Density/Weight [2.04]

4.8.3 Additives [2.04]
4.8.4 *Contamination and Deterioration [2.05]
4.8.5 Tank Location [2.06]
4.8.6 Venting [2.06]
4.8.7 Fuel Line — Filters/Drains [2.06]
4.8.8 Induction Manifold [2.06]
4.8.9 *Detonation — Causes/Effects [2.04]
4.8.10 Vapour Lock [2.04]
4.8.11 Primers [2.06]
4.8.12 Fuel Management — Ground/Air [2.05]
4.8.13 *Fuel Handling — Fuelling Aircraft [2.05]
4.8.14 *Grounding/Bonding [2.05]

4.9. OTHER AIRCRAFT SYSTEMS

4.9.1 Oxygen [2.13]
4.9.2 Vacuum [4.06]

5. Theory of Flight

5.1. PRINCIPLES OF FLIGHT

5.1.1 Bernoulli's Theorem [3.03]
5.1.2 Newton's Laws [3.03]

5.2. FORCES ACTING ON AN AEROPLANE

5.2.1 *Lift [3.03]
5.2.2 *Drag — Induced/Parasite/Profile [3.03]
5.2.3 *Relationship of Lift and Drag to Angle of Attack [3.03]
5.2.4 Thrust [3.03]
5.2.5 Weight [3.03]
5.2.6 Equilibrium [3.03]
5.2.7 Centre of Pressure (C of P) [3.04]
5.2.8 Centrifugal/Centripetal [3.05]
5.2.9 *Forces Acting on an Aircraft during Manoeuvres [3.05]

5.3. AEROFOILS

5.3.1 Pressure Distribution about an Aerofoil [3.05]
5.3.2 *Relative Airflow and Angle of Attack [3.05]
5.3.3 Downwash [3.05]
5.3.4 Wing Tip Vortices [3.03]
5.3.5 Angle of Incidence [3.01]

5.4. PROPELLERS

5.4.1 Propeller Efficiency at Various Speeds [3.06]
5.4.2 Fixed/Variable Pitch [3.06]
5.4.3 Torque/Slipstream/Gyroscopic Effect/ Asymmetric Thrust [3.06]

5.5. DESIGN OF THE WING

5.5.1 Wing Planform [3.01]
5.5.2 Area/Span/Chord [3.01]
5.5.3 Aspect Ratio [3.01]
5.5.4 Streamlining [3.01]
5.5.5 Camber [3.01]
5.5.6 Laminar Flow [3.01]
5.5.7 Dihedral/Anhedral [3.01]
5.5.8 Wash-in/Wash-out [3.01]
5.5.9 Slots/Slats [3.01]
5.5.10 Wing Fences/Stall Strips [3.01]
5.5.11 Spoilers [3.01]
5.5.12 *Flaps [3.01]
5.5.13 Canards [3.07]

5.6. LOAD FACTOR

5.6.1 Centrifugal Force/Weight [3.05]
5.6.2 Load Factor — Linear/Turns [3.05]
5.6.3 *Relationship of Load Factor to Stalling Speed [3.05]
5.6.4 *Structural Limitations [3.05]
5.6.5 Gust Loads [3.05]

5.7. STABILITY

5.7.1 *Longitudinal/Lateral/Directional [3.02]
5.7.2 Inherent Stability [3.02]
5.7.3 Methods of Achieving Stability [3.02]

5.8. FLIGHT CONTROLS

5.8.1 Aeroplane Axes and Planes of Movement [3.07]
5.8.2 Functions of Controls [3.07]
5.8.3 Relationship Between Effects of Yaw and Roll [3.07]
5.8.4 Adverse Yaw/Aileron Drag [3.07]
5.8.5 Static/Dynamic Balancing of Controls [3.07]
5.8.6 Trim/Trimming Devices [3.07]

6. Flight Instruments

6.1. PITOT STATIC SYSTEM

6.1.1 *Pitot [4.01]
6.1.2 *Static [4.01]
6.1.3 Anti-icing [4.01]
6.1.4 Alternate Static — Sources/Errors [4.01]

6.2. AIRSPEED INDICATOR

6.2.1 Principles of Operation [4.02]
6.2.2 *Errors [4.02]
6.2.3 *Markings [4.02]
6.2.4 Definitions (IAS/CAS/TAS) [4.02]

6.3. VERTICAL SPEED INDICATOR

6.3.1 Principles of Operation [4.03]
6.3.2 Errors [4.03]
6.3.3 Lag [4.03]

6.4. ALTIMETER/ENCODING ALTIMETER

6.4.1 Principles of Operation [4.04]
6.4.2 Errors [4.04]

6.5 DIRECT READING MAGNETIC COMPASS

6.5.1 Principles of Operation [4.05]
6.5.2 *Magnetic Dip [4.05]
6.5.3 *Turning and Acceleration Errors [4.05]
6.5.4 *Reading the Compass [4.05]
6.5.5 Factors Adversely Affecting Compass Operation [4.05]
6.5.6 Deviation [4.05]
6.5.7 Compass Correction Card [4.05]
6.5.8 Turning and Acceleration Errors [4.05]
6.5.9 Compass and Serviceability Checks [4.05]
6.5.10 Checking Compass Heading on the Ground and in Flight [4.05]

6.6. GYROSCOPE

6.6.1 Principles of Operation [4.06]
6.6.2 Inertia [4.06]
6.6.3 Precession [4.06]

6.7. HEADING INDICATOR

6.7.1 *Principles of Operation [4.08]
6.7.2 *Errors [4.08]
6.7.3 Limitations [4.08]
6.7.4 Power Sources [4.08]

6.8. ATTITUDE INDICATOR

6.8.1 *Principles of Operation 4.07]
6.8.2 *Errors [4.07]
6.8.3 Limitations [4.07]
6.8.4 Power Sources [4.07]

6.9 TURN AND BANK INDICATOR/TURN COORDINATOR

6.9.1 Principles of Operation [4.09]
6.9.2 Errors [4.09]
6.9.3 Limitations [4.09]
6.9.4 Power Sources [4.09]

6.10. INSTRUMENT FLYING

6.10.1 Loss of Visual Reference [4.11]
6.10.2 The Control and Performance Instruments [4.11]
6.10.3 Instrument Scan/Interpretation [4.11]
6.10.4 Aircraft Control [4.11]
6.10.5 Unusual Attitudes/Recoveries [4.11]

7. Flight Operations

7.1. GENERAL

7.1.1 Pilot-in-Command Responsibilities [7.06]
7.1.2 Winter Operations [7.01]
7.1.3 Thunderstorm Avoidance [7.03]
7.1.4 Mountain Flying Operations [7.09]
7.1.5 Collision Avoidance — Use of Landing Lights [7.05]
7.1.6 *Runway Numbering [7.08]
7.1.7 Airport Rotating Beacon [1.38]
7.1.8 VASIS/PAPI [7.25, 7.26]
7.1.9 Approach, Runway and Aerodrome Marking/Lighting [1.38, 7.22]
7.1.10 Obstruction Markings/Lighting [6.03]
7.1.11 Units of Measurement and Conversion [7.21]
7.1.12 Radio Communications (as per Section 1.)
7.1.13 *Aerodrome Operations (including Procedures for the Prevention of Runway Incursions) [1.38]
7.1.14 *Wheelbarrowing [7.11]
7.1.15 *Hydro-planing [7.35]
7.1.16 *Taxiing [7.15]
7.1.17 *Effects of Wind/Wind Shear [7.04]
7.1.18 *Side-slips [7.17]

7.2. AIRCRAFT PERFORMANCE

7.2.1 *Effects of Critical Surface Contamination [7.01]
7.2.2 Lift/Drag Ratio [7.17]

7.2.3 *Effects of Density Altitude/Humidity [7.17]
7.2.4 Attitude Plus Power Equals Performance — Climb/Descent/Level Flight [7.17]
7.2.5 Normal/Short/Soft and Rough Field Take-offs and Landings [7.17]
7.2.6 *Ground Effect [7.12]
7.2.7 *Best Angle of Climb (Vx) [7.17, 7.30]
7.2.8 *Best Rate of Climb (Vy) [7.17, 7.30]
7.2.9 *Manoeuvring Speed (Va) [7.17]
7.2.10 Safe Operating Speed (Vno) [7.17]
7.2.11 Never Exceed Speed (Vne) [7.17]
7.2.12 Maximum Flap Speed (Vfe) [7.17]
7.2.13 Maximum Gear Operating Speed (Vlo) [7.17]
7.2.14 Gliding for Range [7.17]
7.2.15 Flying for Range [7.17]
7.2.16 Flying for Endurance [7.17]
7.2.17 *Slow Flight [7.17]
7.2.18 *Stalls [3.04, 7.17]
7.2.19 *Indicated and True Stalling Speed [3.04]
7.2.20 *Stall Speed vs Altitude [3.04]
7.2.21 *Spins [7.14]
7.2.22 *Spirals [7.13]
7.2.23 Recommended Safe Recovery Altitudes [7.14]
7.2.24 Bank/Speed vs Rate/Radius of Turn [7.23]
7.2.25 *Effect of Change of Weight or Centre of Gravity (C of G) on Performance [7.16]
7.2.26 Use of Aircraft Flight Manual (Including Approved and Unapproved Operational Information) [7.17]

7.3. USE OF PERFORMANCE CHARTS

7.3.1 *Take-off Charts [7.17]
7.3.2 *Cross-wind Charts [7.17]
7.3.3 *Cruise Charts [7.17]
7.3.4 Fuel Burn Charts [7.17]
7.3.5 *Landing Charts [7.17]
7.3.6 Performance (V) Speeds — Va, Vno, Vfe, Vlo, Vne, Vs, Vx, Vy [7.17]
7.3.7 Effect of Ice/Snow/Frost/Slush/Water on Take-off and Landing Run [7.17]
7.3.8 Effect of Various Runway Surfaces on Take-off and Landing Run [7.17]
7.3.9 Upslope/Downslope Runway [7.17]

7.4. WEIGHT AND BALANCE

7.4.1 Terms (e.g. Datum/Arm/Moment) [7.16]
7.4.2 *Locating C of G [7.16]
7.4.3 *C of G Limits [7.16]
7.4.4 Weights (e.g. Empty/Gross) [7.16]
7.4.5 Load Adjustment [7.16]
7.4.6 Cargo Tie-down/Passenger Loading [7.16]
7.4.7 *Normal/Utility Category [7.16]

7.5. WAKE TURBULENCE

7.5.1 *Causes [7.10]
7.5.2 *Effects [7.10]
7.5.3 *Avoidance [7.10]

7.6. SEARCH AND RESCUE (SAR) (AIP Canada — SAR Information)

7.6.1 Types of Service Available [7.27]
7.6.2 *ELT (exclude Categories) [7.27]
7.6.3 Aircraft Emergencies [7.27]
7.6.4 Survival — Basic Techniques [7.27]

7.7. AIRCRAFT CRITICAL SURFACE CONTAMINATION

7.7.1 Clean Aircraft Concept [7.01]
7.7.2 Frozen Contaminants [7.01]
7.7.3 Cold Soaking Phenomenon [7.01]
7.7.4 Pre-take-off Inspection [7.01]

8. Human Factors

8.1. AVIATION PHYSIOLOGY

8.1.1 *Hypoxia/Hyperventilation [7.24]
8.1.2 Gas Expansion Effects [7.24]
8.1.3 Decompression (Including SCUBA Diving) [7.24]
8.1.4 *Vision/Visual Scanning Techniques [7.05]
8.1.5 Hearing [7.24]
8.1.6 *Orientation/Disorientation (Including Visual/Vestibular Illusions) [7.31]
8.1.7 Positive/Negative "G" [7.24]
8.1.8 Sleep/Fatigue [7.24]
8.1.9 Anaesthetics/Blood Donations [7.24]

8.2. THE PILOT AND THE OPERATING ENVIRONMENT

8.2.1 Personal Health/Fitness [7.24]
8.2.2 Diet/Nutrition [7.24]
8.2.3 *Medications (Prescribed, over-the-counter) [7.24]
8.2.4 Substance Abuse (Alcohol/Drugs) [7.24]
8.2.5 Pregnancy [7.24]
8.2.6 Heat/Cold [7.24]
8.2.7 Noise/Vibration [7.24]
8.2.8 Effects of Smoking [7.24]
8.2.9 *Toxic Hazards (Including Carbon Monoxide) [7.24]

8.3. AVIATION PSYCHOLOGY

8.3.1 The Decision-making Process [7.34]
8.3.2 *Factors that Influence Decision-making [7.34]
8.3.3 *Situational Awareness [7.34]
8.3.4 *Stress [7.34]
8.3.5 *Managing Risk [7.34]
8.3.6 Attitudes [7.34]
8.3.7 Workload — Attention and Information Processing [7.34]

8.4. PILOT — EQUIPMENT/MATERIALS RELATIONSHIP

8.4.1 Controls and Displays — Errors in Interpretation and Control [7.34]
8.4.2 Errors in the Interpretation and Use of Maps/Charts [7.34]
8.4.3 Correct Use of Check-lists and Manuals [7.34]

8.5. INTERPERSONAL RELATIONS

8.5.1 Communication with — Maintenance Personnel/Air Traffic Services/Passengers [7.34]
8.5.2 Operating Pressures — Family Relationships/Peer Group [7.34]

TRANSPORT CANADA SUBJECT CODES: RECREATIONAL PILOT PERMIT

INSTRUCTIONS: Transport Canada indicates the subject material for study and review in preparation for the RPP in the publication *Study and Reference Guide Recreational Pilot Permit - Aeroplane, Third Edition June, 1998 - TP 12467E.* Following your completion of your RPPAE written, you will receive on your scoring results sheet a listing of any subjects scored incorrectly utilizing these Transport Canada subject codes. All Transport Canada subject codes are listed below, and following each Transport Canada code, is a number in brackets which represents the corresponding code used in this course. To cross reference from a Transport Canada subject number to this manual, first locate the subject from the numbers below, note our corresponding subject reference number (in brackets), then look up the subject in this manual. EXAMPLE: To cross reference Transport Canada number **1.3.3.2** (Issuance, Renewal and Validity Period of Medical Certificate), refer to subject **1.02** in this course. Subjects with an asterisk (*) are essential knowledge areas that will receive particular emphasis on the Transport Canada written.

1. Air Law and Procedures

1.1. CARs PART I - GENERAL PROVISIONS

1.1.1. 101 - Interpretation
1.1.1.1 101.01 - Definitions [1.01]

1.1.2. 103 - Administration and Compliance
1.1.2.1 103.02 - Inspection of Aircraft, Requests for Production of Documents and Prohibitions [1.01]
1.1.2.2 103.03 - Return of Canadian Aviation Documents [1.01]
1.1.2.3 103.04 - Record Keeping [1.01]

1.2. CARs PART III - AERODROMES AND AIRPORTS

1.2.1. 300 - Interpretation
1.2.1.1 300.01 - Interpretation [1.38]

1.2.2. 301 - Aerodromes
1.2.2.1 301.01 - Application [1.38]
1.2.2.2 *301.04 - Markers and Markings [1.38]
1.2.2.3 *301.06 - Wind Direction Indicator [1.38]
1.2.2.4 301.07 - Lighting [1.38]
1.2.2.5 301.08 - Prohibitions [1.38]
1.2.2.6 301.09 - Fire Prevention [1.38]

1.2.3. 302 - Airports
1.2.3.1 302.10 - Prohibitions [1.38]
1.2.3.2 302.11 - Fire Prevention [1.38]

1.3. CARs PART IV - PERSONNEL LICENSING AND TRAINING

1.3.1. 400 - General
1.3.1.1 400.01 - Definitions [1.01]

1.3.2. 401 - Flight Crew Permits, Licences and Ratings
1.3.2.1 401.03 - Requirements to Hold a Flight Crew Permit, Licence or Rating [1.11]
1.3.2.2 401.04 - Flight Crew Members of Aircraft Registered in Contracting States other than Canada [1.11]
1.3.2.3 *401.05 - Recency Requirements [1.11]
1.3.2.4 401.08 - Personal Logs [1.11]
1.3.2.5 *401.26 - Pilot Permit - Recreational - Aeroplane - Privileges [1.05]
1.3.2.6 401.28 - Reimbursement of Costs Incurred in Respect of a Flight [1.12]

1.3.3. 404 - Medical Requirements
1.3.3.1 404.03 - Requirement to Hold a Medical Certificate (MC) [1.02]
1.3.3.2 404.04 - Issuance, Renewal and Validity Period of MC [1.02]
1.3.3.3 *404.06 - Prohibitions Regarding Exercise of Privileges [1.02]
1.3.3.4 404.18 - Permission to Continue to Exercise the Privileges of a Licence or Rating [1.02]

1.4. CARs PART VI - GENERAL OPERATING AND FLIGHT RULES

1.4.1. 600 - Interpretation
1.4.1.1 600.01 - Definitions [1.13]

1.4.2. 601 - Airspace Structure, Classification and Use
1.4.2.1 601.01 - Airspace Structure [1.16]
1.4.2.2 601.02 - Airspace Classification [1.16]
1.4.2.3 601.03 - Transponder Airspace [1.19]
1.4.2.4 *601.04 - IFR or VFR Flight in Class F Special Use Restricted Airspace or Class F Special Use Advisory Airspace [1.16]
1.4.2.5 601.06 - VFR Flight in Class A Airspace [1.16]
1.4.2.6 601.07 - VFR Flight in Class B Airspace [1.16, 1.33]
1.4.2.7 *601.08 - VFR Flight in Class C Airspace [1.16, 1.33]
1.4.2.8 *601.09 - VFR Flight in Class D Airspace [1.16]
1.4.2.9 601.15 - Forest Fire Aircraft Operating Restrictions [1.39]
1.4.2.10 601.16 - Issuance of NOTAM for Forest Fire Aircraft Operating Restrictions [1.39]

1.4.3. 602 - Operating and Flight Rules
GENERAL
1.4.3.1 *602.01 - Reckless or Negligent Operation of Aircraft [1.13]
1.4.3.2 *602.02 - Fitness of Flight Crew Members [1.13]
1.4.3.3 *602.03 - Alcohol or Drugs - Crew Members [1.13]
1.4.3.4 602.04 - Alcohol or Drugs - Passengers [1.13]
1.4.3.5 602.05 - Compliance with Instructions [1.13]
1.4.3.6 602.06 - Smoking [1.13]
1.4.3.7 602.07 - Aircraft Operating Limitations [1.13]
1.4.3.8 602.08 - Portable Electronic Devices [1.13]
1.4.3.9 *602.10 - Starting and Ground Running of Aircraft Engines [1.13]
1.4.3.10 *602.11 - Aircraft Icing [1.13]
1.4.3.11 *602.12 - Overflight of Built-up Areas or Open-Air Assemblies of Persons during Take-offs, Approaches and Landings [1.13]
1.4.3.12 *602.13 - Take-offs, Approaches and Landings within Built-up Areas of Cities and Towns [1.13]
1.4.3.13 602.14 - Minimum Altitudes and Distances [1.13]
1.4.3.14 602.15 - Permissible Low Altitude Flight [1.13]
1.4.3.15 *602.19 - Right of Way - General [1.13]
1.4.3.16 602.20 - Right of Way - Aircraft Maneuvering on Water [1.13]
1.4.3.17 *602.21 - Avoidance of Collision [1.13]
1.4.3.18 602.22 - Towing [1.13]
1.4.3.19 602.23 - Dropping of Objects [1.13]
1.4.3.20 602.24 - Formation Flight [1.13]
1.4.3.22 602.26 - Parachute Descents [1.13]
1.4.3.23 602.27 - Aerobatic Manoeuvres - Prohibited Areas and Flight Conditions [1.13]
1.4.2.24 602.28 - Aerobatic Maneuvres with Passengers [1.13]
1.4.2.25 *602.31 - Compliance with Air Traffic Control Instructions and Clearances [1.13]
1.4.2.26 *602.34 - Cruising Altitudes and Cruising Flight levels [1.37]
1.4.2.27 *602.35 - Altimeter Setting and Operating Procedures in the Altimeter-Setting Region [1.17]
1.4.2.28 *602.36 - Altimeter Setting and Operating

Procedures in the Standard Pressure Region [1.17]
1.4.2.29 602.37 - Altimeter Setting and Operating Procedures in Transition Between Regions [1.17]

OPERATIONAL AND EMERGENCY EQUIPMENT REQUIREMENTS
1.4.3.30 602.58 - Prohibition [1.35]
1.4.3.31 602.59 - Equipment Standards [1.35]
1.4.3.32 602.60 - Requirements for Power-Driven Aircraft [1.35]
1.4.3.33 602.61 - Survival Equipment - Flights over Land [1.35]
1.4.3.34 *602.62 - Life Preservers and Personal Flotation Devices [1.35]
1.4.3.35 *602.63 - Life Rafts and Survival Equipment - Flights over Water [1.35]

FLIGHT PREPARATION, FLIGHT PLANS & FLIGHT ITINERARIES
1.4.3.36 602.70 - Definitions [1.20]
1.4.3.37 602.71 - Pre-Flight Information [1.20]
1.4.3.38 602.72 - Weather Information [1.20]
1.4.3.39 *602.73 - Requirements to file a Flight Plan or Flight Itinerary [1.20]
1.4.3.40 602.74 - Contents of a Flight Plan or Fight Itinerary [1.20]
1.4.3.41 602.75 - Filing of a Flight Plan or Fight Itinerary — Sub-sections (3), (4) [1.20]
1.4.3.42 602.76 - Changes in the Flight Plan — Subsections (3) and (4) [1.20]
1.4.3.43 *602.77 - Requirement to File an Arrival Report [1.20]
1.4.3.44 602.78 - Contents of an Arrival Report [1.20]
1.4.3.45 602.79 - Overdue Aircraft Reports [1.20]

PRE-FLIGHT AND FUEL REQUIREMENTS
1.4.3.46 *602.88 - Fuel Requirements [1.21]
1.4.3.47 602.89 - Passenger Briefings [1.21]

OPERATIONS AT OR IN THE VICINITY OF AN AERODROME
1.4.3.48 *602.96 - General [1.32]
1.4.3.49 *602.97 - VFR and IFR Aircraft Operations at Uncontrolled Aerodromes within a Mandatory Frequency (MF) Area [1.32]
1.4.3.50 *602.98 - General MF Reporting Procedures [1.32]
1.4.3.51 602.99 - MF Reporting Procedures before Entering Maneuvring Area [1.32]
1.4.3.52 602.100 - MF Reporting Procedures on Departure [1.32]
1.4.3.53 602.101 - MF Reporting Procedures on Arrival [1.32]
1.4.3.54 602.102 - MF Reporting Procedures when Flying Continuous Circuits [1.32]
1.4.3.55 602.103 - Reporting Procedures when Flying through an MF Area [1.32]

VISUAL FLIGHT RULES
1.4.3.56 *602.114 - Minimum Visual Meteorological Conditions for VFR Flight in Controlled Airspace [1.15]
1.4.3.57 *602.115 - Minimum Visual Meteorological Conditions for VFR Fight in Uncontrolled Airspace [1.15]
1.4.3.58 *602.117 - Special VFR Flight [1.15]

RADIOCOMMUNICATIONS
1.4.3.59 602.136 - Continuous Listening Watch [1.15]
1.4.3.60 *602.138 - Two-way Radiocommunication Failure in VFR Flight [1.15]

EMERGENCY COMMUNICATIONS AND SECURITY
1.4.3.61 602.143 - Emergency Radio Frequency Capability [1.28]
1.4.3.62 602.144 - Interception Signals, Interception of Aircraft and Instructions to Land [1.28]
1.4.3.63 602.145 - ADIZ [1.36]
1.4.3.64 602.146 - SCATANA Plan [1.36]

1.4.5. 605 - Aircraft Requirements

GENERAL
1.4.4.1 605.03 - Flight Authority [1.14]
1.4.4.2 605.04 - Availability of Aircraft Flight Manual [1.14]
1.4.4.3 605.05 - Markings and Placards [1.14]
1.4.4.4 605.08 - Unserviceable and Removed Equipment - General [1.14]

AIRCRAFT EQUIPMENT REQUIREMENTS
1.4.4.5 605.14 - Power-Driven Aircraft - Day VFR [1.15]
1.4.4.6 605.22 - Seat and Safety-Belt Requirements [1.26]
1.4.4.7 605.24 - Shoulder Harness Requirements [1.26]
1.4.4.8 *605.25 - General Use of Safety Belts and Restraint Systems — Sub-sections (1) and (2) [1.26]
1.4.4.9 605.28 - Child Restraint System — Sub-section 1(a) [1.26]
1.4.4.10 605.29 - Flight Control Locks [1.18]
1.4.4.11 605.31 - Oxygen Equipment and Supply [1.27]
1.4.4.12 *605.32 - Use of Oxygen [1.27]
1.4.4.13 *605.35 - Transponder and Automatic Pressure Altitude Reporting Equipment — Sub-sections (1) and (3) [1.19]
1.4.4.14 605.38 - ELT [1.29]
1.4.4.15 *605.40 - ELT Activation [1.29]

AIRCRAFT MAINTENANCE REQUIREMENTS
1.4.4.16 605.84 - General [1.14]
1.4.4.17 605.85 - Maintenance Release and Elementary Work [1.14]
1.4.4.18 605.86 - Maintenance Schedule [1.14]
1.4.4.19 605.88 - Inspection after Abnormal Occurrences [1.14]

TECHNICAL RECORD
1.4.4.20 605.92 - Requirements to Keep Technical Records — Sub-section (1) [1.14]
1.4.4.21 605.93 - Technical Records - General [1.14]
1.4.4.22 605.94 - Journey Log Requirements [1.14]
1.4.4.23 605.95 - Journey Log - Carrying on Board [1.14, 1.40]
1.4.4.24 605.97 - Transfer of Records [1.14]

1.4.5. Miscellaneous
1.4.5.1 606.02 - Liability Insurance — Sub-section (9) [1.43]

1.5. AIP CANADA
1.5.1 AIP Canada Supplements [1.23]
1.5.2 Aeronautical Information Circulars [1.23]
1.5.3 Aviation Notices [1.23]
1.5.4 AIRAC Canada [1.23]

1.6. TRANSPORTATION SAFETY BOARD OF CANADA (TSB) (AIP Canada GEN 3.0) [1.24]

1.7. AIR TRAFFIC SERVICES AND PROCEDURES
1.7.1 *Air Traffic and Advisory Services [1.30]
1.7.2 *Flight Service Stations [1.30]
1.7.3 *Communications Procedures [1.30]
1.7.4 *Radar Service — Clock Position System [1.30]
1.7.5 *ATC Clearances/Instructions [1.30]
1.7.6 Wake Turbulence Separation [1.30]
1.7.7 *Airport/Aerodrome Operations — Controlled [1.31]
1.7.8 *Airport/Aerodrome Operations — Uncontrolled [1.32]
1.7.9 Mandatory & Aerodrome Traffic Frequencies [1.32]
1.7.10 VFR Enroute Procedures [1.33]
1.7.11 VFR Holding Procedures [1.34]
1.7.12 *VFR/IFR Traffic Mix at Uncontrolled Airports/Aerodromes [1.34]
1.7.13 *Simultaneous Intersecting Runway Operations (SIRO) [1.32]
1.7.14 *Procedures for the Prevention of Runway Incursions [1.31]

2. Navigation and Radio Aids

2.1. DEFINITIONS
2.1.1 Meridian [6.01]
2.1.2 Equator [6.01]
2.1.3 Latitude [6.01]
2.1.4 Great Circle [6.01]
2.1.5 Variation [6.01]
2.1.6 Isogonal [6.01]
2.1.7 Agonic Line [6.01]
2.1.8 Deviation [6.01]
2.1.9 Track [6.01]
2.1.10 Heading [6.01]
2.1.11 Airspeed [6.01]
2.1.12 Ground Speed [6.01]
2.1.13 Ground Position [6.01]
2.1.14 Wind Velocity [6.01]
2.1.15 Drift [6.01]

2.2. MAPS AND CHARTS
2.2.1 VTA — Transverse Mercator Projection [6.02]
2.2.2 VNC — Lambert Conformal Conic Projection [6.02]
2.2.3 *Topographical Symbols [6.02]
2.2.4 *Elevation and Contours (Relief) [6.02]
2.2.5 *Aeronautical Information [6.02]
2.2.6 *Scale and Units of Measurement [6.02]
2.2.7 *Locating Position by Latitude and Longitude [6.05]
2.2.8 Navigational Aids [6.03]

2.3. TIME AND LONGITUDE
2.3.1 24 Hour System [6.04]
2.3.2 Time Zones and Relation to Longitude [6.04]
2.3.3 Conversion of UTC to Local and Vice Versa [6.04]

2.4. PILOT NAVIGATION
2.4.1 Use of Aeronautical Charts [6.06, 6.08]
2.4.2 *Measurement of Track and Distance [6.06]
2.4.3 Map Reading [6.06]
2.4.4 Setting Heading — Visual Angle of Departure [6.07]
2.4.5 Check-points and Pin-points [6.13]
2.4.6 *Ground Speed Checks and ETA Revisions [6.14]
2.4.7 *Variation/Deviation [6.01, 6.14]
2.4.8 *True Track/Magnetic Track [6.01, 6.14]
2.4.9 *Track Made Good
2.4.10 Determining Drift by 10° Lines [6.08, 6.15]
2.4.11 Double Track Error Method to Regain Track [6.15]
2.4.12 Sum of Opening and Closing Angles to Destination [6.15]
2.4.13 Visual Alteration Method of Correcting Track [6.15]
2.4.14 Diversion to Alternate [6.16]
2.4.15 *Return to Departure Point (Reciprocal Track) [6.17]
2.4.16 *Procedures When Lost [6.18]
2.4.17 Indicated/Calibrated Airspeed (IAS, CAS) [6.12]

2.5. TRIANGLE OF VELOCITIES
2.5.1 True Airspeed and Heading [6.01]
2.5.2 Wind Velocity [6.01]
2.5.3 Ground Speed and Track [6.01]

2.6. NAVIGATION COMPUTERS
2.6.1 *Heading and True Airspeed [6.09]
2.6.2 *Applying the Wind [6.09]
2.6.3 *True Track and Ground Speed [6.09]
2.6.4 *Magnetic Heading and Magnetic Track [6.09]
2.6.5 *Pressure/Density Altitudes [4.04]
2.6.6 *Time/Ground Speed/Distance [6.10]
2.6.7 *Fuel Consumption and Conversions [6.11]

2.7. PRE-FLIGHT PREPARATION
2.7.1 *Factors Affecting Choice of Route [6.13]
2.7.2 Map Preparation [6.08, 6.13]
2.7.3 *Meteorological Information [5.16 - 5.26]
2.7.4 *NOTAM [1.22, 6.13]
2.7.5 Selection of Check-points [6.13]
2.7.6 *Fuel Requirements [6.11]
2.7.7 *Weight and Balance [7.16]
2.7.8 *Use of Canada Flight Supplement [7.22]
2.7.9 Documents to be Carried in Aircraft [1.40]
2.7.10 *Flight Plans/Itineraries [1.20]
2.7.11 Aircraft Serviceability [1.20

2.8. RADIO THEORY
2.8.1 Characteristics of Very High Frequency Radio Waves [6.24]
2.8.2 Frequency Bands Used in Navigation and Communication [6.24]
2.8.3 Reception Limitations [6.24]

2.9. RADIO NAVIGATION - BASIC USE AND LIMITATIONS
2.9.1 VHF Omni-direction Range (VOR) [6.19]
2.9.2 Automatic Direction Finder (ADF) [6.20]
2.9.3 Long Range Area Navigation (LORAN C) [6.27]
2.9.4 Global Navigation Satellite System (GNSS/GPS) [6.25]

2.10. OTHER RADIO AND RADAR AIDS — BASIC PRINCIPLES AND USE
2.13.1 *Transponder [1.19]
2.13.2 *Emergency Locator Transmitter (ELT) [1.29]
2.13.3 *VHF Direction Finding (DF) Assistance [6.21]
2.13.4 Airport Surveillance Radar (ASR) Primary/Secondary [6.23]

3. Meteorology

3.1. THE EARTH'S ATMOSPHERE
3.1.1 The Standard Atmosphere [5.01]
3.1.2 Density and Pressure [5.01]
3.1.3 Mobility [5.01]
3.1.4 Expansion and Compression [5.01]

3.2. ATMOSPHERIC PRESSURE
3.2.1 Pressure Measurements [5.01]
3.2.2 Station Pressure [5.01]
3.2.3 Sea Level Pressure [5.01]
3.2.4 Effects of Temperature [5.03]
3.2.5 Isobars [5.01]
3.2.6 *Horizontal Pressure Differences [5.01]

3.3. METEOROLOGICAL ASPECTS OF ALTIMETRY
3.3.1 *Pressure Altitude [5.02]
3.3.2 *Density Altitude [5.02]
3.3.3 *Altimeter Settings [5.02]
3.3.4 *Considerations When Flying from High to Low Pressure and Temperature Areas, and Vice Versa [5.02]

3.4. TEMPERATURE
3.4.1 Temperature Scale — Fahrenheit/Celsius [5.03]
3.4.2 *Temperature Variations with Altitude [5.03]
3.4.3 *Inversions [5.03]
3.4.4 Isothermal Layers [5.03]

3.5. MOISTURE
3.5.1 *Relative Humidity/Dewpoint [5.04]
3.5.2 Sublimation/Condensation [5.04]
3.5.3 Cloud Formation [5.04]
3.5.4 Precipitation [5.04]

3.6. STABILITY AND INSTABILITY
3.6.1 *Characteristics of Stable/Unstable Air [5.05]
3.6.4 Surface Heating/Cooling [5.05]
3.6.5 Lifting Processes [5.05]

3.7. CLOUDS
3.7.1 Formation and Structure [5.06]
3.7.2 *Types and Recognition [5.06]
3.7.3 *Associated Precipitation and Turbulence [5.06]

3.8. TURBULENCE
3.8.1 Convection [5.07]
3.8.2 Mechanical [5.07]
3.8.3 Orographic [5.07]
3.8.4 *Wind Shear [5.07]

3.9. WIND
3.9.1 *Pressure Gradient [5.08]
3.9.2 Low Level Winds — Variation in Surface Wind [5.08]
3.9.3 *Veer and Back [5.08]
3.9.4 Squalls and Gusts [5.08]
3.9.5 Diurnal Effects [5.08]
3.9.6 Land and Sea Breezes [5.08]
3.9.7 *Topographical Effects [5.08]
3.9.8 *Wind Shear — Types and Causes [5.08, 7.04]

3.10. AIR MASSES
3.10.1 Definition and Characteristics [5.09]
3.10.2 Factors that Determine Weather [5.09]
3.10.3 Seasonal/Geographic Effects [5.09]
3.10.4 Air Masses Affecting North America [5.09]

3.11. FRONTS
3.11.1 Structure [5.10]
3.11.2 *Types [5.10]
3.11.3 Formation [5.10]

3.12. FRONTAL WEATHER
3.12.1 *Cold Front [5.10]
3.12.2 *Warm Front [5.10]
3.12.3 TROWAL and Upper Fronts [5.10]

3.13. AIRCRAFT ICING
3.13.1 In-flight — Freezing Rain [5.11]
3.13.2 Hoar Frost [5.11]

3.14. THUNDERSTORMS
3.14.1 Requirements for Development [5.12]
3.14.2 Structure/Development [5.12]
3.14.3 Types — Air Mass/Frontal [5.12]
3.14.4 *Hazards — Updrafts/Downdrafts/Gust Fronts/Downbursts/Microbursts/Hail/Lightning/Tornadoes [5.12]
3.1.4.5 Squall lines [5.13]

3.15. SURFACE BASED LAYERS
3.15.1 *Fog Formation [5.15]
3.15.2 *Fog Types [5.15, 5.20]
3.15.3 Haze/Smoke [5.15]
3.15.4 Blowing Obstructions to Vision [5.15]

3.16. METEOROLOGICAL SERVICES AVAILABLE TO PILOTS
3.16.1 Aviation Weather Information Service (AWIS) [5.16]
3.16.2 Aviation Weather Briefing Service (ABWS) [5.16]
3.16.3 Flight Service Stations (FSS) [5.16]
3.16.4 Pilot's Automatic Telephone Weather Answering Service (PATWAS) [5.16]
3.16.5 Transcribed Weather Broadcasts (TWB) [5.16]
3.16.6 Automatic Terminal Information Service (ATIS) [5.16]

3.17. AVIATION WEATHER REPORTS
3.17.1 *Aviation Routine Weather Report (METAR) - decoding [5.20]
3.17.2 Automated Weather Observation Station (AWOS) [5.20]
3.17.3 Pilot Reports (PIREP) [5.22]

3.18. AVIATION WEATHER FORECASTS
3.18.1 *Times Issued and Period of Coverage [5.17, 5.18, 5.19]
3.18.2 *Decoding [5.17, 5.18, 5.19]
3.18.3 *Area Forecasts (FA) [5.17]
3.18.4 *Aerodrome Forecasts (TAF) [5.18]
3.18.5 Upper Level Winds and Temperature Forecasts (FD) [5.19]
3.18.6 ~~Airman's~~ Pilot's Meteorological Advisory (AIRMET) [5.25]
3.18.7 Significant In-flight Weather Warning Message (SIGMET) [5.21]

3.21. WEATHER MAPS AND PROGNOSTIC CHARTS
3.21.1 Times Issued and Validity Periods [5.23]
3.21.2 Symbols and Decoding [5.23]
3.21.3 Surface Weather Map [5.23]
3.21.4 Prognostic Surface Chart [5.23]

4. Engines and Systems

4.1. LANDING GEAR, BRAKES AND FLAPS
4.1.1 Mechanical [2.01, 2.02]
4.1.2 Hydraulic [2.01, 2.02]
4.1.3 Electrical [2.01, 2.02]

4.2. ENGINES
4.2.1 Two/Four Stroke Cycle [2.03]
4.2.2 Methods of Cooling [2.03]
4.2.3 *Dual Ignition [2.07]
4.2.4 Exhaust Systems [2.12]
4.2.5 Ancillary Controls [2.09]
4.2.6 *Effect of Density Altitude/Humidity [7.17]
4.2.7 Limitations and Operations [7.17]
4.2.8 Instruments [2.03]

4.3. CARBURATION
4.3.1 *Mixture Controls [2.09]
4.3.2 *Carburettor Icing [2.09]
4.3.3 *Use of Carb Heat and its Effects on Mixture [2.09]

4.4. ELECTRICAL SYSTEM
4.4.1 Generator/Alternator/Battery [2.10]
4.4.2 Master/Alternator/Generator Switches [2.10]
4.4.3 Ammeter/Load Meter [2.10]
4.4.4 Circuit Breakers/Fuses [2.10]
4.4.5 Grounding/Bonding [2.10]

4.5. LUBRICATING SYSTEMS AND OILS
4.5.1 *Types — Viscosity/Grades/Seasonal Use [2.08]
4.5.2 *Purposes [2.08]
4.5.3 Venting [2.08]
4.5.4 Filters [2.08]

4.6. FUEL SYSTEM AND FUELS
4.6.1 Types — Colour and Properties [2.04]
4.6.2 Density and Weight [2.04]
4.6.3 Additives [2.04]
4.6.4 *Contamination and Deterioration [2.05]
4.6.5 Tank Location [2.06]
4.6.6 Venting [2.06]
4.6.7 Fuel Line — Filters/Drains [2.06]
4.6.8 *Detonation — Causes/Effects [2.04]
4.6.9 Vapour Lock [2.04]
4.6.10 Primers [2.06]
4.6.11 Fuel Management — Ground/Air [2.05]
4.6.12 *Fuel Handling — Fuelling Aircraft [2.05]
4.6.13 *Grounding/Bonding [2.05]

5. Theory of Flight

5.1. PRINCIPLES OF FLIGHT
5.1.1 Bernoulli's Theorem [3.03]
5.1.2 Newton's Laws [3.03]

5.2. FORCES ACTING ON AN AEROPLANE
5.2.1 Lift [3.03]
5.2.2 Drag — Induced/Parasite/Profile [3.03]
5.2.3 Relationship of Lift and Drag to Angle of Attack [3.03]
5.2.4 Thrust [3.03]
5.2.5 Weight [3.03]
5.2.6 Equilibrium [3.03]
5.2.7 Centre of Pressure (C of P) [3.04]
5.2.8 Centrifugal/Centripetal [3.05]
5.2.9 *Forces Acting on an Aircraft during Manoeuvres [3.05]

5.3. AEROFOILS
5.3.1 Pressure Distribution about an Aerofoil [3.05]
5.3.2 *Relative Airflow and Angle of Attack [3.05]
5.3.3 Downwash [3.05]
5.3.4 *Wing Tip Vortices [3.03]
5.3.5 Angle of Incidence [3.01]

5.4. PROPELLERS
5.4.1 Fixed Pitch [3.06]
5.4.2 Torque/Slipstream/Gyroscopic Effect/Asymmetric Thrust [3.06]

5.5. DESIGN OF THE WING
5.5.1 Area/Span/Chord [3.01]
5.5.2 Aspect Ratio [3.01]
5.5.3 Camber [3.01]
5.5.4 Laminar Flow [3.01]
5.5.5 Dihedral/Anhedral
5.5.6 Wash-in/Wash-out [3.01]
5.5.7 Slots/Slats [3.01]
5.5.8 Stall Strips [3.01]
5.5.9 Spoilers [3.01]
5.5.10 *Flaps [3.01]
5.5.11 Canards [3.07]

5.6. LOAD FACTOR
5.6.1 Centrifugal Force/Weight [3.05]
5.6.2 Load Factor — Linear/Turns [3.05]
5.6.3 *Relationship of Load Factor to Stalling Speed [3.05]
5.6.4 *Structural Limitations [3.05]
5.6.5 Gust Loads [3.05]

5.7. STABILITY
5.7.1 *Longitudinal/Lateral/Directional Stability [3.02]
5.7.2 Inherent Stability [3.02]
5.7.3 Methods of Achieving Stability [3.02]

5.8. FLIGHT CONTROLS
5.8.1 Aeroplane Axes and Planes of Movement [3.07]
5.8.2 Functions of Controls [3.07]
5.8.3 Relationship Between Yaw and Roll [3.07]
5.8.4 Adverse Yaw/Aileron Drag [3.07]
5.8.5 Static/Dynamic Balancing of Controls [3.07]
5.8.6 Trim/Trimming Devices [3.07]

6. Flight Instruments

6.1. PITOT STATIC SYSTEM
6.1.1 *Pitot [4.01]
6.1.2 *Static [4.01]
6.1.3 Anti-icing [4.01]
6.1.4 Alternate Static — Sources/Errors [4.01]

6.2. AIRSPEED INDICATOR
6.2.1 *Principles of Operation [4.02]
6.2.2 Errors [4.02]
6.2.3 *Markings [4.02]
6.2.4 Definitions (IAS/CAS/TAS) [4.02]

6.3. **VERTICAL SPEED INDICATOR**
6.3.1 *Principles of Operation [4.03]
6.3.2 Errors [4.03]

6.4. **ALTIMETER/ENCODING ALTIMETER**
6.4.1 Principles of Operation [4.04]
6.4.2 Errors [4.04]

6.5 **DIRECT READING MAGNETIC COMPASS**
6.5.1 *Reading the Compass [4.05]
6.5.2 *Magnetic Dip [4.05]
6.5.3 *Turning and Acceleration Errors
6.5.4 Compass Serviceability Checks [4.05]
6.5.5 Compass Correction Card [4.05]
6.5.6 Checking Compass Heading on the Ground and in Flight [4.05]

6.6. **GYROSCOPE**
6.6.1 Principles of Operation [4.06]
6.6.2 Inertia [4.06]
6.6.3 *Precession [4.06]

6.7. **HEADING INDICATOR**
6.7.1 *Principles of Operation [4.08]
6.7.2 *Errors [4.08]
6.7.3 Limitations [4.08]
6.7.4 Power Sources [4.08]

6.8. **ATTITUDE INDICATOR**
6.8.1 *Principles of Operation 4.07]
6.8.2 *Errors [4.07]
6.8.3 Limitations [4.07]
6.8.4 Power Sources [4.07]

6.9. **TURN AND BANK INDICATOR/TURN COORDINATOR**
6.9.1 *Principles of Operation [4.09]
6.9.2 Errors [4.09]
6.9.3 Limitations [4.09]
6.9.4 Power Sources [4.09]

7. Flight Operations

7.1. **GENERAL**
7.1.1 Pilot-in-Command Responsibilities [7.06]
7.1.2 Winter Operations [7.01]
7.1.3 Thunderstorm Avoidance [7.03]
7.1.4 Mountain Flying Operations [7.09]
7.1.5 Collision Avoidance — Use of Landing Lights [7.05]
7.1.6 *Runway Numbering [7.08]
7.1.7 VASIS/PAPI [7.25, 7.26]
7.1.8 Approach, Runway and Aerodrome Marking/Lighting [1.38, 7.22]
7.1.9 Obstruction Markings [6.03]
7.1.10 Units of Measurement /Conversion [7.21]
7.1.11 Radio Communications (as per Sec. 1.)
7.1.12 *Aerodrome Operations (including Procedures for the Prevention of Runway Incursions) [1.38]
7.1.13 *Wheelbarrowing [7.11]
7.1.14 *Hydro-planing [7.35]
7.1.15 *Taxiing [7.15]
7.1.16 *Effects of Wind [7.04]
7.1.17 *Side-slips [7.17]

7.2. **AIRCRAFT PERFORMANCE**
7.2.1 *Effects of Critical Surface Contamination [7.01]
7.2.2 *Effects of Density Altitude/Humidity [7.17]
7.2.3 Normal/Short/Soft and Rough Field Take-offs and Landings [7.17]
7.2.4 *Ground Effect [7.12]
7.2.5 *Best Angle of Climb (Vx) [7.17, 7.30]
7.2.6 *Best Rate of Climb (Vy) [7.17, 7.30]
7.2.7 *Manoeuvring Speed (Va) [7.17]
7.2.8 Safe Operating Speed (Vno) [7.17]
7.2.9 Never Exceed Speed (Vne) [7.17]
7.2.10 Maximum Flap Speed (Vfe) [7.17]
7.2.11 Gliding for Range [7.17]
7.2.12 Flying for Range [7.17]
7.2.13 Flying for Endurance [7.17]
7.2.14 *Slow Flight [7.17]
7.2.15 *Stalls [3.04, 7.17]
7.2.16 *Indicated and True Stalling Speed [3.04]
7.2.17 *Stall Speed vs Altitude [3.04]
7.2.18 *Spins [7.14]
7.2.19 *Spirals [7.13]
7.2.20 Recommended Safe Recovery Altitudes [7.14]
7.2.21 *Effects of Change of Weight or Centre of Gravity (C of G) on Performance [7.16]
7.2.22 Use of Aircraft Flight Manual (Including Approved and Unapproved Operational Information) [7.17]

7.3. **USE OF PERFORMANCE CHARTS**
7.3.1 *Take-off Charts [7.17]
7.3.2 *Cross-wind Charts [7.17]
7.3.3 *Cruise Charts [7.17]
7.3.4 Fuel Burn Charts [7.17]
7.3.5 Landing Charts [7.17]
7.3.6 Effect of Ice/Snow/Frost/Slush/Water on Take-off and Landing Run [7.17]
7.3.7 *Effect of Various Runway Surfaces on Take-off and Landing Run [7.17]
7.3.9 Upslope/Downslope Runway [7.17]

7.4. **WEIGHT AND BALANCE**
7.4.1 Terms (e.g. Datum/Arm/Moment) [7.16]
7.4.2 *Locating C of G [7.16]
7.4.3 *C of G Limits [7.16]
7.4.4 Weights (e.g. Empty/Gross) [7.16]
7.4.5 Load Adjustment [7.16]
7.4.6 Cargo Tie-down/Passenger Loading [7.16]
7.4.7 *Normal/Utility Category [7.16]

7.5. **WAKE TURBULENCE**
7.5.1 *Causes [7.10]
7.5.2 *Effects [7.10]
7.5.3 *Avoidance [7.10]

7.6. **SEARCH AND RESCUE (SAR) (AIP Canada — SAR Information)**
7.6.1 Types of Service Available [7.27]
7.6.2 ELT (Exclude Categories) [7.27]
7.6.3 Aircraft Emergencies [7.27]
7.6.4 Survival — Basic Techniques [7.27]

7.7. **AIRCRAFT CRITICAL SURFACE CONTAMINATION**
7.7.1 *Clean Aircraft Concept [7.01]
7.7.2 *Frozen Contaminants [7.01]
7.7.3 Cold Soaking Phenomenon [7.01]
7.7.4 *Pre-take-off Inspection [7.01]

8. Human Factors

8.1. **AVIATION PHYSIOLOGY**
8.1.1 *Hypoxia/Hyperventilation [7.24]
8.1.2 Gas Expansion Effects [7.24]
8.1.3 Decompression (Including SCUBA Diving) [7.24]
8.1.4 *Vision/Visual Scanning Techniques [7.05]
8.1.5 Hearing [7.24]
8.1.6 *Orientation/Disorientation (Including Visual/Vestibular Illusions) [7.31]
8.1.7 Positive/Negative “G” [7.24]
8.1.8 Sleep/Fatigue [7.24]
8.1.9 Anaesthetics/Blood Donations [7.24]

8.2. **THE PILOT AND THE OPERATING ENVIRONMENT**
8.2.1 Personal Health/Fitness [7.24]
8.2.2 Diet/Nutrition [7.24]
8.2.3 *Medications (Prescribed, over-the-counter) [7.24]
8.2.4 Substance Abuse (Alcohol/Drugs) [7.24]
8.2.5 Pregnancy [7.24]
8.2.6 Heat/Cold [7.24]
8.2.7 Noise/Vibration [7.24]
8.2.8 Effects of Smoking [7.24]
8.2.9 *Toxic Hazards (Including Carbon Monoxide) [7.24]

8.3. **AVIATION PSYCHOLOGY**
8.3.1 The Decision-making Process [7.34]
8.3.2 *Factors that Influence Decision-making [7.34]
8.3.3 *Situational Awareness [7.34]
8.3.4 *Stress [7.34]
8.3.5 *Managing Risk [7.34]
8.3.6 Attitudes [7.34]
8.3.7 Workload — Attention and Information Processing [7.34]

8.4. **PILOT — EQUIPMENT/MATERIALS RELATIONSHIP**
8.4.1 Controls and Displays — Errors in Interpretation and Control [7.34]
8.4.2 Correct Use of Check-lists and Manuals [7.34]

8.5. **INTERPERSONAL RELATIONS**
8.5.1 Communication with — Maintenance Personnel/Air Traffic Services/Passengers [7.34]
8.5.2 Operating Pressures — Family Relationships/Peer Group [7.34]

NOTES

INDEX

NOTES

NOTES

CULHANE Aviation Training Manuals - 1999 order form

Please send: Cost:

____ copy (ies) **Private Pilot and Recreational Pilot Ground School Course ($69.95)** . . . . ________

____ copy (ies) **Commercial Pilot Ground School Course ($69.95)** . . . . ________

____ copy (ies) **Instrument Rating Ground School Course ($69.95)** . . . . ________

____ copy (ies) **Airline Transport Ground School Course ($69.95)** . . . . ________

____ copy (ies) **Instructor Rating Ground School Course ($69.95)** . . . . ________

____ copy (ies) **Helicopter Pilot Ground School Course ($69.95)** . . . . ________

____ copy (ies) **Helicopter Airline Transport Ground School Course ($69.95)** . . . . ________

____ copy (ies) **PSTAR and Student Pilot Written Test Book ($29.95)** . . . . ________

____ copy (ies) **Private Pilot and Recreational Pilot Written Test Book ($49.95)** . . . . ________

____ copy (ies) **Commercial Pilot Written Test Book ($49.95)** . . . . ________

____ copy (ies) **Instrument Rating Written Test Book ($49.95)** . . . . ________

____ copy (ies) **Airline Transport Written Test Book ($49.95)** . . . . ________

____ copy (ies) **Instructor Rating Written Test Book ($49.95)** . . . . ________

____ copy (ies) **Air Traffic Controller Aptitude Test Written Test Book ($49.95)** . . . . ________

____ copy (ies) **AME Regulations Written Test Book ($49.95)** . . . . ________

____ copy (ies) **AVIATION LAW: Legal Guide for Canadian Owners and Pilots ($39.95)** . . . . ________

SHIPPING: ($7.50 flat rate, allow 2 - 7 days) ________

INTERNATIONAL SHIPPING: ($15.00 flat rate USA, $30.00 flat rate all other nations, 1 - 4 weeks) ________

SUBTOTAL: (Books plus shipping) ________

GST/HST: (For Canadian destinations, add 7% of subtotal for GST, 15% HST in Maritime provinces). ________

TOTAL PAYMENT REQUIRED: (Subtotal plus GST/HST) ________

Payment:

❑ Cheque/Money Order enclosed (payable to Accelerated Aviation Training - personal cheques subject to 14 day delay for bank clearance).

❑ VISA Card No. ______________________ Exp. ______

❑ MASTERCARD No. ______________________ Exp. ______

Authorizing signature: ______________________

Name: ______________________

Address: ______________________

______________ **Country - if other than Canada:** ______________

Postal Code: ____________ **Phone**: (____)______________

Mail or FAX this form to:

Accelerated Aviation Training
5980 Forsythe Crescent
Richmond BC V7C 2C3 CANADA
Ph. 604-279-0179 FAX 604-279-0161
E-mail: info@culhanebooks.com

Come and visit our website:

www.culhanebooks.com